SiBCN Ceramics and Composites Prepared by Mechanical Alloying

Daxin Li · Dechang Jia · Zhihua Yang · Yu Zhou

SiBCN Ceramics and Composites Prepared by Mechanical Alloying

Springer

Daxin Li
School of Material Science and Engineering
Harbin Institute of Technology
Harbin, Heilongjiang, China

Dechang Jia
School of Material Science and Engineering
Harbin Institute of Technology
Harbin, Heilongjiang, China

Zhihua Yang
School of Material Science and Engineering
Harbin Institute of Technology
Harbin, Heilongjiang, China

Yu Zhou
School of Material Science and Engineering
Harbin Institute of Technology
Harbin, Heilongjiang, China

ISBN 978-981-97-3028-5 ISBN 978-981-97-3029-2 (eBook)
https://doi.org/10.1007/978-981-97-3029-2

This Springer imprint is published by the registered company Springer Nature Singapore Pte Ltd.
The registered company address is: 152 Beach Road, #21-01/04 Gateway East, Singapore 189721, Singapore

Preface

The development of novel multifunctional thermal protection materials operated under harsh environments, typical of high-temperature oxidation, severe thermal shock, ablation by combustion gas flow, etc., is one of the urgent needs in modern aerospace industry. The amorphous Si-based multicomponent ceramics, such as SiCN, SiBN, SiBC, SiBCN, SiBON, SiBOCN, etc. have received much attention in recent years owing to their highly stable amorphous structures and specific properties different from the crystalline counterparts. Coincidently, the quaternary SiBCN system stands out from all recent materials offering great potentials at high temperatures due to their higher microstructural stability and excellent high-temperature properties including resistance to oxidation, thermal shock, creep and ablation.

In past 30 years, significant progress on the synthesis and fabrication of SiBCN ceramics through the precursor-derived ceramics (PDCs) route as an appropriate synthesis method for synthesizing multi-component materials have been made. The PDCs approach based on well-defined precursors can offer precise control over chemical composition, microstructure and ceramic shapes that have provided access to a vast number of SiBCN ceramics with interesting properties. In this aspect, the PDCs route is mostly technologically suitable for preparing ceramic films, coatings, fibers, porous structure or ceramic matrix composition (CMCs), with extremely homogeneous distribution of the constituting elements without near-range order.

By using inorganic powders (such as Si, graphite, boron, h-BN, etc.) instead of the organic precursor as raw materials, the inorganic processing route based on mechanical alloying (MA), one of the nonequilibrium processing techniques, coupled with sequential sintering, although apparently very "hard" compared to the "soft" PDCs method, is actually a simple and effective way to prepare monoliths with uniform microstructure and superior properties. It has been used to obtain dense SiBCN monoliths and structural parts stable at high-temperatures providing new experimental data and therefrom a more detailed understanding of the intrinsic properties of metastable SiBCN materials, benefitting progress toward engineering applications.

Herein this book mainly provides a comprehensive treatment of the MA-derived SiBCN (MA-SiBCN) ceramics and their CMCs, aspect of theme content including material design and preparation, microstructural feature and evolutionary process, mechanical and thermophysical properties, resistance to oxidation, thermal shock and ablation, and the mechanisms of oxidation, ablation and crystallization of SiBCN-based materials. Chapter 1 introduces the motivation, structure characteristics, performance advantages and prospects of the SiBCN materials derived from PDCs or MA routes. Chapter 2 describes the basis of mechanical alloying, and outlines the historical prospective and principle of the non-equilibrium phases formation, and the alloying mechanism and amorphization kinetics of SiBCN. Chapter 3 includes the crystallization behavior and thermal stability of MA-derived amorphous SiBCN powders and also focuses on the effects of metals and/or ceramics particles on the microstructural evolution of the mixing inorganic powders. Chapter 4 introduces the densification kinetics and crystallization mechanisms of MA-SiBCN materials. This section also involves the comparison of sintering methods on densification and microstructural/morphological evolution of PDCs or MA-derived SiBCN. Chapter 5 deals with mechanical behavior and strengthening/toughening mechanisms in MA-SiBCN materials, focusing on the intrinsic toughness and brittle fracture of nanocrystalline or amorphous monoliths, and "additive" strengthened/toughened MA-SiBCN composite ceramics and CMCs. Chapter 6 outlines the thermal damage behavior and damage mechanisms of MA-SiBCN materials.

We wish to acknowledge all the contributing authors for their great enthusiasm in reviewing portions of the draft manuscript and providing photographs in their respective area of expertise. We also thank the financial supports from the National Key Research and Development Program (No. 2017YBB0310400), the National Natural Science Foundation-Innovation Research Group Project (No. 51621091), Outstanding Youth Funding (No. 51225203), Key Project Funding (Nos. 51832002, 52232004), General Funding (Nos. 52172068, 52372059, 51272300, 51072041), Youth Funding (Nos. 52002092, 50902031), and Heilong Jiang Natural Science Fund for Young Scholars (No. YQ2021E017). The financial supports by the Postdoctoral Innovation Talent Support Program (No. BX20190095), Heilongjiang Touyan Team Funding and Advanced Talents Scientific Research Foundation of Shenzhen: Yu Zhou, are also acknowledged.

At Science Park of Harbin
Institute of Technology
August 2024

Daxin Li
Dechang Jia

Contents

Chapter 1
Introduction

1.1 The Processing Route: Motivation, Principle and History

The aerospace industry is continually searching for lighter high-strength materials that perform under more and more extreme conditions. In particular there is considerable demand for structural materials and components that are stable to combustion gas flows, as leading edges and ablation shields. Such extreme conditions encountered in these applications mandate materials that integrate multiple properties including excellent mechanical properties (especially high tolerance to thermal and mechanical stress), high thermal stability, and superior resistance to thermal shock, oxidation and ablation to ensure their safety and reliability in service [1–6].

Currently, binary ceramics such as SiC and Si_3N_4 are used widely for high temperature applications due to their low density, high hardness, excellent mechanical properties and high thermo-chemical stability. However, the mechanical properties and oxidation resistance of SiC ceramics generally degrade in air above 1500 °C due to softening of protective SiO_2 oxide scale. Commercial Si_3N_4 components offer upper use temperatures of no more than 1300 °C in oxidizing environments [7–10]. Additionally, the high-temperature structural applications of the single-phase ultra-high temperature ceramics (UHTCs) such as ZrB_2 and HfB_2 are limited by their high density as well as poor damage tolerance detrimental to thermal shock resistance, although their unique combined properties, such as high melting points, high chemical inertness against molten metals, and good resistance to ablation, allow them to be potential candidates for nose cone and wing leading edges as thermal protection for reentry systems [11–15].

Consequently, there is a continuing search for superior materials building on what is currently state-of-the-art. These include carbide and nitride ceramic matrix composites (CMCs), multicomponent ceramics and their CMCs, such as C_f/(C, SiC, Si_3N_4) [16–20], SiC_f/(SiC, ZrB_2) [21–24], SiC_p/(Si_3N_4, ZrB_2, HfB_2) [25–28],

D. Li et al., *SiBCN Ceramics and Composites Prepared by Mechanical Alloying*,
https://doi.org/10.1007/978-981-97-3029-2_1

Fig. 1.1 Relationship between using temperature and specific strength of the commonly used structural ceramic materials [4]

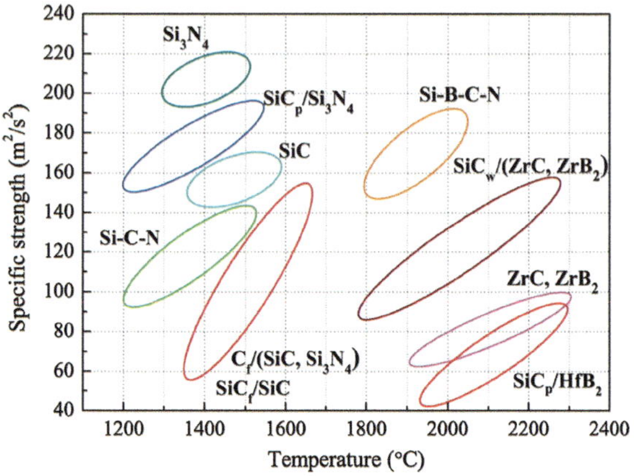

$SiC_w/(ZrC, ZrB_2)$ [29–31], SiCN [32–34], $(C_f, SiC_f)/SiCN$ [35–37], SiBCN(-Zr, Hf, Al) [38–44], $(C_f, SiC_f)/SiBCN$ [45–48], etc. Among them, metastable silicoboron carbonitride (SiBCN) ceramics have attracted considerable attention due to their exceptional microstructural stability, relatively higher specific strength and use temperature (Fig. 1.1) and satisfying high-temperature property retention.

Precursor-derived SiBCN fibers, one of the very first described SiBCN materials, remain amorphous at least up to 1800 °C, and do not undergo microstructural changes at this temperature. At temperatures up to 1600 °C, SiBCN ceramics show oxidation resistance superior to those of SiC and Si_3N_4. On exposure to air at 1500 °C for 50 h, the tensile strength of oxidized SiBCN fibers still reaches 2.5 GPa, as high as that (3.0 GPa) of the as-prepared fibers within the experimental error. In contrast, the residual tensile strength of oxidized SiC fibers with an overcoat of cristobalite approaches zero. Additionally, the resulting SiBCN ceramic fibers offer high elastic modulus at 1400 °C/Ar, a low coefficient of thermal expansion, good flexibility and low density [49, 50]. Interestingly, SiBCN fibers are the only fibers that seem to fulfill the rather ambitious requirements the potential users (European turbine producers) stated for so-called third generation fibers.

The development of quaternary SiBCN ceramics can be traced back to the precursor-derived SiC and Si_3N_4 [51, 52]. In the early '60s, Ainger [53] and Chantrell [54] have reported the synthesis of non-oxide ceramics derived from molecular precursors. In the early '70s, Verbeek [55, 56], Winter [57] and Yajima [58–61] developed sequentially the practical production of Si_3N_4/SiC and SiC ceramic fibers by transformation of polyorganosilicon compounds. Since then, the preparation of precursor-derived non-oxide ceramics has gained significant interest. Many efforts have been made to produce the classical binary ceramics (SiC and Si_3N_4) and multi-component materials (SiCN, SiBN, SiBCN). In 1985, Takamizawa [62–64] first synthesized the SiBCN precursors. Until the '90s, new breakthroughs were made in the development of precursor-derived ceramic monoliths with tailored microstructures and excellent properties. In the late '90s, Riedel et al. [42] reported processing of

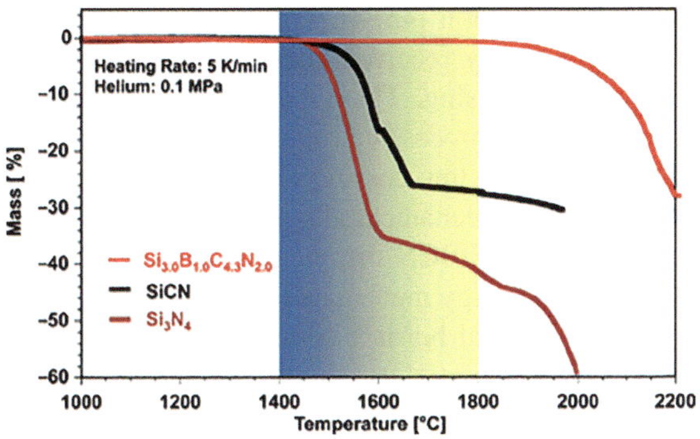

Fig. 1.2 The thermogravimetric curves of the precursor-derived SBCN, SiCN and Si_3N_4 ceramics when heated to 2000–2200 °C in an inert atmosphere [42]

SiBCN monoliths by precursor pyrolysis together with Si_3N_4, Si_3N_4/SiC and SiCN monoliths [65–67] (Fig. 1.2).

The development of SiBCN and SiBCN-Me (Me=Al, Ti, Zr, Hf, etc.) materials including precursor synthesis and ceramic processing has received much attention from researcher working on the chemistry and material science [68–71]. The preparation process can be generally divided into five steps: (1) Synthesis of the precursor by standard Schlenk techniques [72, 73]; (2) Crosslinking or curing; (3) Warm pressing (200–400 °C) of ground SiBCN precursor powders; (4) Thermolysis (1000–1400 °C) of warm pressed precursor shapes to obtain amorphous SiBCN ceramic shapes; (5) Post-annealing (temperatures up to 2000 °C) of amorphous ceramics to obtain polycrystalline nanocomposites [43, 74].

The chemical composition, microstructures and properties of precursor-derived ceramics SiBCN (PDCs-SiBCN) can be tailored effectively by controlling the initial reagent structures, chemical reactions and processing parameters. The thermolysis process generally proceeds at relatively low temperatures (1000–1400 °C) and the resulting polycrystalline SiBCN provides clean grain boundaries. The precursor-derived ceramic route (hereinafter referred to as the organic method) is the only mechanism for producing SiBCN fibers, films, monoliths and their ceramic matrix composites with addition of continuous fibers [75–82]. The investigation of PDCs-SiBCN materials over the last three decades continues to confirm their excellent properties, as noted above.

Unfortunately, the organic method also has some shortcomings that limit its widely commercial utility. The raw materials are usually expensive thus leading to high-cost products. Some organic solvents are harmful to human health and the environment. They are also flammable and present waste disposal problems. This processing route is also low-yield and complex. The microstructures and properties of final products are greatly influenced by multiple factors. The gas release and mass loss during pyrolysis requires careful monitoring to avoid porous structures and to carefully control shrinkage (even microcracks). These issues can be countermanded by addition of passive or active fillers (like ceramic or metal powders) to minimize shrinkage during pyrolysis [41, 83–85]. The organic method can be problematic when used to fabricate dense SiBCN monoliths and structural parts, although the vacuum-assisted polymer infiltration [45, 46], pressure-assisted pyrolysis [86], and sintering [87–89] have recently received attention and can offer better processing than simple precursor processing. Thus, there is considerable motivation to develop new and complementary processing routes.

As a relatively low-cost and simple approach to monoliths, Jia et al. [90] proposed in 2004 using mechanical alloying (MA) as a novel inorganic processing route to fabricate dense SiBCN ceramics. Amorphous SiBCN powders are synthesized by simple, high energy mechanical alloying using inorganic elemental powders (cubic silicon, hexagonal boron nitride, graphite, boron, aluminum, zirconium, hafnium, etc.) as starting materials, and then sintered by hot pressing (HP), or spark plasma sintering (SPS) or hot-isostatic pressing (HIP) to produce fully dense monoliths. The only drawback to this approach is the need for high sintering temperatures up to 1700–2000 °C and crystallization of the intermediate amorphous powders which cannot be avoided. So, nanocrystalline microstructures with residual amorphous phases are obtained eventually similar to those of nanocrystalline SiBCN materials developed by precursor pyrolysis after higher temperature treatment.

Note that recent use of high-pressure synthesis technique provides quite dense and completely amorphous SiBCN monoliths the first to be prepared at high pressures and low temperatures. The inorganic processing route has proven simple and effective for preparing SiBCN monoliths and structural parts with the uniform microstructures and good properties. This processing route readily provides centimeter-sized samples convenient for evaluating basic properties and thus complements the precursor route especially for producing SiBCN monoliths. Alternatives to the precursor and inorganic methods include chemical vapor deposition (CVD) [91–94], reactive magnetron sputtering [75–77] and sol–gel processing [95]. The development of PDCs-SiBCN ceramics processing has been reviewed by Jansen [50] and Baldus [49, 96], Kroke and Li [97], Bill [98], Riedel [74] and Colombo [42], and Viard [99]. Coincidently, the phase equilibria and material thermodynamics of PDCs-SiBCN coupled with the ternary systems (SiBN, SiBC, SiCN) were summarized by Aldinger and Seifert [100].

1.2 Crystallization and Phase Equilibrium in Amorphous SiBCN

The outstanding structural stability at high-temperatures, and good resistance to oxidation (Fig. 1.3) and creep are the main features of the amorphous PDCs-SBCN materials. Apparently, the temperature stability of the PDCs-SBCN is limited and strongly depends on the particular ceramic composition and presence of strong 3-D amorphous covalent network structures [101–103].

Generally, these amorphous PDCs-SiBCN materials contain amorphous Si–C-N domains related with regions of amorphous domains composed of stacked layers containing carbon, boron, and nitrogen (turbostratic B-N-C domains) [104]. Further

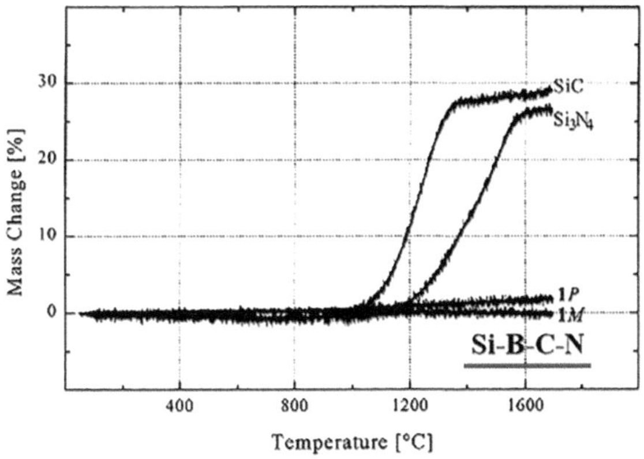

Fig. 1.3 The thermogravimetric curves of the amorphous PDCs-SiBCN, crystalline SiC and crystalline Si_3N_4 ceramics when heated to 1650 °C in flowing air [81]

annealing of the as-thermolyzed PDCs-SiBCN at temperatures $T \geq 1400$ °C leads to the formation of SiC nanocrystallites [105, 106], as the primary crystalline phase, within the amorphous microstructure. Si_3N_4 nanocrystallites emerge at significantly higher temperatures, depending on the boron content. In this connection, the increase in boron content up to 8–9 at% retards the crystallization of Si_3N_4 [107], whereas the crystallization of SiC is promoted with addition of the boron content (Fig. 1.4). The available thermodynamic model of amorphous Si-C-N domains, nanocrystalline Si_3N_4, and nanocrystalline SiC treated as the separated phases with variable dimensions were proposed by Tavakoli [108] to the as-forementioned kinetic crystallization process. Note that this thermodynamic model approach comprises of both nucleation and growth processes, which are connected with kinetic aspects of crystallization.

Clear information on phase equilibria in PDCs-SiBCN system at constant temperatures can be derived from calculated quaternary isothermal sections. Concentration sections through the 1673 and 2273 K tetrahedra at 10 at% boron were calculated by Seifert et al. [100] (Fig. 1.5). In both cases the (regular) concentration tetrahedron is divided into different distorted tetrahedra indicating the four-phase equilibria at the specific temperature. The phase equilibria and composition of some high temperature ($T > 2273$ K) stable PDCs-SiBCN PMBS(m), PHBS(p), PHBS(m), PMBS(p)

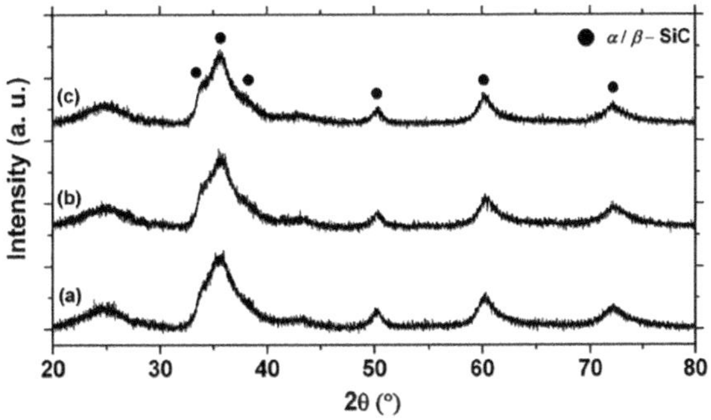

Fig. 1.4 XRD patterns of the PDCs-SiBCN system: **a** 3.7 at% B; **b** 6.0 at% B; **c** 8.3 at% B prior to the crystallization of Si_3N_4 after isothermal heat treatment at $T = 1700$ °C for 2, 6, and 15 h, respectively [107]

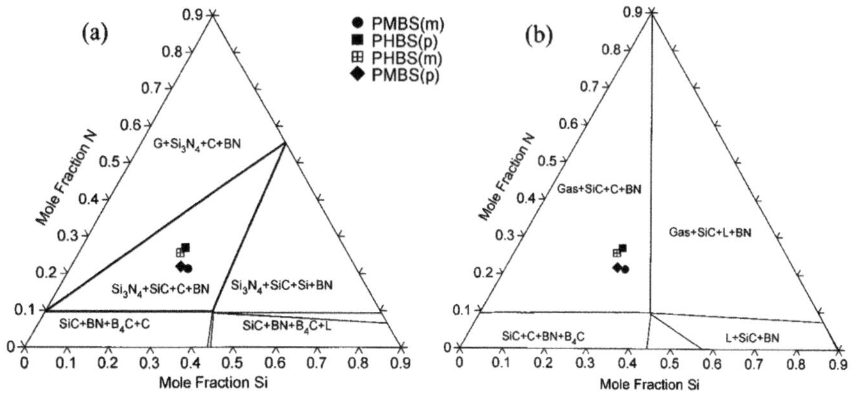

Fig. 1.5 Concentration sections in the PDCs-SiBCN system at constant boron content (~ 10 at%) and temperature: **a** $T = 1673$ K; **b** $T = 2273$ K [100]

are indicated (all with ~ 10 at% boron content). At $T = 1673$ K, they are all located in the phase field $Si_3N_4 + SiC + C + BN$. This phase equilibrium remains stable as low as room temperature. However, at high temperatures (e.g., 2273 K) the composition is located in the phase field Gas + SiC + C + BN, and some decomposition is expected.

The phase fraction diagrams computed by Thermo-Calc software [109] for the composition of the PDCs-SiCN and PDCs-SiBCN have been compared (Fig. 1.6) [101]. In these calculations, the hydrogen content of the ceramics was ignored, assuming hydrogen exists in the form of H_2 gas at temperatures $T > 1050$ °C [99]. The equilibrium state of the PDCs-$SiC_{1.4}N_{0.9}$ consists of three phases SiC, Si_3N_4 and C at low temperatures ($T < 1484$ °C), and its thermal stability is limited by the carbothermal reduction of Si_3N_4 with C at $T = 1484$ °C. As expressed, this reaction results in the formation of SiC and a mass loss of the solid phase due to the gas (N_2) escape. Consequently, the new phase equilibrium is obtained at $T >$ 1484 °C, including SiC, C and N_2. This equilibrium state remains unchanged up to the maximum temperature of calculation $T = 2000$ °C.

The main difference between the phase fraction diagrams computed for the PDCs-$SiC_{1.4}N_{0.9}$ and PDCs-$SiC_{1.5}N_{1.0}B_{0.05}$ materials is the emergence of the additional phase BN and the consequential formation of the four-phase equilibrium SiC + Si_3N_4 + C + BN for the boron-containing ceramics [101]. However, the same reaction as explained above for PDCs-$SiC_{1.4}N_{0.9}$ also determines the stability of PDCs-$SiC_{1.5}N_{1.0}B_{0.05}$. Since BN is very stable within the computed temperature range, the amount of BN phase remains unchanged and consequently, the three-phase equilibrium including SiC, C and BN is anticipated at the temperature range between 1484 and 2000 °C for the PDCs-$SiC_{1.5}N_{1.0}B_{0.05}$. Note that BN phase regarded as non-interacting with other phases in this PDCs-SiBCN system to generate the computed phase fraction diagrams may be considered insufficiently, since the amorphous

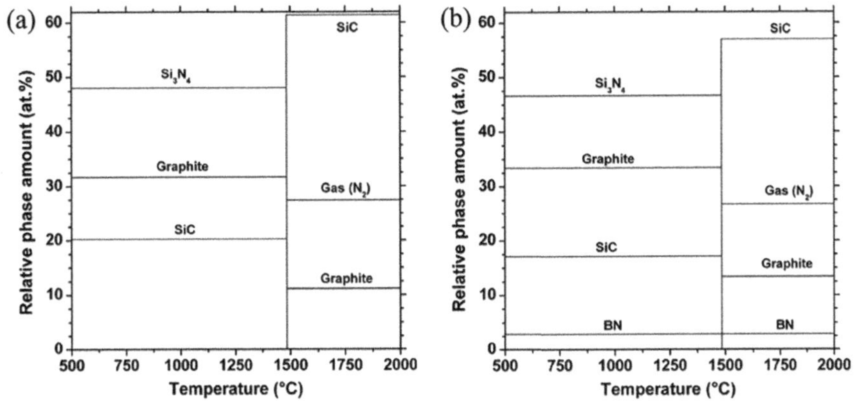

Fig. 1.6 The computed phase fraction diagrams: **a** PDCs-$SiC_{1.4}N_{0.9}$; **b** PDCs-$SiC_{1.5}N_{1.0}B_{0.05}$ (hydrogen content in ceramics was ignored) [101]

domains of Si-C-N and B-C-N must interact on each other during crystallization process.

1.3 Microstructure and Performance Characteristics of SiBCN

1.3.1 Overview of Microstructural Features

The microstructural features of the mechanical alloying SiBCN (MA-SiBCN) nanocrystalline ceramics are extremely similar to those of the precursor-derived counterparts. The chemical composition of the studied MA-Si_2BC_3N is approximately located on the triple-phase line of SiC, BN and C. Consequently, crystalline MA-Si_2BC_3N ceramics should consist mainly of SiC and BN(C), confirmed experimentally. The PDCs-SiBCN nanocrystalline ceramics consist of the same phases especially when the chemical composition is located in the four-phase field of SiC + BN + C + B_4C. Coincidently, the encapsulation of SiC by BN(C), the turbostratic structure of BN(C) and "clean" grain boundaries are common features to the two "types" of crystalline SiBCN materials.

The only difference is that other phases including Si_3N_4, SiC_xN_y, B_4C, Si likely form in PDCs-SiBCN system depending on initial chemical composition and processing. For example [110], a PDCs-SiBCN ceramic particle with a chemical composition of $Si_{24}B_8C_{44}N_{24}$ obtained by hydroboration of oligovinylsilazane is composed of Si_3N_4 and SiC nanocrystallites which are embedded in a turbostratic BNC_x matrix after being annealed at 1800 °C for 10 h in N_2 (Fig. 1.7). However, another PDCs-SiBCN ceramic monoliths pyrolyzed at 1400 °C for 2 h in Ar derived

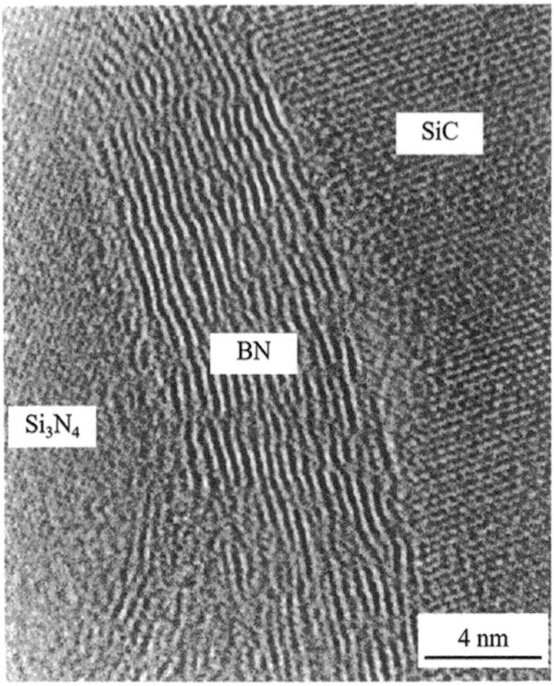

Fig. 1.7 The typical microstructure composed of the nano SiC, Si_3N_4 and turbostratic BN(C) for a crystallized PDCs-SBCN material [110]

from a boron-modified polyvinylsilazane remain amorphous but form α-SiC crystals of a mean grain size of ~ 100 nm embedded in a turbostratic BNC_x matrix after annealing at 1800 °C for 3 h in N_2.

Note that, nitrogen cannot be added independently in MA so that the composition design is less flexible [111]. Additionally, it is likely difficult to produce powders by MA with extremely uniform elemental distribution and strong chemical bonds, like that in PDCs route [112]. As a result, MA derived amorphous SiBCN powders exhibit a relatively low initial temperature of crystallization. However, the dense MA-SiBCN monolithic ceramics with a relatively large dimension can be easily prepared by this inorganic route thus facilitating the material property evaluation.

Both β-SiC and α-SiC form in hot-pressed MA-Si_2BC_3N at 1900 °C. The SiC grains are nearly equiaxed with average grain size (AGS) of ca.100 nm [113–115]. Most β-SiC grains are well developed with a high degree of crystallinity. A few β-SiC grains contain stacking faults, which is common in β-SiC single crystals, whiskers, nanowires and monoliths. These defects are usually found in the (111) plane acting as the primary fault plane. The α-SiC content is relatively low compared with β-SiC. They have similar shape, size and stacking faults. However, α-SiC grains show hexagonal wurtzite structure and their (0002) planes contain most of the stacking faults (Fig. 1.8).

BN(C) grains have no fixed shape and are uniformly distributed around SiC grains [114]. Most are 15–50 nm thick. A few larger ones with AGS of ca. 100 nm also

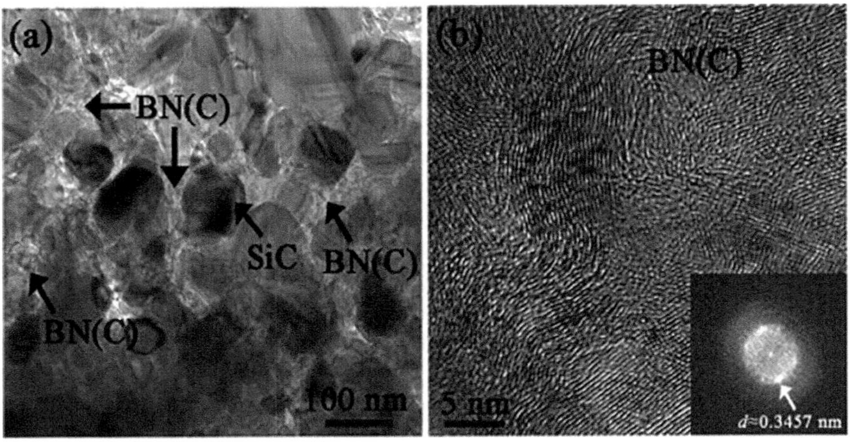

Fig. 1.8 Microstructural feature of MA-Si_2BC_3N monoliths hot-pressed at 1900 °C: **a** Bright filed image showing the shape, size and the distribution of each phase; **b** HRTEM image of turbostratic BN and corresponding fast Fourier transform (FFT) pattern [115]

appear in selected regions. BN(C) grains exhibit a turbostratic structure, as found in PDCs-SiBCN crystalline ceramics. They consist of layers similar to graphite or h-BN, rotated by a random angle around the *c* axis or perpendicular to *c* axis. Therefore, the interplanar spacings along the c axis in turbostratic-BN(C) are generally enlarged by different degrees. The detected (0002) peak of BN(C) shifts towards lower angle by ~0.7° resulting in the expansion of *d*(0002) by 0.009 nm. Coincidently, the fluctuation in interlayer spacings in graphite or h-BN is generally accompanied by rotation, translation or curvature of single layers. The structural distortion and peak shift result in the formation of turbostratic structure with a relatively low three-dimensional order degree. Turbostratic BN(C) is most likely composed of heterogeneously distributed t-carbon layers, t-BN layers and B-doped t-carbon layers.

The formation of turbostratic BN(C) can be attributed to the easy-to-strip character of graphite and h-BN. The fierce collisions during MA break h-BN and graphite into tiny fragments, mix them as uniformly as possible and activate mechanochemical effects leading to formation of B-N-C and B-C-N bonds. Coincidently, h-BN and graphite have similar crystal structure and lattice parameters thus providing chance to the solid solution of carbon in t-BN, or boron and nitrogen in t-carbon. However, the atomic arrangement and chemical bonding states of BN(C) in SiBCN materials developed both by inorganic processing route and by polymer precursor method are still not clearly understood thus requiring further investigations.

The atomic arrangements in SiC and BN(C) phases at and near the interfacial regions are quite different [115]. SiC atoms exhibit long-range order, while BN(C) species have only short- or at best medium-range order. The interfacial regions between the two phases are limited to three or four atomic layers, 0.8–1.1 nm thick.

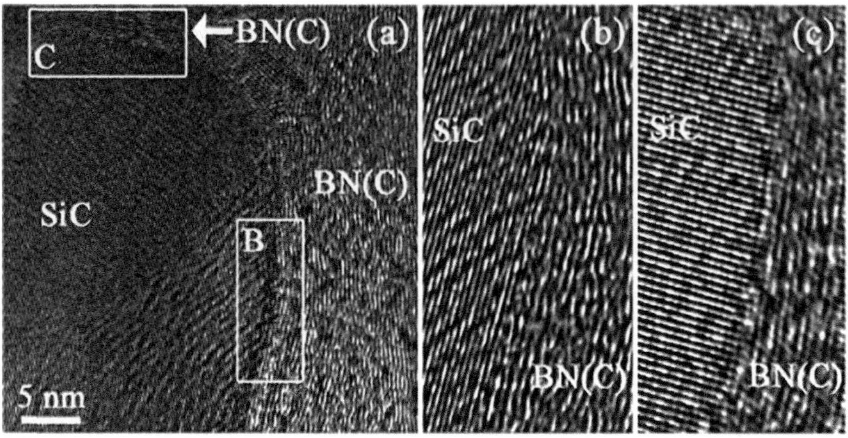

Fig. 1.9 Microstructural features of the interfacial region between SiC and BN(C) in MA- Si_2BC_3N ceramics hot-pressed at 1900 °C: **a** HRTEM image; **b**, **c** inverse FFT images of the areas marked by the white rectangles B and C in (**a**), respectively [115]

No impurity or any other phases like low-melting-point phases form in pure MA-SiBCN monoliths (Fig. 1.9). The "clean" interfacial regions coupled with capsule-like structures hinder long-range atomic diffusion thus limiting the grain growth. This is favorable to the structural stability, resistance to creep and oxidation of the SiBCN materials at high temperatures.

1.3.2 Mechanical Properties

Due to the limits of the sample sizes, only hardness and elastic modulus tests can be performed for the PDCs-SiBCN monoliths, in some cases toughness can also be measured using fracture-mechanics methods, which evaluate the critical value of a crack-driving force [116–122]. Compared to the SiC and Si_3N_4 crystalline ceramics, amorphous PDCs-SiBCN have lower density and Young's modulus, but higher microhardness. The tensile strength of amorphous PDCs-SiBCN ceramic fibers at room temperature is up to 3.2–4.0 GPa and the elastic modulus is in range of 200–210 GPa (Table 1.1). The maximum using temperature of the PDCs-SiBCN fibers in the air can reach to ~ 1500 °C, while their tensile strength at 1500 °C is obtained at ~ 2.3 GPa. The elastic modulus of the PDCs-SiBCN fibers at 1400 °C can still maintain 80–90%. The average thermal expansion coefficient of the PDCs-SiBCN ceramic fibers is 3.5×10^{-6} °C^{-1} [52, 82, 117, 118].

MA-Si_2BC_3N monoliths exhibit excellent mechanical properties including room temperature flexural strength to 400 and 240 MPa at 1400 °C in air, and elastic modulus of 150 GPa. The flexural strength and fracture toughness of MA-Si_2BC_3N are equivalent to pure SiC values. The elastic modulus is 1/3 ~ 1/2 lower than SiC, and

Table 1.1 Comparison of the mechanical properties for the different Si-based ceramic fibers [43]

Properties	SiC	SiOC	SiCTiO	SiN	SiBC	SiBCN	SiBCNO
Room temperature tensile strength (GPa)	3.0–4.0	3.0	2.9	2.1	3.4–4.8	3.2–4.0	4.0
1500 °C tensile strength (GPa)	–	–	–	–	–	1.5	–
Room temperature Young's modulus (GPa)	300–420	190–250	200	220	430	200–210	400
Fracture elongation	0.6–1.0	–	–	–	–	–	–
Thermal expansion coefficient ($\times 10^{-6}$ °C^{-1})	3.3–4.0	–	–	–	–	3.5	–
Density (g/cm^3)	2.74–3.10	2.5	2.5	2.4	3.1	1.8	–
Diameter (μm)	10–14	10–20	8–12	–	8–10	10–12	~ 10

the hardness is ~4/5 lower. The combination of high flexural strength and relatively low elastic modulus contributes to excellent thermal shock resistance [123]. Furthermore, a relative low hardness coupled with the turbostratic structure of BN(C) phase is beneficial to machinability. An increasing ratio of Si: C results in an increasing relative content of α/β-SiC in the as-sintered MA-SiBCN ceramics, leading to a nearly linear increase in mechanical properties. The flexural strength, elastic modulus, fracture toughness and Vickers' hardness of the MA-SiBCN ceramics with a Si: C ratio of 3: 4 are 511 MPa, 157 GPa, 5.64 MPa·m$^{1/2}$ and 5.9 GPa, respectively [124].

The mechanical properties of the as-sintered monoliths are also correlated with the sintering atmosphere [124]. The hot-pressed MA-Si_2BC_3N monoliths sintered in 1 bar N_2 atmosphere show satisfying bulk density, flexural strength, elastic modulus and fracture toughness, higher than those sintered in Ar. For instance, the bending strength of the MA-Si_2BC_3N ceramics sintered in N_2 and Ar, is 526 MPa and 422 MPa, respectively. The Si_2BC_3N-Al monoliths show flexural strength of 527 MPa and fracture toughness of 5.25 MPa m$^{1/2}$. A high N_2 partial pressure inhibits decomposition of some phases during sintering of Si_2BC_3N-Al ceramics. Consequently, the values of density, flexural strength, elastic modulus and fracture toughness of the Si_2BC_3N-Al ceramics sintered in N_2 are much higher than those of sintered counterparts in Ar.

Increases in a boron content results in formation of B_xC and belt-like BN(C) improving the hardness and fracture toughness [4]. Hot-pressed MA-$Si_2B_{1.5}C_2N$ bulk ceramics at 1900 °C provide better mechanical properties: flexural strength of 413 ± 13 MPa, fracture toughness of 5.41 ± 0.91 MPa·m$^{1/2}$, elastic modulus of 213 ± 7 GPa, Vickers' hardness of 7.66 ± 0.32 GPa. Hot-pressed MA-Si_2BC_xN ($2 \leq x \leq 3$) ceramics at the same hot-pressed condition also provide desirable mechanical properties. But an under or over dose of carbon will deteriorate their mechanical properties.

The addition of reasonable Al, Zr, Al–Zr, AlN, ZrO_2, ZrB_2, HfB_2, TaC, ZrC, TaC, Ta_4HfC_5, MgO–ZrO_2–SiO_2 (MZS), MgO–Al_2O_3–SiO_2 (MAS), multi-walled carbon nanotubes (MWCNTs), or graphene platelets (GPLs), etc. leads to varying improvement on mechanical properties of the MA-Si_2BC_3N materials [43]. Few ZrO_2 addition results in the optimal reinforcing effects, while MZS addition results in the optimal toughening effects. Additionally, a desirable strengthening together with toughening effect is also achieved by adding metal Al particle [125] or MWCNTs. For example, the Si_2BC_3N-Al monoliths show flexural strength of 527 MPa and fracture toughness of 5.25 MPa $m^{1/2}$ (Table 1.2).

The as-sintered MA-SiBCN bulk ceramics exhibit brittle fracture characteristics when loaded at room temperature; however, the C_f/Si_2BC_3N composites display pseudoplastic fracture behavior suggesting potential applications in thermal shock environments [126, 127]. On this basis, the introduction of few ZrO_2 in C_f/Si_2BC_3N composites effectively promotes their densification with moderate interfacial bonding, thereby guaranteeing the fiber pullout and bridging [123]. Nevertheless, the introducing SiC_f cannot improve the fracture toughness of MA-Si_2BC_3N, which still show brittle fracture behavior. However, once h-BN is coated on SiC_f surfaces, the fracture toughness of the SiC_f/SiBCN composites is significantly improved; however, although fiber pullout and crack deflection are observed, and the resulting composites still show catastrophic fracture. Yet if both short C_f and SiC_f are incorporated in MA-Si_2BC_3N matrix, the $(C_f\text{-}SiC_f)/Si_2BC_3N$ composites show pseudoplastic fracture characteristics even though lower relative densities and room temperature mechanical properties are obtained. The mechanical properties of the $(C_f\text{-}SiC_f)$/SiBCN composites decrease after introducing Pyrolytic carbon (PyC) or h-BN weak interfaces, but still maintain pseudoplastic fracture characteristics.

1.3.3 Oxidation Resistance and Oxidation Mechanisms

(1) **Oxidation resistance property**

MA derived Si_2BC_3N monoliths both show excellent oxidation resistance in dry air and moist air. Only a thin oxide scale less than 10 μm thick forms on monolithic surfaces oxidized at 1200 °C for 85 h in flowing dry air. The oxide scale thickness is about 200 μm, though the oxidation rate is markedly accelerated in moist air with an absolute humidity of 0.8 g/cm^3. The ternary chemical bonds including C-B-N and B-C-N can effectively lower diffusion of oxygen inward thus leading to improved oxidation resistance superior to SiC-BN monolithic ceramics.

Too much addition of oxygen-sensitive carbon with fine grain sizes, whether free carbon or carbon imbedded into interlayers of h-BN will weaken the oxidation resistance of MA-Si_2BC_xN ($x = 1$–4) monoliths. The excessive introduction of boron into MA-$Si_2B_yC_3N$ ($y = 1.5$–4) is also detrimental to the oxidation property. Too much boron addition causes the rapid consumption of BN(C), B_xC and loss of B_2O_3 thereby promoting the oxidation rate at high temperatures (≥1500 °C).

The Al, Zr, ZrO_2, AlN and MZS etc. additives also have a significant influence on oxidation resistance of MA-$Si_2B_2C_2N$ monolithic ceramics [128–135]. ZrO_2 and AlN

Table 1.2 The bulk density and mechanical properties of the MA-SiBCN monoliths and their ceramic matrix composites prepared by various sintering techniques

Material composition	Density (g/cm^3)	Flexural strength (MPa)	Young's modulus (GPa)	Fracture toughness ($MPa \cdot m^{1/2}$)	Hardness (GPa)	Fracture modes
[a]SiBCN	2.12–2.82	66.3–423.4	30.6–139.4	1.25–3.09	2.4–5.7	Brittle
[b]SiBCN	2.64–2.83	–	220.0–291.0	3.60–4.25	20.0–29.4	Brittle
[c]SiBCN	2.52–2.76	331.0–459.5	120.0–139.4	3.28–4.77	5.7–11.0	Brittle
[d]SiBCN	2.49–2.84	200.0–511.0	133.0–220.0	2.80–5.64	1.5–6.0	Brittle
SiBCN-Zr	3.22–4.08	202.0–400.0	142.1–252.4	2.34–3.16	3.3–9.6	Brittle
SiBCN-Al	2.77–2.90	422.0–527.0	174.0–222.0	3.40–5.25	11.6–12.7	Brittle
SiBCN-(Zr-Al)	2.70–2.75	480.1–590.2	120.4–122.6	4.93–5.60	5.2–5.8	Brittle
SiBCN-Al_4SiC_4	2.65–2.66	345.1–451.5	119.4–128.4	5.17–5.44	6.7–6.9	Brittle
SiBCN-ZrC	2.33–2.64	153.9–229.5	113.4–141.7	1.96–2.25	2.5–3.4	Brittle
SiBCN-TaC	2.61–2.79	351.3–399.5	141.4–164.1	4.04–4.26	5.0–7.2	Brittle
SiBCN-Ta_4HfC_5	2.68–2.79	310.2–344.1	–	3.67–4.52	–	Brittle
[e]SiBCN-ZrB_2	2.82–3.53	315.0–411.0	133.0–172.0	3.50–5.10	3.1–7.8	Brittle
[f]SiBCN-ZrB_2	2.78–2.99	512.3–559.6	163.6–178.9	6.71–6.77	5.2–5.7	Brittle
[g]SiBCN-ZrB_2	3.67–3.74	343.1–386.1	232.8–242.4	4.56–4.84	4.3–4.6	Brittle
SiBCN-HfB_2	3.29–4.91	158.2–176.1	173.5–270.1	3.88–4.17	6.9–7.2	Brittle
SiBCN-(TiB_2-TiC)	2.03–2.80	158.4–311.2	76.1–141.3	1.85–3.92	1.5–7.7	Brittle
SiBCN-ZrO_2	2.83	575.4 ± 73.7	159.2 ± 21.7	3.67 ± 0.01	6.7 ± 0.7	Brittle
SiBCN-AlN	2.74	415.7 ± 147.3	148.4 ± 8.3	4.08 ± 1.18	6.4 ± 1.2	Brittle
SiBCN-MZS	2.78	394.2 ± 41.7	152.9	5.86 ± 0.06	8.3 ± 0.6	Brittle
SiBCN-MAS	2.44–2.58	274.4–425.1	–	–	4.2–6.2	Brittle
SiBCN-$C_{f(PyC)}$	1.92–2.18	30.4–70.5	20.3–55.6	2.24–2.38	0.9–2.3	Pseudoplastic

(continued)

Table 1.2 (continued)

Material composition	Density (g/cm^3)	Flexural strength (MPa)	Young's modulus (GPa)	Fracture toughness (MPa·m$^{1/2}$)	Hardness (GPa)	Fracture modes
SiBCN-(C_f-ZrO_2)	2.39	112.4 ± 12.1	111.1 ± 23.3	2.94 ± 0.25	–	Pseudoplastic
SiBCN-(C_f-MAS)	2.38–2.56	256.9–400.2	–	3.3–4.5	4.2–6.2	Brittle
SiBCN-(C_f-ZrB_2)	2.20–3.36	221.2–405.7	108.8–176.4	3.21–5.25	2.6–4.5	Pseudoplastic
[h]SiBCN-SiC_f	2.35–2.57	70.2–284.3	64.1–183.5	1.04–3.66	–	Brittle
[i]SiBCN-$SiC_{f(BN)}$	2.14–2.20	79.0–101.2	–	1.63–1.94	2.9–4.7	Brittle
[i]SiBCN-($SiC_{f(BN)}$-MAS)	2.69–2.76	318.7–436.6	–	5.87–6.81	9.1–11.4	Brittle
[i]SiBCN-(SiC_f-MAS)	2.69–2.76	363.1–453.3	–	5.50–6.15	10.1–11.2	Brittle
[j]SiBCN-$SiC_{f(BN)}$	2.62–2.70	149.4–208.0	92.9–114.3	2.50–3.92	–	Brittle
[j]SiBCN-(C_f-SiC_f)	2.24	97.2 ± 2.6	83.8 ± 2.5	3.51 ± 0.22	–	Brittle
[j]SiBCN-($C_{f(BN)}$-$SiC_{f(BN)}$)	2.18	69.2 ± 8.5	72.9 ± 8.3	2.60 ± 0.21	–	Pseudoplastic
[j]SiBCN-($C_{f(PyC)}$-$SiC_{f(PyC)}$)	2.04	59.4 ± 5.1	54.0 ± 5.0	2.04 ± 0.31	–	Pseudoplastic
SiBCN-GPLs	2.17–2.45	135.3–196.6	94.6–150.1	3.04–5.40	2.4–5.4	Brittle
SiBCN-MWCNTs	2.58	462 ± 50	115 ± 2	5.54 ± 0.6	5.1 ± 0.2	Brittle
SiBCN-$MWCNTs_{(SiC)}$	2.59–2.61	390.9–532.1	142.5–144.0	5.80–6.66	–	Brittle

[a] Hot-pressed sintering; [b] High-pressure sintering; [c] Hot-isostatic sintering; [d] Spark plasma sintering; [e] Introduction of sol–gel derived ZrB_2; [f] Introduction of MA derived nano ZrB_2; [g] Introduction of Zr and B mixture; [h] Introduction of unmodified SLFC1 SiC_f; [i] Introduction of KD-SA SiC_f; [j] Introduction of KD-II SiC_f

are detrimental to oxidation resistance with oxidation initiated below 1000 °C, and accelerated above 1200 °C. A large number of pores form in the oxide scale ~ 100 μm away from the oxidized surfaces. This could be due to the high diffusion rate of oxygen in ZrO_2 or AlN providing the extra inward diffusion paths for oxygen during composite oxidation. In contrast, MZS significantly improves oxidation resistance. The oxide scale thickness of MZS/$Si_2B_2C_2N$ composite ceramics oxidized at 1500 °C for 10 h is only 10 ~ 20 μm, while it is up to 40 ~ 100 μm for the pure MA-$Si_2B_2C_2N$ oxidized at 1400 °C with a dwell time of several hours.

Significantly, the addition of ultrahigh temperature ceramics (UHTCs) [131, 132], such as ZrB_2, HfB_2, ZrC, Ta_4HfC_5 and TaC, etc. shows higher oxidation resistance beyond 1500 °C due to the formation a three-dimensional oxide skeleton of UHTCs based oxides fulfilled within an amorphous silica glass. For example, the oxide layer thickness of 10 wt.% nano Ta_4HfC_5 containing MA-$Si_2B_2C_2N$ monoliths is only ~ 14.5 μm after oxidation at 1650 °C for 5 h, much lower than ~ 91.9 μm of pure MA-$Si_2B_2C_2N$.

(2) **Oxidation thermodynamics**

As noted above, the as-sintered MA-$Si_2B_2C_2N$ monoliths are composed of α/β-SiC and BN(C) sometimes together with residual amorphous phase. BN(C) consists of heterogeneously distributed t-carbon, t-BN and B-doped t-carbon layers. Therefore, the oxidation of MA-$Si_2B_2C_2N$ likely consists of oxidative attack on the amorphous phases, SiC, h-BN and C.

The oxidation of carbon results in the release of CO and/or CO_2 and coincidently, the oxidation of BN results in the B_2O_3 formation and nitrogen release. The formed liquid-B_2O_3 layer with a melting point of 450 °C uniformly covers the sample surfaces and starts to evaporate at ~ 900 °C. Above 1200 °C, the rapid evaporation of B_2O_3 occurs due to its high vapor pressure [136]. The high temperature oxidation of SiC may be either passive or active depending on the local oxygen concentration. Silica usually forms as a dense protective oxide scale via the passive oxidations at oxygen pressures close to one bar. The reaction can start below 780 °C leading to a net increase in the mass. The SiO_2 scale protects SiC from further catastrophic oxidation to SiO occurring at higher temperatures. Such protective action can theoretically continue up to the melting point of SiO_2 (~ 1720 °C) [137].

However, in cases where the oxidant pressure is less than one bar, the pressure is insufficient to form SiO_2 and SiO form resulting in rapid consumption of SiC via the active oxidations at higher temperatures [138]. At 1200–1300 °C, B_2O_3 can react thermodynamically with SiO_2 forming borosilicate glass [139]. The borosilicate will discompose leading to evaporation of B_2O_3 and even SiO_2 at much higher temperatures. Therefore, the formation of bubbles and loose, porous oxide scale are due primarily to the release of CO, CO_2, N_2 and the evaporation of B_2O_3.

(3) **Oxide scale microstructure**

The oxide scale of MA-$Si_2B_2C_2N$ monoliths show mono- or dual-layer structures depending mainly on the fabrication processing (Fig. 1.10). The hot-pressed MA-$Si_2B_2C_2N$ oxidized at 1300 °C shows a monolayer structural oxide scale, adhered

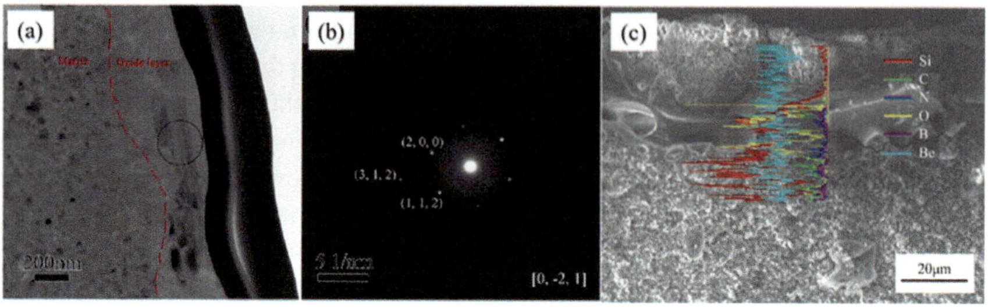

Fig. 1.10 The oxide scale microstructure of the oxidized MA-Si_2BC_3N monoliths consolidated by **a**, **b** HP or **c** SPS: **a**, **b** Oxidized at 1300 °C/1 h in dry flowing air; **c** oxidized at 1200 °C dry air for 85 h [124, 133]

tightly to the ceramic matrix [133]. The oxide scale exhibiting a low degree of crystallinity consists of amorphous SiO_2 and few cristobalite. The MA-$Si_2B_2C_2N$ counterparts prepared by SPS show a dual-layer structural oxide scale [124]. The outer layer is loose and porous which is rich in silicon and oxygen, while the relatively dense inner layer is rich in silicon, oxygen and carbon, together with few boron and nitrogen. The inner layer hinders effectively the inward diffusion of oxygen thus slowing down the oxidation rate.

The oxidation investigation of hot-pressed Zr-containing MA-$Si_2B_2C_2N$ monoliths also indicates a monolayer oxide scale consisting mainly of SiO_2 glass forms on the surfaces at 1500 °C/5 h/air. In contrast, the PDCs-SiBCN materials usually exhibit dual- or multi-layer oxide scale most likely contributing to improved oxidation resistance [82]. It seems reasonable to suggest that thin amorphous oxide scales with a dual-layer structures form on a PDCs-$SiBN_3C$ fiber surfaces in air at high-temperatures responsible for excellent oxidation resistance. The outer layer is composed of silica containing some boron and carbon, while the inner layer is turbostratic B-N phase containing some silicon and oxygen. The PDCs-$SiBN_3C$ fiber shows similar strength to those of polycrystalline SiC fibers, and resistance to oxidation and creep superior to polycrystalline fibers, although the reasons why such microstructure provides better oxidation resistance than single passivating layer of silica are not completely clear.

In another case [140], a tri-layer α-cristobalite/SiO_2/SiBCNO (with BN precipitates) scale forms on a PDCs-$SiBC_{0.8}N_{2.3}$ fiber surfaces heated at 1500 °C for 2 h in air (Fig. 1.11). The unreacted fiber core retains its amorphous nature. The outermost layer oxidized at 1500 °C has devitrified to cristobalite. The second layer consists of essentially pure silica glass. The oxidation layer next to the core consists of amorphous SiBCNO and nanosized turbostratic BN, which evolves into a more oxygen-rich glass with hexagonal and turbostratic BN grains dispersed throughout nearer the surface. This interlayer does not disappear on heating above 1450 °C and is suggested to be the reason for offering 10 times more oxidation resistance than similar coatings of Si_3N_4 and SiC. However, the multilayer oxide scale cannot ensure an expected high-temperature strength which degrades significantly after only short

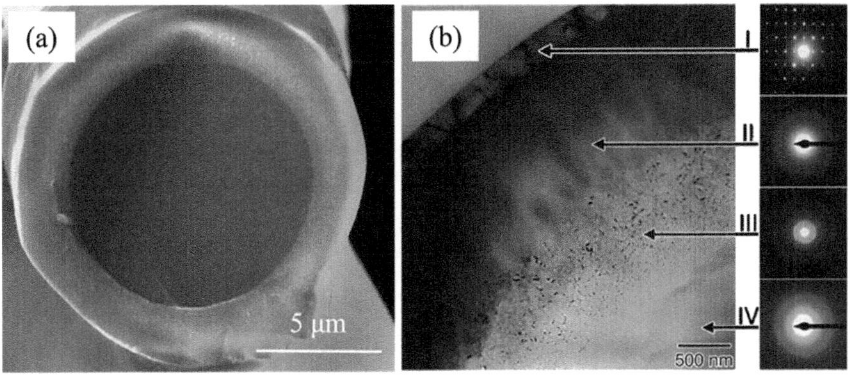

Fig. 1.11 Microstructure of the oxide layer for the amorphous PDCs-$SiBC_{0.8}N_{2.3}$ ceramic fibers after oxidation at 1500 °C for 2 h under dry flowing air condition: **a** SEM cross-section image; **b** bright-field image and corresponding SAED pattern [140]

exposures at temperatures of 1300–1400 °C. Similar oxide scale structure is found for the a PDCs-SiBCN fiber reinforced SiCN composites after air exposure at 1100 °C.

Investigation on a highly dense and amorphous MA-$Si_2B_2C_2N$ monolith prepared by high pressure sintering shows oxidation resistance superior to SiC and Si_3N_4 above 1600 °C, and retains significant resistance to oxidation up to 1800 °C [134]. The microstructure observed for a cross section of the oxidized MA-$Si_2B_2C_2N$ at 1700 °C for 8 h can be divided into three zones from outside to inside: the outermost layer of amorphous N-containing SiO_2; the loose interlayer; the innermost unoxidized amorphous matrix with some nanosized precipitates of SiC and BN(C) (Fig. 1.12). No complex oxide scale like the previously reported two- and tri-layer structures forms here indicating that the good oxidation resistance is not necessarily related to the increasing complexity of oxide scale. No obvious interfaces are observed indicative of strong adhesion between them. Some atomic clusters, which are most likely to be cristobalite, observed in the partial zone close to and beneath the amorphous SiO_2 layer. The loose interlayer mainly includes cristobalite, and unoxidized SiC and BN(C) precipitates.

The formation of dense, passivating surface layers of N-containing amorphous SiO_2 is thus primarily responsible for the excellent oxidation resistance of these amorphous MA-$Si_2B_2C_2N$. Consequently, no necessary correlation exists most likely between the oxidation resistance and the microstructural complicacy of oxide scale, which has been proved by all the investigations up to now.

The phase composition and microstructure of the oxide layers in PDCs-SiBCN ceramic fibers are also impacted by the in-service environment [141]. For example, a PDCs-$SiB_{0.9}C_{1.6}N_{2.4}$ ceramic fiber oxidized at 1400 °C in static air exhibits the three distinct zones with different microstructural feature: (i) Outermost layer of a glass-like structure; (ii) intermediate layer with dispersive nanocrystals; and (iii) innermost homogeneous and amorphous fiber core. All the three zones are dense and intact with no clear boundaries (Fig. 1.13).

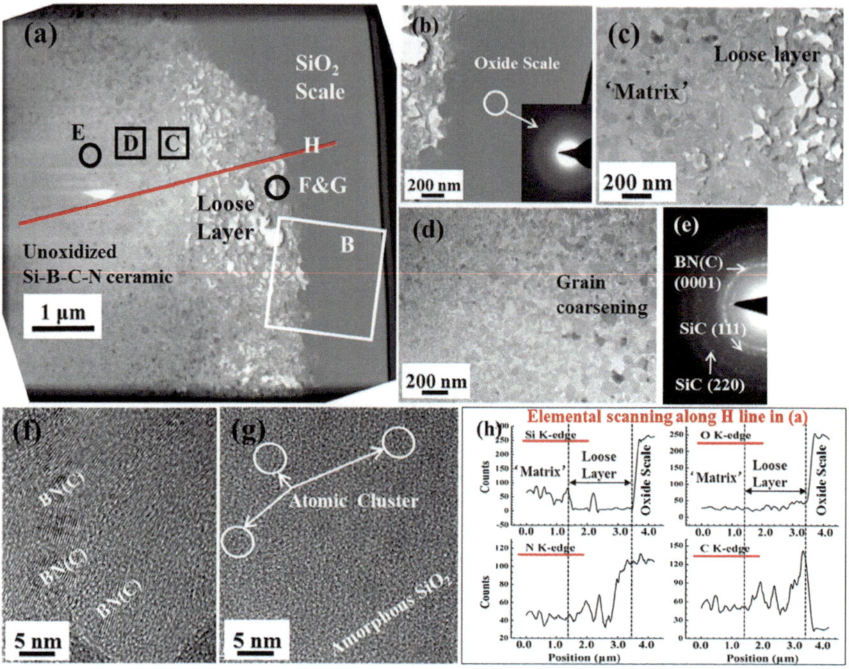

Fig. 1.12 Oxide scale microstructure of a highly dense completely amorphous MA-Si_2BC_3N monolith oxidized at 1700 °C for 8 h: **a** Overview of a cross section; **b–d** magnification of boxes "B, C and D" in (**a**), respectively; **e** SAED pattern of circle "E" in (a); **f**, **g** HRTEM images of circle "F&G" in (**a**); **h** element distribution along the red line "H" in (**a**) [134]

For the PDCs-$SiB_{0.9}C_{1.6}N_{2.4}$ ceramic fiber oxidized at 1400 °C in a simulated combustion environment ($P_{H_2O}/P_{O_2}/P_{N_2}$ = 14:8:78) for 2 h, it also exhibits the three distinct zones: two-layer oxide scale and unoxidized fiber core, which are similar to the former. The outermost layer shows only Si and O elements illustrating the SiO_2. However, its thickness of the outmost layer is ~ 160 nm, which is 300 nm thinner than that of the former (~ 470 nm). Interestingly, the high percentage of water vapor in simulated combustion environment triggers the formation of abundant α-cristobalite nanoparticles dispersing in the outermost amorphous SiO_2 layer, which are absent in fibers corroded in air. The formation of h-BN grains in the interlayer of the two oxide scales can be ascribed to the sacrificial effects of free carbon clusters, Si-C, and Si-N units reacting with oxygen diffusing inward, and thereby improve corrosion resistance of the PDCs-$SiB_{0.9}C_{1.6}N_{2.4}$ fibers under an oxygen- and water-rich atmosphere at high-temperatures.

(4) **Oxidation kinetics**

The oxidation activation energy for amorphous MA-Si_2BC_3N powder (< 0.1 μm) varies from 211.2 (one-step milling) to 561.3 kJ/mol (two-step milling), which are much higher than that of 0.2 μm-sized SiC powder (82.6 kJ/mol) and Si_3N_4 powder (<

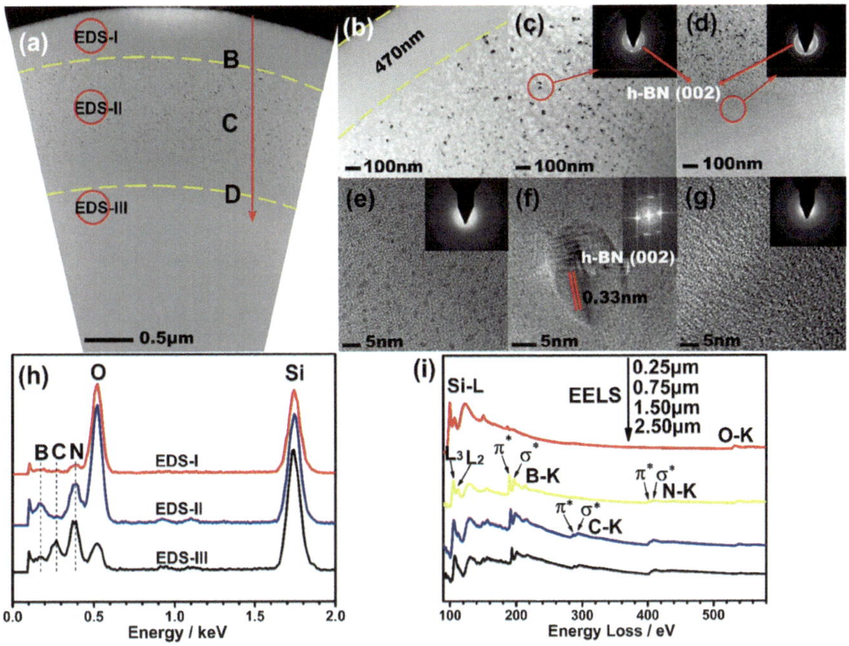

Fig. 1.13 Microstructure of a cross section of the PDCs-$SiB_{0.9}C_{1.6}N_{2.4}$ fiber annealed at 1400 °C in static air for 2 h: **a** Bright-field image; **b**, **d** HRTEM images of the zones marked by "B, C, and D" in (**a**), respectively; **e**, **g** HRTEM images from the glass-like area in (**b**), **c** and amorphous area in **d**; **h** corresponding EDS spectra (I), (II), and (III); **i** EELS spectra with the distance of 0.25, 0.75, 1.50, and 2.50 μm from fiber surface to the inner in (**a**) [141]

100 kJ/mol) [142]. However, the additive of Al results in a poor oxidation resistance. The oxidation activation energy of various Al containing MA-Si_2BC_3N powder (100–200 nm) is 150–220 kJ/mol, much lower than that of pure MA-Si_2BC_3N powder, but still higher than those SiC and Si_3N_4 powders [143]. Besides, the oxidation activation energy decreases with the increasing Al content in MA-Si_2BC_3N powder. In detail, the MA-$Si_2BC_3N_{1.6}Al_{0.6}$ powder (using AlN as Al source) provides oxidation activation energy of 189.0 kJ/mol, higher than that 177.5 kJ/mol of MA-$Si_2BC_3NAl_{0.6}$ (using Al as Al source).

Coincidently, hot-pressed MA-$Si_2BC_3N_{1.6}Al_{0.6}$ monoliths show a low mass change and thin oxide scale compared with $Si_2BC_3NAl_{0.6}$ counterparts under the same oxidation condition (Fig. 1.14). For example, after oxidation at 1400 °C for 80 h, the hot-pressed MA-$Si_2BC_3N_{1.6}A1_{0.6}$ and MA-$Si_2BC_3NAl_{0.6}$ gain weight of ~ 0.58 mg/cm^2 and ~ 0.70 mg/cm^2, respectively. Moreover, the mass change of the two Al-containing materials is higher than that of pure MA-Si_2BC_3N monoliths. That is to say, MA-$Si_2BC_3N_{1.6}Al_{0.6}$ monolithic ceramics show an oxidation resistance superior to $Si_2BC_3NAl_{0.6}$, but inferior to MA-Si_2BC_3N.

Increasing SiC content due to the increasing Si: C ratio in spark plasma sintered monoliths leads to relatively thin and dense oxide scale indicative of superior

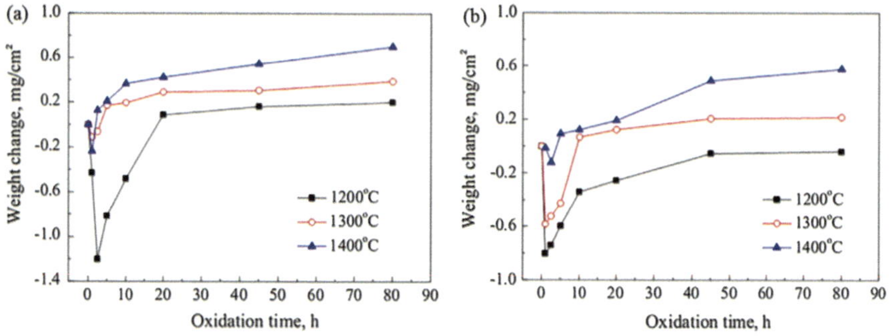

Fig. 1.14 Oxidation kinetic curves of the two hot-pressed Al-containing monoliths oxidized at 1200–1400 °C in flowing air: **a** MA-$Si_2BC_3NAl_{0.6}$ (using Al as the Al source); **b** MA-$Si_2BC_3N_{1.6}Al_{0.6}$ (using AlN as the Al source) [128]

oxidation resistance [43]. Hot-pressed Ta_4HfC_5/Si_2BC_3N composite ceramics have a lower oxidation kinetic constant ~ 215.97 $\mu m^2/h$ at 1600 °C/5 h [144, 145] compared with the MA-Si_2BC_3N counterparts. Nevertheless, the oxide scale growths of GPLs/Si_2BC_3N materials are controlled by a linear kinetic rate law in ranges of 1100–1500 °C indicating a rapid oxidation ratio [146].

The scale growths for the high-pressure sintered amorphous MA-Si_2BC_3N monoliths oxidized at 1500 and 1600 °C fit well to the parabolic laws [135], with kinetics constants of 32.5 $\mu m^2/h$ and 86.1 $\mu m^2/h$, respectively (Fig. 1.15). However, oxidation at 1700 °C becomes complex. The scales grow obeying approximately linear growth for 8 h (with a linear rate constant of 4.0 $\mu m^2/h$), and complying roughly with a parabolic raw within 4 h (with a parabolic rate constant of 79.4 $\mu m^2/h$). They then grow irregularly with oxidation times longer than 8 h. These results suggest the diffusion of oxidant through the growing oxide scales is the rate-controlling process at 1500 and 1600 °C, while the oxidation rate at 1700 °C is co-controlled by the rates of oxidant diffusion through the oxide scale and the reactions on scale/ceramic matrix interfaces. The oxidation activation energy can be deduced to be ~ 116 kJ/mol at 1500–1600 °C below 16 h of oxidation [147]. Note that the values for the apparent activation energies are estimated from the upper and lower extremes of the Arrhenius plot.

For the high-pressure sintered partially crystalline MA-Si_2BC_3N monoliths, their fitted results suggest that at 1500–1600 °C the growth of oxide scales also follows a parabolic law as the oxidation time increases from 0.5 to 16 h, while it is a little bit complicated at 1700 °C [148]. The scale growth follows a linear law within 8 h and also a parabolic trend within 4 h. The faster scale growth for partially crystalline ceramics can be well verified quantitatively by the fitted oxidation rate constant. The aforementioned fitting processing provides parabolic constants of 42.5 $\mu m^2/h$ for 1500 °C/$\le$ 16 h, 172.2 $\mu m^2/h$ for 1600 °C/$\le$ 16 h, and 120.4 $\mu m^2/h$ for 1700 °C/$\le$ 4 h, respectively, for oxidation of partially crystalline MA-Si_2BC_3N bulk ceramic, each of which is larger than those for the completely amorphous counterparts.

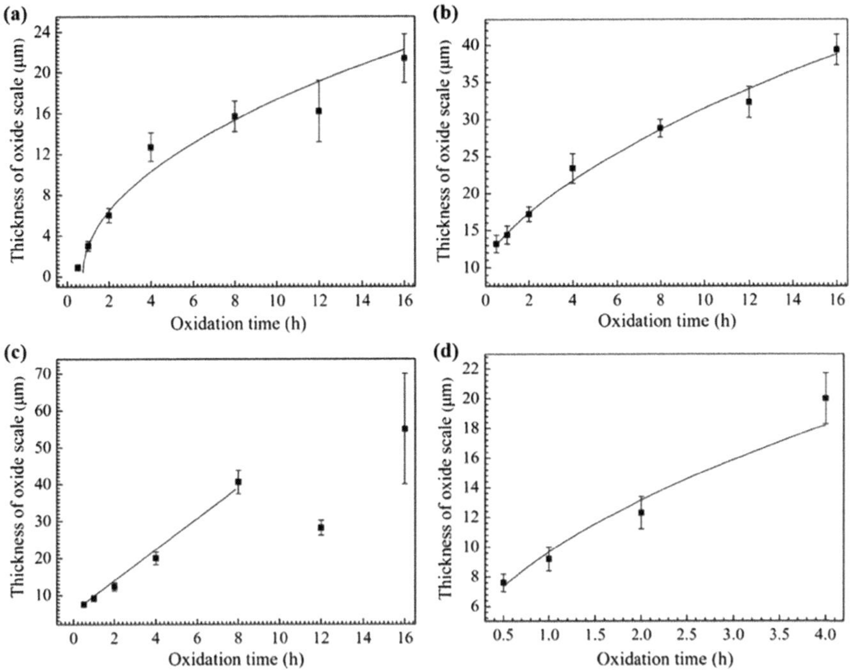

Fig. 1.15 Oxidation kinetic curves of high-pressure sintered amorphous MA-Si_2BC_3N monoliths oxidized in a flowing dry air atmosphere: **a** 1500 °C/≤ 16 h; **b** 1600 °C/≤ 16 h; **c** 1700 °C/≤ 16 h; **d** 1700 °C/≤ 4 h [135]

Oxidation behavior investigation of the amorphous PDC-$Si_{4.46}BC_{7.32}N_{4.40}$ (AMF2p), PDCs-$Si_{2.7}BC_{4.51}N_{2.69}$ (AMF3p) and PDCs-$Si_{3.08}BC_{4.39}N_{2.28}$ (T2/1p) shows that these particles develop thinner scales than pure SiC or Si_3N_4, at least at 1500 °C within 100 h [149]. Scales at 1500 °C in air show a parabolic growth with rate constants of 0.0599 m^2/h (AMF3p) and 0.0593 m^2/h (T2/1p) (Fig. 1.16). The oxidation rate constants are smaller than the determined parabolic rate constants at 1500 °C in oxygen for CVD-SiC (0.095 m^2/h) and CVD-Si_3N_4 (0.113 m^2/h) [150]. Oxidation of the T2/1 particles at 1500 °C for 15 h results in a dense, bubble-free scale. For longer oxidation times of 24 h and even up to 100 h, bubbles form almost completely limited to the so-called "edges" of the thermolyzed particles indicating good oxidation resistance.

Note that the kinetic data of the amorphous AMF2p particles do not obey a parabolic equation. In particular, their oxide scale thicknesses during the short-time experiments are noticeably above the parabolic fit. Nickel et al. have measured their mass changes occurring during the oxidation; however, the obtained results are not easily interpreted. In another work, Müller et al. [151] have reported the parabolic rates for amorphous PDCs-$SiC_{1.6}N_{1.1}B_{0.37}Al_{0.13}$, higher than those of PDCs-SiBCN materials.

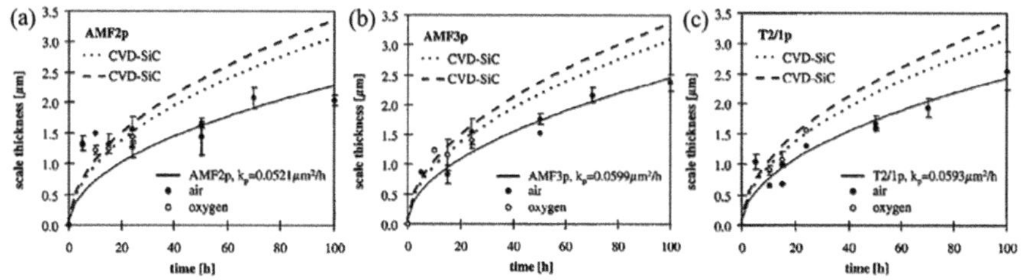

Fig. 1.16 Oxide scale thickness as a function exposing time for the **a** amorphous PDC-$Si_{4.46}BC_{7.32}N_{4.40}$ (AMF2p), **b** PDCs-$Si_{2.7}BC_{4.51}N_{2.69}$ (AMF3p) , **c** PDCs-$Si_{3.08}BC_{4.39}N_{2.28}$ (T2/1p) particles oxidized at 1500 °C [149]

Interestingly, the warm-pressed and subsequently thermolyzed T2/1p ceramic bodies exhibit an oxidation resistance inferior to T2/1 particles counterparts. They show beginning bubble formation already after oxidation at 1100 °C for 24 h. Oxidation at 1500 °C for the same time span results in completely foamy scales (Fig. 1.17), possibly due to the quasi-grain boundaries acting as diffusion paths of oxygen between amorphous particles. For AMF3 particles oxidized at 1500 °C for 100 h, some bubble forms accompanied with cristobalite precipitates within the oxide scales. In contrast, bubble-free regions of the oxide scales of all B-containing ceramic bodies are amorphous most likely due to the fact that the incorporation of B in the oxide scales obviates crystallization. The inhibiting effect of B on the crystallization of amorphous SiO_2 oxide layer is also observed in an oxide scale of reactive magnetron sputtered SiBCN film. The oxide scale is amorphous consisting mainly of SiO_2 coupled with a small amount of B, C and N so that reactive magnetron sputtered SiBCN films remain amorphous after annealing at 1700 °C in air.

However, some other early investigations suggested exceptionally low oxidation rate of amorphous PDCs-SiBCN materials, which is most likely underestimated for several reasons, e.g., the low oxide/ceramic volume ratio, the borosilicate viscous flow and/or the B_2O_3 volatilization. Nickel et al. [152] found that the distinct oxidation of amorphous PDCs-SiBCN materials occurs at 1000 °C, at least 100 °C lower than the initial temperature of PDCs-SiCN oxidation. Zhang et al. [153] proposed an amorphous PDCs-SiBCN ceramics with merely 2 at% B, and suggested these ceramics should have some similarity with amorphous PDCs-SiCN. Coincidently, the PDCs-SiCN exhibits similar oxidation behavior as SiC and Si_3N_4 once the free carbon has been burned out [154]. The parabolic rate constants of PDCs-SiCN ceramics calculated by Mocaer et al. [155] are lower than those observed for SiC but higher than Si_3N_4, and the calculated activation energy is 170 kJ/mol comparable to SiC.

Nickel [149] and Müller [151] have both pointed out that a low oxide growth rate may not be a necessary assessment for high oxidation resistance, which is better evaluated by recession rates of a material. For example, though highly dense amorphous MA-Si_2BC_3N monoliths show higher oxidation rates and lower activation energies, they provide superior oxidation resistance to the SiC above 1600 °C. In a word, the

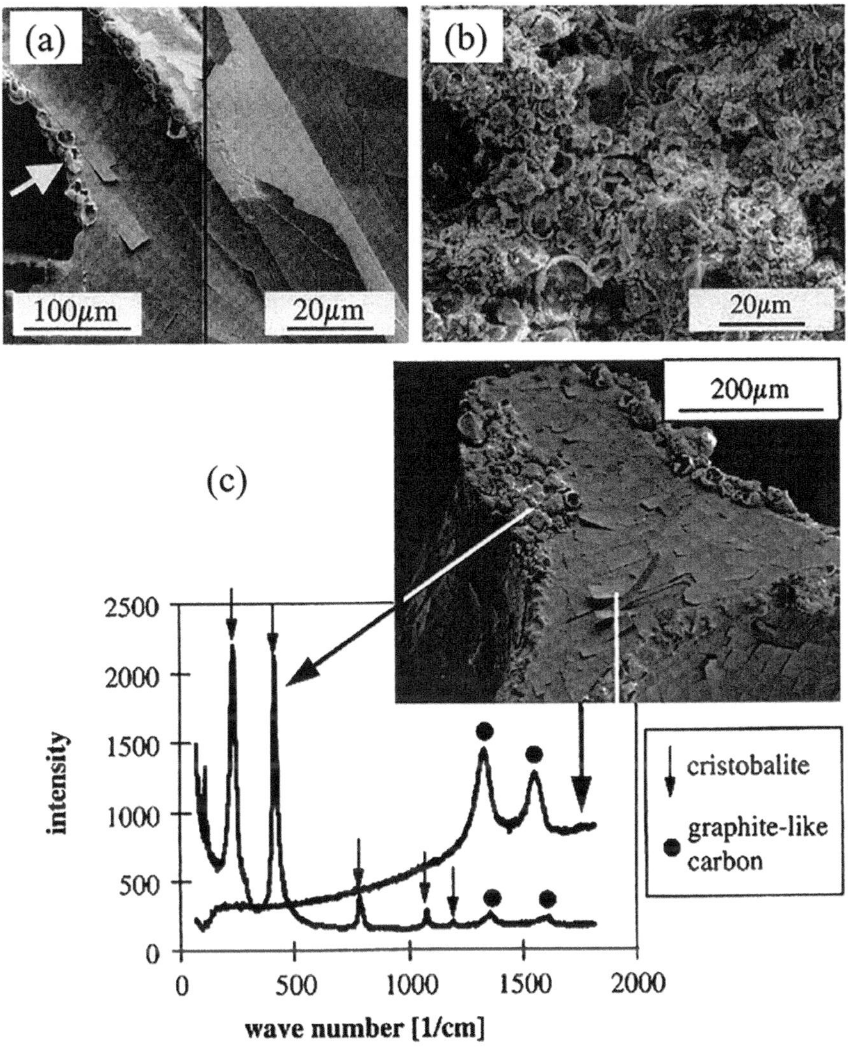

Fig. 1.17 Oxide scale morphologies of the as-thermolyzed T2/1 (oxidized at 1500 °C for 24 h in air) and AMF3 ceramic materials (oxidized 1500 °C for 100 h in air): **a** T2/1 particles; **b** T2/1 ceramic body with the formation of bubble-rich oxide scale; **c** bubble formation and crystallization of the oxide scale on the surfaces of AMF3 particles [149]

disagreement about oxidation behavior and mechanisms of amorphous SiBCN materials still exists due to its complicated variable structure depending on the starting precursors and process parameters used by the different research groups. Densities of ceramic monoliths and oxidation conditions (like oxidation temperature, time and atmosphere, gas flux, partial pressure of oxygen, the rates of ramping and cooling, etc.) also have a role on oxidation behavior leading to the divergence of data.

A Comparison of oxidation activation energy and oxidation kinetic constant for the SiBCN and SiC materials oxidized at different condition are summarized in Table 1.3. Narushima [156] reported during the oxidation of CVD SiC, the crystallization of SiO_2 to β-cristobalite changed the oxygen diffusion rates in SiO_2 scales thereby leading to two-step parabolic oxidation behavior over 1550 ~ 1675 °C in dry O_2. The boron or aluminum addition causes increases in oxidation rates of hot-pressed SiC at lower temperatures (1000–1200 °C), and the presence of free Si in CVD SiC can also increase oxidation rates, although SiC materials usually obey a parabolic oxidation [157, 158]. Costello et al. [159] investigated the oxidation kinetics of SiC monoliths in dry oxygen over 1200–1500 °C which fitted to the linear-parabolic model. The corresponding apparent activation energies increased from 120 kJ/mol below 1400 °C to 300 kJ/mol above this temperature, also induced by the crystallization of oxide scales.

Table 1.3 Comparison on oxidation activation energy and oxidation kinetic constant of the oxidizing SiBCN and SiC materials prepared by different method [156–159]

Ceramics	Preparation method	Oxidation condition	Oxidation kinetic constant K_p (μm^2/h)	Oxidation activation energy E_a (kJ/mol)
Si_2BC_3N bulk	MA + high-pressure sintering	1500 °C/16 h/dry air	32.50	116
		1600 °C/16 h/dry air	86.10	
SiBCN powder	Precursor pyrolysis	1500 °C/100 h/air	0.06	–
SiBCNAl powder	Precursor pyrolysis	1500 °C/16 h/dry O_2	0.57	–
SiC/SiBCN bulk	Precursor pyrolysis + hot-pressing	1400 °C/2 h/air	0.36	87
		1500 °C/2 h/air	0.65	
SiC bulk	Hot-pressing	1400 °C/100 min/air	0.07	155–498
		1500 °C/100 min/air	0.54	
SiC bulk	Hot-pressing	1400 °C/24 h/dry O_2	0.07	120
		1500 °C/24 h/dry O_2	0.20	300
SiC film	CVD	1400 °C/100 h/dry O_2	0.06	118
		1500 °C/100 h/dry O_2	0.10	
SiC film	CVD	1550–1675 °C/O_2	–	345–387

1.4 Development Tendency and Prospect

MA route is an indispensable way other than the organic precursor pyrolysis method. The inorganic processing route, i.e., MA coupled with sequential sintering, is a combination of top-down and bottom-up ways for preparation of nano-structural or amorphous materials. It is a “hard processing” which requires the assist of external mechanical force or energy. Moreover, it is simple for highly densified ceramic monoliths and structural parts. The starting materials needed are cheap, easily available and environment- or human-friendly.

In contrast, the polymer precursor route belonging to bottom-up way is a “soft processing” without the external energy input. It is suitable for fabrication of nano or amorphous SiBCN materials, especially for fibers, films, porous structures. But the starting organic materials are expensive and poisonous thus economy- and environment-unfriendly and harmful to human health. The complex and demanding process procedure makes it much more difficult to control the microstructures and properties of the final products, especially when higher density is needed.

Significantly, the inorganic processing route can reach the same goal as the organic precursor pyrolysis approach, and thus is proved to be a successful way to expand the SiBCN system. The two processing methods have their advantages and make significant contribution to the development of SiBCN materials. They are mutually complementary and cannot be replaced by each other. Though encouraging developments and fascinating achievements about the SiBCN ceramics and their ceramic matrix composites developed by the inorganic processing route have been achieved as documented above, the relevant study is only on the half way or even in its infancy to some aspects. Many basic scientific issues about SiBCN system materials are still not completely clear and thus require extensive further study.

These topics involve (1) the solid-state amorphization process and mechanisms of MA derived SiBCN powders; (2) the more flexible design of chemical composition of SiBCN-Me (Me=Al, Zr, Hf, Ta, etc.) as well as SiCN and SiBN systems; (3) the atomic conformation in amorphous SiBCN phases; (4) the ceramic densification mechanisms and the crystallization-densification coupling behavior during sintering process; (5) the influence behavior and mechanisms of chemical composition changes on the amorphization of SiBCN-Me composite powders during MA, the thermal stability (such as crystallization process during high-temperature sintering or annealing), and the properties; (6) the harsh environments (such as high-temperature creep, thermal shock, oxidation and ablation) responding behavior and damage mechanisms; (7) fatigue property and fatigue damage mechanisms; (8) the basic issues relevant to hot-isostatic pressing (HIP) and high-pressure sintering (HPS), such as the influence of (ultra-)high pressure and post-HIP treatment on material microstructural evolution and properties; (9) the rational co-design of the vast range of complex composition and multiscale structure, perhaps developed a combination of “soft” (organic route) and “hard” (inorganic route) synthetic technique.

Overall, the vastness of composition optimization and multiscale structure design gives the chance to tailor the performance of the SiBCN materials but also raises new challenges. These difficulties have suggested the development of specific approaches to produce new structure, new high-throughput calculation tools for composition optimization and structure design, high-throughput experiments to rapidly assess performance of the structural parts. It is believed that more extensive progress on the basic scientific issues about the metastable SiBCN system ceramics can be achieved and more types of novel materials with unique microstructure and properties can be developed and promising engineering applications will be found in the near future.

References

1. Riedel R, Kleebe HJ, Schonfelde H, Aldinger F. A covalent micro-nano-composite resistant to high-temperature oxidation. Nature. 1995;374:526–8.
2. Riedel R, Kienzle A, Dressler W, Ruwisch L, Bill J, Aldinger F. A silicoboron carbonitride ceramic stable to 2000 °C. Nature. 1996;382:796–8.
3. Liang B, Yang Z, Li Y, Yuan J, Jia D, Zhou Y. Ablation behavior and mechanism of SiC_f/C_f/SiBCN ceramic composites with improved thermal shock resistance under oxyacetylene combustion flow. Ceramic Int. 2015;41:8868–77.
4. Jia D, Liang B, Yang Z, Zhou Y. Metastable Si-B-C-N ceramics and their matrix composites developed by inorganic route based on mechanical alloying: fabrication, microstructures, properties and their relevant basic scientific issues. Prog Mater Sci. 2018;98:1–67.
5. Opeka MM, Talmy IG, Zaykoski JA. Oxidation-based materials selection for 2000 °C plus hypersonic aerosurfaces: theoretical considerations and historical experience. J Mater Sci. 2004;39:5887–904.
6. Van Wie DM, Drewry DG, King DE, Hudson CM. The hypersonic environment: required operating conditions and design challenges. J Mater Sci. 2004;39:5915–24.
7. Roy J, Chandra S, Das S, Maitra S. Oxidation behavior of silicon carbide—A review. Rev Adv Mater Sci. 2014;38:29–39.
8. Huang QW, Jin ZH. The high temperature oxidation behavior of reaction-bonded silicon carbide. J Mater Process Technol. 2001;110:142–5.
9. Goto T. High-temperature oxidation behavior of chemical vapor deposited silicon carbide. J Ceram Soc Jpn. 2002;110:884–9.
10. Klemm H, Herrmann M, Reich T, Hermel W. High temperature properties of Si_3N_4 materials. J Eur Ceram Soc. 1991;7:315–8.
11. Guo SQ. Densification of ZrB_2-based composites and their mechanical and physical properties: a review. J Eur Ceram Soc. 2009;29:995–1011.
12. Opeka MM, Talmy IG, Wuchina EJ, Zaykoski JA, Causey SJ. Mechanical, thermal, and oxidation properties of refractory hafnium and zirconium compounds. J Eur Ceram Soc. 1999;19:2405–14.
13. Squire TH, Marschall J. Material property requirements for analysis and design of UHTC components in hypersonic applications. J Eur Ceram Soc. 2010;30:2239–51.
14. Gasch M, Ellerby D, Irby E, Beckman S, Gusman M, Johnson S. Processing, properties and arc jet oxidation of hafnium diboride/silicon carbide ultra high temperature ceramics. J Mater Sci. 2004;39:5925–37.
15. Xiang LY, Cheng LF, Shi L, Yin XW, Zhang LT. Mechanical and ablation properties of laminated ZrB_2–SiC/BN ceramics. J Alloy Compd. 2015;638:261–6.
16. Zhang Y, Zhang L, Liu Y, Liu X, Chen B. Oxidation effects on in-plane and interlaminar shear strengths of two-dimensional carbon fiber reinforced silicon carbide composites. Carbon. 2016;98:144–56.
17. Yang J, Dong S, Xu C. Mechanical response and microstructure of 2D carbon fiber reinforced ceramic matrix composites with SiC and Ti_3SiC_2 fillers. Ceramic International. 2016;42:3019–27.

18. Li L. Fatigue hysteresis of carbon fiber-reinforced ceramic-matrix composites at room and elevated temperatures. Appl Compos Mater. 2016;23:1–27.
19. Han Y, Yan Q, Li X. Thermal/mechanical properties of short carbon fibre/SiC co-reinforced graphite matrix composites produced by low temperature hot pressing. Nano-Micro Letters. 2015;10:263–6.
20. Wang HY, Liu Q, Zhou Y, Zhou ZZ, Liu GH. Preparation and properties of carbon fiber/Si_3N_4 composites. J Inorg Mater. 2014;29:1003–8.
21. Jin E, Du S, He H. The evolution of mechanical and structural properties at the fiber/matrix interphase of SiC/SiC composites. Comput Mater Sci. 2015;104:84–91.
22. Yoshida K, Akimoto H, Yano T, Kotani M, Ogasawara T. Mechanical properties of unidirectional and crossply SiC_f/SiC composites using SiC fibers with carbon interphase formed by electrophoretic deposition process. Prog Nucl Energy. 2015;82:148–52.
23. Silvestroni L, Sciti D, Hilmas GE, Fahrenholtz WG, Watts J. Effect of a weak fiber interface coating in ZrB_2 reinforced with long SiC fibers. Mater Des. 2015;88:610–8.
24. Zoli L, Medri V, Melandri C, Sciti D. Continuous SiC fibers-ZrB_2 composites. J Eur Ceram Soc. 2015;35:4371–6.
25. Rezaie A, Fahrenholtz WG, Hilmas GE. Effect of hot pressing time and temperature on the microstructure and mechanical properties of ZrB_2–SiC. J Mater Sci. 2007;42:2735–44.
26. Kasiarova M, Tatarko P, Burik P, Dusza J, Sajgalik P. Thermal shock resistance of Si_3N_4 and Si_3N_4–SiC ceramics with rare-earth oxide sintering additives. J Eur Ceram Soc. 2014;34:3301–8.
27. Jiang Y, Wu LE, Sun WZ. Sintering behavior and properties of SiC/Si_3N_4 composite. Rare Met. 2015;34:95–100.
28. Ye F, Liu S, Hu S, Yang H, Liu Q, Zhang B. A route to increase fracture toughness of layered Si_3N_4/SiC composite using interlocked interfaces. Ceramic Int. 2015;41:10331–5.
29. Li J, Li L, Wang Z, Liu Q. Effect of SiC Whisker content on the properties and microstructure of the ZrC matrix ultrahigh temperature ceramics. Rare Metal Materials and Engineering. 2013;42:238–40.
30. Sciti D, Pienti L, Fabbriche DD, Guicciardi S, Silvestroni L. Combined effect of SiC chopped fibers and SiC whiskers on the toughening of ZrB_2. Ceramic Int. 2014;40:4819–26.
31. Lin J, Huang Y, Zhang H. Crack healing and strengthening of SiC whisker and ZrO_2 fiber reinforced ZrB_2 ceramics. Ceramic Int. 2014;40:16811–5.
32. Galusek D, Riley FL, Riedel R. Nanoindentation of a polymer-derived amorphous silicon carbonitride ceramic. J Am Ceram Soc. 2001;84:1164–6.
33. Liew LA, Saravanan RA, Bright VM, Dunn ML, Daily JW, Raj R. Processing and characterization of silicon carbon-nitride ceramics: application of electrical properties towards MEMS thermal actuators. Sens Actuators A-Phys. 2003;103:171–81.
34. Vishnyakov VM, Ehiasarian AP, Vishnyakov VV, Hovsepian P, Colligon JS. Amorphous boron containing silicon carbo-nitrides created by ion sputtering. Surf Coat Technol. 2011;206:149–54.
35. Ziegler G, Richter I, Suttor D. Fiber-reinforced composites with polymer-derived matrix: processing, matrix formation and properties. Compos A Appl Sci Manuf. 1999;30:411–7.
36. Niihara K, Nakano K, Sekino T, Yasuda E. High temperature ceramic matrix composites. 1999;103–04.
37. Herrmann M, Schoenfeld K, Klemm H, Lippmann W, Hurtado A, Michaelis A. Laser-supported joining of SiC-fiber/SiCN ceramic matrix composites fabricated by precursor infiltration. J Eur Ceram Soc. 2014;34:2913–24.
38. Yan XB, Gottardo L, Bernard S, Dibandjo P, Brioude A. Ordered mesoporous silicoboron carbonitride materials via preceramic polymer nanocasting. Chem Mater. 2008;20:6325–34.
39. Wang ZC, Aldinger F, Riedel R. Novel silicon-boron-carbon-nitrogen materials thermally stable up to 2200 °C. J Am Ceram Soc. 2001;84:2179–83.
40. Li WH, Wang J, Xie ZF, Wang H. A novel polyborosilazane for high-temperature amorphous Si-B-N-C ceramic fibres. Ceramic International. 2012;38:6321–6.
41. Bill J, Aldinger F. Precursor-derived covalent ceramics. Adv Mater. 1995;7:775–87.

42. Colombo P, Mera G, Riedel R, Soraru GD. Polymer-derived ceramics: 40 years of research and innovation in advanced ceramics. J Am Ceram Soc. 2010;93:1805–37.
43. Zhang P, Jia D, Yang Z, Duan X, Zhou Y. Progress of a novel non-oxide Si-B-C-N ceramic and its matrix composites. J Adv Ceram. 2012;1:157–78.
44. Ye D, Jia D C, Yang Z H, Duan X M, Zhou Y. Structural and microstructural characterization of $SiB_{0.5}C_{1.5}N_{0.8}Al_{0.3}$ powders prepared by mechanical alloying using aluminum nitride as aluminum source. Ceramic Int. 2011;37:2937–940.
45. Lee SH, Weinmann M, Aldinger F. Processing and properties of C/Si-B-C-N fiber-reinforced ceramic matrix composites prepared by precursor impregnation and pyrolysis. Acta Mater. 2008;56:1529–38.
46. Lee SH, Weinmann M, Gerstel P, Aldinger F. Extraordinary thermal stability of SiC particulate-reinforced polymer-derived Si-B-C-N composites. Scripta Mater. 2008;59:607–10.
47. Lee SH, Weinmann M. C fiber/SiC filler/Si-B-C-N matrix composites with extremely high thermal stability. Acta Mater. 2009;57:4374–81.
48. Wang J, Duan X, Yang Z, Jia D, Zhou Y. Ablation mechanism and properties of SiC_f/SiBCN ceramic composites under an oxyacetylene torch environment. Corros Sci. 2014;82:101–7.
49. Baldus HP, Jansen M. Novel high-performance ceramics—Amorphous inorganic networks from molecular precursors. Angewandte Chemie-Int Edn. 1997;36:329–43.
50. Jansen M, Jongermann H. A new class of promising ceramics based on amorphous inorganic networks. Curr Opin Solid State Mater Sci. 1997;2:150–7.
51. Jansen M. Highly stable ceramics through single source precursors. Solid State Ion. 1997;101:1–7.
52. Cooke TF. Inorganic fibers—A literature review. J Am Ceram Soc. 1991;74:2959–78.
53. Ainger FW, Herbert JM. The preparation of phosphorus-nitrogen compounds as non-porous solids. Spec Ceram. 1960;661:168–82.
54. Chantrell PG, Popper P. Inorganic polymers for ceramics. Spec Ceram. 1965;67.
55. Verbeek W. Production of shaped articles of homogeneous mixtures of silicon carbide and nitride. US Pat No 3853567; 1973.
56. Verbeek W, Winter G. Ger Pat No 2236078; 1974.
57. Winter G, Verbeek W, Mansmann M. Ger Pat No 2243527; 1974.
58. Yajima S, Hayashi J, Omorj M, Okamura K. Development of a silicon carbide fibre with high tensile strength. Nature. 1976;261:683–5.
59. Yajima S, Hasegawa Y, Okamure K, Matsuzawa T. Development of high tensile strength silicon carbide fibre using an organosilicon polymer precursor. Nature. 1978;273:525–7.
60. Yajima S, Hayashi J, Omorj M. Continuous silicon carbide fiber of tensile strength. Chem Lett. 1975;9:931.
61. Yajima S, Okamura K, Matsuzawa T, Hasegawa Y, Shishido T. Anomalous characteristics of the microcrystalline state of SiC fibres. Nature. 1979;279:706–7.
62. Takamizawa M, Kobayashi T, Hayashida A. Organoborosilicon polymer and a method for the preparation thereof. US Pat No 4550151; 1985.
63. Takamizawa M, Kobayashi T, Hayashida A. US Pat No 4604367; 1986.
64. Takamizawa M, Kobayashi T, Hayashida A, Y T. Dtsch Pat No 3444306; 1986.
65. Riedel R, Passing G, Schonfelder H, Brook RJ. Synthesis of Dense silicon-based ceramics at low-temperatures. Nature. 1992;355:714–7.
66. Riedel R, Kleebe HJ, Schonfelde H, Aldinger F. A covalent micro/nano-composite resistant to high-temperature oxidation. Nature. 1995;374:526–8.
67. Funayama O, Nakahara H, Okoda M, Okumura M, Isoda T. Conversion mechanism of polyborosilazane into silicon nitride-based ceramics. J Mater Sci. 1995;30:410–6.
68. Long X, Shao C, Wang H, Wang J. Synthesis and characterization of a polyborosilazane/Cp_2ZrCl_2 hybrid precursor for the Si-B-C-N-Zr multinary ceramic. Dalton Trans. 2015;44:15463–9.
69. Yuan J, Hapis S, Breitzke H, Xu Y, Fasel C. Single-source-precursor Synthesis of hafnium-containing ultrahigh-temperature ceramic nanocomposites (UHTC-NCs). Inorg Chem. 2014;53:10443–55.

70. Yuan J, Galetz M, Luan XG, Fasel C, Riedel R, Ionescu E. High-temperature oxidation behavior of polymer-derived SiHfBCN ceramic nanocomposites. J Eur Ceram Soc. 2016;36:3021–8.
71. Zeman P, Zuzjaková Š, Mareš P, Čerstvý R, Zhang M. Superior high-temperature oxidation resistance of magnetron sputtered Hf-B-Si-C-N film. Ceram Int. 2016;42:4853–9.
72. Weinmann M, Schuhmacher J, Kummer H, Prinz S, Peng J. Synthesis and thermal behavior of novel Si-B-C-N ceramic precursors. Chem Mater. 2000;12:623–32.
73. Tavakoli AH, Gerstel P, Golczewski JA, Bill J. Effect of boron on the crystallization of amorphous Si-(B-)C-N polymer-derived ceramics. J Non-Cryst Solids. 2009;255:2381–9.
74. Riedel R, Mera G, Hauser R, Klonczynski A. Silicon-based polymer-derived ceramics: synthesis properties and applications—A review. J Ceram Soc Jpn. 2006;114:425–44.
75. Kalaš J, Vernhes R, Hřeben S, Vlček J, Klemberg-Sapieha JE, Martinu L. High-temperature Stability of the mechanical and optical properties of Si-B-C-N films prepared by magnetron sputtering. Thin Solid Films. 2009;518:174–9.
76. Zeman P, Čapek J, Čerstvý R, Vlček J. Thermal stability of magnetron sputtered Si-B-C-N materials at temperatures up to 1700 °C. Thin Solid Films. 2010;519:306–11.
77. Hayashi N, Choi J, Kumagai T, Kato T, Kawaguchi M. Thermal stability of SiBCN films. Tribology Online. 2008;3:254–8.
78. Kumar R, Rixecker G, Aldinger F. Anelasticity of precursor derived Si-B-C-N ceramics. J Eur Ceram Soc. 2007;27:1475–80.
79. Dressler W, Riedel R. Progress in silicon-based non-oxide structural ceramics. Int J Refract Metal Hard Mater. 1997;15:13–47.
80. Tang Y, Wang J, Li X, Li W, Wang H, Wang X. Thermal stability of polymer derived SiBNC ceramics. Ceram Int. 2009;35:2871–6.
81. Weinmann M, Kamphowe TW, Schuhmacher J, Muller K, Aldinger F. Design of polymeric Si-B-C-N ceramic precursors for application in fiber-reinforced composite materials. Chem Mater. 2000;12:2112–22.
82. Baldus P. Ceramic fibers for matrix composites in high-temperature engine applications. Science. 1999;285:699–703.
83. Raj R, Riedel R, Soraru GD. Introduction to the special topical issue on ultrahigh-temperature polymer-derived ceramics. J Am Ceram Soc. 2001;84:2158–9.
84. Seibold M, Greil P. Thermodynamics and microstructural development of ceramic composite formation by active filler-controlled pyrolysis (AFCOP). J Eur Ceram Soc. 1993;11:105–13.
85. Zern A, Mayer J, Janakiraman N, Weinmann M, Bill J, Ruhle M. Quantitative EFTEM study of precursor-derived S-B-C-N ceramics. J Eur Ceram. 2002;22:1621–9.
86. Bharadwaj L, Fan Y, Zhang LG, Jiang DP, An LN. Oxidation Behavior of a fully dense polymer-derived amorphous silicon carbonitride ceramic. J Am Ceram Soc. 2004;87:483–6.
87. Christ M, Thurn G, Weinmann M, Bill J, Aldinger F. High-temperature mechanical properties of Si-B-C-N-precursor-derived amorphous ceramics and the applicability of deformation models developed for metallic glasses. Journal of the American Ceramics Society. 2000;83:3025–32.
88. Lu B, Zhang Y. Densification behavior and microstructure evolution of hot-pressed SiC-SiBCN ceramics. Ceram Int. 2015;41:8541–51.
89. Bechelany MC, Salameh C, Viard A, Guichaoua L, Rossignol F. Preparation of polymer-derived Si-B-C-N monoliths by spark plasma sintering technique. J Eur Ceram Soc. 2015;35:1361–74.
90. Yang ZH, Jia DC, Zhou Y. Amorphous and nanocrystalline SiBCN ceramic composite and its synthesis method. China Paten No. 2005100014927; 2005.
91. Hegemann D, Riedel R, Oehr C. PACVD-derived thin films in the system Si-B-C-N. Chem Vap Deposition. 1999;5:61–5.
92. Ko CH, Kuan TM, Zhang K, Tsai G, Seutter SM. A novel CVD-SiBCN low-K spacer technology for high-speed applications. Symposia VLSI technology and circuits; 2008: 108–9.
93. Wilden J, Wank A, Bykava A. DC thermal plasma CVD synthesis of coatings from liquid single source SiBCN and SiCNTi precursors. Surf Coat Technol. 2005;200:612–5.

94. Sun X, Liu H, Li J, Cheng H. Effects of CVD SiBCN interphases on mechanical and dielectric properties of SiC_f/SiC composites fabricated via a PIP process. Ceram Int. 2016;42:82–9.
95. Widgeon S, Mera G, Gao Y, Sen S, Navrotsky A. Effect of precursor on speciation and nanostructure of SiBCN polymer-derived ceramics. J Am Ceram Soc. 2013;96:1651–9.
96. Jansen M, Jaschke B, Jaschke T. Amorphous multinary ceramics in the Si-B-N-C system. Struct Bond. 2002;101:137–91.
97. Kroke E, Li YL, Konetschny C, Lecomte E, Fasel C, Riedel R. Silazane derived ceramics and related materials. Mater Sci Eng R-Rep. 2000;26:97–199.
98. Bill J, Kamphowe TW, Muller A, Wichmann T, Zern A. Precursor-derived Si-(B-)C-N ceramics: thermolysis, amorphous state and crystallization. Appl Organomet Chem. 2001;15:777–93.
99. Viard A, Fonblanc D, Lopez-Ferber D, Schmidt M, Lale A, Durif C, Balestrat M, Rossignol F, Weinmann M, Riedel R, Bernard S. Polymer derived Si-B-C-N ceramics: 30 years of research. Adv Eng Mater. 2018;20:1800360.
100. Seifert H J, Aldinger F. Phase Equilibria in the Si-B-C-N system. In: High performance non-oxide ceramics I. Structure and bonding, vol. 101. Berlin, Heidelberg: Springer; 2002.
101. Tavakoli AH, Gerstel P, Golczewski JA, Bill J. Kinetic effect of boron on the thermal stability of Si-(B-)C-N polymer-derived ceramics. Acta Mater. 2010;58(18):6002–11.
102. Gao Y, Mera G, Nguyen H, Morita K, Kleebe HJ, Riedel R. Processing route dramatically influencing the nanostructure of carbon-rich SiCN and SiBCN polymer-derived ceramics. Part I: low temperature thermal transformation. J Eur Ceram Soc. 2012;32(9):1857–66.
103. Muller A, Zern A, Gerstel P, Bill J, Aldinger F. Boron-modified poly(propenylsilazane)-derived Si-B-C-N ceramics: preparation and high temperature properties. J Eur Ceram Soc. 2002;22(9–10):1631–43.
104. Janakiraman N, Weinmann M, Schuhmacher J, Müller K, Bill J, Aldinger F. Thermal stability, phase evolution, and crystallization in Si-B-C-N ceramics derived from a polyborosilazane precursor. J Am Ceram Soc. 2002;85:1807–14.
105. Müller A, Gerstel P, Weinmann M, Bill J, Aldinger F. Correlation of boron content and high temperature stability in Si-B-C-N ceramics. J Eur Ceram Soc. 2000;20:2655–9.
106. Müller A, Gerstel P, Weinmann M, Bill J, Aldinger F. Correlation of boron content and high temperature stability in Si-B-C-N ceramics II. J Eur Ceram Soc. 2001;21:2171–7.
107. Tavakoli AH, Gerstel P, Golczewski JA, Bill J. Kinetic effect of boron on the crystallization of Si_3N_4 in Si-B-C-N polymer-derived ceramics. J Mater Res. 2010;26(04):600–8.
108. Tavakoli AH, Gerstel P, Golczewski JA, Bill J. Effect of boron on the crystallization of amorphous Si-(B-)C-N polymer-derived ceramics. J Non-Cryst Solids. 2009;355:2381–9.
109. Sundman B, Jansson B, Anderson JO. The thermo-calc databank system. Calphad. 1985;9:153–90.
110. Jalowiecki A, Bill J, Aldinger F, Mayer J. Interface characterization of nanosized B-doped Si_3N_4/SiC ceramics. Compos A Appl Sci Manuf. 1996;27(9):717–21.
111. Yang ZH, Jia DC, Zhou Y, Yu CQ. Fabrication and characterization of amorphous Si-B-C-N powders. Ceram Int. 2007;33(8):1573–7.
112. Zhang PF, Jia DC, Yang ZH, Duan XM, Zhou Y. Influence of ball milling parameters on the structure of the mechanically alloyed SiBCN powder. Ceram Int. 2013;39(2):1963–9.
113. Zhang PF, Jia DC, Yang ZH, Duan XM, Zhou Y. Physical and surface characteristics of the mechanically alloyed SiBCN powder. Ceram Int. 2012;38(8):6399–404.
114. Zhang PF, Jia DC, Yang ZH, Duan XM, Zhou Y. Crystallization and microstructural evolution process from the mechanically alloyed amorphous Si-B-C-N powder to the hot-pressed nano SiC/BN(C) ceramic. J Mater Sci. 2012;47(20):7291–304.
115. Zhang PF, Jia DC, Yang ZH, Duan XM, Zhou Y. Microstructural features and properties of the nano-crystalline SiC/BN(C) composite ceramic prepared from the mechanically alloyed SiBCN powder. J Alloy Compd. 2012;537(19):346–56.
116. Liang B, Yang ZH, Chen QQ, Wang SJ, Duan XM, Jia DC, Zhou Y, Luo K, Yu DL, Tian YJ. Crystallization behavior of amorphous Si_2BC_3N ceramic monolith subjected to high pressure. J Am Ceram Soc. 2015;98(12):3788–96.

117. Bernard S, Weinmann M, Gerstel P, Miele P, Aldinger F. Boron-modified polysilazane as a novel single-source precursor for SiBCN ceramic fibers: synthesis, melt-spinning, curing and ceramic conversion. J Mater Chem. 2005;15(2):289–99.
118. Li WH, Wang J, Wang ZF, Wang H. A novel polyborosilazane for high-temperature amorphous Si-B-C-N ceramic fiber. Ceram Int. 2012;38(8):6321–6.
119. Sujith R, Kumar R. Experimental investigation on the indentation hardness of precursor derived Si-B-C-N ceramics. J Eur Ceram Soc. 2013;33(13–14):2399–405.
120. Janakiraman N, Aldinger F. Fabrication and characterization of fully dense Si-C-N ceramics from a poly(ureamethylvinyl)silazane precursor. J Eur Ceram Soc. 2009;29(1):163–73.
121. Shah R, Raj R. Mechanical properties of a fully dense polymer derived ceramic made by a novel pressure casting process. Acta Mater. 2002;50(16):4093–103.
122. Soraru GD, Dallapiccola E, Andrea GD. Mechanical characterization of sol-gel derived silicon oxycarbide glasses. J Eur Ceram Soc. 1996;79(8):2074–80.
123. Zhang PF. Hot pressing behavior and high-temperature properties of mechanically alloyed 2Si-B-3C-N ceramic. Ph.D. thesis. Harbin: Harbin Institute of Technology; 2013.
124. Yang ZH. Microstructure and high-temperature properties of the Si-B-C-N MA-powder and ceramics. Ph.D. thesis. Harbin: Harbin Institute of Technology; 2008.
125. Liang B, Yang ZH, Rao JC, Cai DL, Duan XM, Jia DC, Zhou Y, Yu DL, Tian YJ. Highly dense amorphous Si_2BC_3N monoliths with excellent mechanical properties prepared by high pressure sintering. J Am Ceram Soc. 2015;98(12):3782–7.
126. Li YT. Mechanical properties, thermal shock and ablation resistance of (C_f-SiC_f)/SiBCN composites. Master thesis. Harbin: Harbin Institute of Technology; 2014.
127. Pan LJ. The surface coating on C_f and the preparation and properties of C_f/SiBCN composite. Master thesis. Harbin: Harbin Institute of Technology; 2012.
128. Ye D. Microstructure and oxidation resistance of the mechanically alloyed Si-B-C-N-Al powders and ceramics. Ph.D. thesis. Harbin: Harbin Institute of Technology; 2012.
129. Miao Y. Preparation, oxidation resistance and ablation mechanisms of ZrB_2/SiBCN composites. Ph.D. thesis. Harbin: Harbin Institute of Technology; 2017.
130. Zhao Y. Microstructure and properties of the ZrC/SiBCN multiphase ceramics by hot pressing. Master thesis. Harbin: Harbin Institute of Technology; 2016.
131. Liao XQ. Microstructure and properties of the (TiB_2+TiC)/SiBCN composites. Bachelor thesis. Harbin: Harbin Institute of Technology; 2014.
132. Li DX, Yang ZH, Mao ZB, Jia DC, Cai DL, Liang B, Duan XM, He PG, Rao JC. Microstructures, mechanical properties and oxidation resistance of SiBCN ceramics with the addition of MgO, ZrO_2 and SiO_2 (MZS) as sintering additives. RSC Adv. 2015;5:52194–205.
133. Zhang M, Li DX, Hong YZ, Niu ZB, Yang ZH, Jia DC, Zhou Y. Study on the oxygen diffusion in the oxide layers of SiBCN ceramics by SIMS. J Eur Ceram Soc. 2022;42:1341–7.
134. Liang B, Yang ZH, Jia DC, Rao JC, Li YD, Tian YJ, Li Q, Miao Y, Zhu QS, Zhou Y. Amorphous silicoboron carbonitride monoliths resistant to flowing air up to 1800 °C. Corros Sci. 2016;109:162–73.
135. Liang B, Yang ZH, Miao Y, Zhu QS, Liao XQ, Yang ZH, Zhou Y. High temperature oxidation kinetics of amorphous silicoboron carbonitride monoliths and silica scale growth mechanisms determined by SIMS. Corros Sci. 2017;122:100–7.
136. Patel M, Janardhan Reddy J, Bhanu Prasad VV, Jayaram V. Strength of hot pressed ZrB_2–SiC composite after exposure to high temperatures (1000–1700 °C). J Eur Ceram Soc. 2012;32:4455–67.
137. Li YJ, Zhang Y, Wang YS. Structure and oxidation behavior of high temperature ZrB_2-SiBCN ceramics with polyborosilazane as a sintering additive. J Ceram Soc Jpn. 2013;121:520–3.
138. Jacobson NS, Lee KN, Fox DS. Reactions of SiC and SiO_2 at elevated temperature. J Am Ceram Soc. 1992;75:1603–11.
139. Borisov VG, Yudin BF. Reaction thermodynamics in the SiO_2–SiC system. Refract Ind Ceram. 1968;9:162–5.
140. Cinibulk MK, Parthasarathy TA. Characterization of oxidized polymer-derived SiBCN fibers. J Am Ceram Soc. 2001;84(10):2197–202.

141. Ji XY, Wang SS, Shao CW, Wang H. High-temperature corrosion behavior of SiBCN fibers for aerospace applications. ACS Appl Mater Interfaces. 2018;10:19712–20.
142. Zhang XH, Hu P, Han JC, Du SY. Study on thermal shock resistance and oxidation resistance of ultra-high temperature ceramics. Materials China. 2011;30(1):27–31.
143. Ye D, Jia DC, Yang ZH, Sun ZL, Zhang PF. Microstructure and thermal stability of amorphous SiBCNAl powders fabricated by mechanical alloying. J Alloy Compd. 2010;506(1):88–92.
144. Wang BZ, Li DX, Yang ZH, Jia DC, Guan JY, Cai DL, He PG, Duan XM, Zhou Y. Study on oxidation resistance and oxidative damage mechanism of SiBCN-Ta_4HfC_5 composite ceramics. Corros Sci. 2022;197: 110049.
145. Wang BZ, Li DX, Yang ZH, Cai DL, Jia DC, Guan JY, Zhou Y. Oxidation Kinetic behavior of SiBCN-Ta_4HfC_5 ceramics at 1600 °C. Corros Sci. 2022;207: 110552.
146. Li DX, Yang ZH, Jia DC, Wu DX, Zhu QS, Liang B, Wang SJ, Zhou Y. Microstructure, oxidation and thermal shock resistance of graphene reinforced SiBCN ceramics. Ceram Int. 2016;42:4429–44.
147. Liang B. Crystallization and high temperature oxidation mechanisms of amorphous Si_2BC_3N ceramics by high pressure sintering. Harbin: Harbin Institute of Technology; 2017.
148. Liang B, Zhu QS, Liao XQ, Liu KW, Yang ZH, Jia DC, Zhou Y. Crystallinity Dependence of high-temperature oxidation of silicoboron carbonitride monoliths. Corros Sci. 2021;187: 109473.
149. Butchereit E, Nickel KG, Müller A. Precursor-derived Si-B-C-N ceramics: oxidation kinetics. J Am Ceram Soc. 2001;84(10):187:2184–188.
150. Ogbuji LUJT, Opila EJ. A comparison of the oxidation kinetics of SiC and Si_3N_4. J Electrochem Soci 1995;142:925–30.
151. Müller A, Gerstel P, Butchereit E, Nickel KG, Aldinger F. Si/B/C/N/Al precursor-derived ceramics: synthesis, high temperature behavior and oxidation resistance. J Eur Ceram Soc. 2004;24:3409–17.
152. Butchereit E, Nickel KG. Oxidation behaviour of precursor derived Si-(B)-C-N-ceramics. J Mater Process Manuf Sci. 1998;7:15–21.
153. Dai J, Zhang Y, Gao D, Song Y, Shi X. Oxidation behavior of amorphous Si-(B)-C-N ceramic in ambient air. Rare Met. 2011;30:557–62.
154. Bahloul D, Pereira M, Goursat P. Silicon carbonitride derived from an organometallic precursor-influence of the microstructure on the oxidation behavior. Ceram Int. 1992;18:1–9.
155. Mocaer D, Chollon G, Pailler R, Filipuzzi L, Naslain R. Si-C-N Ceramics with a high microstructural stability elaborated from the pyrolysis of new polycarbosilazane precursors. J Mater Sci. 1993;28:3059–68.
156. Narushima T, Goto T, Hira T. High-temperature passive oxidation of chemically vapor deposited silicon carbide. J Am Ceram Soc. 1989;72:1386–90.
157. Schlichting J, Kriegesmann J. Oxidation behavior of hot-pressed silicon carbide. Berichte der Deutschen Keramischen Gesellschaft. 1979;56:72–5.
158. Singhal SC, Lange FF. Effect of alumina content on the oxidation of hot-pressed silicon carbide. J Am Ceram Soc. 1975;58:433–5.
159. Costello JA, Tressler RE. Oxidation kinetics of silicon carbide crystals and ceramics: I, in dry oxygen. J Am Ceram Soc. 1986;69:674–81.

Chapter 2
Principle and Kinetic Mechanisms of Mechanical Alloying Derived Non-equilibrium Materials

Mechanical alloying (MA) has been extended to synthesize a variety of metastable materials, such as supersaturated solid solutions, intermetallic compounds, nanocrystalline, quasi-crystalline, and amorphous phases, which otherwise cannot be synthesized by conventional routes. As an alternative synthesis route, though apparently very "hard" compared to the "soft" precursor-derived ceramics (PDCs) method, MA has been proven to produce completely amorphous SiBCN powder due to the advantages including the avoidance of external heating and potentially harmful organic solvents, the shorter processing time, and the simpler operation.

MA process has been shown to be critically dependent on various parameters, details about the raw materials, type of millers, milling containers, grinding mediums, ball-to-powder mass ratio, milling speed, milling time, and atmosphere, etc., which in turn have a strong influence on the sinterability of the as-milled powders and on the microstructures and properties of sintered bodies. Here recent advances in these synthesis areas have been summarized after discussing the alloying mechanism, thermodynamics and kinetics of solid-state amorphization, and process variables involved in MA. The alloying mechanism and amorphization kinetics of MA derived SiBCN are also introduced. Wherever possible, advantages of MA derived SiBCN route have been compared with PDCs counterparts.

2.1 Historical Perspective and Nomenclature

MA technology is a powder processing technology that allows fabrication of homogeneous composition materials starting from blended elemental or pre-alloyed powder mixtures [1]. The MA technique was first proposed by Benjamin et al. [2] in the early 1970s, and was originally used to prepare a nickel-based oxide-dispersion strengthened (ODS) superalloy for the gas turbine applications, which was expected to combine the two effects of precipitation hardening (intermediate-temperature

D. Li et al., *SiBCN Ceramics and Composites Prepared by Mechanical Alloying*,
https://doi.org/10.1007/978-981-97-3029-2_2

strength of γ' precipitates) and dispersion hardening (high-temperature strength of oxide dispersions) [3]. Coincidentally, the MA derived nickel-based ODS powders with each of the powder particles contain a dispersion of oxide particles. Benjamin et al. thus believed that the oxide layers formed on the powder particle surfaces were fractured and further incorporated into the solid powder particles through cold welding during high-energy milling. Benjamin and the co-workers called this milling process as "mechanical alloying", but was not widely accepted by the scientific communities at that time.

Until the early 1980s, the synthesis of an amorphous phase in a Y–Co intermetallic compound [4] and an amorphous Ni–Nb phase in a Ni–Nb system [5] was reported, both confirming the true alloying occurring during milling. These discoveries thereby greatly contribute the recognition of MA as a potential non-equilibrium processing technique, and the term "mechanical alloying" has gained widespread acceptance. Inspired by this new understanding, considerable investigations [6–8] have been carried out to produce a variety of amorphous or nanostructured powders, and metal-based or ceramic-based nanocomposites. In the late 1980s, another exciting area: mechanochemical processing, a combinatory of mechanical milling and chemical reactions, has been observed in a Ca–CuO system [9]. MA is, therefore, sometimes referred to as mechanochemical synthesis [10]. In the early 1990s, efforts were conducted to understand the processing fundamentals of MA through modeling studies [11]. Nevertheless, the accurate quantification and modelling of MA is still not available.

In the late 1990s, MA technique to synthetize novel alloy phases or new ODS materials has received widespread recognition and the main themes mainly focused on three fields: (1) Alloying two or more pure metals or alloys to form a new alloy phase, (2) destabilizing intermetallic compounds and element materials to form a nonequilibrium phase, (3) mechanically activated to induce chemical reactions, i.e., mechanochemical reactions at room temperature or at least at much lower temperatures than normally required of the traditional powder metallurgy methods. In the early twenty-first century, the MA theory and technology develop continuing, and now has been used to produce new multifunctional materials, such as magnetic materials, high-temperature materials, superconducting materials, superplastic alloy materials, composite materials, hydrogen storage materials, etc. [10–14]. Up to present, a deep understanding of the base alloying and solid-state amorphization mechanisms related to the MA process has not yet established, which deserves the close attention of all material scientists and engineers deeply engrossed in the area of MA processing non-equilibrium materials. The outstanding contributors and some important milestones in MA field have been summarized by Suryanarayana [1].

The term, MA, is used to describes the process when mixtures of element powders or alloys/compounds are milled together [1]. In this case, material transfer is involved to obtain a homogeneous composition alloy, thus the process is referred to as MA. In another case, material transfer is not required for homogenization when ball-milled uniform (often stoichiometric) composition powders, the process can be termed mechanical milling (MM) [10]. If the destruction of long-range order in intermetallics is needed to produce either a disordered intermetallic (solid solution) or an amorphous phase, this process is referred to as mechanical disordering (MD) [15]. The main advantage of MM/MD over MA is that the time required for processing is shorter [16], since only particle size refinements or other transformations are needed to be induced mechanically.

In case of powder process involving chemical reactions, structural changes, and phase transformations that are activated by inputting mechanical energy, then the process can be termed mechanochemical processing (MCP) [10]. Some other terms are also used in literature on mechanical alloying, including cryomilling [17], reactive milling (RM) [18], rod milling [15], mechanically activated annealing (M2A) [19, 20], double mechanical alloying (dMA) [21, 22], and mechanically activated self-propagating high-temperature synthesis (MASHS), etc. [23, 24].

Occasionally, be necessarily ball-milled associated with a liquid medium, this process is referred to as wet grinding [1]; otherwise, it is referred to as dry grinding. One of the main advantages for the wet grinding is that it seems more suitable to gain finer-ground products than that of dry grinding, since the solvent molecules are adsorbed on the newly formed surfaces of the particles lowering their surface energy. Another is the less-agglomerations of the powder particles in the wet milling conditions. The chief disadvantage of the wet grinding, however, is the often-vexing problem of powder contamination resulting from the milling medium. For this reason, most of the MA/MM processes are carried out under dry conditions to produce high-quality powders with high purity, superfine and homogeneous particle size distribution.

2.2 The Process of Mechanical Alloying

The microstructures and properties of the final milling products can be tailored by adjusting the interdependent processing parameters. Obviously, factors like the raw materials, type of millers, and the process variables hold the key to control the chemical and physicochemical transformations of the substances during MA process.

2.2.1 Raw Materials

Generally, the commercially available pure powder with particle sizes of 1–200 μm is suitable for MA as the raw materials [10]. The powder particle size is not strictly limited in that range, but should be smaller than the grinding ball size. This is because the powder particle size decreases exponentially as milling time progresses (only a few minutes of milling), finally reaching a small value of a few microns. The multiple types of raw materials including pure metals, master alloys, pre-alloyed powders, non-metallic powders and refractory compounds are commonly used. For example, amorphous systems of Si/C [25], Si/Ge [26, 27], BN/C [28] and SiBN [14] usually require raw materials of pure metals and/or non-metallic powders, while the quaternary amorphous SiBCN system [29] are ball milled from pure metals (c–Si), non-metallic powders (boron and graphite) and refractory compounds (SiC, Si_3N_4, B_4C, h–BN, etc.).

2.2.1.1 Type of Millers

MA derived amorphous and/or nanocrystalline powder starts with mixing of the inorganic powders in reasonable ingredient. Initially, the powder mixtures are loaded into the millers containing some grinding medium (e.g., steel balls, WC balls, ZrO_2 balls and Si_3N_4 balls, etc.) with the right ball-to-powder ratio. The milling processing is then conducted for the desired length of time at certain milling intensity until reaching to a steady state. The as-milled powders with the pre-designed composition and structure, is then compacted into a heating body to obtain the satisfying microstructure and properties.

Different types of high-energy milling equipment are used to produce mechanically alloyed powders [30]. They differ in their capacity, efficiency of milling and additional arrangements for cooling, heating, etc. At present, three commonly millers, such as attritor millers, shaker millers and planetary ball millers are used in laboratory (Fig. 2.1).

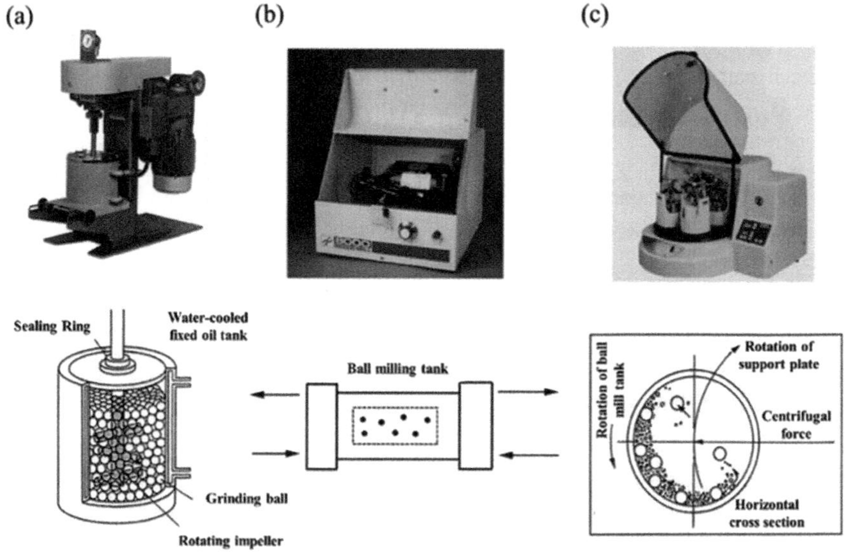

Fig. 2.1 The commonly used milling equipment and its schematic diagrams: **a** Attritor millers; **b** shaker millers; **c** planetary ball miller [1]

(1) **Attritor millers**

This type of millers is suitable for large-scale industrial production and can generally produce 0.5–40 kg powders at a time. The ordinary attritor millers have a horizontal placed drum half-filled with small steel balls [1]. As the drum begins to rotate, the balls drop on the powder mixtures that is being ground. Once the drum reaches a certain speed, the centrifugal force will bring the steel balls to the higher places and subsequently falling down to achieve the purpose of effective milling. At higher speeds, however, the balls are pinned to the drum wall by centrifugal force, which is unfavorable for mechanical alloying (At this point the grinding action stops).

(2) **Shaker millers**

The shaker millers, such as SPEX 800 millers with a small volume capacity, are generally used for alloy screening purposes and for the laboratory investigations [1]. The traditional shaker millers have only one vial, containing the powder mixtures and grinding balls, secured in the clamp and swung energetically back and forth several thousand times per minute. On average, the average ball velocities are on the order of 5 m/s, thereby leading to the unusually great impact force of the balls. Therefore, these millers are regarded as the high-energy variety and can generate 10–20 g powders each time.

(3) **Planetary ball millers**

The planetary ball millers are the popular millers for laboratory applications, in which dozens to several hundred grams of the finer powders can be obtained at a time [1]. To start with, the vias are fixed tightly on a rotating support disk and subsequently rotated oppositely to that of the disk. Consequently, the centrifugal forces alternately act in like and opposite directions. This causes the high-frequency friction and impact effects among the grinding balls, raw materials and inside wall, leading to the material being ground. In the modern versions (e.g., Fritsch Pulverisette 5, 6 and 7), the rotating velocities of the disk and the vials can be independently controlled [31, 32], and various milling balls, such as silicon nitride, zirconia, sintered corundum, chrome steel, Cr–Ni steel, tungsten carbide, and plastic polyamide are available for different milling purposes.

2.2.2 *Process Variables*

The process variables primarily include milling speed/time, weight ratio of ball-to-powder, milling atmosphere, milling temperature, type and size of grinding media and process control agent, etc. Taking the processing of amorphous SiBCN powder as an example, flat-ended stainless-steel vials with high purity Si_3N_4 linings are often used as the milling container. Simultaneously, high purity Si_3N_4 balls ($\sim$10 mm in diam.) and argon are essential to MA [33]. The variables such as milling speed, weight ratio of ball-to-powder, milling time, filling quantity, milling atmosphere, milling temperature and other factors are briefly introduced below.

(1) **Milling speed**

Below the maximum speed the miller allows (miller design limitations), higher milling speeds lead to higher energy input. Coincidently and as expected, high milling speeds result in higher system temperatures (grinding vials, balls and starting powders) having both positive and negative influences on final products. Increasing temperatures promote homogenization and powder alloying. However, transformation processes are accelerated resulting in the decomposition of formed supersaturated solid solutions and other metastable phases. Additionally, higher temperatures increase average particle sizes and decrease internal strain due to dynamic recrystallization [1]. During MA of Si_2BC_3N, milling speeds of 600 rpm with a ball-to-powder ratio of 20:1 provide completely amorphous phase. At 800 rpm, nanosized SiC precipitates in as-milled amorphous powder [34].

(2) **Weight ratio of ball-to-powder**

Weight ratio of the ball-to-powder (also ball-to-powder ratio, BPR) is one of the most important processing parameters. The BPR varies from as low as 1:1 to as high as 220:1. A 10:1 ratio is most commonly used in small capacity millers, while higher BPR up to 50:1 or even 100:1 is often used for large capacity millers. Ratios between 10:1 and 50:1 are most suitable for planetary millers equipped with grinding vials (500 mL capacity) [35, 36]. BPR has a significant influence on the desirable milling time. At higher BPR, the number of collisions/unit-time is higher as is input energy improving alloying rates thereby shortening milling time. However, higher BPR generates more heat, which may be detrimental to forming target products as noted just above. According to previous reports on the Zr–Co [37] and Cu–In–Ga–Se [38] systems, "soft" conditions (e.g., a combination of low BPR and low milling speed) produce non-equilibrium metastable phases, whereas "hard" conditions (e.g., combination of high BPR and high milling speed) produce equilibrium phases. In MA of Si_2BC_3N, milling at 600 rpm with a BPR of 20:1 produces desirable amorphous phase, while a BPR of $\geq$30:1 results in nanosized SiC precipitates [36].

(3) **Milling time**

The milling time depends mainly on a combination of milling speed, the weight ratio of ball-to-powder, etc. A reasonable milling time is usually chosen to achieve a balance between fracturing and cold welding of powder particles. It should be noted that the expected phases cannot be obtained with insufficient milling time. Longer milling times may increase contamination (due to the abrasion of balls and vial wall, etc.) and the formation of undesirable phases [12]. For example, insufficient time (<20 h) at 600 rpm and a BPR of 20:1 cannot provide amorphous Si_2BC_3N powder. At $\geq$ 25 h amorphous Si_2BC_3N forms, perhaps with a little free silicon remaining due to the insufficient mechanochemical reaction between the blended elements. Amorphous Si_2BC_3N powder can be obtained by combinations of a milling speed of 600 rpm, BPR of 20:1 and a milling time of 40 h. At times longer than 40 h SiC nanocrystallites form [34]. Similar results are obtained for MA of other systems such as Si–C [26], Ni–Nb [39], Fe–Co–Ni [40], Fe–Al [41], etc.

(4) **Mill atmosphere**

The choice of milling atmosphere depends on specific purposes since types of atmospheres affect the nature of the as-milling phases [42, 43]. If nitrides products [44, 45] are required, nitrogen or ammonia atmospheres are desirable; however, hydrogen atmosphere is used to produce hydrides [46]. For oxides, milling conducted in air is feasible. But in case of powders are easily reacted with the surrounding nature an inert atmosphere is suggested. Actually, nitrogen can react with many metals to form nitrides, which cannot be used to prevent N-contamination during milling. The high-purity argon or helium atmospheres are the common choice to prevent oxidation and/or N-contamination of the resulting powders. Note that the ambient conditions sensitive powder should be operated under an inert gas during production, transportation, storage and conversion. In all cases, MA derived amorphous or nanocrystalline SiBCN based powders are conducted under high purity Ar.

(5) **Milling container**

In MA derived amorphous Si_2BC_3N [14], grinding medium of Si_3N_4 balls and ZrO_2 bowls can cause ZrO_2 contamination in the resulting powders, while no trace of Si_3N_4 is found when both Si_3N_4 balls and bowls are adopted. In case of the two materials (powder being milled and grinding medium adopted) are the same, the chemistry of the final products may be changed unless proper precautions have been taken to minimize the risk [46].

(6) **Grinding medium**

Selecting desirable grinding balls for specific purposes. Most commonly in MA of SiBCN, the grinding medium (balls) used mainly include the SiC, ZrO_2 and Si_3N_4, etc. Note that the density or the hardness of the grinding balls should be high enough to produce effective impact on the powder particles during MA process. Otherwise, the grinding medium contamination may occur during long-term milling. The milling efficiency is confirmed to be closely corelated with the grinding medium sizes [1, 10]. For MA of Si_2BC_3N a larger size of the Si_3N_4 grinding balls ~ 8 mm promotes the solid-state amorphization of the powder mixtures. In MA derived SiBCN system, it is always desirable to have the grinding container and the grinding medium made of the same material so as to avoid cross contamination.

The above factors are not independent of each other; the optimum milling time strongly depends on the type of millers, ball sizes, milling speed, milling temperature and ball-to-powder ratio. For example, completely amorphous Si_2BC_3N is impossible below 10 h of high-energy milling, though higher BPR 30:1, higher milling speed 800–1000 rpm, and both Si_3N_4 grinding balls ~ 10 mm and Si_3N_4 vessels are adopted.

2.3 Mechanism of Mechanical Alloying

As noted above, MA is a complex process involving many interdependent variables. Thus, accurate quantification and modeling of MA are still not accessible, despite many efforts on multiple aspects of the process. During MA, the complex process can be simplified and divided into three stages [14]: (1) At an early stage, high-speed milling shears and plastically deforms powder particles leading to work hardening and fracture. The newborn particle surfaces weld particles (especially ones with high ductility) together resulting in initial increases in particle sizes. (2) In the second stage, work hardening and fatigue failure lead to increases in fracture, which now dominates over cold welding leading to decreases in particle size without forming strong agglomerates. (3) In the final stage, steady-state equilibrium is attained balancing the rates of welding and fracturing. A narrow particle size distribution is eventually reached by the size reduction of larger particles and size increase of smaller particles. Significantly, each particle contains all the starting ingredients in the proportion they are mixed together indicating the occurrence of "atomic mixing". Chemical compounds can be directly established between elements for some special systems.

Three types of system cover most examples of the MA processing.

(1) Ductile-ductile system (e.g., Cu–Ag, Al–Ni, Al–Ti, Ni–Cr). True alloying arises from repeated cold welding and fracture of powder particles. Cold welding in ductile particles under strong impact occurs readily during MA. Benjamin et al. [47, 48] suggest that in the early stages, ductile components are flattened to platelet shapes by micro-forging. Then the flattened particles cold weld forming lamellar structures and causing increases in particle sizes. They eventually approach a steady-state processing stage. At this stage, work hardening leads to particle fragmentation to more equiaxed particles. Hardness and particle sizes eventually reach equilibrium. Decreases in interlamellar spacing (helpful to mass diffusion) and increases in lattice defect density indicate initiation of alloying. The rate of structural refinement depends on the rates of mechanical energy input and work hardening. With further milling, layer spacing decreases and disappears. The powders attain "atomic mixing" producing new phases.
(2) Ductile–brittle system (ODS nickel- and iron-based alloys; Ni–$NiZr_2$). MA of ductile–brittle components requires both fragmentation of brittle particles facilitating short-range diffusion, and good solid solubility in the ductile matrix. If the brittle component has a negligible solid solubility, alloying is unlikely to occur, e.g., boron in iron, and oxide-dispersion strengthened (ODS) alloys. Taking the originally developed ODS alloys (brittle oxide particles are dispersed in a ductile matrix) as an example [1], in the initial stage, the ductile powder particles get flattened by the impact arising from ball-powder-ball and ball-powder-vial wall collisions. Coincidently, the brittle particles fragment and tend to be trapped in the ductile particles. The brittle constituent is close spaced along the interlamellar spacings. With further milling, the lamellae become convoluted and refined, interlamellar spacings decrease, and the brittle particles (if insoluble)

become uniformly dispersed in the ductile matrix (e.g., the dispersion of Er_2O_3 in α-titanium aluminide). The composition of each particle converges towards the overall compositions of the starting blended elemental powders. In contrast, if the brittle particles have good solubility in the ductile component, chemical homogeneity results in true alloying (e.g., the formation of amorphous phase by MA of ductile Zr and brittle $NiZr_2$ intermetallic powders).

(3) Brittle-brittle system (e.g., Si–Ge, Mn–Bi, Si–C, SiBCN). All particles fragment and their particle size are reduced with continuous ball milling. The harder particles (with higher brittleness) embed in the softer components. Additionally, atomic diffusion is likely essential to true alloying in all three types of systems. Pseudo-plastic deformation (not genuine plastic deformation, but having the appearance of plastic deformation) is the most likely mechanism contributing to mass transport during high-energy ball milling of brittle-brittle system. It is likely driven by a combination of local temperature rises, micro-deformation in defect-free volumes, deformation on particle surface and hydrostatic stress state in powders produced during ball milling [27]. Note that MA does not occur in brittle-brittle systems at sub-ambient temperatures probably due to long diffusion times. In contrast, alloying occurs in ductile-ductile and ductile–brittle systems due to diffusion paths improved by severe plastic deformation of ductile components.

2.4 Thermodynamics and Kinetics of Amorphization

The underlying mechanisms of amorphization by MA are not yet been clearly understood. The MA process consists of a balance of both thermodynamic and kinetic conditions [49, 50]. Amorphous phase is most easily obtained for equiatomic composition and an illustration of the process is exhibited in Fig. 2.2. The α, β, γ and mixtures of these phases are thermodynamically stable, while L is metastable. The free energy of the starting equiatomic powder mixture is G_c corresponding to point 1. The γ phase shows the lowest free energy, G_i, corresponding to point 3. Even though G_i is the lowest free energy state based on thermodynamics, the formation of γ phase can be prevented kinetically when the free energy G_a (point 2) is reached indicative of formation of an amorphous phase.

One can prevent the occurrence of reaction c → i while favoring c → a by choosing proper starting powders, reaction temperature and time. Here a, c, and i represent amorphous phase, starting powder mixture and intermetallic, respectively. Johnson et al. [51] suggest two conditions required to form an amorphous phase. The two starting components must first have a large negative heat of mixing in liquid/amorphous state to provide the driving force for thermodynamic reaction c → a. This criterion has proven to be important for forming amorphous phases in the Cu–Ti, Ni–Ti, and Co–Ti systems (a large negative mixing heat leads to a completely amorphous phase), and in the Fe–Ti, Mn–Ti and Cr–Ti systems (a small mixing heat leads to a partially amorphous product). A large negative heat of mixing is an essential

Fig. 2.2 Thermodynamics of an amorphous phase formation: **a** A hypothetical binary phase diagram indicating some solid solubility on either end and also exhibiting the presence of an intermetallic phase; **b** Gibbs free energy versus composition at a certain temperature (Points 1, 2, 3 correspond to the starting powder mixtures, amorphous phase and intermetallic, respectively) [1]

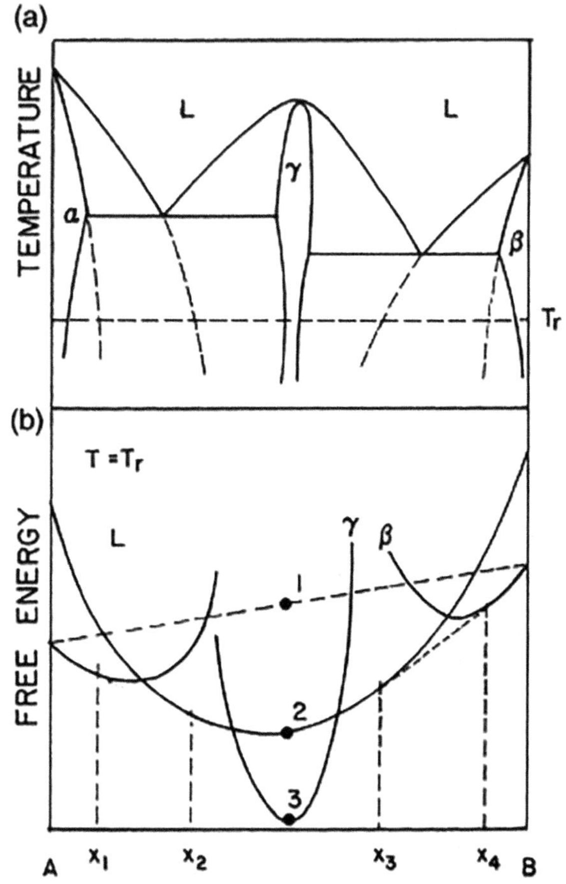

condition for glass formation by rapid solidification processing, although amorphous phase formation by MA can be achieved in a few alloy systems with a positive mixing heat. The other criterion is that the two starting components must exhibit significantly different diffusion rates into each other and in the new amorphous phase during MA. This ensures reaction c → a instead of the kinetic c → i.

As illustrated in Fig. 2.3, an appropriate time must not only ensure formation of an amorphous phase, but also prevent transformation to equilibrium phases. To prevent formation of an intermetallic, the time scale for formation of the amorphous phase from the starting powder mixture, referred to as $t_{c \to a}$, should be much shorter than $t_{c \to i}$. Coincidently, the time for transformation of a metastable amorphous phase into an intermetallic, $t_{a \to i}$, should be much longer than $t_{c \to a}$ ensuring higher stability for the amorphous phase. The two time conditions co-determine the kinetic formation of an amorphous phase.

In recent years, studies indicate that increases in temperature during MA are insufficient to cause powder particles to melt. However, amorphization is not purely a

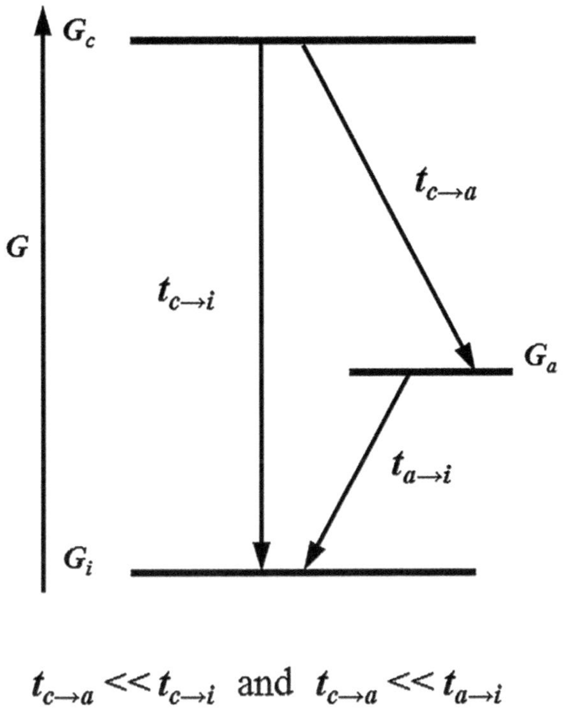

Fig. 2.3 Schematic free energy diagram indicating the criteria to be met for solid state amorphization. It shows kinetics of amorphous phase formation [1]

mechanical process or a simple solid-state reaction. Most likely, crystalline phases are destabilized due to an accumulation of structural defects including vacancies, dislocations, grain boundaries, and anti-phase boundaries. Increases in grain boundary area (due to continuously deceasing grain size), lattice expansion and lattice distortion can also contribute to stored energy. The two aspects lead to the free energy of intermetallic that is higher than that of amorphous phases thereby driving the formation of an amorphous phase.

There have been extensive attempts to clarify the MA mechanisms [26, 52–54]. Shaw et al. [26] investigated bond formation and structural changes of Si in Si-graphite powder mixtures during MA by NMR, TEM and XRD finding that high-energy milling does not result in in-situ formation of crystalline SiC, most Si and C atoms are intimately mixed at atomic levels in the amorphous phase and the nominal composition of the amorphous phase is 1:1 (Si:C). It is proposed that amorphization of Si is primarily induced by crystallite refinement, and mixing of Si and C at atomic levels is assisted by micro-diffusion during MA.

Studies on the amorphization of Ni–Nb and Ni–Zr binary alloys reported by Enayati et al. [39] showed that MA leads to refinement of grain sizes and increases in internal strain, followed by amorphization. An ultrafine layered structure forms containing nano-scaled grains prior to amorphization suggesting that numerous high-speed diffusion paths are necessary to allow high diffusion rates at low temperatures permitting kinetic amorphization reaction. Amorphous phase formation begins along interfacial layers via solid-state processes involving diffusion/alloying between A/B layers. The reduction in crystallite size only aids amorphization by accelerating diffusion and/or providing suitable interfacial sites.

2.5 MA Processing and Amorphization of SiBCN

2.5.1 Basic Process of MA

In fields of structure ceramics, MA or MM techniques are generally used to refine ceramic particles or synthesize nanoceramic powders. Up to now, the completely amorphous ceramics and their composite materials induced by MA or MM are rarely reported. Because most ceramic systems tend to form nanocomposites or amorphous-nanocrystalline hybrid structure instead of completely amorphous phases after MA process.

MA of SiBCN is not purely a mechanical action, which involves complex physico-chemical reactions of grain refinement and fragmentation, cold sintering, short-range diffusion of elemental atoms [55]. During MA, destabilization of the crystalline phases of the raw materials, such cubic Si, graphite, boron, h–BN, SiC, B_4C, Si_3N_4, etc., tends to occur by the accumulation of structural defects [56]. The annihilation of crystalline volume accompanied with defects formation changes the local atomic configuration thus raising the free energy of the powder mixture to a level higher than that of the amorphous phase. This provides the best information possible to its solid-sate amorphization. When the driving energy of nanocrystals destabilization has accumulated so greatly that the crystalline structure of the raw materials becomes unstable, a uniform distribution of all elements is eventually formed and consequently, amorphous SiBCN powder is obtained at the end of milling. MA offers an opportunity for the atoms and/or molecules to interact with each other, and break of old chemical bonds and the establishment of new ones in a cyclical manner. However, during MA, not all atomic interactions are valid, only "intimate" atoms can bind to form new chemical bonds. For example, substantial new bonds of Si–C, C–B–N and C–N–B form at the end of milling in final amorphous Si_2BC_3N powder, but no trace of Si–N and Si–B bonds is observed or below the limits of detection for the given designed composition.

2.5.2 Alloying Mechanism of SiBCN

According to the difficult level of alloying, three types of milling systems, "ductile"-"ductile", "ductile"-brittle, and brittle-brittle systems, are also divided covering the most examples for MA derived SiBCN, SiBON, SiCN, SiBC, SiBN, etc. (Fig. 2.4).

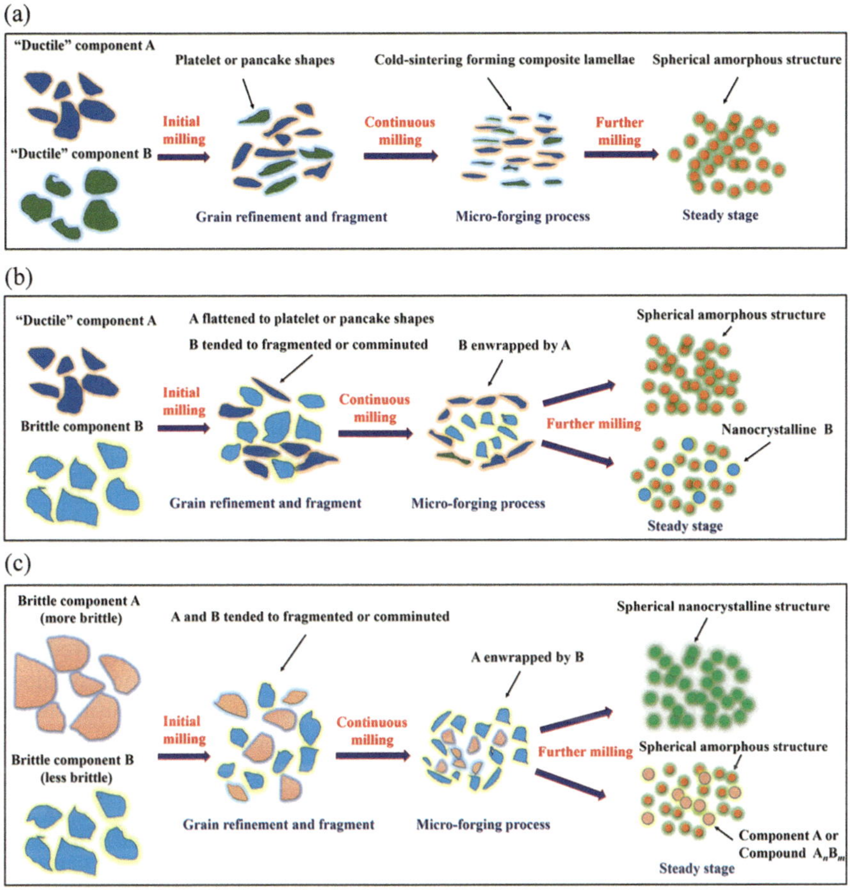

Fig. 2.4 Three models of the structural evolution schematic for MA-SiBCN, MA-SiBON, MA-SiCN, MA-SiBC, MA-SiBN, etc. covering the most examples: **a** "Ductile"-"ductile" system; **b** "ductile"-brittle system; **c** brittle-brittle system

The raw materials of cubic Si, graphite, h-BN for SiBCN system, and SiO_2, h-BN for SiBON system can be regarded as the "ductile" components during MA. At initial stage of milling, these "ductile" components begin to flattened to platelet shapes by a micro-forging process. Continued milling, these flattened particles get repeated cold sintering and fracture forming a composite lamellar structure and therefore particle size refinement. After approaching a steady-state processing stage, the hardness and particle size eventually reach equilibrium and alloying can occur at this stage due to the combination of decreasing short-range diffusion distances, increasing lattice defect density and local high-temperatures. Further ball-milling, true alloying arises at an atomic level leading to the formation of amorphous phases in SiBCN or SiBON systems.

The formation of a hybrid structure of amorphous and nanocrystalline phases on milling a powder mixture of Si_3N_4 (brittle) and graphite ("ductile") is a typical "ductile"-brittle example for SiCN system. This is the same situation for SiBN (e.g., using brittle Si_3N_4 and "ductile" boron as raw materials) and SiBC (e.g., using "ductile" cubic Si and brittle B_4C as raw materials) systems. MA of the SiCN, SiBN and SiBC "ductile"-brittle systems also requires both fragmentation of brittle components (e.g., Si_3N_4, B_4C) and their good solid solubility in a "ductile" matrix (e.g., boron, cubic Si, graphite). In the initial stage of milling, the "ductile" ceramic particles get flattened by the effective impact of ball-powder-ball and ball-powder-vial wall collisions, during which the brittle components get fragmented and/or comminuted. Coincidently, at this stage, the refined brittle particles tend to become occluded by the "ductile" ingredients and trapped in the "ductile" matrix. Continued milling, the "ductile" ceramic particles get work hardening, while the lamellae get further convoluted and refined. Under stable state, the composition of each particle eventually converges toward the overall composition of the starting blended elemental powders. In case the brittle component is soluble, true alloying tends to appear and chemical homogeneity can be achieved. However, if the brittle component has a negligible solid solubility, then any alloying is unlikely to occur.

It is unlikely that alloying can appear in the brittle-brittle system from an intuitive standpoint. This is based on Benjamin's suggestion that at least 15% of a "ductile" component is required for achieving possible alloying [47, 48]. This has proved to be true when MA on a blend of brittle Si_3N_4 and brittle B_4C components only results in constituents' refinement without any trace of alloying. During milling of brittle-brittle ceramic systems, all the particles occur to fragment and their particle sizes get reduced continuously. Continued milling, the harder (more brittle) component gets further fragmented and consequently, gets embedded in the softer (less brittle) matrix. For example, the softer Si_3N_4 particles are embedded in the harder SiC matrix for SiCN system. Note that, at very small particle sizes the more brittle particles can behave in a "ductile" fashion, but further reduction in size is not possible (comminution limit in current technical conditions). At an atomic or molecular scale, these brittle particles are unlikely to contact intimately so that the possible material migration driven by short-range diffusion is limited and thus, true alloying is unlikely to take place.

2.5.3 Amorphization Kinetics of SiBCN

Zhang [12] has pointed out that, amorphous phases can be produced by three routes: Route 1, MA two or more elemental phases; route 2, MA one elemental phase and another intermetallic compound, or two or more intermetallic compounds; route 3, MM a single-phase elemental or intermetallic compound powder. Exemplify the most common route 1, when the elemental phases of A and B react, the final product phases are those of an amorphous phase, A(B) and B(A) solid solutions, or A_xB_y intermetallic compounds. In cease the kinetic condition is unfavorable to the formation of A(B) or B(A) solid solution, and A_xB_y intermetallic compounds, an amorphous phase often nucleates. As milling further progresses, amorphous phase increases continually through the growth of the original amorphous domains and/or through the nucleation and growth of newly generating amorphous area. Zhang thus believes that, milling induced an amorphous phase is very similar to the annealing of multi-layered thin films composed of alternating layers of different phases. But if the A and B elements have a negative heat of mixing, then they tend to be alloyed into compounds or solid solutions. In this case, the kinetic condition is the only factor determining whether an amorphous phase can form.

However, in another case, Schwarz and Johnson [58] have proposed two conditions needed to be satisfied for an amorphous phase formation produced from a blended elemental powder mixture. First, the two constituents must have a large negative heat of mixing in the liquid or amorphous state, thus it can provide the thermodynamic driving force for the reaction to occur. The $\Delta H_{\mathrm{AB}}^{\mathrm{mix}}$ (kJ/mol) calculated by Miedema's model [59] for atomic pairs in SiBCN system, such as Si–B (−14 kJ/mol), Si–C (−39 kJ/mol), Si–N (−81 kJ/mol), B–C (−10 kJ/mol), B–N (−28 kJ/mol) and C–N (−2 kJ/mol), are negative. The mixing heat for SiBCN system cannot be obtained directly; however, it can be reasonably inferred that values of $\Delta H_{\mathrm{SiBCN}}^{\mathrm{mix}}$ may be negative. In experimental findings [60, 61], SiBCN powder prepared from graphite, h-BN and cubic silicon, or SiBON powder using fused SiO_2 and h-BN as starting materials can be completely amorphized by MA; however, SiBN system derived from Si_3N_4 and boron, or SiCN using graphite and Si_3N_4 as raw materials can be only partially amorphized. In practical terms, alloying systems with a positive heat of mixing have also been amorphized by MA in recent years [47, 48].

The second criterion is that the two constituents must have largely different diffusion rates into each other and in the amorphous phase. This will contribute to the reaction kinetics of path c → a over that of reaction path c → i. This asymmetry in the diffusivities is strongly correlated with the difference in atomic sizes for Si, B, C and N elements which determines the composition range in which a solid solution is stable.

MA often introduces some ineluctable impurities, such as substituted impurities like aluminum, iron and nickel from raw ceramic powders, grinding medium like ZrO_2, and interstitial impurities like nitrogen, carbon, oxygen either from the atmosphere or the process control agents [55–60]. Substituted impurities, of course, can dilate the lattice; however, interstitial impurities can either dilate or contract the lattice depending on the atomic sizes of these elements with respect to the constituent elements. So, another possible solid-state amorphization process of MA derived SiBCN is also proposed in which amorphization is assumed to be realized through interstitial impurity formation during MA. However, this gives a hypothesis that MA induced amorphization occurs when the impurity atoms can dissolve into interstitial sites leading to local lattice distortion. When the local lattice distortion energy reaches a critical threshold, crystalline phases are most likely destroyed due to an accumulation of structural defects in lattice and increase in grain boundary area leading to the free energy of nanocrystals that is higher than that of amorphous phases thereby driving the formation of an amorphous phase in SiBCN.

The reaction temperature should be considered during MA of SiBCN, since the diffusion processes involved in the formation of alloying phases irrespective of whether the final product phase is a solid solution, intermetallics, nanocrystals, or an amorphous phase. It is expected that the temperature of milling will have a feeble effect in any alloying system. But the formation of an amorphous phase is unlikely occurring once the system powder being milled at high-temperatures. This is because raising milling temperature results in to a situation such that the actual temperature in local region is above the crystallization temperature of the amorphous phase. In this case, the preforming amorphous phases will inevitably transfer to nanocrystals. This has been confirmed to be true in SiBCN and SiBON quaternary systems when higher milling speeds and/or higher ball-to-powder ratios (refers to higher local temperature) employed leads to the formation of some nano SiC or SiO_2, respectively [59, 60]. It thus suggests that milling should be conducted at a relatively low temperature and an enough time interval of milling should not be ignored if the final product of an amorphous SiBCN is required.

The milling time of MA, mainly depending on a combination of the milling speed and ball-to-powder ratio, is also important because not only should one allow the formation of the amorphous phase, but also maintain it without changing to the equilibrium nanocrystalline phases [62]. To prevent the thermostable nanocrystals from forming, as mentioned above, the time scale $t_{\mathrm{c}\rightarrow\mathrm{a}}$ for the formation of amorphous SiBCN from the blended elemental mixture should be much shorter than $t_{\mathrm{c}\rightarrow\mathrm{i}}$. More significantly, since amorphous phase is thermodynamically metastable, $t_{\mathrm{a}\rightarrow i}$ should be much longer than $t_{\mathrm{c}\rightarrow\mathrm{i}}$, so that the stability of amorphous SiBCN phase can be retained. Note that prolonging the milling time can favor the nucleation and growth of possible nanocrystals since it can promote the effective collision frequency among the atoms and/or molecules. Therefore, the kinetic conditions for the amorphous phase formation in MA derived SiBCN system can also be summarized as: $t_{\mathrm{c}\rightarrow\mathrm{a}} \ll t_{\mathrm{c}\rightarrow\mathrm{i}}$ and $t_{a\rightarrow\mathrm{i}} \gg t_{\mathrm{c}\rightarrow\mathrm{a}}$.

2.5.4 Advantages and Challenges of MA Derived SiBCN

The precursor-derived ceramics (PDCs) route can provide high-purity amorphous SiBCN with a homogenous elemental distribution at atomic scales via the pyrolysis of cross-linked infusible precursors [63, 64]. This organic method allows the variegated design of the skeleton of the preceramic network thereby tailoring the physical, chemical, and high-temperature properties of the final ceramic products. However, the operational handling is complex and requires strict water-free and oxygen-free conditions. The expensive and toxic raw materials including the organic monomers and solvents may lead to increases in the product cost, human health damage, and environmental pollution (Fig. 2.5).

Unlike PDCs route, MA is a non-equilibrium solid-state powder processing technique that enables the production of intimately mixed, fine-scale amorphous SiBCN from a wide range of blended elemental or pre-alloyed powders. MA can be applied

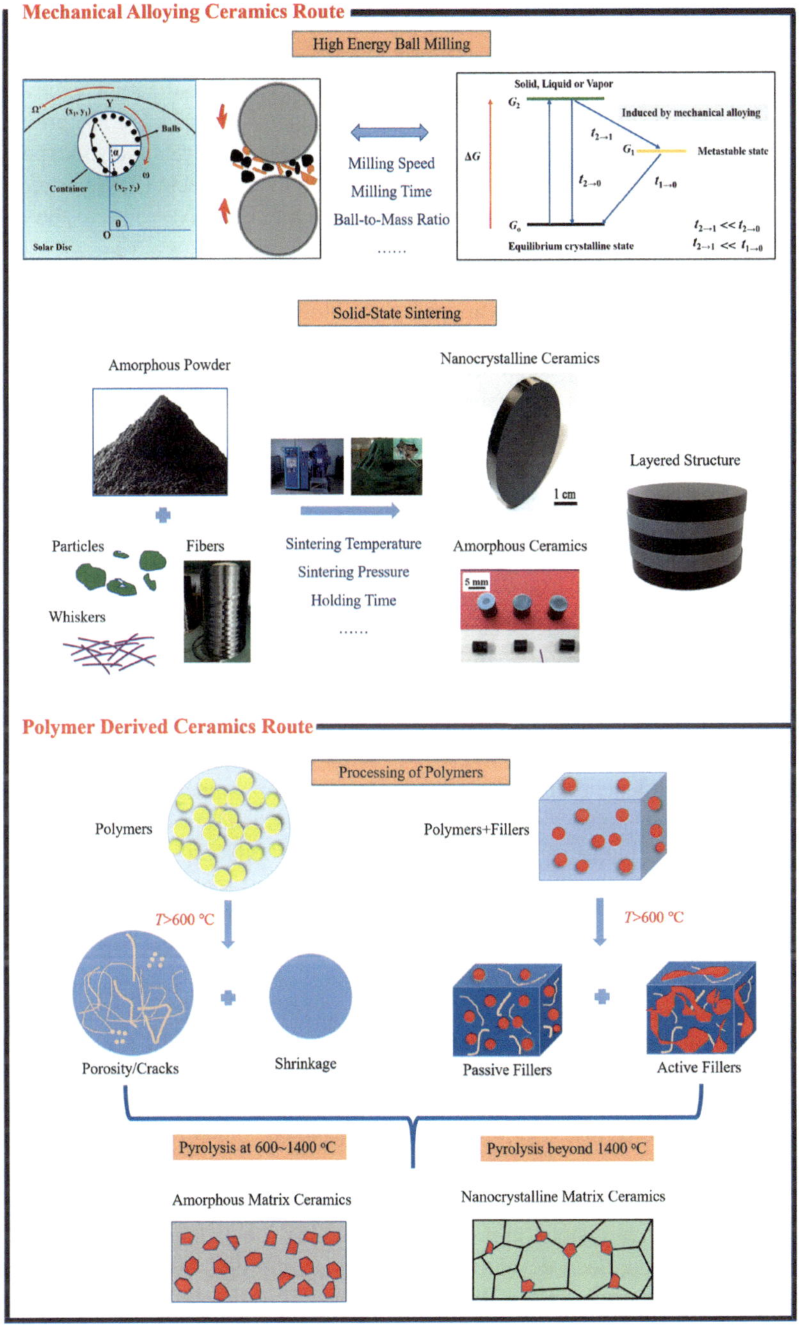

Fig. 2.5 Schematic illustration of the SiBCN materials processing by mechanical alloying plus pressure-assisted sintering method or precursor-derived ceramics route [63, 64]

to the ductile and brittle SiBCN systems, and mixtures thereof. The high mechanical energy achieved during MA of SiBCN can activate mechanochemical effects thereby attaining atomic level mixing. MA combined with powder densification achieved by pressure-assisted sintering, can offer several potential advantages: (1) The process is relatively green because it uses readily available elemental and relatively inexpensive inorganic starting powders compared with the precursor approach; (2) yields are high without essentially by-products; (3) it is a simple and effective route to the dense monoliths and structural parts that avoids the problems of excessive gas evolution (mass loss) and volume changes that occur in the precursor processing approach; (4) production on an industrial scale appears technologically feasible.

The MA derived amorphous SiBCN powders typically must be densified to monoliths using mechanical pressure assisted sintering techniques, such as hot-pressing (HP), hot-isotactic pressing (HIP), spark plasm sintering (SPS) and high-pressure sintering (HPS), due to the low atomic diffusivity of covalently bonded ceramics. Just like PDCs route, MA derived amorphous SiBCN powder remains amorphous at relatively low-temperatures and exhibits homogeneous structure due to the atomic mixing. The crystallization of amorphous phase at high-temperatures leads to the nanostructures with higher homogeneity than counterparts prepared by the conventional processing using reactive sintering of powders. As noted just above, the microstructures and properties can be tailored by controlling the starting powder types and judicious choice of chemical composition.

In MA processing of SiBCN, the changes in chemical bonding, degree of crystallinity of the starting powder mixtures, and microstructural/morphological evolution (such as particle shape and size) still require more studies. Additionally, the influence of chemical composition changes on amorphization of SiBCN-Me (Me = Al, Zr, Hf, Ta, etc.) powder also requires further investigation. Much help can be got from the combination of various characterization methods, especially the advanced analytical methods including SAXS (small angle X-ray scattering), WAXS (wide angle X-ray scattering) and XANES (X-ray absorption near edge structure) based on synchrotron radiation technique. For example, SAXS provides information on microstructures at nano-scales while WAXS provides information on short-range ordered microstructures at the atomic-scales. Both offer insights into basic scientific information about microstructural evolution during MA of SiBCN-Me powder.

Note that the intrinsic brittleness of SiBCN ceramics themselves frequently leads to catastrophic failure severely limiting their engineering applications; however, design criteria defined by Raj [65] for "superceramics" identify room-temperature fracture toughness and thermal shock resistance as being crucial to successful microstructural designs. Efforts to improve toughness remain one of the most important topics in SiBCN system. In precursor processing route, damage-tolerant PDCs-SiBCN CMCs have been developed by precursor impregnation of fiber preforms (sometimes with particle fillers). In the inorganic approach, chopped fibers are most suitable for processing CMCs as MA derived powders are not particularly suitable for impregnation or infiltration into continuous fiber preforms.

One can envision using short/chopped carbon or SiC fibers as reinforcements to develop short carbon fiber C_{sf}, SiC_{sf} or C_{sf}/SiC_{sf} toughened SiBCN CMCs. Coincidently, particle fillers (such as refractory Zr, Hf, Ta, borides, nitrides and carbides, graphene, etc.) can be incorporated into SiBCN matrix to effectively improve their high-temperature properties, especially resistance to oxidation and ablation. Attention should be paid to the interfaces between the secondary phase (including the short fibers and particle fillers) and SiBCN matrix when it comes to CMCs. The interface design strategies such as the introduction of proper h-BN and/or pyrolytic carbon coating (PyC) on the fiber surfaces can provide appropriate interfacial strength thereby substantially strengthening and toughening the SiBCN CMCs.

References

1. Suryanarayana C. Mechanical alloying and milling. Prog Mater Sci. 2001;46:1–184.
2. Benjamin JS. Dispersion strengthened superalloys by mechanical alloying. Metall Trans. 1970;1:2943–51.
3. Benjamin JS. Mechanical alloying. Sci Am. 1976;234:40–8.
4. Ermakov AE, Yurchikov EE, Barinov VA. Magnetic properties of amorphous powdered Y-Co system alloys, obtained by mechanical griding. Fizika Metallovi Metallovedenie. 1981;52:1184–93.
5. Koch CC, Cavin OB, McKamey CG, Scarbrough JO. Preparation of "'Amorphous"' $Ni_{60}Nb_{40}$ by mechanical alloying. Appl Phys Lett. 1983;43:1017–9.
6. Suryanarayana C. Recent advances in the synthesis of alloy phases by mechanical alloying/milling. Met Mater Int. 1996;2:195–209.
7. Lv L, Lai MO. Mechanical alloying. Springer Science & Business Media; 1997.
8. Murty BS, Ranganathan S. Novel materials synthesis by mechanical alloying/milling. Int Mater Rev. 1998;43:101–41.
9. Schaffer GB, Mccormick PG. Reduction of metal oxides by mechanical alloying. Appl Phys Lett. 1989;55:45–6.
10. Suryanarayana C, Al-Aqeeli N. Mechanically alloyed nanocomposites. Prog Mater Sci. 2013;58:383–502.
11. Maurice DR, Courtney TH. The physics of mechanical alloying: a first report. Metall Trans A. 1990;21:289–303.
12. Zhang DL. Processing of advanced materials using high-energy mechanical alloying. Prog Mater Sci. 2004;49:537–60.
13. Rairden JR, Habesch EM. Low-pressure-plasma-deposited coating formed from mechanical alloying powders. Thin Solid Films. 1981;83:353–60.
14. Jia DC, Liang B, Yang ZH, Zhou Y. Metastable Si-B-C-N ceramics and their matrix composites developed by inorganic route based on mechanical alloying: fabrication, microstructures, properties and their relevant basic scientific issues. Prog Mater Sci. 2018;98:1–67.
15. El-Eskandarany MS, Aoki K, Suzuki K. Rod milling for solid-state formation of Al30Ta70 amorphous alloy powder. ChemInform. 1990;167(1):113–8.
16. Weeber AW, Bakker H, de Boer FR. The preparation of amorphous Ni-Zr powder by grinding the crystalline alloy. Europhys Lett. 1986;2:445.
17. Luton MJ, Jayanth CS, Disko MM, Matras MM, Vallone J. Multicomponent ultrafine microstructures. Materials Research Society, 1989. p. 79–86.
18. Jangg G, Kuttner F, Korb G. Production and properties of dispersion hardened aluminum. Aluminium 1975; 51:641–645.

19. Gaffet E, Malhouroux N, Abdellaoui M. Far from equilibrium phase transition induced by solid-state reaction in the Fe-Si system. J Alloy Compd. 1993;194(2):339–60.
20. Gaffet E, Malhouroux-Gaffet N. Nanocrystalline $MoSi_2$ phase formation induced by mechanically activated annealing. J Alloy Compd. 1994;205:27–34.
21. Froyen L, Delaey L, Niu X-P, LeBrun P, Peytour C. Synthesizing aluminum alloys by double mechanical alloying. JOM. 1995;47:16–9.
22. Le Brun P, Froyen L, Delaey L. The modelling of the mechanical alloying process in a planetary ball mill: comparison between theory and in-situ observations. Mater Sci Eng, A. 1993;157:79–88.
23. Charlot F, Gaffet E, Zeghmati B, Bernard F, Niepce JC. Mechanically activated synthesis studied by X-ray diffraction in the Fe-Al system. Mater Sci Eng, A. 1999;262:279–88.
24. Gauthier V, Josse C, Bernard F, Gaffet E, Larpin JP. Synthesis of niobium aluminides using mechanically activated self-propagating high-temperature synthesis and mechanically activated annealing process. Mater Sci Eng, A. 1999;265:117–28.
25. Davis RM, Koch CC. Mechanical alloying of brittle components: silicon and germanium. JOM. 1987;39:60–1.
26. Xie X, Yang Z, Ren R, Shaw LL. Solid state ^{29}Si magic angle spinning NMR: investigation of bond formation and crystallinity of silicon and graphite powder mixtures during high energy milling. Mater Sci Eng, A. 1998;255:39–48.
27. Davis RM, Mcdermott B, Koch CC. Mechanical alloying of brittle materials. Metall Trans A. 1988;19:2867–74.
28. Zhang YF, Tang YH, Lee CS, Bello I, Lee ST. Nanocrystalline C-BN synthesized by mechanical alloying. Diamond Related Mater. 1999;8:610–3.
29. Zhang PF, Jia DC, Yang ZH, Duan XM, Zhou Y. Influence of ball milling parameters on the structure of the mechanically alloyed SiBCN powder. Ceram Int. 2013;39:1963–9.
30. Suryanarayana C. Powder metal technologies and applications. ASM Handbook. 1998;7:80–90.
31. Yamada K, Koch CC. The influence of mill energy and temperature on the structure of the TiNi intermetallic after mechanical attrition. J Mater Res. 1993;8:1317–26.
32. ThuÈmmler F, Oberacker R. Introduction to powder metallurgy. The Institute of Materials; 1993.
33. Zhang PF, Jia DC, Yang ZH, Duan XM, Zhou Y. Physical and surface characteristics of the mechanically alloyed SiBCN powder. Ceram Int. 2012;38:6399–404.
34. Yang ZH, Jia DC, Zhou Y, Yu CQ. Fabrication and characterization of amorphous Si-B-C-N powders. Ceram Int. 2007;33:1573–7.
35. Zhang PF, Jia DC, Yang ZH, Duan XM, Zhou Y. Microstructural features and properties of the nano-crystalline SiC/BN(C) composite ceramic prepared from the mechanically alloyed SiBCN powder. J Alloy Compd. 2012;537:346–56.
36. Zhang PF, Jia DC, Yang ZH, Duan XM, Zhou Y. Crystallization and microstructural evolution process from the mechanically alloyed amorphous Si-B-C-N powder to the hot-pressed nano SiC/BN(C) ceramic. J Mater Sci. 2012;47:7291–304.
37. Gerasimov KB, Gusev AA, Ivanov EY, Boldyrev VV. Tribochemical equilibrium in mechanical alloying of metals. J Mater Sci. 1991;26:2495–500.
38. Suryanarayana C, Ivanov E, Noufi R, Contreras MA, Moore JJ. Phase selection in a mechanically alloyed Cu-In-Ga-Se powder mixture. J Mater Res. 1999;14:377–83.
39. Enayati MH, Dastanpoor E. Comparative study of mechanical alloying induced nanocrystallization and amorphization in Ni-Nb and Ni-Zr systems. Metall and Mater Trans A. 2013;44:3984–98.
40. Duan YP, Zhang YH, Wang TM, Gu SC, Xin L, Lv XJ. Evolution study of microstructure and electromagnetic behaviors of Fe-Co-Ni alloy with mechanical alloying. Mater Sci Eng, B. 2014;185:86–93.
41. Fan RH, Sun JT, Gong HY, Sun KN, Wang WM. Structural evolution of mechanically alloyed nanocrystalline Fe-28Al powders. Powder Technol. 2005;149:121–6.
42. Kimura H, Kimura M. Solid state powder processing. TMS;1990.

43. Lee CH, Mori M, Fukunaga T, Mizutani U. Effect of ambient temperature on the MA and MG processes in Ni-Zr alloy system. Jpn J Appl Phys. 1990;29:540–4.
44. Miki M, Yamasaki T, Ogino Y. Preparation of nanocrystalline NbN and (Nb, Al)N powders by mechanical alloying under nitrogen atmosphere. Mater Trans, JIM. 1992;33:839–44.
45. Calka A, Williams JS. Synthesis of nitrides by mechanical Alloying. In: Materials science forum, vol. 88. Trans Tech Publications Ltd.; 1992. p. 787–794.
46. Chen Y, Williams JR. Hydriding reactions induced by ball milling. In: Materials science forum, vol. 225. Trans Tech Publications Ltd.; 1996. p. 881–888.
47. Benjamin JS. Fundmentals of mechanical alloying. Mech Alloying. 1992;88:1–17.
48. Benjamin JS. Mechanical alloying-history and future potential. Adv Powder Metallic Part Mater. 1992;7:155–68.
49. Johnson WL. Thermodynamic and kinetic aspects of the crystal to glass transformation in metallic materials. Progr Mater Sci. 1986;30:81–134.
50. Gilman PS, Benjamin JS. Mechanical alloying. Annu Rev Mater Res. 1983;13:279–300.
51. Johnson WL. Crystal-to-glass transformation in metallic materials. Mater Sci Eng. 1988;97:1–13.
52. Chakk Y, Berger S, Weiss BZ, Brook-Levinson E. Solid state amorphization by mechanical alloying–an atomistic model. Acta Metall Mater. 1994;42(11):3679–85.
53. Omuro K, Miura H. Amorphization of mechanically alloyed Fe-C and Fe-N materials with additive elements and their concentration dependence. In: Materials science forum, vol. 179–181. Trans Tech Publications, Ltd.; 1995. p. 273–278.
54. Zhang H. The quantification of criteria for predicting glass-formation of binary transition-metals by mechanical alloying. J Phys: Condens Matter. 1993;5:1337–42.
55. Liang B, Yang ZH, Chen QQ, Wang SJ, Duan XM, Jia DC, Zhou Y, Luo K, Yu DL, Tian YJ. Crystallization behavior of amorphous Si_2BC_3N ceramic monolith subjected to high pressure. J Am Ceram Soc. 2015;98:3788–96.
56. Zhang PF, Jia DC, Yang ZH, Duan XM, Zhou Y. Progress of a novel non-oxide Si-B-C-N ceramic and its matrix composites. J Adv Ceram. 2012;1:157–78.
57. Schwarz RB, Johnson WL. Formation of an amorphous alloy by solid-state reaction of the pure polycrystalline metals. Phys Rev Lett. 1983;51:415.
58. Bakker H. Enthalpies in alloys, Miedema's semi-empirical model. Trans. Tech: Publications; 1998.
59. Yang ZH, Zhou Y, Jia DC, Meng QC. Microstructures and properties of $SiB_{0.5}C_{1.5}N_{0.5}$ ceramics consolidated by mechanical alloying and hot pressing. Mater Sci Eng A 2008;489:187–192.
60. Li Q, Yang ZH, Miao Y, Liang B, Cai DL, Wang SJ, Duan XM, Jia DC, Zhou Y. Effect of the BN content on the thermal shock resistance and properties of BN/SiO_2 composites fabricated from mechanically alloyed SiBON powders. RSC Adv. 2017;7:48994–9003.
61. Li Q, Yang ZH, Zhong J, Sun YS, Jia DC, Duan XM, Zhou Y. Effect of the starting AlN content on the phase formation and property of the novel in-situ fabricated X-SiAlON/BN composites. J Eur Ceram Soc. 2019;39(4):934–43.
62. Riedel R, Kleebe HJ, Schonfelde H, Aldinger F. A covalent micro/nano-composite resistant to high-temperature oxidation. Nature. 1995;374:526–8.
63. Colombo P, Mera G, Riedel R, Soraru GD. Polymer-derived ceramics: 40 years of research and innovation in advanced ceramics. J Am Ceram Soc. 2010;93:1805–37.
64. Li D, Jia D, Yang Z, Zhou Y. Principles, design, structure and properties of ceramics for microwave absorption or transmission at high-temperatures. Int Mater Rev. 2022;67(3):266–97.
65. Rishi R. Fundamental research in structural ceramics for service near 2000 °C. J Am Ceram Soc. 1993;76:2147–74.

Chapter 3
Solid-State Amorphization, Crystallization Behavior and Thermal Stability of MA-SiBCN Powders

Mechanical alloying (MA) usually involves repeated welding, fracturing, and rewelding of powder particles which promotes mechanochemical effect making possible the homogeneous mixture at atomic levels and new chemical bonding states. MA-driven mechanochemical reactions have some advantages over the conventional powder processing techniques. For example, MA can induce completely amorphous SiBCN powder with higher thermal stability while only partial amorphous phase is synthesized by powder metallurgy or sol–gel processing ways. The appropriate ball milling parameters of MA, is critical to the preparation of amorphous SiBCN powder, though their atomic structures still remain open to debate despite considerable experimental studies and computer simulation.

MA derived SiBCN powder is characterized by their morphologies (including size and shape), surface area, phase constitution, microstructural features, chemical bonding states and even transformation behavior on further treatment. Additions of fine second phases like metal particles, ceramic particles, whiskers, graphene, etc. to amorphous SiBCN matrix generally hinder the solid-state amorphization process and lower their structural stability at high-temperatures. Such fine reinforcement particles can be most effectively and homogeneously distributed in a powder matrix through the process of MA or sol–gel.

3.1 MA Induced Amorphization and Valence Bond Changes

Figure 3.1 summarizes the typical XRD patterns of MA-Si_2BC_3N powder blends after milling from 0.5 to 40 h [1]. The characteristic peaks of the constituent elements gradually decrease in intensity, and increase in width (full width at half maximum, FWHM) finally disappearing beyond 40 h of milling. In detail, the diffraction intensities of the raw materials such as h-BN and graphite drop dramatically after 0.5

D. Li et al., *SiBCN Ceramics and Composites Prepared by Mechanical Alloying*,
https://doi.org/10.1007/978-981-97-3029-2_3

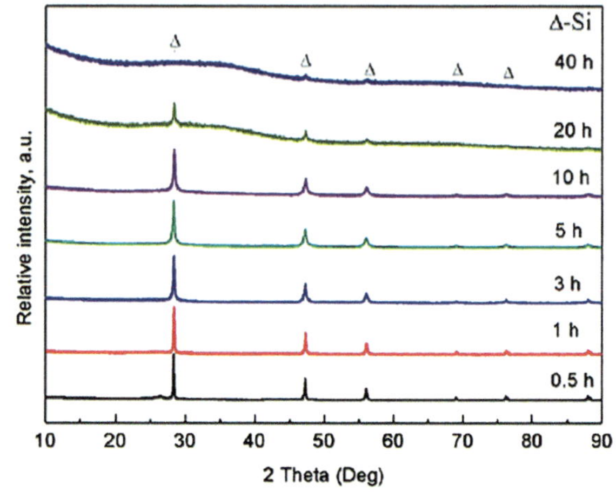

Fig. 3.1 Representative XRD patterns of MA-Si_2BC_3N powder milled at different time (milling speed of 600 rpm; ball-to-powder mass ratio of 20:1) [1]

h of milling and increase in FWHM to finally disappear beyond 1 h of milling. Continued milling up to 5 h leads to a small diffuse halo coexisting with unprocessed starting silicon, indicating the formation of few amorphous phase. However, no distinct features are detected other than a vibration of broadening and intensity for diffraction lines of remaining c-Si for all investigated powder blends with prolonging milling time. Note that for carbon-lean blends, such as MA-$Si_2BC_{0.1}N$ and MA-$Si_2BC_{0.5}N$, the c-Si remains till 40 h of milling, an indicative of carbon contribution to amorphization.

Since the increase of the milling time beyond 40 h of milling cannot change the mechanical energy input into the blends, microstructure of MA derived powder with certain constituent show nothing different, but differs in the crystallite content and crystalline size. The average particle size of the as-milled MA-Si_2BC_3N powder is 6.9 ~ 3.4 mm with the median particle size of 6.6 ~ 3.2 mm (Fig. 3.2). Interestingly, further milling decreases the crystallite size but results in an increase in the particle size (Table 3.1). Silicon crystallite sizes decrease with prolonging milling, suggesting effective introduction of internal strain. The carbon-rich blends, like MA-$Si_2BC_{3.5}N$ and MA-Si_2BC_4N, show smaller grain size compared with the carbon-lean counterparts.

The typical HRTEM images of the as-milled MA-Si_2BC_3N powder particle on edge are provided in Fig. 3.3. Evidently, particle edge of the powder is predominantly amorphous with a small volume fraction of nano-crystalline phase [2]. The interlayer spacing measured is 0.249 ± 0.02 nm, matched well with $d(102) = 0.251$ nm of α-SiC or $d(111) = 0.251$ nm of β-SiC. In contrast, a typical composite microstructure of both amorphous structure and nanocrystalline phases is found in the center area of the particles. In detail, some very fine crystallites of SiC 4 ~ 5 nm embedded in an amorphous matrix. Besides, some structural defects form within the SiC crystallites.

Previous ^{29}Si solid-state NMR studies have outlined the spectra difference of SiC prototype [3–5]. Typically, a single resonance line is observed in the spectrum of β-SiC while up to three closely spaced resonances characterize the α-SiC. Figure 3.4

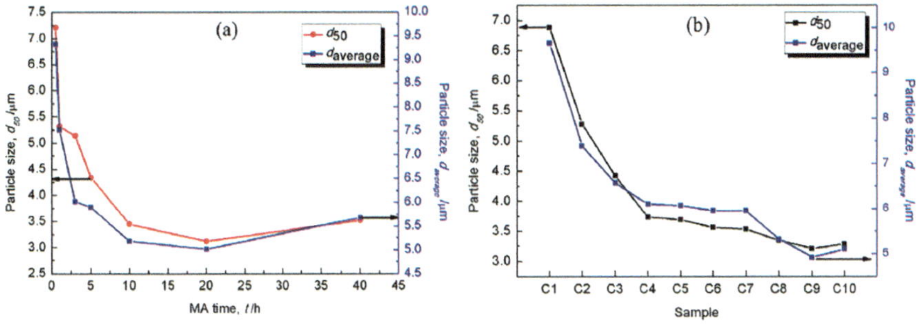

Fig. 3.2 The particle size distribution of ceramic powders: **a** MA-Si_2BC_3N powder milled at different time; **b** various carbon content containing powders milled for 20 h [1]

Table 3.1 Variation of the characteristic parameters for MA-Si_2BC_3N at various milling time [1]

Milling time (h)	Interplanar spacing of c-Si (111) (nm)	Crystallite size (nm)	Lattice strain (%)	Average size of powder (μm)
0.5	0.3136 ± 0.0001	952 ± 35	0.032 ± 0.002	9.33
1	0.3138 ± 0.0002	542 ± 27	0.096 ± 0.004	7.52
3	0.3139 ± 0.0001	132 ± 16	0.13 ± 0.01	6.02
5	0.3141 ± 0.0002	102 ± 15	0.19 ± 0.01	5.90
10	0.3142 ± 0.0003	85 ± 12	0.22 ± 0.02	5.18
20	0.3143 ± 0.0002	61 ± 11	0.26 ± 0.01	5.02

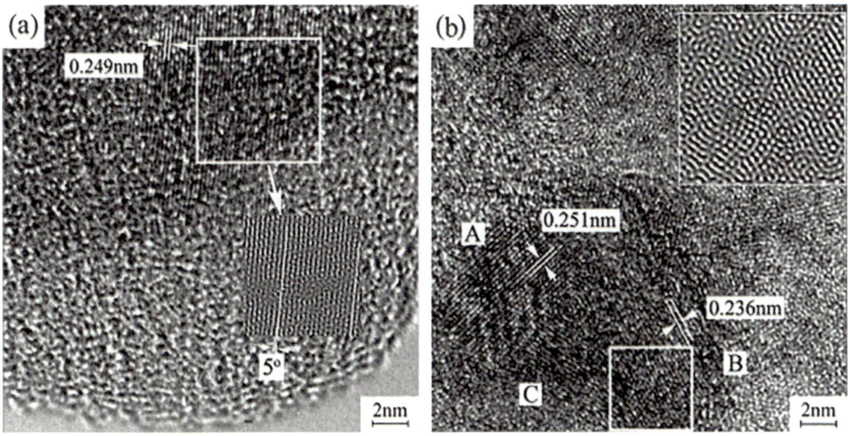

Fig. 3.3 HRTEM images of the as-milled MA-Si_2BC_3N powder: **a** The edge of powder particle; **b** the center of powder particle. The insets are the inverse FFT images taken from the white square highlights some structural defects [2]

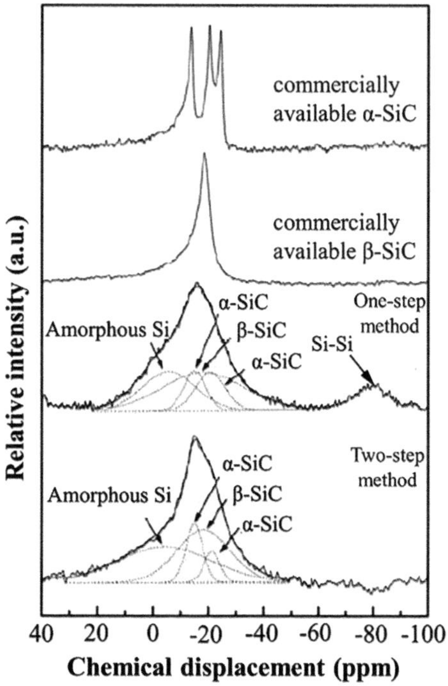

Fig. 3.4 Solid-state ^{29}Si NMR spectra of the MA-Si_2BC_3N ceramic powder [2]

shows the ^{29}Si NMR spectra for CS (MA-Si_2BC_3N, one-step milling, the mixtures of three starting materials are simultaneously milled in a vessel for 20 h) and CD (MA-Si_2BC_3N, two-step milling, both c-Si and partial graphite powders in mole ratio of 1:1 are firstly milled for 15 h, and then the residual graphite and h-BN powders are added into the same vessel for another 5 h milling) powders. Three closely spaced resonance lines at −13.8, −20.3, and −24.4 ppm indicate the α-SiC polytype, while a single resonance line (−18.4 ppm) is observed for the β-SiC.

For CS sample, a broad peak at −80 ppm is attributed to the starting crystalline Si material due to the insufficient milling. The broad ^{29}Si peaks position at range of −50 ~ 30 ppm for both CS and CD samples, implying the starting c-Si have been reacted with C to form crystalline and/or amorphous SiC. After spectra fitting by Gaussian–Lorentzian multi-peaks fitting procedure (Table 3.2), four components (P1, P2, P3, and P4) have been observed and the P1 resonance peak with a large FWHM is assigned to the SiC amorphous part. The signals at about -—18 ppm (−18.1 for CD and −17.9 ppm for CS) can been assigned to Si sites of the β-polytype. The peaks P3 (at about −14.9 ppm) and P4 (at about −21 ppm) should be responded to the α-polytype [6].

The deconvoluted Si 2p spectra of starting silicon powder and as-milled powders are presented in Fig. 3.5. The following elemental sensitivity factors: $S_{Si} = 0.27$, $S_B = 0.13$, $S_C = 0.25$, and $S_N = 0.42$ are used. For as-milled CS, the Si 2p spectra consist of two peaks located at ~ 99.5 and ~ 101.6 eV, identified as those due to

Table 3.2 The relative content of cubic silicon, SiC crystallites, and amorphous SiC, deduced from the ^{29}Si solid-state NMR spectra [2]

MA-Si_2BC_3N powder	Si (%)	α-SiC (%)	β-SiC (%)	Amorphous SiC (%)	β-SiC/α-SiC
CS	8.3	25.3	39.8	26.6	1.57
CD	0	19.1	36.6	44.3	1.92

CS: one-step milling, the mixtures of three starting materials are simultaneously milled in a vessel for 20 h. CS: two-step milling, both c-Si and partial graphite powders in mole ratio of 1:1 are firstly milled for 15 h, and then the residual graphite and h-BN powders are added into the same vessel for another 5 h milling.

vibration of Si–Si and Si–C bonds, respectively [7–9]. But for the CD sample, only one peak centered at ~ 100.9 eV corresponded to Si–C bonds can be deconvoluted.

A distinct change occurs for B 1s spectra, in which the B 1s spectra of CS and CD are deconvoluted into two distinguishable peaks (Table 3.3). The component centered at 190.0 ~ 190.8 eV is assigned to B-N bonds of h-BN, while that with higher binding energy ~ 192.1 eV for sample CS or ~ 192.7 eV for sample CD is attributed to B–C–N bonds [10]. For the graphite raw materials, the C–C bonds position at ~ 284.3 eV [11, 12] and C–O bonds locate at ~ 286.7 eV [13, 14]. After high-energy ball milling, the C 1s spectra are deconvoluted into three peaks. The peaks at ~ 284.9 eV and ~ 287 eV correspond to the C–C and C–O/C–B–N bonds, respectively. The small peak ~ 283.7 eV for CS and ~ 283.8 eV for CD both corresponds to the C–Si bonds [8]. For starting h-BN raw materials, a peak ~ 398.4 eV is assigned to the N–B bonds of h-BN [10]. After MA, another new distinct peak at ~ 400 eV is regarded as the C–B–N bonds [15].

3.2 Crystallization Behavior of Amorphous MA-SiBCN Powder

3.2.1 *Effect of Annealing Atmosphere*

In 1 bar N_2, the CS powder (one-step milling) show sustained mass gain, mainly resulting from reactions between residual Si and N_2. In Ar or N_2 atmosphere, weight loss of CD powder occurs (Fig. 3.6). The mass loss should be attributed to the reactions between the surface oxides and as-milled components [16, 17].

$$SiO_2(s) + 2SiC(s) = 3Si(s) + 2CO(g) \tag{3.1}$$

$$SiO_2(s) + 1/2SiC(s) = 3/2SiO(g) + 1/2CO(g) \tag{3.2}$$

$$SiO_2(s) + 3C(s) = SiC(s) + 2CO(g) \tag{3.3}$$

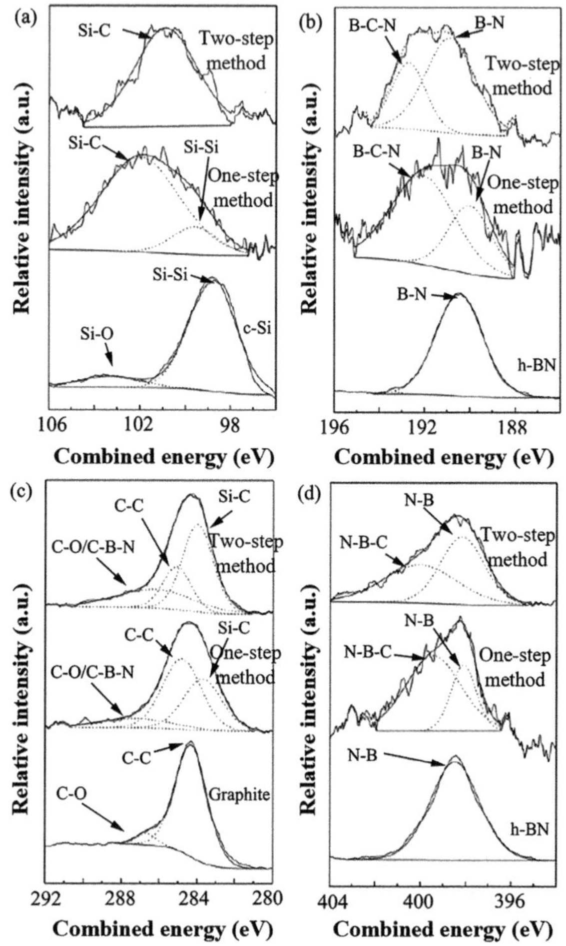

Fig. 3.5 XPS spectra of graphite, h-BN, cubic Si, CS and CD as-milled powders: **a** Si 2p; **b** B 1s; **c** C 1s; **d** N 1s [2]

$$SiO_2(s) + 2C(s) = SiC(s) + CO_2(g) \quad (3.4)$$

Oxidation of SiC and Si_3N_4 seems occurring at ~ 900 and ~ 1100 °C (Fig. 3.7). The final mass gain reaches 30.7% for SiC and 20.3% for Si_3N_4 at 1400 °C. For the CS and CD powders, three stages for mass change can be divided. The first stage of weight loss occurs at 485 °C for CS and 398 °C for CD, respectively. At 600 °C, the mass loss −1.97% for CS and −8.95% for CD are detected. The mass loss at the first stage corresponds to the BN(C) oxidation, confirmed by the detection of C, CO_2, N_2, NO and NO_2.

$$2C(s) + O_2(g) = 2CO(g) \quad (3.5)$$

Table 3.3 Results of deconvolution and the relative content of chemical bonds for the raw materials, CS, and CD [2]

Element lines	Bond types	Binding energy (eV)			Relative content of the bonds (%)	
		Starting materials	CS	CD	CS	CD
Si 2p	Si–Si	~ 98.8	~ 99.5	–	3.9	–
	Si–C	–	~ 101.6	~ 100.9	22.6	6.2
	Si–N	~ 103. 3	–	–	–	–
B 1s	B–N	~ 190.5	~ 190.0	~ 190.8	4.7	14.6
	C–B–N	–	192.1	~ 192.7	9.2	7.0
	B–O	~ 193.4	–	–	–	–
C 1s	C–Si	–	~ 283.7	~ 283.9	19.3	28.7
	C–C	~ 284.3	~ 284.8	~ 285.0	26.6	12.9
	C–O/C–N	~ 286.7	~ 287.3	~ 286.2	6.6	15.4
N 1s	N–B/N–C	~ 398.4	~ 398.3	~ 398.2	2.0	7.8
	C–B–N	–	~ 399.4	~ 400.0	5.0	7.4

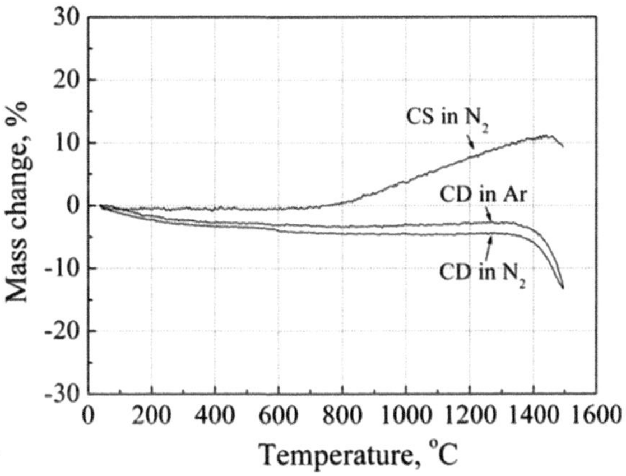

Fig. 3.6 TG curves for the as-milled MA-Si_2BC_3N powder treated in 1 bar N_2 or Ar [2]

$$C(s) + O_2(g) = CO_2(g) \tag{3.6}$$

$$4/3BN(s) + O_2(g) = 2/3B_2O_3(s, l) + 2/3N_2(g) \tag{3.7}$$

Reaction (3.7) leads to a mass gain, and others result in the mass loss. Therefore, the mass loss of MA-Si_2BC_3N powder at this initial stage should be contributed to the oxidation of free C in BN(C). The second stage of weight gain increases from 680 to 920 °C. In this temperature range, the ion current for C and CO_2 decreases, while the N, NO and NO_2 show a small releasing peaks at about 810 °C in CS. But a sharp releasing peaks for these five ions appear at ~ 830 °C in CD. So, the mass

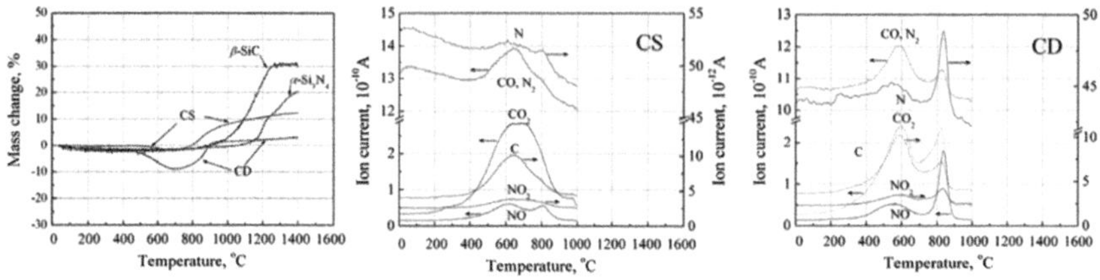

Fig. 3.7 TG curves of pure β-SiC, α-Si_3N_4, and as-milled CS, CD powders in flowing air: (MS curves for the CS and CD are conducted in flowing Ar + O_2 up to 1000 °C) [2]

gain of CS in this stage mainly come from the oxidation of Si, and the mass gain of CD can be attributed to the oxidation of BN and SiC [18]:

$$SiC(s) + O_2(g) = SiO_2(s, l) + C(s) \tag{3.8}$$

Part of C, resulting from SiC oxidation, is further oxidized to CO and CO_2. The third stage occurs above 920 °C, during which the oxides form a continuous passivating scale on the surfaces of the powder particles leading to a decreasing mass gain rate. The mass gains are 6.06 and 2.11% for the CS and CD powders from 920 to 1400 °C, respectively.

As described above, amorphous MA-Si_2BC_3N powder regardless of the milling method show unsatisfying oxidation resistance below 1000 °C due to the oxygen-sensitive C, BN and/or BN(C). But the mass loss of CS is only −1.9% at this stage. For CS and CD powders, the only difference is the milling time of h-BN and graphite. To investigate the effect of milling time of BN-C system on the oxidation of MA-Si_2BC_3N powder, the BCN5 (h-BN and graphite co-milled for 5 h) and BCN15 (h-BN and graphite co-milled for 15 h) samples are heated to 1500 °C in air (Fig. 3.8). At temperature below 800 °C, mass loss of BCN15 sample is about 6.2%, lower than that of BCN5. Beyond 800 °C, the mass of BCN15 remains almost unchanged from 915 to 1370 °C, and BCN5 shows unchanged mass in range of 1025 ~ 1214 °C. Generally, BN(C) become more stable at high-temperatures with increasing MA time. Beyond 1000 °C, CS and CD powders are more stable in air than pure SiC and Si_3N_4 [19–22].

3.2.2 *Effect of Chemical Composition*

The amorphous MA-$Si_2B_yC_2N$ ($y = 1 \sim 4$) powders show a similar weight loss accompanied with a strong exothermic peak at ~ 1400 °C [23]. Below 400 °C, all the ceramic powders show continuous mass loss; however, the weight loss keeps almost constant at 400 ~ 1300 °C. Besides, the weight loss rate decreases as boron content increases. The final weight loss at 1500 °C for the MA-$Si_2B_4C_2N$, MA-$Si_2B_3C_2N$, MA-$Si_2B_2C_2N$, MA-$Si_2B_{1.5}C_2N$ and MA-Si_2BC_2N powders is 3.21, 3.69, 4.12,

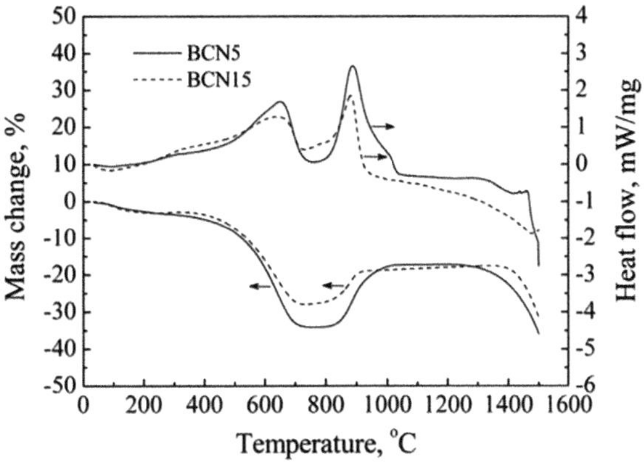

Fig. 3.8 TG-DSC spectra of the BCN5 (h-BN and graphite co-milled for 5 h) and BCN15 (h-BN and graphite co-milled for 15 h) powders heated to 1500 °C in air [2]

3.31, and 3.71 wt.%, respectively. Note that the summit of the exothermic peaks shifts to lower temperatures with increasing boron content, although it is limited. It thus suggests that increasing boron content reduces the mass loss, but promotes the crystallization (Fig. 3.9).

In Air atmosphere, development of all the TG-DSC curves is similar. The as-milled ceramic powders begin to gain weight at ~ 800 °C accompanied with an exothermic peak at ~ 840 °C. The boron and h-BN are very sensitive to the oxygen, which can be oxidized to B_2O_3(l) and/or N_2(g) at ~ 450 °C. B_2O_3 products begin to evaporate violently beyond 900 °C leading to mass loss. Besides, other weight gain reactions must also occur. The Si and SiC can oxidize to SiO_2(s,l) or SiO(g) via on the passive or active reactions. For the continuous flux of oxygen, Si and/or SiC are expected to oxidize through an active manner. In addition, some B_2O_3 can react with SiO_2 to form borosilicate glass that reduces weight loss. Furthermore, the evaporation of SiO(g) can also transform to SiO_2(s,l) at elevated oxygen gradient. In a whole, the more boron content, the worse anti-oxidation ability is expected.

As indicated, both SAED patterns on heating at 1000 °C in 1 bar Ar present a diffuse elastically scattered ring halo, typical of amorphous nature (Fig. 3.10). Few contrast variations are observed for the MA-Si_2BC_2N powder annealed at 1200 °C, implying strong amorphous retention ability. However, a faint halo with bright-spots indicative of nanocrystals is detected for MA-$Si_2B_{1.5}C_2N$ at 1200 °C. Coincidently, some randomly distributed atomic clusters are observed in amorphous matrix, see HRTEM image. Further annealing at 1400 °C, the broad band assigned to the first amorphous halo transfers into a much narrower ring. Besides, the second band splits into two propounded rings overlapped with a wide diffuse background. These changes are indicative of an increasing short-range order and an onset of crystallization. At this annealing temperature, however, the crystal nuclei with very small sizes develop incompletely. Even some changes of the microstructures occur, which can be reflected in the diffraction pattern. It is, therefore, still very hard to identify what the crystal nuclei are until 1600 °C.

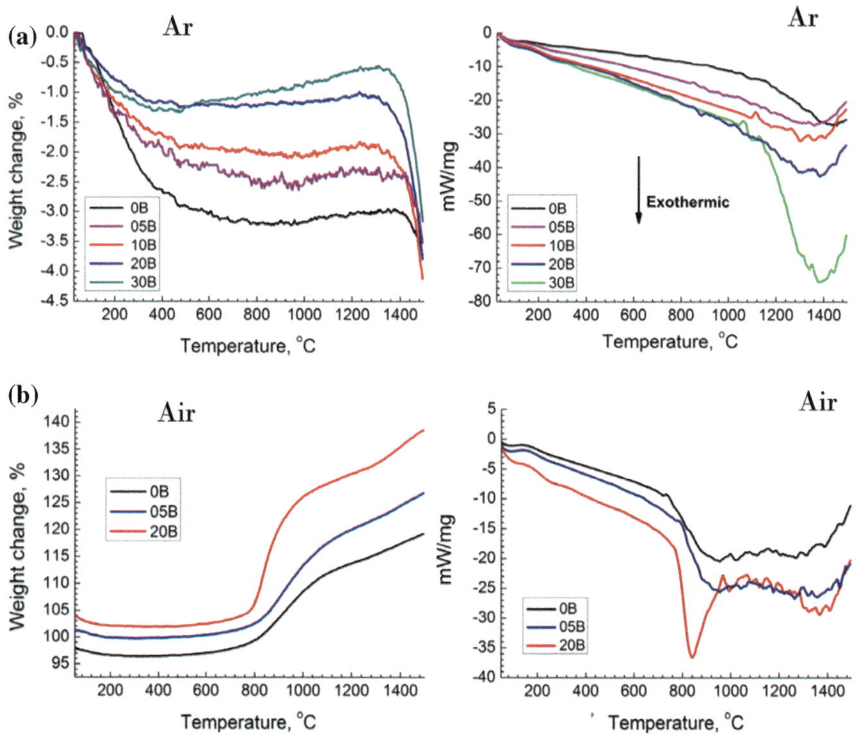

Fig. 3.9 The TG-DSC profiles of the amorphous MA-Si_2BC_2N, MA-$Si_2B_{1.5}C_2N$ and MA-$Si_2B_3C_2N$ heated to 1500 °C in **a** Ar and **b** Air, performed with a heating rate of 20 K/min [23]

For the as-milled amorphous powders annealed at 1600 °C, both HRTEM images and SAED patterns clearly confirm the formation of some nanocrystals. The rings in both SAED patterns are relatively well-defined and have now become solid lines which suggest the increasing volume fraction of the nanocrystals. However, a broad but faint halo is still observed, overlapped with the outer rings, implying some residual amorphous structures. Besides, lattice fringes and Moire firings are found in SiC and BN(C) grains, respectively. The size of the nanocrystals is in a range of 5–10 nm for MA-Si_2BC_2N; however, the size of the nanocrystals in MA-$Si_2B_{1.5}C_2N$ is 10–20 nm. The *d*-spacings measured from SAED patterns are 0.2521 nm, 0.1541 nm, and 0.1312 nm, respectively, corresponded well with SiC (111), (220), and (310). Some nano-SiC grains almost have perfect structures. In addition, BN(C) crystallites are ill-developed with defects, typical of turbostratic structures.

The above results have symmetrically provided investigations by different characterization techniques that give information about the pathway of crystallization in MA derived SiBCN powder. In the as-milled state, these investigations resolve characteristics of a typical amorphous structure. After 2 h of annealing at 1000

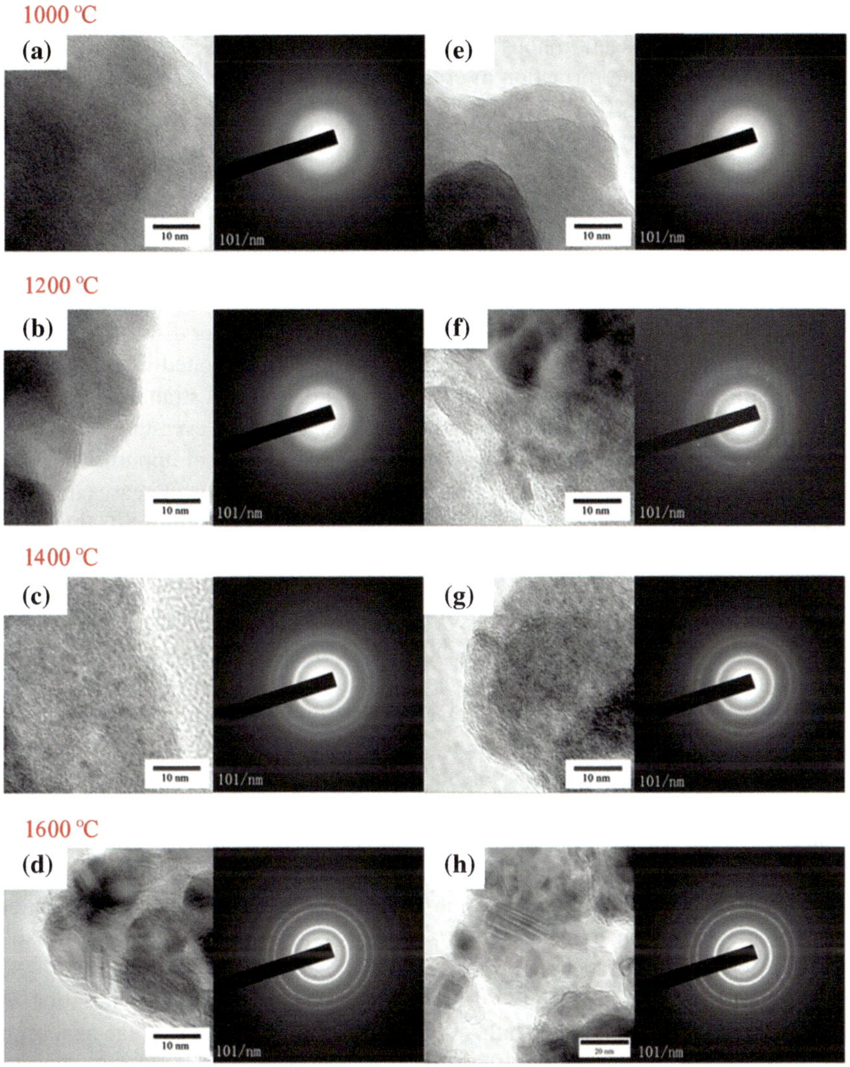

Fig. 3.10 The TEM microstructure of the as-milled amorphous powders heated at different temperature for 2 h in Ar: **a–d** MA-Si_2BC_2N; **e–h** MA-$Si_2B_{1.5}C_2N$ [23]

°C, XRD does not resolve crystal formation. This is associated with the formation of an amorphous microstructure observed by HRTEM. However, compared to the as-milled state, the corresponding SAED pattern shows the broad band of the first amorphous halo transformed into a much narrower ring. Moreover, the second broad band appears although it is very faint. An attempt to image the structural changes by STEM-EDS is not successful due to the small changes in concentration at this early stage. Thus, the possibility of phase separation at this temperature cannot be excluded. After 2 h of annealing at 1200 °C, for MA-Si_2BC_2N, XRD and TEM results both confirm its amorphous feature. For MA-$Si_2B_{1.5}C_2N$, SAED pattern with bright-spots verifies the formation of crystalline phase. Further annealing at higher temperatures (1400–1700 °C), XRD and HRTEM both identify the formation of nano-SiC and turbostratic BN(C).

The crystallization of a crystal in an amorphous body composes of nucleation and growth, which are usually dependent on atomic diffusion. For MA derived SiBCN system, three kinds of structural evolution should be considered during annealing, namely strain relaxation, reordering and grain growth, which are closely related to the diffusivity of atoms in an amorphous matrix. The microstructural evolution may take place in the following ways: atoms in the disordered or distorted positions rearrange and/or move into more thermodynamically equilibrated and ordered or perfect lattice positions, enhancing the degree of ordering and strain release in the matrix. At the same time, the grain grows to a critical size. However, these structural evolutions are accompanied with the simultaneous segregation of impurity atoms to grain boundaries pinning the grain growth [24]. Accordingly, the increases in boron content should effectively inhabit the reordering and grain growth of SiC due to its lowest self-diffusivity. However, this is in contradiction with current investigations. Therefore, it is reasonable assumed that the role of boron on the crystallization kinetics of SiC is mainly owing to the effect of boron on the nucleation stage [25].

Besides, we should consider the MA induced SiC nanoclusters since the preformed SiC can offer nucleation sites driving crystallization. In this case, the introduction of inorganic boron powder acted as extra boron source provides a new chemical environment around B atoms (BC_3 unit). The BC_3 unit possibly forms at the interfaces between B–N–C and Si–C–N nanodomains. This will significantly reduce the hindering effects of B–N–C nanodomains that retard SiC crystallization. It should be noted that the effects of boron on promoting SiC crystallization are limited. Since the nano-grain growth depends on the diffusion of atoms of the elements that constitute the phase. Besides, the competition of SiC and BN(C) crystallization for the limited carbon also constrains the development of abutting phase. Moreover, the soft impingement of the diffusion fields can also limit the growth of nanocrystals.

The sequence of crystallization depends on the driving force for nucleation. A smaller activation energy of crystallization promotes the nucleation of the nanocrystals, and a following increase in the activation energy if helpful for impeding its growth during crystallization process [26, 27]. Such a crystallization behavior (an easy nucleation and a difficult growth for a crystal) is required for the crystal to form a uniform microstructure with fine grain. As documented, the self-diffusivity of the four elements is extremely low allowing only short-range diffusion [28]. Especially, boron has the lowest diffusion coefficient. Nucleation depends on diffusion while diffusion is concerned with the number and type of neighboring atoms. Since the nucleation and growth of the primary phase SiC in amorphous SiBCN system require less solute diffusion. Therefore, SiC occurs at lower temperatures during annealing.

It is accepted that the degradation of amorphous PDCs-SiBCN occurs mainly due to the carbothermal reactions between Si_3N_4 and carbon as well as Si_3N_4 self-decomposition at high temperatures [29]. Crystalline segregation of Si_3N_4, however, is possibly protected from decomposition by a turbostratic BN(C) [30]. In a previous report [31], the final phase composition of MA-Si_2BC_3N (using c-Si, graphite and h-BN as raw materials) is within the limits of four-phase region Si–SiC–BN–C, without any trace of Si_3N_4. It thus suggests that the decomposition in MA-SiBCN system leading to weight loss is not caused by Si_3N_4 self-decomposition or carbothermal reaction. The mass loss at lower temperatures is most possibly resulting from desorption of gases on surfaces and/or evaporation of H_2O molecules. Nevertheless, the reactions between carbon and surface oxidizing species at higher temperatures are also likely.

Up to now, the influence of chemical composition on the structural stability of MA derived SiBCN materials remains insufficient. A preliminary study indicates that increasing the ratio of Si: C in MA-SiBCN powder increases the grain sizes of β-SiC crystallites in hot-pressed monoliths at high temperatures. Coincidently, increases in thermal conductivity and coefficient of thermal expansion together with the mechanical properties are observed with increasing Si:C ratios. Houska et al. suggests that increasing the Si:C ratio increases the network density, i.e., the coordination number of atoms thus inhibiting diffusion and improving the structural stability.

For the PDCs-SiBCN system, some researcher believed increasing boron content promotes nucleation and growth of SiC, while most considered boron additives inhibit the crystallization of SiC and/or Si_3N_4 [25]. Results also found the increasing boron or carbon contents inhibit Si_3N_4 decomposition thus improving the thermal stability of PDCs-SiBCN materials [29]. The disordered structures in amorphous 3-D networks increase the activation energy for Si_3N_4 crystallization and reaction of Si_3N_4 with carbon thus improving thermal stability, in accord with reports on PDCs-SiCN and PDCs-SiOC ceramics. The results obtained by molecular dynamics simulation and experiential studies both suggest that the self-diffusion coefficient of silicon in PDCs-SiBCN is over an order of magnitude lower than that in PDCs-SiCN. Increasing carbon content lowers the coefficient of self-diffusion of silicon, and increasing boron content lowers the diffusion coefficient of nitrogen and boron, beneficial to the structural stability of PDCs-SiBCN system.

For the PDCs-SiBCN system, a high carbon content coupled with incorporated rigid structural elements like borazine rings leads to a maximum temperature durability to at least 2000 °C. Carbon bonds to four neighbors via strong covalent bonds instead of three in the case of nitrogen. The rigid borazine rings with strong p_π–p_π interactions between boron and nitrogen make the covalent SiBCN network stiffer and stronger thus significantly suppressing the conformational changes as well as reorientation of atoms which always precede decomposition and improving eventually the high temperature stability. On the other hand, the "free" carbon in SiC or $BN(C)_x$ is beneficial to high temperature stability, while graphite-like carbon leads to decreases in mechanical properties. The extra nitrogen bonds to silicon also decrease structural stability. These results suggest that high Si:C ratios, N:(Si + B + C) ratios and boron contents are usually beneficial to structural stability at high-temperatures. The control of chemical composition can tailor the microstructures thus adjusting the mechanical and thermophysical properties of the final SiBCN products.

3.3 Microstructure Evolution of MA Derived SiBCN-Me Composite Powders

(1) SiBCN-Zr powder

For one-step ball milling derived MA-$Si_2B_5C_3NZr_2$ powder [32], limited reactions among Zr, B, h-BN and C particles occur after 1 h of MA, since substantial Zr and Si are still observed in the powder blends (Fig. 3.11). Only a small amount of SiC is detected. The absence of other components such as boron, graphite and h-BN are due to their amorphous states. Interestingly, peaks of raw materials decrease in intensity dramatically and broaden with increasing milling time. Further ball milled for 3 h, ZrB_2 compounds are obtained. Continuous deformation of powder particles through prolonged milling time results in crystallite refinement and increase in lattice strain. After ball milling for 20 h, the MA-$Si_2B_5C_3NZr_2$ ceramic powder can be regarded as completely amorphous structure. Since prolonged milling time cannot change the mechanical energy input into powder, the microstructures and crystallite content in blends may have little variation after 40 h of MA.

For two-step ball milling, ZrB_2 forms with residual Zr in powder blends after 1 h of milling (Fig. 3.12). However, amorphous powder is obtained after ball milling for 5 h. Afterwards, the resulting amorphous Zr-B powder is added to a jar that contains c-Si, graphite and h-BN. After milling for 20 h and more, as-prepared MA-$Si_2B_5C_3NZr_2$ powder is in well amorphous states.

The as-milled MA-$Si_2B_5C_3NZr_2$ powder exposed to air instantaneously enables complete solid-state reactions of themselves with oxygen, and the formation of

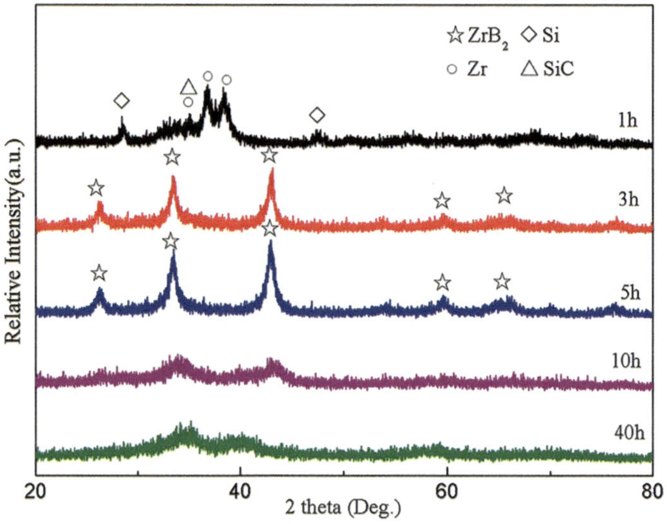

Fig. 3.11 XRD patterns of MA-$Si_2B_5C_3NZr_2$ composite ceramic powders prepared by one-step milling [32]

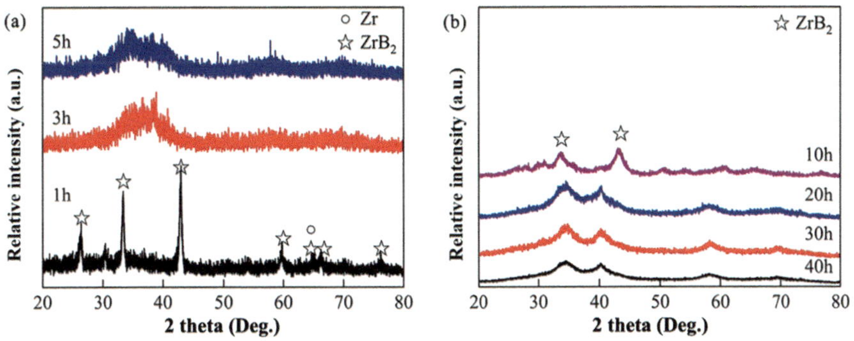

Fig. 3.12 XRD patterns of MA-$Si_2B_5C_3NZr_2$ powder prepared by two-step milling: **a** 1–5 h of milling; **b** 10–40 h of milling [32]

ZrN, ZrB_2, SiC and ZrO_2 are confirmed via the self-propagating high-temperature synthesis (SHS) process [33–35] (Fig. 3.13). The as-milled MA-$Si_2B_5C_3NZr_2$ powder further heated at 1200 °C for 60 min in N_2, and some ZrN, ZrB_2, SiC and a small amount of ZrO_2 also form. The formation of ZrO_2 in N_2 atmosphere may come from the impurity of N_2 or the pre-existing oxides on powder surfaces.

The Gibbs free energy changes of all the predicted reactions are negative which mean they may take place and therefore they are thermodynamically feasible at temperatures between 300 and 2000 K (Fig. 3.14). Moreover, all the reactions are exothermic implying heat release during SHS or sintering process. The preferential formation of ZrB_2 and ZrN rather than ZrC and B_4C are duo to their higher Gibbs free energy change than others.

According to Tsuchida's work [36], the self-ignition reactions are due to the oxidation of MA induced disordered carbon in air. This is because the reactivity of the nanosized particles, such as disordered carbon, is extremely high. Once the highly active amorphous powder contact with oxygen before cooling, the disordered carbon

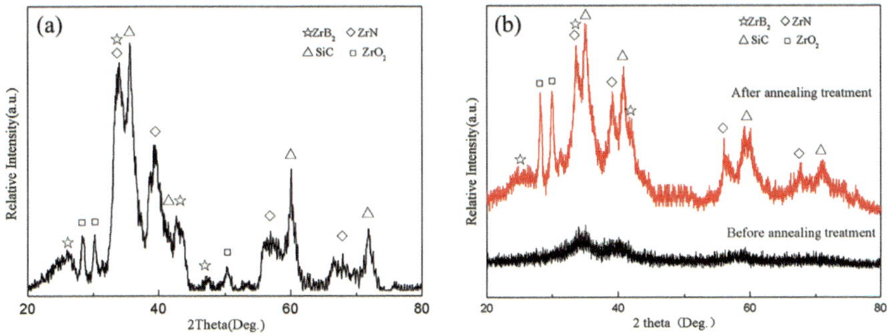

Fig. 3.13 XRD patterns of the MA-$Si_2B_5C_3NZr_2$ composite powder under different treatment condition: **a** Directly exposed to air after high energy ball milling; **b** as-milled powder heated to 1200 °C for 1 h in N_2 atmosphere with a rate of 20 °C/min [32]

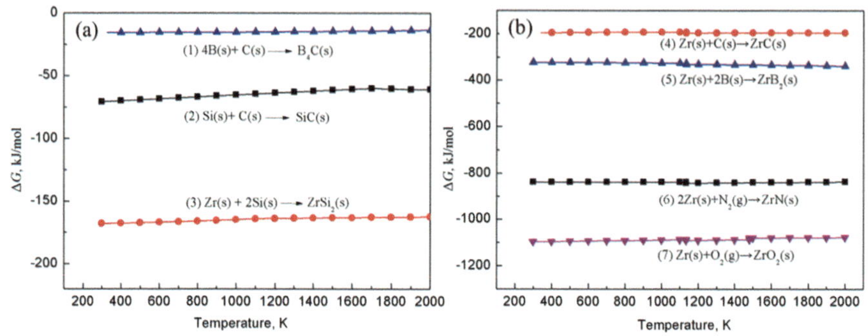

Fig. 3.14 Gibbs free energy change versus temperature curves of the reactions predicted occurred during MA or sintering process in N_2 atmosphere: **a** -200~0 kJ/mol; **b** -1200~-200 kJ/mol [32]

will react with oxygen to release heat first. Despite the oxidation of Zr particles in air are expected to occur preferentially based on the thermodynamic date, but Zr particles are enwrapped with finely divided particle of disordered carbon and amorphous boron after MA process, therefore its oxidation process must be retarded. For homogeneously amorphous powder is obtained by MA, the heat release from disordered carbon oxidation is transferred to the nearby powder effectively.

Hence SHS process in MA-$Si_2BC_3NZr_2$ powder system is induced. For both reactions in Fig. 3.14 are exothermic, these reactions become self-sustaining and propagate through overall reactant mixture. However, the formation of SiC in MA-$Si_2B_5C_3NZr_2$ powder after exposing to air may result from the pre-existing MA derived SiC nuclei, and the absence of ZrC is due to the exhaustion of disordered carbon and Zr. In summary, mechanical alloying can be employed to prepare amorphous Zr containing MA-Si_2BC_3N powder system with fine microstructure. The exposure of mechanical activation disordered carbon to air results in an ignition reaction between disordered carbon and oxygen, therefore SHS process can be induced and sustains. The formation of ZrN possibly results from the reactions between residual Zr and N_2.

After milling for 20 h, the two MA-$Si_2B_2C_3NZr_{0.5}$ (by one-step or two-step milling method) powders both mainly consist of near-spherical agglomerates, whose diameters are below 500 nm (Fig. 3.15). After exposure to air instantaneously before cooling, some particles seem to react with each other and form larger agglomerates. Trace of oxides is observed on powder surfaces, indicating ZrO_2 formed in blends.

After MA process, the as-milled MA-$Si_2B_2C_3NZr_{0.5}$ powder shows amorphous structure confirmed by two pieces of evidence: the completely disordered atomic arrangement in HRTEM image and corresponded SEAD pattern with large diffraction spot (Fig. 3.16). However, the exposure in air leads to the formation of nanosized grains at local region. These nanocrystalline region are composed of Zr, Si, B, C, N and O elements.

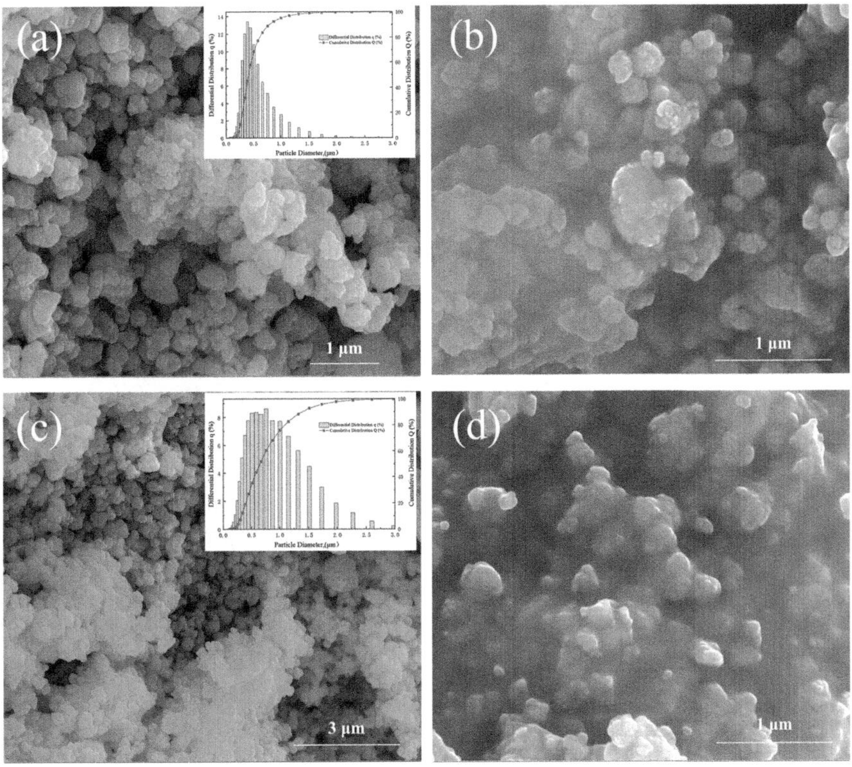

Fig. 3.15 SEM surface morphologies of the as-milled amorphous MA-$Si_2B_2C_3NZr_{0.5}$ powder before and after exposing to air: **a** Powder prepared by one-step milling; **b** powder (one-step milling) exposed to air instantaneously before cooling; **c** powder prepared by two-step milling; **d** powder (two-step milling) exposed to air instantaneously before cooling. Inserted are the corresponded particle size distribution [32]

(2) SiBCN-Al powder

In an attempt to prepare amorphous MA-$Si_2BC_3NAl_{0.6}$ powder (using Al particles as the Al source), a set of experiments with different milling time are conducted [37]. As indicated, the diffraction peaks of h-BN, graphite and Al disappear completely while those of crystalline c-Si still exist beyond 5 h of milling. Solid-state amorphization of h-BN, graphite and Al is easier than that of c-Si. After 15 h of milling, the diffraction peaks of Si obviously decrease in intensity. Prolonging the grinding time to 25 h, a halo peak corresponding to an amorphous phase is observed which can be considered as completely X-ray amorphous (Fig. 3.17). The introduction of metal Al seems accelerate the solid-state amorphization process of MA-Si_2BC_3N system.

The resulting MA-$Si_2BC_3NAl_{0.6}$ powder particles are in sizes of 100–200 nm (Fig. 3.18). The small particles and the edge of the larger particles of the TEM bright-field images are featureless, suggesting the presence of amorphous phases. Such conjecture is consistent with the SAED pattern that exhibits a broad and diffused halo.

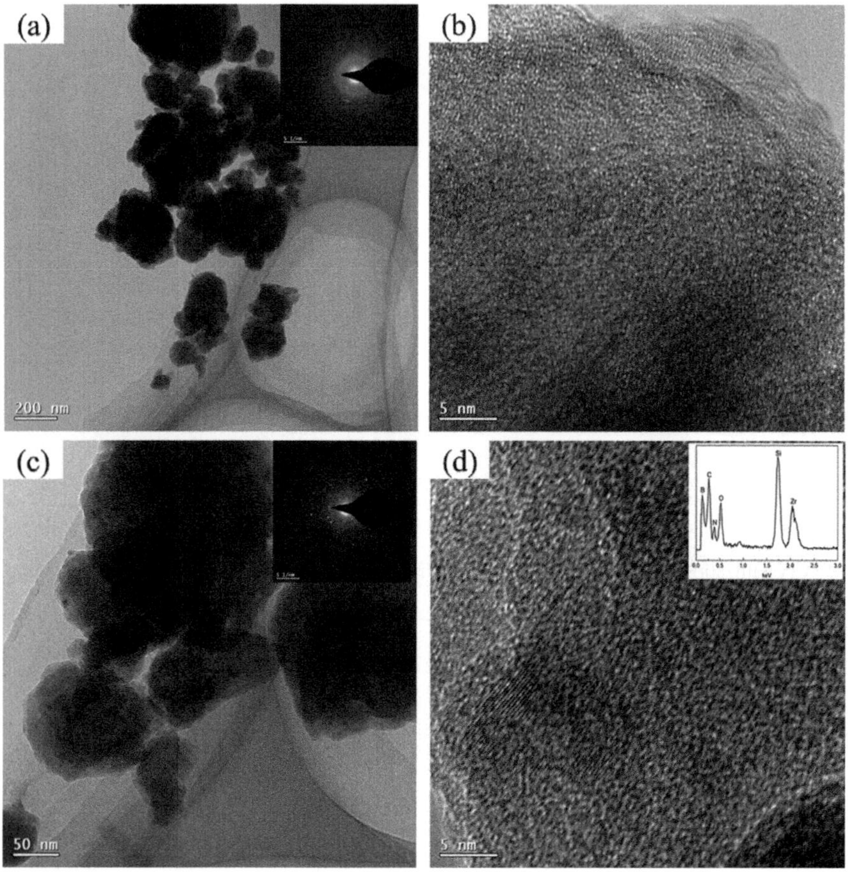

Fig. 3.16 TEM analysis of the MA-$Si_2B_2C_3NZr_{0.5}$ powder (by one-step milling) before and after exposing to air: **a**, **b** As-milled; **c**, **d** exposed to air instantaneously before cooling. Inserted are the corresponded SAED pattern and EDS result [32]

Fig. 3.17 XRD patterns of MA-$Si_2BC_3NAl_{0.6}$ powder using metal Al particles as Al source (milling speed of 600 rpm; ball-to-powder mass ratio of 30: 1) [37]

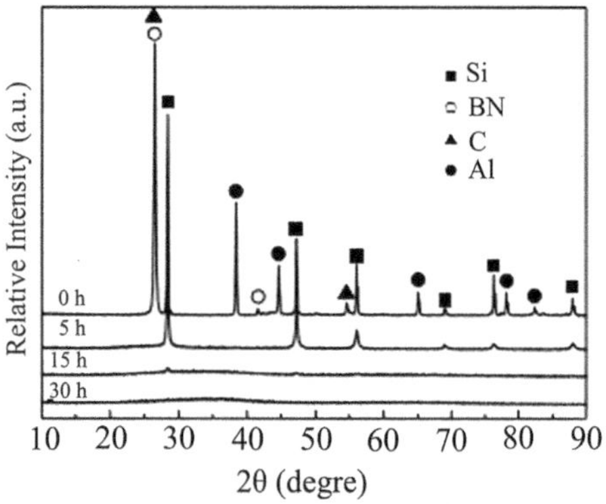

HRTEM images further confirm such particles are completely amorphous (Fig. 3.19). The EFTEM results also show the elemental atoms are uniformly distributed in the whole area without elemental segregation (Fig. 3.20).

The as-milled amorphous MA-$Si_2BC_3NAl_{0.6}$ powder show various bonds such as Si–C, B–C–N, C–C, B–N and Al–N (Fig. 3.21). The characteristic peaks at ~ 100.8 (Si 2p) and ~ 282.5 eV (C 1s) are assigned to the Si–C bonds of SiC [38]. The peaks at ~ 192.2 (B 1s), ~ 286.1 (C 1s) and ~ 398.9 eV (N 1s) can reflect the existence of B–C–N bonds [10]. There is a peak at ~ 284.6 eV for C 1s spectrum, assigned to the C–C bonds of graphite [38]. The peaks at ~ 189.3 (B 1s) and ~ 398.9 eV (N 1s) can

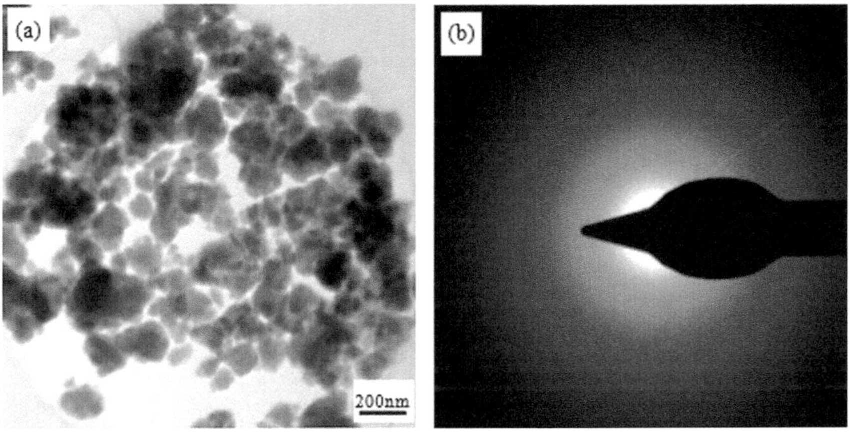

Fig. 3.18 TEM analysis of the MA-$Si_2BC_3NAl_{0.6}$ powder: **a** Bright-field image; **b** SAED pattern [37]

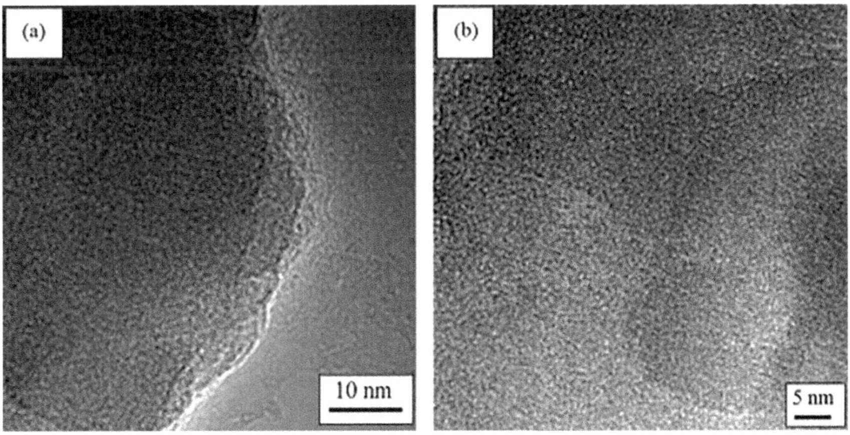

Fig. 3.19 Typical HRTEM images of MA-$Si_2BC_3NAl_{0.6}$ powder particles: **a** The edge of a particle; **b** the center area of a particle [37]

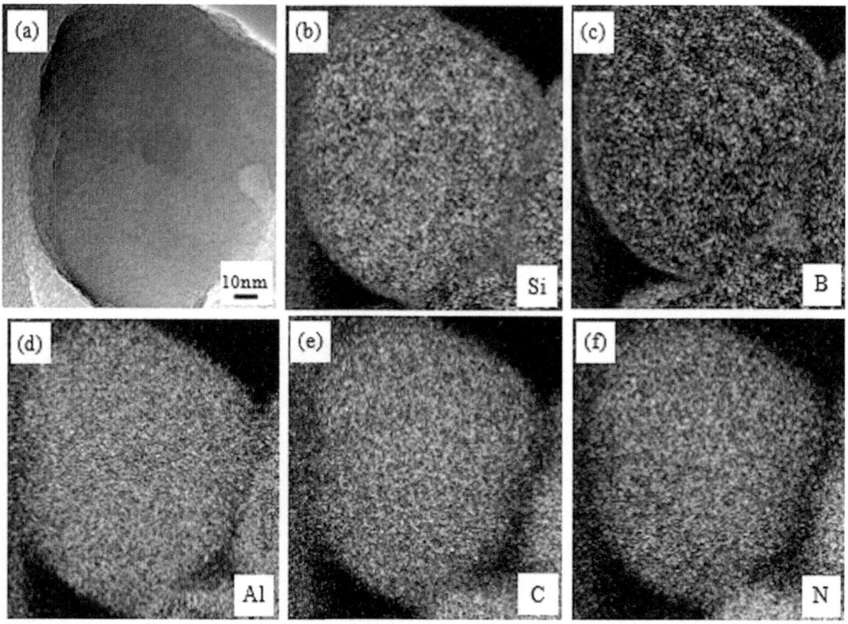

Fig. 3.20 The energy filtration transmission electron microscopy (EFTEM) results of the MA-$Si_2BC_3NAl_{0.6}$ powder after 30 h of milling: **a** Bright-field image; **b** Si mapping; **c** B mapping; **d** Al mapping; **e** C mapping; **f** N mapping [37]

be assigned to the B–N bonds of h-BN [10]. The characteristic peaks at ~ 73.5 and ~ 397 eV are identified as Al 2p and N 1s signals of the Al-N bonds, respectively [39]. The peaks at ~ 75.6 (Al 2p), ~ 531.7 (O 1s) and ~ 401.6 eV (N 1s) may come from the vibration of Al–O–N bonds [40]. The detective Al–O–N bonds may be derived from Al_2O_3 oxides on the surfaces of the Al particles (Table 3.4).

The ^{29}Si NMR spectrum shows a broad peak around −15 ppm characteristic of an amorphous SiC structure based on the SiC_4 sites. A small hump at −85 ppm suggests the presence of a small amount of amorphous Si due to insufficient milling [41]. Three signals centered at 100, 57 and 4 ppm are observed in the ^{27}Al NMR spectrum. The ~ 100 ppm signal corresponds to tetrahedral AlN_4 sites [42], implying the presence of AlN in the as-milled powders. The position of the resonance at ~ 57 ppm can be assigned to pentahedral AlN_5 and/or tetrahedral AlO_4 units [43]. A small peak at ~ 4 ppm should be responded to the octahedral AlN_6 units [44]. Aluminum is presented as mixed AlN_4, AlN_5/AlO_4 and AlN_6 units (Fig. 3.22).

(3) SiBCN-AlN powder

The AlN particles are used as the Al source to prepare amorphous MA-$Si_2BC_3N_{1.6}Al_{0.6}$ powder (Fig. 3.23). Obviously, the diffraction peaks of AlN can still be observed after 30 h of milling [45]. Even after milling time extended to 60 h, the powder mixtures are still not completely amorphous with very small diffraction

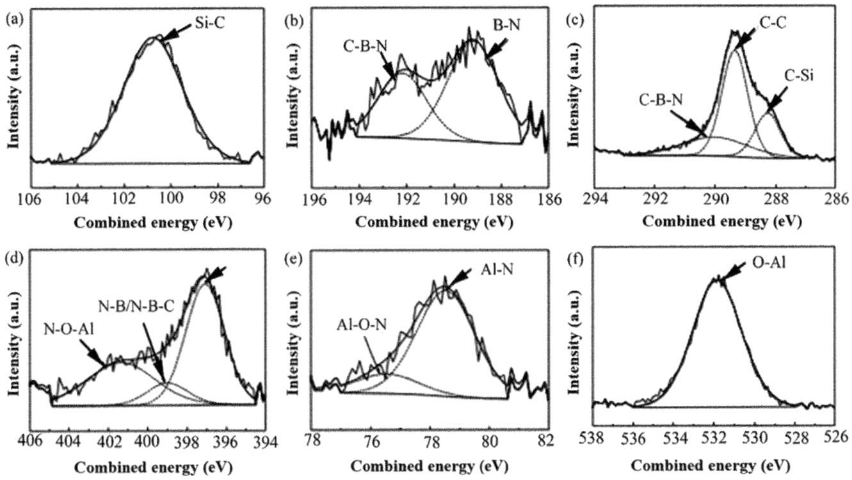

Fig. 3.21 XPS spectra of the MA-$Si_2BC_3NAl_{0.6}$ powder (using Al_p as the Al source) after 30 h of high-energy ball milling: **a** Si 2p; **b** B 1s; **c** C 1s; **d** N 1s; **e** Al 2p; (f) O 1s [37]

Table 3.4 XPS results of the amorphous MA-$Si_2BC_3NAl_{0.6}$ powder using metal Al particles as the Al source [37]

Element lines	Bond types	Binding energy (eV)	Relative content (%)
Si 2p	Si–C	~ 100.8	12.7
B 1s	B–N	~ 189.3	10.0
	B–C–N	~ 192.2	6.1
C 1s	C–Si	~ 282.5	8.3
	C–C	~ 284.6	19.6
	C–B–N	~ 286.1	8.9
N 1s	N–Al	~ 397.0	7.0
	N–B/N–B–C	~ 398.9	1.4
	N–O–Al	~ 401.6	4.4
Al 2p	Al–N	~ 73.5	6.8
	Al–O–N	~ 75.6	1.4
O 1s	O–Al	~ 531.7	13.5

peaks of AlN. The milling efficiency is enhanced as the ball-to-powder (B/P) mass ratio increases (Fig. 3.24). When the B/P increases to 50:1, a halo peak assigned to an amorphous phase is observed. Thus, the amorphous $Si_2BC_3N_{1.6}Al_{0.6}$ powder can be obtained by milled at 600 rpm for 30 h with a B/P of 50:1.

The as-milled $Si_2BC_3N_{1.6}Al_{0.6}$ powder particles are in sizes of about 100–200 nm with some agglomerated particles (Fig. 3.25). HRTEM image further suggests the microstructure of the as-milled powders is predominantly amorphous with a very

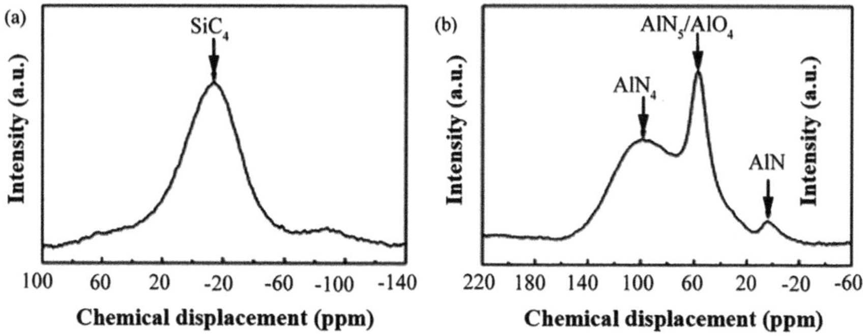

Fig. 3.22 Solid-state NMR spectra of amorphous MA-$Si_2BC_3NAl_{0.6}$ powder after 30 h of ball milling: **a** ^{29}Si; **b** ^{27}Al [37]

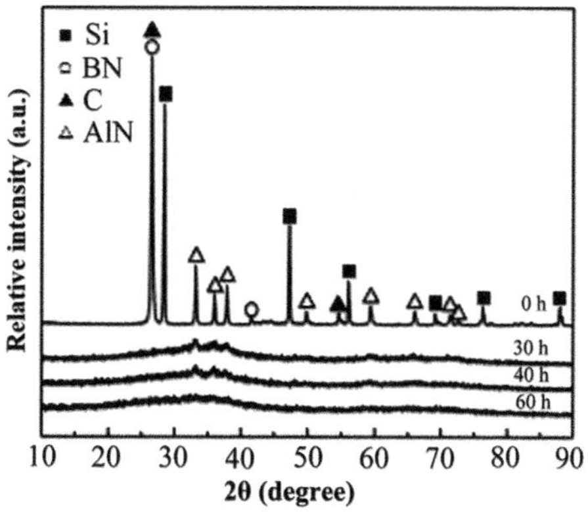

Fig. 3.23 XRD patterns of the MA-$Si_2BC_3N_{1.6}Al_{0.6}$ powder mixture (using AlN_p as the Al source) milled for different time with a B/P of 30:1 (milling speed of 600 rpm) [45]

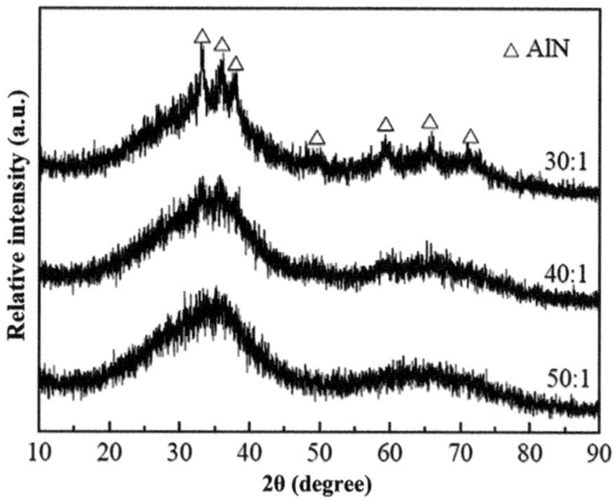

Fig. 3.24 XRD patterns of the MA-$Si_2BC_3N_{1.6}Al_{0.6}$ powder mixture (using AlN_p as the Al source) after 30 h of milling at different BP mass ratio [45]

small volume fraction of nanocrystallites. The interplanar distance measured from the nanocrystallites is 0.250 ± 0.002 nm, matched with $d(111) = 0.251$ nm of β-SiC or $d(002) = 0.251$ nm of AlN.

The o solid reactions among c-Si, h-BN, graphite and AlN, results in the generation of new chemical bonds (Fig. 3.26 and Table 3.5). Compared with the amorphous MA-$Si_2BC_3NAl_{0.6}$ powder (using Al as the Al source), the MA-$Si_2BC_3N_{1.6}Al_{0.6}$ powder shows similar bond types but differs in their relative content. In this case, trace of oxygen-containing bonds is not detected.

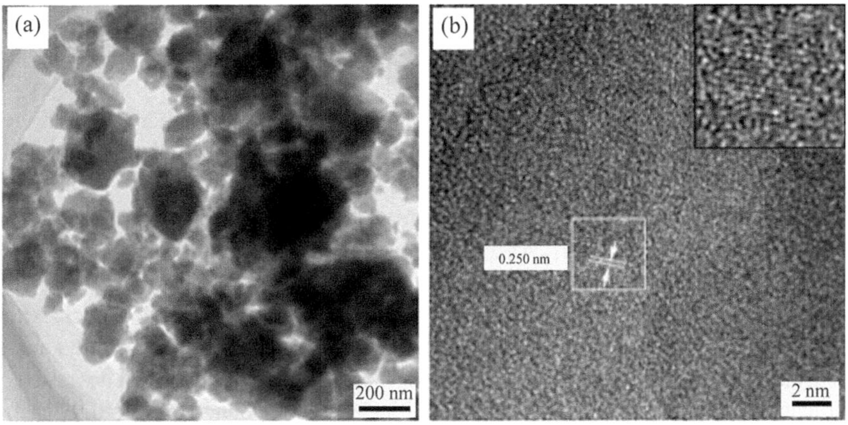

Fig. 3.25 TEM analysis of the MA-$Si_2BC_3N_{1.6}Al_{0.6}$ powder after 30 h of milling (BP mass ratio of 50:1): **a** Bright-field image; **b** HRTEM image with inserted inverse FFT image [45]

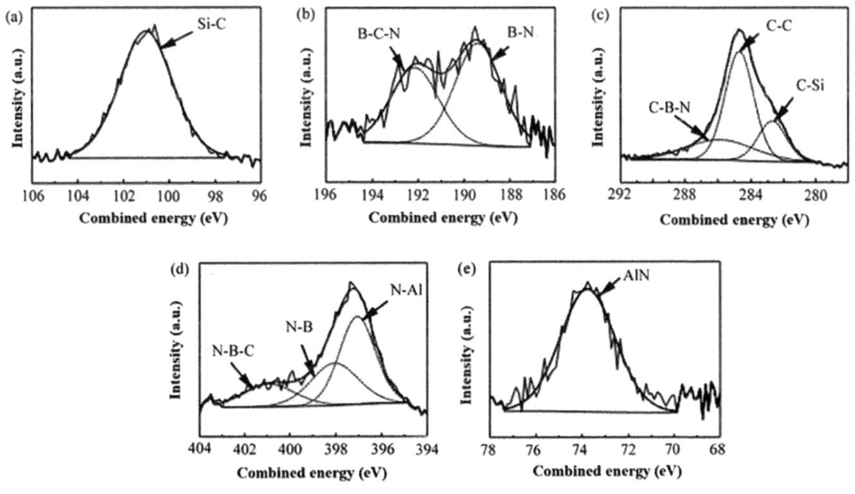

Fig. 3.26 XPS results of the MA-$Si_2BC_3N_{1.6}Al_{0.6}$ powder after 30 h of ball milling: **a** Si 2p; **b** B 1s; **c** C 1s; **d** N 1s; **e** Al 2p [24]

Table 3.5 XPS results of the MA-$Si_2BC_3N_{1.6}Al_{0.6}$ powder using AlN particles as the Al source [45]

Element lines	Bond types	Binding energy (eV)	Relative content (%)
Si 2p	Si–C	~ 101.0	16.6
B 1s	B–N	~ 189.4	11.7
	B–C–N	~ 192.1	9.0
C 1s	C–Si	~ 282.7	9.9
	C–C	~ 284.7	26.7
	C–B–N	~ 285.8	11.9
N 1s	N–Al	~ 397.0	4.5
	N–B	~ 398.1	2.9
	N–B–C	~ 400.9	1.7
Al 2p	Al–N	~ 73.8	5.0

The broad peak at −18.3 ppm of SiC_4 sites indicates the amorphous SiC structure in amorphous MA-$Si_2BC_3N_{1.6}Al_{0.6}$ powder (Fig. 3.27). A small peak at −90.3 ppm is attributed to the residual starting Si material due to insufficient milling. In AlN, the signal at ~ 112.8 ppm suggests that the presence of tetrahedral AlN_4 sites. After milling, the ^{27}Al chemical shifts of MA-$Si_2BC_3N_{1.6}Al_{0.6}$ powder are at ~ 112.9 and ~ 57.0 ppm corresponding to the AlN_4 and AlN_5 units, respectively [13]. These results indicate that the local environment around the Al atoms changes from AlN_4 to the mixture of AlN_4 and AlN_5. As compared to the former (using Al as Al source), the resulting amorphous MA-$Si_2BC_3N_{1.6}Al_{0.6}$ powder presents similar Al atomic environments.

The as-milled powders irrespective of using Al or AlN particles as the Al source, both show similar mass change under Ar atmosphere, and their weight loss is less than 2.5 wt.% at 1200 °C. Annealing temperature higher than 1300 °C, the four powders start to lose weight, ranging from 1.0 to 7.5 wt.%. At certain annealing temperature, the weight loss decreases as Al content increases (Fig. 3.28). The addition of Al or AlN both weakens the high-temperature thermal stability of the amorphous powders. In a flowing air, the as-milled powders gain more weight as Al content increases,

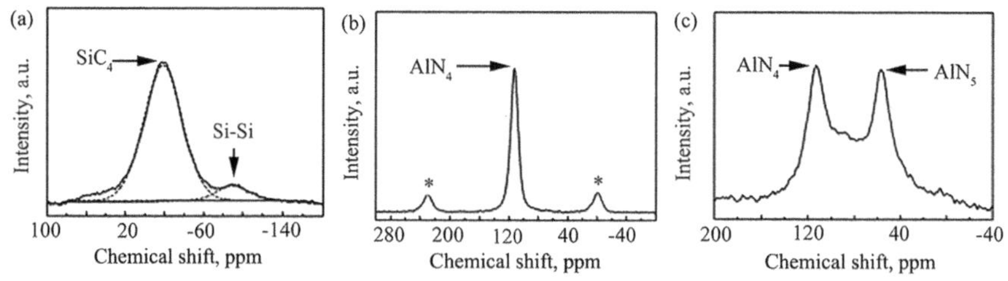

Fig. 3.27 Solid-state NMR spectra of the MA-$Si_2BC_3N_{1.6}Al_{0.6}$ powder after 30 h of ball milling: **a** As-milled, ^{29}Si; **b** AlN_p, ^{27}Al; **c** as-milled, ^{27}Al [45]

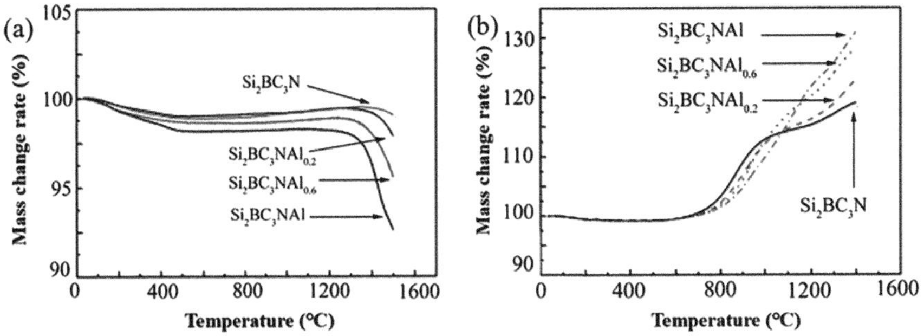

Fig. 3.28 TG curves of amorphous MA-Si_2BC_3N, MA-$Si_2BC_3NAl_{0.2}$, MA-$Si_2BC_3NAl_{0.6}$ and MA-Si_2BC_3NAl (using Al as the Al source) powders when heated to 1500 °C: **a** Ar; **b** Air [24]

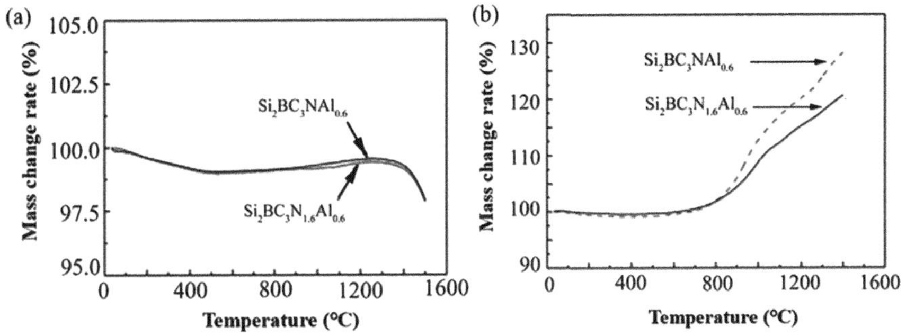

Fig. 3.29 TG curves of the amorphous MA-$Si_2BC_3NAl_{0.6}$ (using Al as the Al source) and MA-$Si_2BC_3N_{1.6}Al_{0.6}$ (using AlN as the Al source) powders when heated to 1500 °C: **a** Ar; **b** Air [24]

and amorphous MA-$Si_2BC_3N_{1.6}Al_{0.6}$ shows better oxidation resistance than their counterparts (Fig. 3.29).

(4) SiBCN-(TiB_2-TiC) powder

The Ti and B powders with a mole ratio of 1:1 are milled for 2 h, and the diffraction peaks of Ti–B mixture appear to be somewhat degree of broadening, but still showing Ti strong diffraction peaks [46]. After 4 h of milling, the diffraction peaks of all crystals show a further broadening, and the intensity of Ti peaks decreases sharply, accompanied with the increasing diffraction peaks of TiB_2 and TiB. After 8 h of milling, XRD pattern shows weak Ti diffraction peaks. Further milling to 16 h, the diffraction peaks assigned to TiB appear; further extending the milling time to 32 h, the diffraction peak intensity of TiB increases while that of TiB_2 decreases slightly (Fig. 3.30). After 16 h of milling, no obvious agglomerations occur but spherical particles are observed. Under the present MA conditions, completely X-ray amorphous TiB and/or TiB_2 powders cannot be prepared from metallic Ti and B powder mixtures.

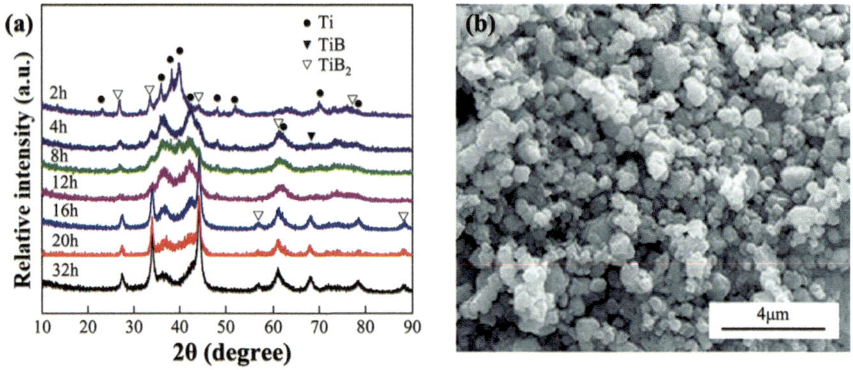

Fig. 3.30 Microstructure of the Ti–B powder system after MA process: **a** XRD pattern; **b** SEM surface morphology after 16 h of milling [46]

Commercially available TiB, TiB_2, TiC powder and amorphous MA-Si_2BC_3N powder are simultaneously ground by high-energy ball milling in order to obtain an amorphous Si_2BC_3N–Ti powder. However, after 3 h of milling, crystalline phases of TiB, TiC, and TiB_2 are still found (Fig. 3.31). Further milling only leads to the particle refinement of TiB, TiC, or TiB_2. The MA-Si_2BC_3N consists of nearly spherical nanoparticle agglomerates, while plate-like TiB_2 and spherical TiC are observed (Fig. 3.32). The Ti containing MA-Si_2BC_3N powder system irrespective of whether elemental powder of Ti or compounds of TiB, TiC, or TiB_2 adopted, is unable to achieve X-ray amorphous structure.

(5) SiBCN-ZrB_2 powder

In sol–gel processing, $Zr(OnPr)_4$ is combined with a mixture of solvents to form a gel. Generally, metal alkoxides react spontaneously with water to form precipitates

Fig. 3.31 XRD patterns of the Si_2BC_3N–(TiB_2–TiC) powder system after 3 h of milling: **a** MA-Si_2BC_3N; **b** + 10 vol.% TiB; **c** + 10 vol.% TiC; **d** + 10 vol.% (TiC–TiB_2), TiB_2/TiC mole ratio of 2:1; **e** + 10 vol.% (TiC–TiB_2), TiB_2/TiC mole ratio of 1:2; **f** + 10 vol.% TiB_2 [46]

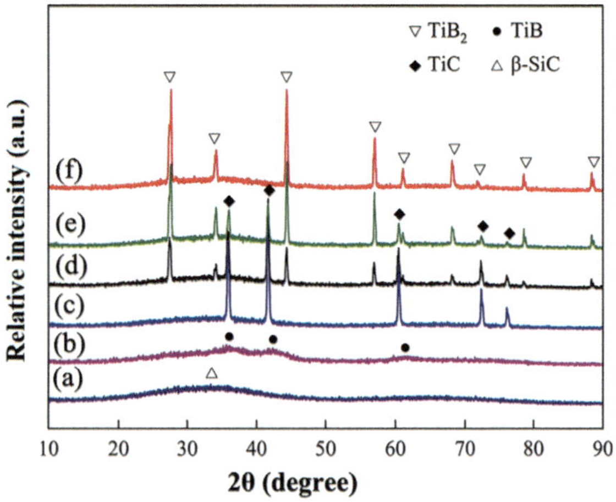

Fig. 3.32 SEM surface morphologies of the Si_2BC_3N–(TiB_2–TiC) powder system after 40 h of milling: **a** MA-Si_2BC_3N; **b** + 10 vol.% TiB; **c** + 10 vol.% TiB_2; **d** + 10 vol.% TiC; **e** + 10 vol.% (TiB_2–TiC), TiB_2/TiC mole ratio of 2:1; **f** + 10 vol.% (TiB_2–TiC), TiB_2/TiC mole ratio of 1:2 [46]

and $Zr(OnPr)_4$ can be modified by acetylacetone (AcAc) as a bridging and chelating ligand to form zirconium acetylacetone by the substitution of AcAc for OPr groups [47]. After sol–gel process, XRD results show the crystalline phases are those of ZrO_2 while MA-Si_2BC_3N still remain amorphous. The ZrO_2 is mainly derived from the decomposition of $Zr(OnPr)_4$ (Fig. 3.33, 3.34 and 3.35).

At SPS temperatures < 1400 °C, the observable peaks are predominant ZrO_2 (Fig. 3.36). Above 1500 °C, reactions are driven between zirconia and carbon, thus ZrC peaks are clearly identified. The rapid heating rate endows the carbothermal reduction reactions starts at a lower temperature than theory predicted. At 1500 °C, some ZrB_2 and SiC appear, and their content increases with annealing temperature.

Fig. 3.33 XRD patterns of as-milled amorphous MA-Si_2BC_3N powder and Si_2BC_3N–ZrO_2 precursor after annealed at 550 °C for 2 h [47]

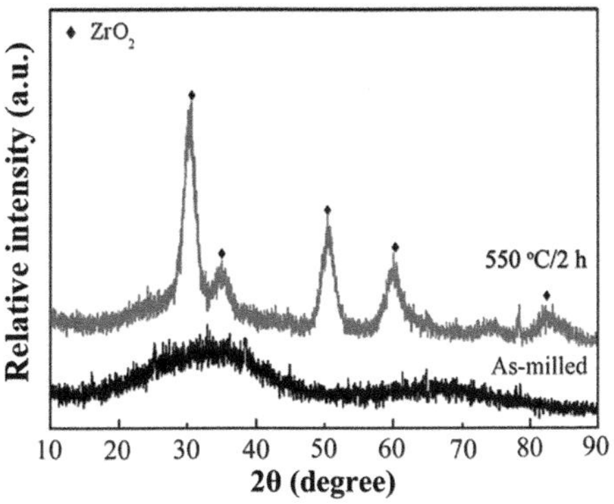

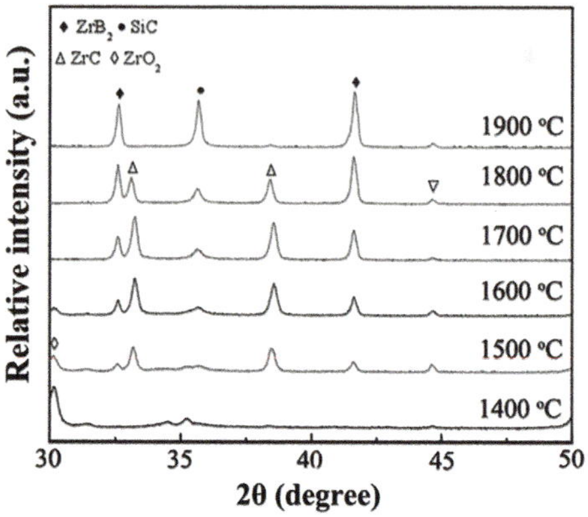

Fig. 3.34 XRD patterns of powder mixture after annealed at 1400–1900 °C for 5 min [47]

Fig. 3.35 SEM images of as-prepared powder annealed at different temperature for 5 min: **a** 1500 °C; **b** 1700 °C; **c** 1900 °C; **d** corresponding EDS [47]

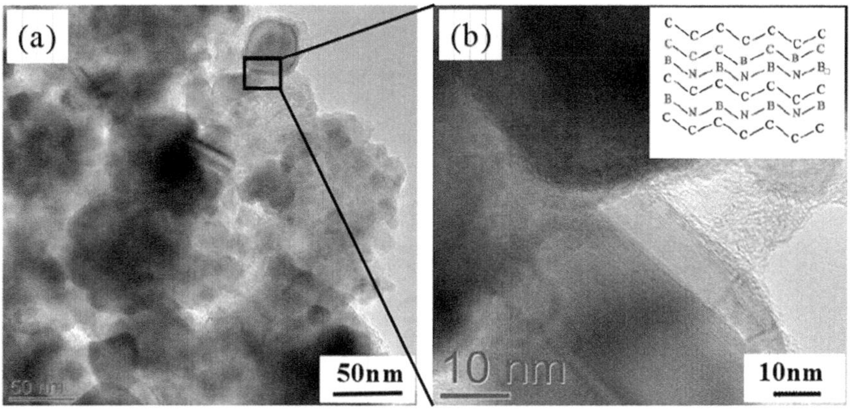

Fig. 3.36 TEM analysis of as-prepared powder annealed at 1500 °C for 5 min: **a** Bright-field image; **b** enlarged area in (**a**) [47]

ZrO_2 finally disappears at 1700 °C due to the sufficient carbothermal reduction. Interestingly, ZrC has all transferred to ZrB_2 at 1900 °C.

Annealing at 1500 °C, the amorphous powder remains aggregated (Fig. 3.35). With increasing annealing temperature, the amorphous MA-Si_2BC_3N matrix starts to crystallize, meanwhile reactions between ZrO_2 and BN(C) occurs. At 1900 °C, the formed ZrB_2 are in sizes of ~ 100 nm, while the larger ones are the Si_2BC_3N particles. After annealing at 1500 °C, SiC and ZrO_2 are surrounded by an amorphous phase. From HRTEM, the lattice fringe spacing ~ 0.51 nm is measured, corresponding to the (001) lattice plane of ZrO_2. At higher annealing temperatures, BN(C) eventually reacts with ZrO_2 forming ZrB_2 (Fig. 3.36).

(6) SiBCN-HfB_2 powder

Hf and B are easily inter-diffused to form HfB_2 during mechanical alloying (Fig. 3.37). After ball milling for 1 h, the diffraction peaks of Hf become weaker and broaden, and finally disappear after only 5 h of milling [48]. The absence of the boron diffraction peaks is attributed to the use of amorphous boron powder. The diffraction peaks of HfB_2 exhibit no obvious change beyond 10 h of milling. Noticeably, Hf gradually refines and tends to become partially amorphous after 3 h of milling. Thus, the final powder mixtures may be composed of main nano HfB_2 and few amorphous phases.

Intuitively, the as-milled HfB_2 powder is composed of near-spherical agglomerates (Fig. 3.38). In detail, these agglomerates are built by the stacking of smaller particles of ~ 500 nm, but the size of the near-spherical agglomerates increases slightly with prolonged milling time. Thus, 10 h of milling of Hf–B mixture is reasonable when both considering the energy consumption and processing efficiency. The amorphous MA-Si_2BC_3N powder will react with the nanosized HfB_2 to form some HfC during annealing or sintering at high-temperatures. In this case, incorporating HfB_2 weakens the thermal stability of amorphous MA-Si_2BC_3N powder.

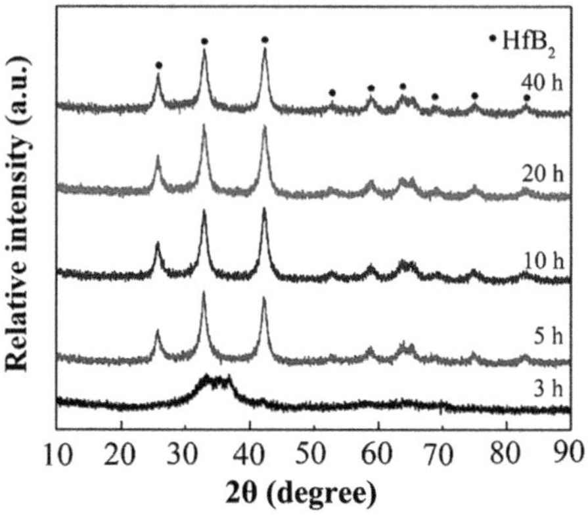

Fig. 3.37 XRD patterns of HfB_2 powder prepared by MA (main speed of 600 rpm; ball-to-powder mass ratio of 20:1) [48]

Fig. 3.38 SEM surface morphologies of HfB_2 powder by MA (milling speed of 600 rpm and BP mass ratio of 20:1): **a**, **b** 3 h; **c**, **d** 10 h; **e**, **f** 20 h; **g**, **h** 40 h [48]

(7) SiBCN-ZrC powder

Using $Zr(PrO)_4$ as zirconium source and furfuryl alcohol ($C_5H_6O_2$) as carbon source, $Zr(PrO)_4$ will converts to ZrO_2 while $C_5H_6O_2$ changes to carbon at high temperatures. Carbothermal reduction reaction will occur between ZrO_2 and carbon which are thermodynamically. After homogeneous mixing of $Zr(PrO)_4$ and $C_5H_6O_2$, the powder mixtures are heated to 550 °C for 2 h in argon [49]. As indicated, the $Zr(PrO)_4$ has changed to ZrO_2 while the MA-Si_2BC_3N remains amorphous. After annealing at 550 °C for 2 h, amorphous MA-Si_2BC_3N powder is covered with some ZrO_2 and C. Interestingly, the as-milled powder has lost the spherical structure (Fig. 3.39).

Further annealing at 1300 °C, the powder mixture consists of two phases, t-ZrO_2 and m-ZrO_2; when annealing at 1350 °C, a small amount of target product ZrC is generated. At 1400–1450 °C, the products are ZrO_2 and ZrC. Increasing the

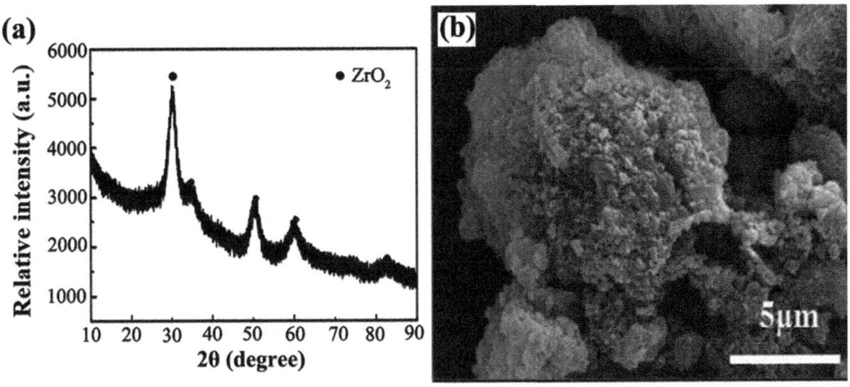

Fig. 3.39 The Zr(PrO)$_4$ and $C_5H_6O_2$ mixture after heated to 550 °C for 2 h: **a** XRD pattern; **b** SEM image [49]

annealing temperature to 1500–1550 °C, ZrO_2 has disappeared instead of forming pure ZrC, indicating the adequate carbothermal reduction reactions. The as-obtained ZrC powder is in sizes of < 1 μm, basically free of agglomerations (Fig. 3.40).

After annealing at 1500 °C for 30 min, the Si_2BC_3N-ZrC powder blends are composed of ZrC, SiC and BN(C) with any trace of ZrO_2. With the increasing ZrC content, the diffraction peak intensity of ZrC gradually enhances. The average sizes of the as-annealed spherical or nearly spherical powder are 100–200 nm without abnormal growth (Fig. 3.41).

(8) SiBCN-TaC powder

Before high-energy milling, TaC has larger particle sizes with ordered crystal structure. After 30 h of milling, the lattice structure of TaC has been destroyed with grains

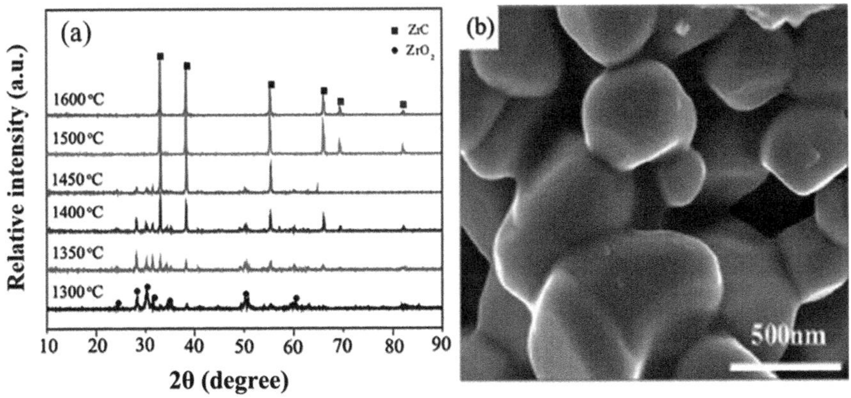

Fig. 3.40 The Zr(PrO)$_4$ and $C_5H_6O_2$ mixture after heated to 1300–1600 °C for 30 min: **a** XRD pattern; **b** SEM image [49]

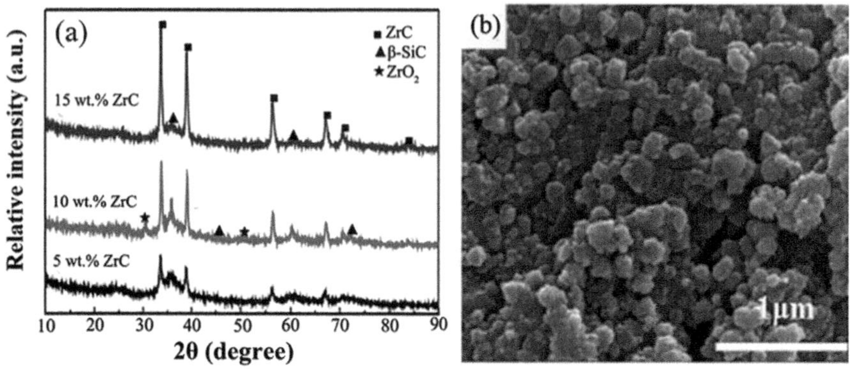

Fig. 3.41 XRD patterns of Si_2BC_3N-ZrC powder blends annealed at 1500 °C for 30 min: **a** XRD patterns; **b** SEM image [49]

refinement [50]. The finial milling products are in the form of nano TaC ~ 5 nm with some amorphous structure (Fig. 3.42).

As expected, the peaks assigned to TaC tend to broaden obviously accompanied with the decreasing diffraction intensities due to the grain refinement induced by MA. After 20 h of milling, the lattice structures of graphite, h-BN and c-Si are completely destroyed, typical of amorphous nature. However, the broad diffraction

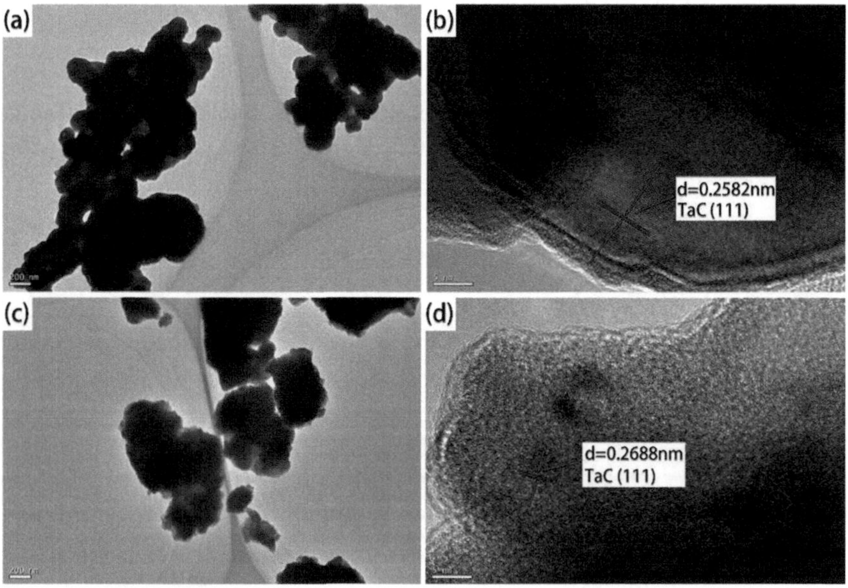

Fig. 3.42 TEM and HRTEM images of TaC powder: **a**, **b** Before milling; **c**, **d** after high-energy ball milling for 30 h [50]

peaks assigned to nano TaC are detected for the Si_2BC_3N–TaC mixture (Fig. 3.43). After mechanical alloying, the Si_2BC_3N–TaC powder mixtures are mainly composed of agglomerated spherical particles 5–10 μm. However, magnified picture further confirms such agglomerates are composed of substantial nearly spherical clusters with a particle diameter of 100–200 nm (Fig. 3.44).

(9) SiBCN–Ta_4HfC_5 powder

Before mechanical alloying, the powder mixtures of the 4TaC–HfC show sharp diffraction peaks [51]. After 0.5 h of milling, the intensities of the corresponding diffraction peaks are reduced and the FWHM becomes wider, resulting from the grain refinement of TaC and HfC. As the milling time further raises to 1–5 h, TaC and HfC gradually undergo a solid solution reaction to form some Ta_4HfC_5. Interestingly, the single-phase Ta_4HfC_5 is obtained after 10 h of MA. Further prolonging the

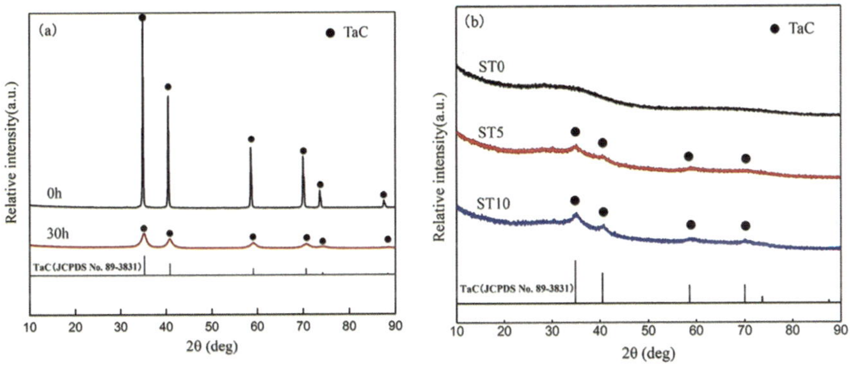

Fig. 3.43 XRD patterns of the TaC and Si_2BC_3N–TaC mixed powder: **a** TaC powder before and after milling for 30 h; **b** Si_2BC_3N–TaC mixed powder after 20 h of milling [50]

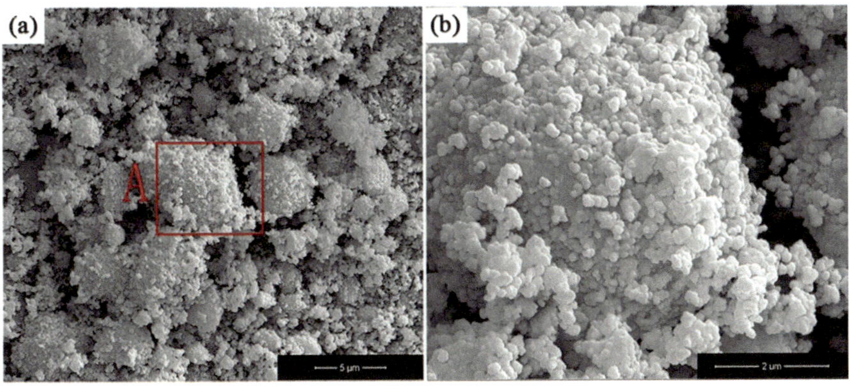

Fig. 3.44 Representative SEM surface images of Si_2BC_3N–TaC mixed powder after ball-milling of 20 h: **a** Low magnification; **b** high magnification of box A [50]

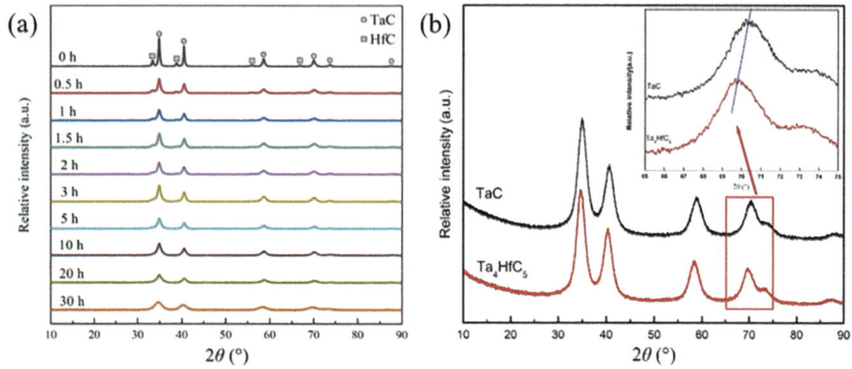

Fig. 3.45 XRD patterns of the 4TaC-HfC powder mixture after mechanical alloying for different time: **a** 0–30 h of milling; **b** magnified image of $2\theta = 65° \sim 75°$ of TaC and Ta_4HfC_5 [51]

milling time to 20–30 h, the diffraction peaks of the as-prepared Ta_4HfC_5 gradually decrease, implying the grain refinement of Ta_4HfC_5 (Fig. 3.45). After 20 h milling, the diffraction peaks of the as-prepared Ta_4HfC_5 shifting to the small angle direction result from the radius difference between Hf and Ta [52, 53]. The radius of Hf atom is bigger than that of Ta. Thus, the short-range diffusion of Hf atoms into the interstitial sites of Ta can lead to the expansion of the TaC lattice.

Before milling, the 4TaC–HfC powder mixtures with larger particle size are observed (Figs. 3.46 and 3.47). After ball milling for only 0.5 h, the grain size of TaC and HfC decreases obviously and some diffraction spots of TaC and HfC have disappeared. Notably, at this milling time, mechanical alloying has induced many structural defects in both TaC and HfC, which can provide convenient channel for Hf atoms diffused inward. With progressing milling of 30 h, only pure Ta_4HfC_5 nanocrystals are obtained without any trace of HfC or below the limits of detection. The Ta, Hf, and C atoms distribute uniformly after 30 h of milling, suggesting a good solid solubility of Hf into TaC. For pure TaC ball-milled for 30 h, SAED pattern with bright-spots and narrow-rings verifies the well-defined crystalline phase of TaC. The interplanar crystal spacing of TaC is slightly smaller than that of Ta_4HfC_5.

After 20 h milling of Si_2BC_3N–Ta_4HfC_5 powder mixtures, the ordered lattice structure of c-Si, h-BN, and graphite has been destroyed to form amorphous MA-Si_2BC_3N. Besides, the broad diffraction peaks of nano Ta_4HfC_5 are observed, and their peak intensities gradually increase with the increasing Ta_4HfC_5 content. Coincidentally, the nano Ta_4HfC_5 does not react with other components and still presents the form of nanocrystals (3–5 nm) in amorphous matrix after MA (Fig. 3.48).

(10) SiBCN-TiCN powder

The Ti and h-BN (mole ratio of Ti:h-BN = 3:2) powder mixtures are initially milled for different time [54]. After 1 h of MA, the h-BN diffraction peaks almost disappear, while obvious peaks assigned to Ti are also found. After 3 h of milling, the diffraction peaks of h-BN have completely disappeared with the deceasing intensity

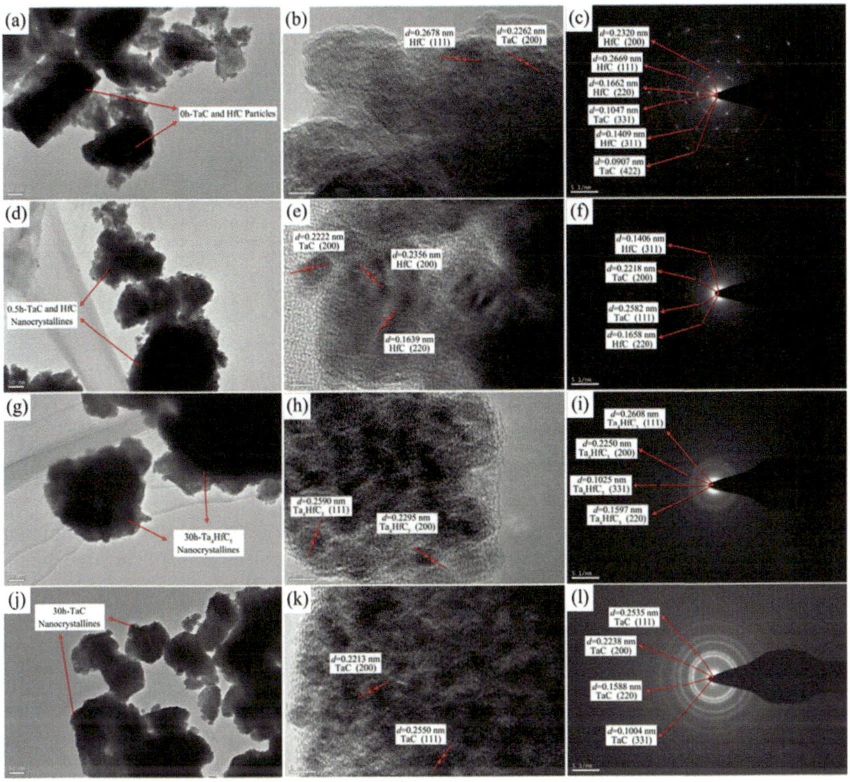

Fig. 3.46 TEM, HRTEM and SAED images of the pure TaC powder and 4TaC-HfC powder mixtures after mechanical alloying for different time: **a–c** 4TaC–HfC, 0 h; **d–f** 4TaC–HfC, 0.5 h; **g–i** 4TaC–HfC, 30 h; **j–i** pure TaC powder, 30 h [51]

of Ti peaks. h-BN is easier to achieve solid-state amorphization than Ti. Further ball milling for 30 h, the diffuse scattering peaks suggest the formation of amorphous phases (Fig. 3.49). It seems that the Ti–BN mixed powders undergo obvious hard agglomerations due to the cold welding, and the powder particle sizes are around 200–500 nm. Almost no nanocrystals appear in the HRTEM image suggesting the almost complete solid-state amorphization of Ti and h-BN mixtures.

For Ti content below 30 wt.%, no trace of crystalline peaks is detected for the Si_2BC_3N–Ti composite powder. However, increasing Ti content to 30 wt.% leads to the emergence of new TiCN and unreacted Ti. The as-milled composite powders are in sizes of 1–5 μm; however, these composite particles are stacked by smaller ones around 100–200 nm. In some cases, some nanocrystals of $TiC_{0.7}N_{0.3}$ about 5–7 nm are observed in an amorphous matrix (Fig. 3.50).

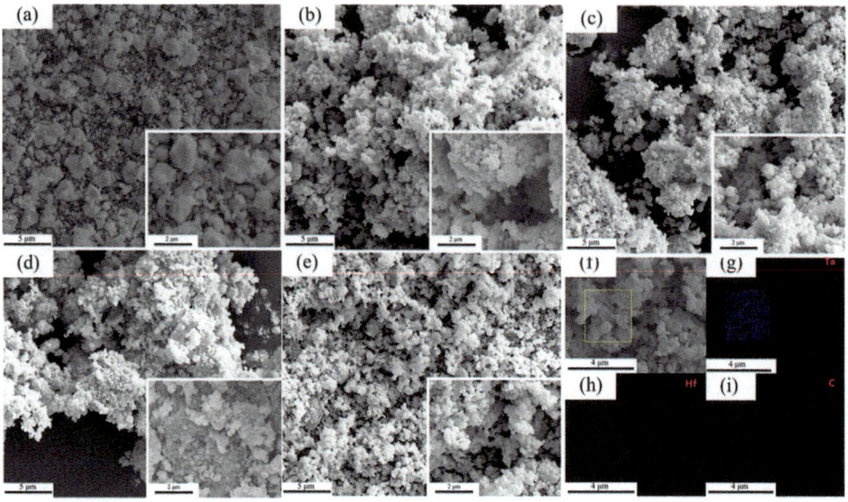

Fig. 3.47 SEM surface images of 4TaC–HfC mixed powders after mechanical alloying for different time. **a** 0 h; **b** 0.5 h; **c** 1.5 h; **d** 20 h; **e**, **f** 30 h; **g**–**i** corresponding element maps (Inserted are the magnifying SEM images) [51]

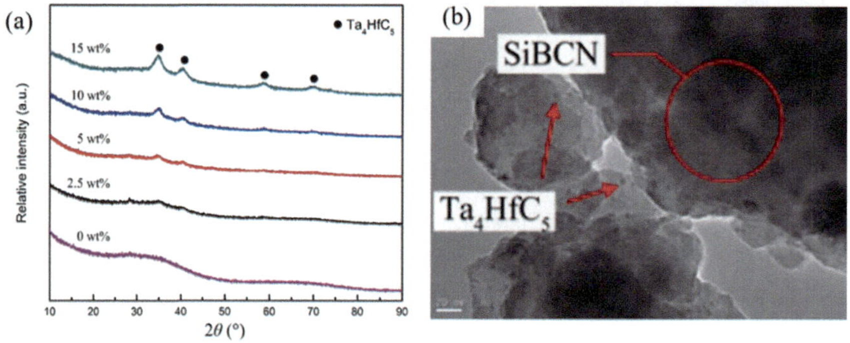

Fig. 3.48 Phase composition and microstructure of Si_2BC_3N–Ta_4HfC_5 powder mixtures after 20 h of MA: **a** XRD patterns; **b** HRTEM image [51]

(11) SiBCN-MWCNTs powder

Commercially available multi-walled carbon nanotubes (MWCNTs) with diameters of ~ 30 nm and lengths of 1–5 μm are mixed with amorphous MA-Si_2BC_3N powder [55]. As indicted, the MWCNTs can be uniformly dispersed among the nanosized MA-Si_2BC_3N ceramic particles, but still a few MWCNTs agglomerates are observed (Fig. 3.51). The SiC coated MWCNTs can improve the dispersion of MWCNTs in amorphous MA-Si_2BC_3N matrix [56].

The MWCNTs with or without nano SiC coating both show smooth surfaces with a coating diameter of 10–30 nm (Fig. 3.52). TG-DSC curves imply the maximum

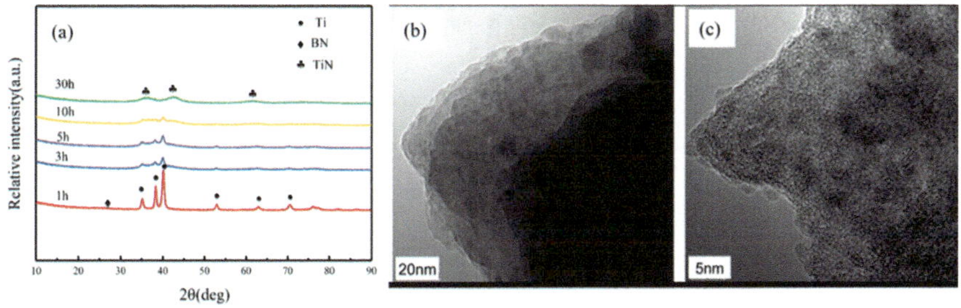

Fig. 3.49 XRD patterns and HRTEM images of the as-milled Ti–BN powder mixture: **a**, **b** phase composition; **b**, **c** partial amorphous Ti–BN [54]

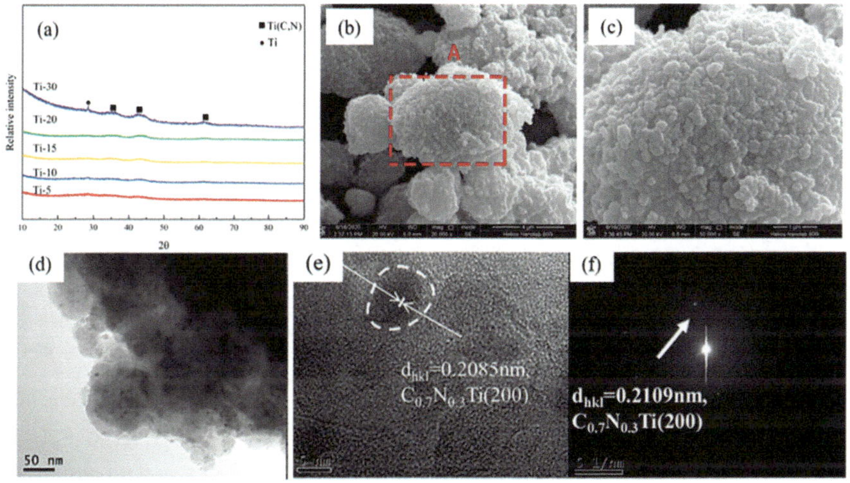

Fig. 3.50 Structure characterization of the as-milled Si_2BC_3N–Ti composite powders: **a**, **b** XRD patterns; **b**, **c** SEM surface morphologies; **d–f** TEM analysis [54]

oxidation rate temperature of MWCNTs increases from 595.2 to 719.1 °C after the coating process. Thus, formation of a uniform SiC coating can effectively improve the oxidation resistance of MWCNTs (Fig. 3.53).

(12) SiBCN-rGO powder

The graphene oxide (GO) is prepared by a modified Hummer's method, and then treated by hydrazine hydrate to obtain reduced graphene oxide (rGO) [57]. The position and intensity of XRD-peaks can directly reflect the oxidation degree of graphite. The increase in layer spacing and the decrease in particle size of oxidized graphite flakes are reflected by the leftward shift of the diffraction peaks and the broadening of the diffraction peaks. The characteristic peak of the natural graphite structure locates at ~ 26°, while the characteristic peak of the oxidized GO positions at ~ 9.4°. This indicates that the oxygen-containing groups have been successfully

Fig. 3.51 SEM surface morphologies of Si_2BC_3N-MWCNTs powder mixtures: **a** Pure MA-Si_2BC_3N; **b** pure MWCNTs; **c** + 1 vol.% MWCNTs; **d** + 2 vol.% MWCNTs [55]

inserted into the graphite (0002) crystal planes. After reduction induced by hydrazine hydrate, the characteristic peak of the rGO appears at ~ 26°, suggesting the removing of a large number of oxygen-containing groups (Fig. 3.54).

The natural graphite flakes are highly oriented and generally shows two characteristic Raman peaks, D-band ~ 1355 cm^{-1} and G-band ~ 1580 cm^{-1}. The G-band is caused by the vibration of sp^2 hybridized carbon in the graphite structure, while the D-band indicates the presence of defects between graphite flakes or free carbon at the edge of graphite flakes. The natural flake graphite has a highly ordered crystal structure, so the intensity of D-band is very weak. The existence of strong D + G peak for both GO and rGO are also found. The laminated structure of rGO is obvious after reduction reaction (Fig. 3.55). The edge of rGO presents crimp structure thus monolayer graphene is obtained (Fig. 3.56). The thickness of GO and rGO are measured to be ~4.5 and ~0.95 nm, respectively, with a similar lateral dimension of 100–400 nm (Fig. 3.57). The rGO and amorphous MA-Si_2BC_3N powder are well-dispersed with each other (Fig. 3.58).

The initial mass loss for all investigated powders below 200 °C likely arises from desorption of adsorbed water, carbon monoxide and carbon dioxide (Fig. 3.59).

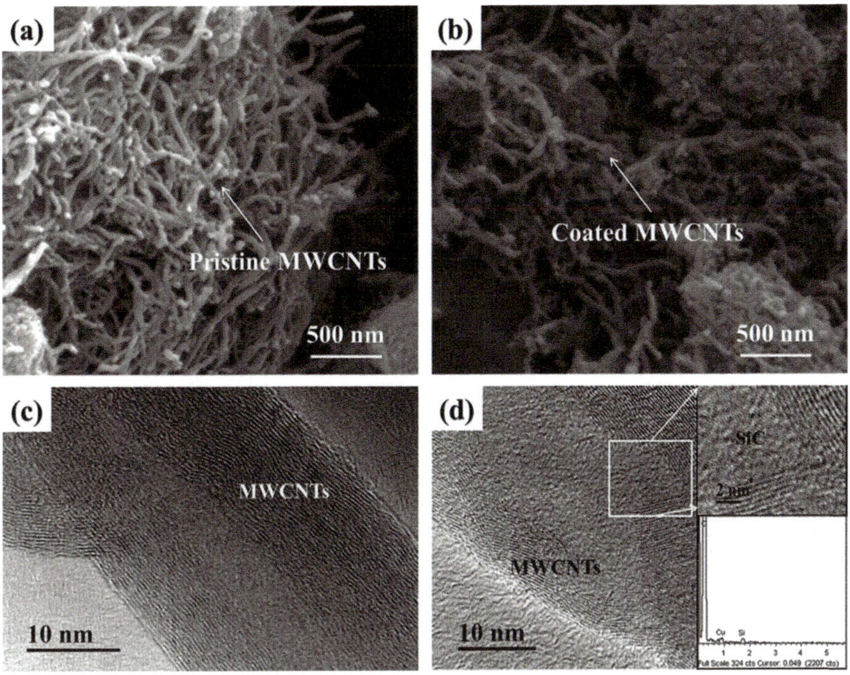

Fig. 3.52 SEM and TEM morphologies of the MWCNTs coated with or without SiC: **a**, **c** Without SiC coating; **b**, **d** with SiC coating (Inserted are the FFT image and corresponding EDS image) [56]

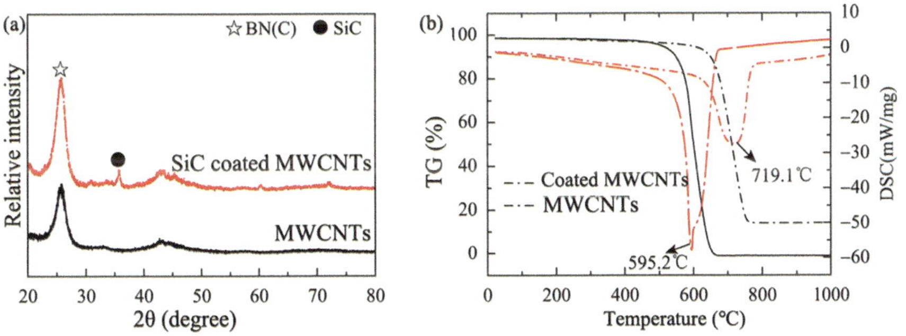

Fig. 3.53 Characterization of MWCNTs coated with or without SiC: **a** XRD patterns; **b** TG-DSC curves [56]

Pure rGO oxidizes dramatically above 600 °C; however, the Si_2BC_3N-rGO powder mixture exhibi mass losses much earlier. In contrast, MA-Si_2BC_3N oxidizes with mass gains beginning only at ~ 600 °C. The mass loss of Si_2BC_3N-rGO powder mixtures oxidized at 1200 °C are ~ 4% (+1 vol.% rGO) and ~ 40% (+5 vol.% rGO), respectively. The mass gain for pure amorphous MA-Si_2BC_3N powder at 1200 °C is ~ 11%. Generally, amorphous MA-Si_2BC_3N powder incorporating with some

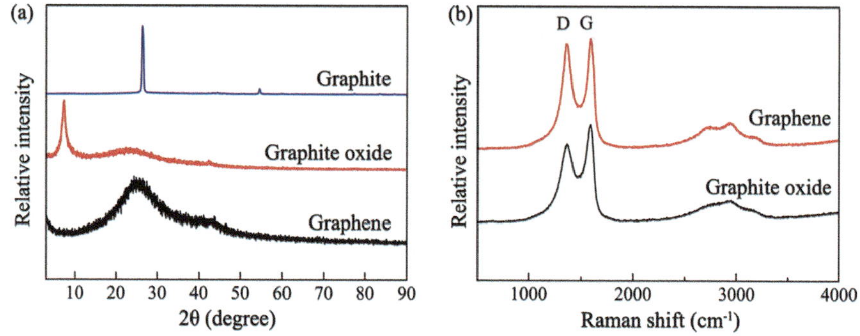

Fig. 3.54 The structural characterization of the natural graphite flakes, graphene oxide and reduced graphene oxide: **a** XRD patterns; **b** Raman spectra [57]

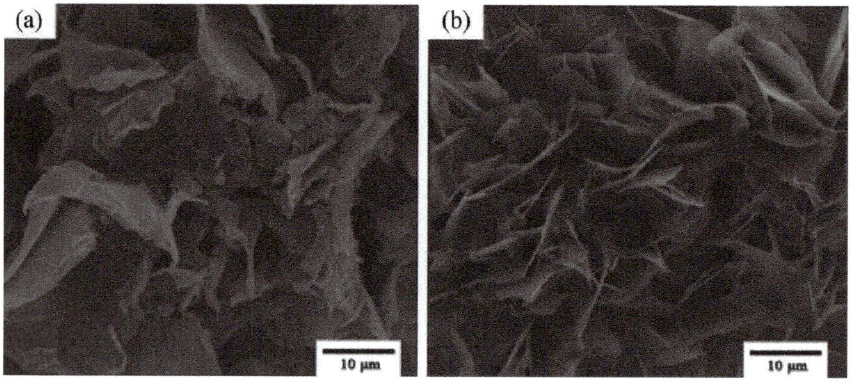

Fig. 3.55 SEM surface morphologies of GO and rGO: **a** GO by improved Hummer's method; **b** rGO by hydrazine hydrate reduction [57]

rGO results in a weak oxidation resistance accompanied with the occurring of three exothermal reactions at relatively low-temperatures.

3.4 Comparison of Amorphous MA-SiBCN Powders with PDCs Counterparts

3.4.1 Microstructure Features

The amorphous MA-Si_2BC_3N from the blended elemental powders including c-Si, graphite, boron and h-BN powders, shows one optimum combination of processing parameters is the milling speed of grinding vials (typically of 600 rpm) coupled with a ball-to-powder (BP) weight ratio of 20:1 and milling time of 30 h. A "harder"

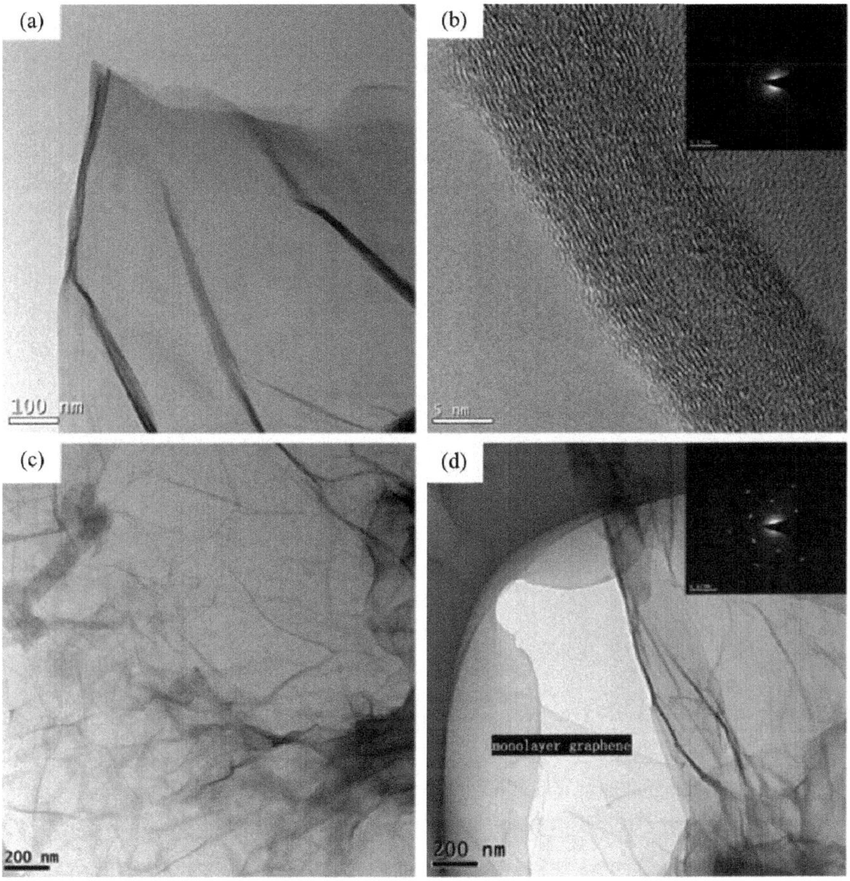

Fig. 3.56 TEM and HRTEM images: **a**, **b** GO; **c**, **d** rGO (Insets are the SAED patterns) [57]

combination (e.g., higher than 800 rpm, 20:1 and 40 h) facilitates formation of SiC nanocrystallites in an amorphous matrix. A "softer" condition (e.g., lower 600 rpm, 20:1, 25 h) results in insufficient mechanochemical reactions characterized by the presence of residual elemental silicon with substantial amorphous phase. Therefore, the appropriate ball milling parameters during MA, besides the used chemical composition of the starting blended elemental powders, is critical to the amorphous SiBCN processing.

MA derived amorphous Si_2BC_3N aggregated powder (600 rpm, 30 h, 20:1) consists of primary particles with average particle size (APS) of ~200 nm and specific surface areas of ~25 m^2/g. The spherical aggregates show APS of ~5 μm with a broad size distribution. Aggregates larger than 15 μm account for ca. 10% of the particles. The aggregates cannot be broken down easily by pressing, grinding or even ultrasonic vibration in ethanol. Thus, the as-milled amorphous powder packs heterogeneously with a pore size range of 2–50 nm.

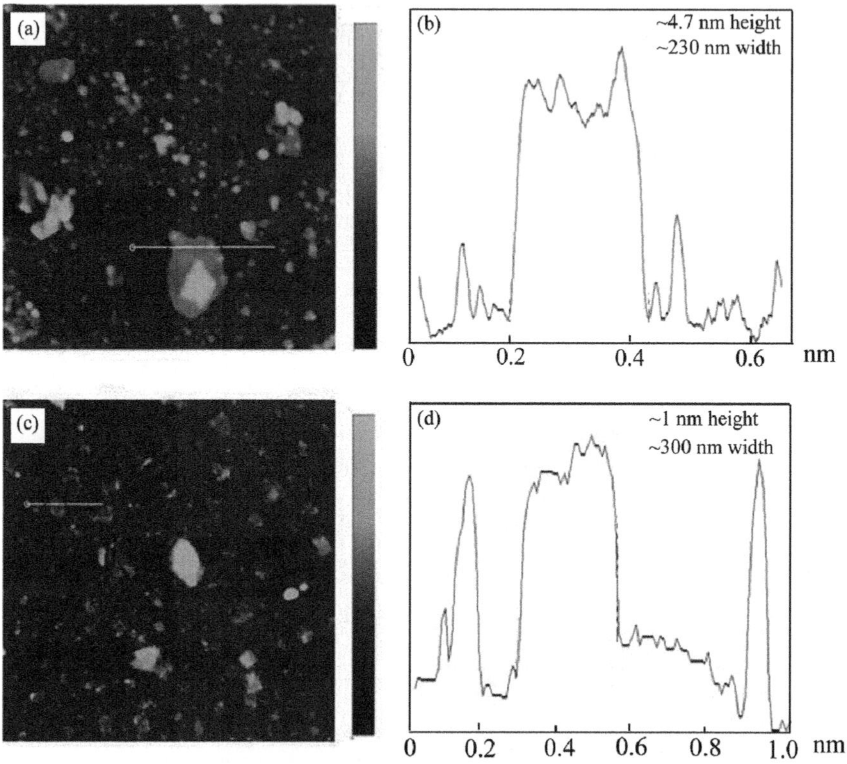

Fig. 3.57 AFM morphologies: **a** GO; **b** corresponding sheet thickness of GO; **c** rGO; **d** corresponding sheet thickness of rGO [57]

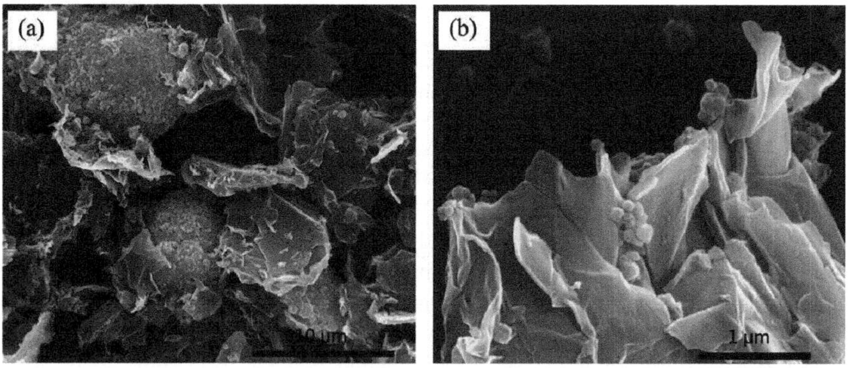

Fig. 3.58 SEM surface morphologies of the Si_2BC_3N-rGO powder mixture after sonication with CTAB aqueous solution: **a** Low magnification; **b** high magnification [57]

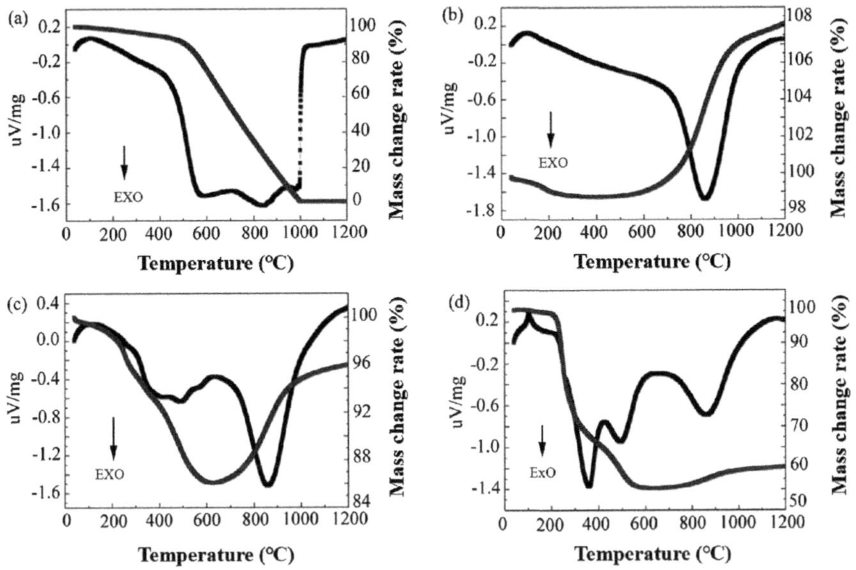

Fig. 3.59 TG-DSC curves of the as-obtained powder mixtures when heated to 1200 °C under air atmosphere with a heating rate of 10 °C/min: **a** Pure rGO; **b** pure MA-Si_2BC_3N; **c** + 1 vol.% rGO; **d** + 5 vol.% rGO [57]

Elemental mappings of as-prepared amorphous MA-Si_2BC_3N powder show uniform elemental homogenization. "Sintering necks" between particles arising as a consequence of MA are the reasons for the observed aggregation. Numerous new chemical bonds of Si–C, C–B–N, B–C, C–N, C–N–B form indicating mechanochemical effects that lead to good-to-excellent alloying and perhaps "atomic mixing" between the starting elements. Additionally, amorphous or partial amorphous Si_2BC_3N–Me (Me = Al, Zr, Ta, Hf, etc.) composite powders are also developed with additions of different types of particles including Al, Zr, Ta, Hf, AlN, ZrO_2, ZrB_2, HfB_2, TaC, Ta_4HfC_5, etc.

The polyborophenylvinylsilylcarbodiimide precursor-derived SiBCN also shows no identifiable diffraction XRD peaks at 1100 °C indicative of long-range disorder [58]. However, not all the as-pyrolyzed PDCs-SiBCN shows completely amorphous structures due to the difference in the starting precursor, chemical composition, synthesis technology and pyrolysis process. For example [59], boron-modified poly(propenylsilazane) derived SiBCN ceramics exhibit small SiC crystallites of 2–4 nm and graphite-like turbostratic BN(C) (Fig. 3.60).

Researcher on early studies [60, 61] hold that the as-pyrolyzed amorphous PDCs-SiBCN consists of a 3-D network with homogeneous chemical composition where the atoms distribute uniformly and form strong chemical bonds taking into consideration that the precursor pyrolysis route attains the material design at atomic level and the well-control of chemical composition and microstructure (Fig. 3.61). Heinemann et al. [62] have investigated the atomic distribution and their chemical bonding states

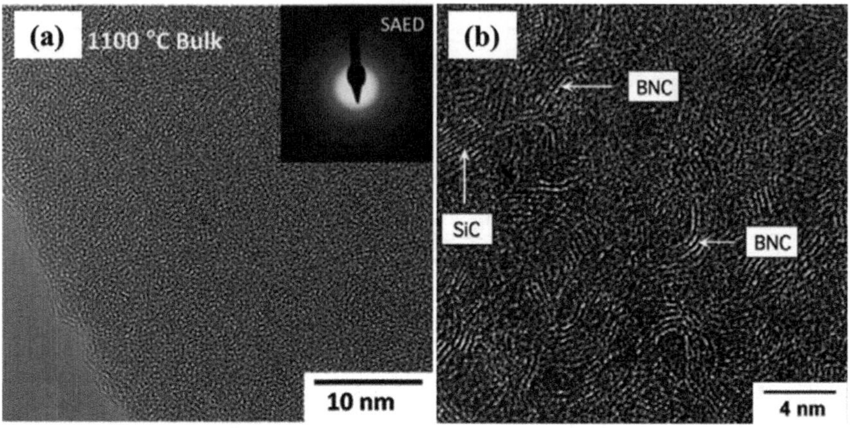

Fig. 3.60 HRTEM images of the as-pyrolyzed PDCs-SiBCN: **a** polyborophenylvinylsilylcarbodiimide derived SiBCN pyrolyzed at 1100 °C with a completely amorphous structure; **b** boron-modified poly(propenylsilazane) derived SiBCN pyrolyzed at 1400 °C with SiC nanocrystallites and turbostratic BN(C) [58, 59]

in the amorphous PDCs-SiB(C)N. HRTEM mappings of Si, B, N, and C reveal a homogeneous elemental distribution on a subnanometer scale without precipitation or separation of atomic clusters. Solid-state NMR together with the pair distribution function derived from electron scattering intensity both indicate short-range order for $Si_3B_3N_7$ consisted primarily of tetrahedral $[SiN_4]$ units and trigonal planar $[BN_3]$ units linked by common nitrogen atoms [63, 64].

Additionally, Franke et al. [60] studied the local atomic structure of an amorphous PDCs-SiBCN. Corresponding XANES characterization shows the formation of SiN_4, SiN_xC_y, BN_3, NSi_xB_y and (C=C) structural units. The relative content of these units

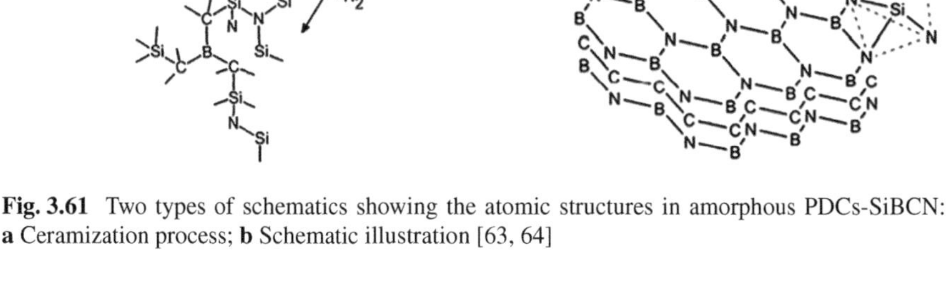

Fig. 3.61 Two types of schematics showing the atomic structures in amorphous PDCs-SiBCN: **a** Ceramization process; **b** Schematic illustration [63, 64]

depends mainly on the chemical constitution of the precursors and processing conditions. NMR, Raman and electron EPR spectroscopies show three major components including h-BN, turbostratic BN (t-BN), and BN_2C groups. Pyrolysis at higher temperatures shows evolution of BN_3 and 'free' carbon with the disappearance of BN_2C groups. The 'free' carbon cluster concentration and the carbon-dangling bond content decrease with increasing pyrolysis temperature [65].

Significantly, new characterization techniques coupled with computer simulation suggests still an additional structure of amorphous SiBCN materials [66, 67]. The presence of amorphous clusters including carbon, BNC_x and SiC_xN_y units implies not all atoms are distributed uniformly. Haug et al. [68, 69] further investigated the atomic structure and phase separation in amorphous PDCs-SiBCN powders using X-ray and neutron diffraction in the wide-angle and small-angle scattering regimes. The short-range order consists mainly of amorphous $Si_{3+(1/4)x}C_xN_{4-x}$ and graphite-like BNC_x phases at scales of 1 nm, although these amorphous units are distributed uniformly in a long-range order.

Just like the PDCs-SiBCN, the atomic structures of completely amorphous or partial amorphous MA-SiBCN powders also deserve attention, and still in its fantasy. They are most likely to be resolved by a combination of experimental efforts and theoretical computation, especially ab initio molecular dynamics simulation coupled with the use of Cs-corrected atomic resolution STEM, NMR and synchrotron radiation spectroscopies.

3.4.2 Structure Stability at High-Temperatures

The crystallization of amorphous structures in MA-Si_2BC_3N powder occurs at 1600 °C/He (0.1 MPa) or 1400 °C/N_2 (0.1 MPa). Higher temperatures enhance crystallization as SiC and BN(C) form. On heating in He or N_2, the amorphous powder exhibits a slight mass gain accompanied by an endotherm (400–1300 °C), and then a rapid mass loss (1350–1500 °C) from loss of CO and CO_2 likely from carbothermal reductions of Si–O surface species. The oxygen contamination derives most likely from surface adsorption of as-milled powders, and the starting Si, h-BN and graphite powders (purity > 99.9%). Carbothermal reductions mean oxygen contamination has negligible influence on the microstructures and properties of the final sintered ceramic materials.

As noted above, amorphous MA-Si_2BC_3N powder shows oxidation resistance superior to SiC and Si_3N_4 counterparts $\geq$ 1000 °C. The oxidation activation energy for amorphous MA-Si_2BC_3N powder (ca. 200 nm in average particle sizes) is about 350 kJ/mol, higher than that of SiC (80 kJ/mol) powder with the similar particle sizes. The addition of Al leads to a decrease in oxidation activation energy. The oxidation behavior of amorphous Si_2BC_3N–Al powder decreases to 150–220 kJ/mol, but it is still higher than that of SiC. Both the content and type of Al appear to influence oxidation resistance according to the TGA curves and oxidation kinetics. For example, introduction of Al as AlN provides oxidation resistance superior to counterparts with direct Al addition attributed to the higher oxidation activation energy of the former (190 kJ/mol) than the latter (178 kJ/mol).

The PDCs-SiBCN also shows structural stability superior to pure carbide and nitride counterparts such as SiC, Si_3N_4 and SiCN. No distinct mass change is detected on heating to 2000 °C/He or 1600 °C/air, while SiC, Si_3N_4 and SiCN show large mass losses under the same conditions. Polyborosilazane-derived $Si_{3.0}B_{1.0}C_{4.3}N_{2.0}$ powder can withstand thermal degradation up to 2100 °C/He (0.1 MPa). The slight continuous mass loss of 6.0 wt.% can be attributed to loss of oxygen-containing products like B_2O_3, SiO and CO, consistent with the investigation of the as-milled powders. At temperatures higher than 2100 °C decomposition rates result in mass losses and composition change, while Si_3N_4 and SiCN show severe mass losses at temperatures higher than 1400 °C. At 1700 °C/air, the PDCs-$Si_{3.0}B_{1.2}C_{4.9}N_{3.1}$ and PDCs-$Si_{3.0}B_{1.1}C_{5.3}N_{3.0}$ powders show no mass change and a slight increase, respectively, whereas pure SiC and Si_3N_4 heated at 1000 °C and 1130 °C/air exhibit large mass gains up to 30 wt.%. As noted above, the thin and dense oxide layer greatly retards further oxidation.

Annealing PDCs-SiBCN powder at higher temperatures, as pointed out above, can drive crystallization (Fig. 3.62). Thus, annealing PDCs-$Si_{3.0}B_{1.0}C_{4.3}N_{2.0}$ powder at 1700 °C/50 h/Ar generates SiC, Si_3N_4 and turbostratic BN, while only SiC is detected at 2100 °C [70]. The PDCs-$Si_{3.0}B_{1.1}C_{5.3}N_{3.0}$ powder crystallizes between 1700 and 1750 °C forming SiC and Si_3N_4, in accord with thermodynamic calculations. The intensity of the Si_3N_4 peaks in XRD patterns does not decrease below 1950 °C. SiC can still be detected and its intensities don't decrease due to any decomposition reaction below 2000 °C [71].

As noted above, the as-milled powders exhibit completely amorphous (Si_2BC_3N) or partial amorphous (some Si_2BC_3N-Me) structures with elemental homogenization. Numerous chemical bonds (Si–C, B–N–C, C–N, C–B–N, Al–N, Zr–B, Hf–B, etc.) indicate sufficient mechanochemical effects and "atomic mixing". Crystallization of amorphous MA-Si_2BC_3N powder at high temperatures results in formation

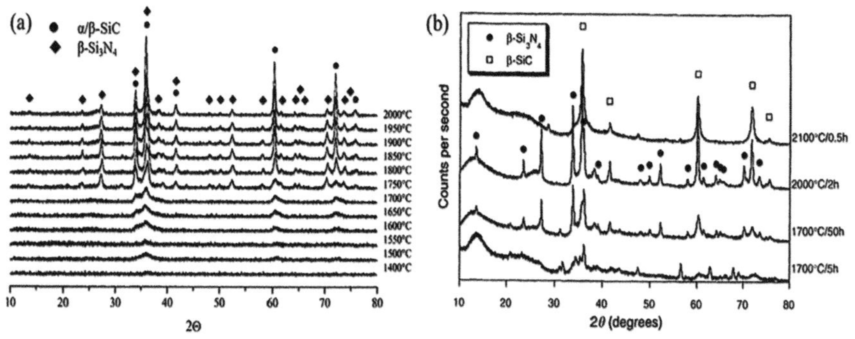

Fig. 3.62 XRD patterns of the PDCs-SiBCN materials: **a** $Si_{3.0}B_{1.0}C_{4.3}N_{2.0}$ annealed at 1400–2000 °C for 3 h under N_2; **b** $Si_{3.0}B_{1.0}C_{4.3}N_{2.0}$ annealed at 1700–2100 °C under Ar [70, 71]

of α/β-SiC and t-BN(C) nanocrystallites. The structural features are very similar to PDCs-SiBCN powder contributing to high-temperature stability. As all roads lead to Rome, these results confirm strongly the scientific rationality of MA technique and its experimental feasibility in developing nonequilibrium or even complete amorphous materials for SiBCN system.

It is well accepted that the extraordinary structural stability of SiBCN at high temperatures can be ascribed to the constraints caused by ternary chemical bonds (like B–N–C or B–C–N) and the BN(C) encapsulation of SiC and/or Si_3N_4 phases. Turbostratic BN can bind free carbon to produce BN(C) with formation of ternary bonds thus decreasing its reactivity and inhibiting the reactions with silicon nitride (At 1484 °C under 1 bar N_2, $Si_3N_4 + 3C \rightarrow 3SiC + 2N_2\uparrow$). Coincidently, turbostratic BN(C) is distributed uniformly around SiC and/or Si_3N_4 phases resulting in a capsule-like structure acting as effective diffusion barriers thus inhibiting atomic diffusion. The encapsulation structure most likely increases significantly internal nitrogen partial pressure thus raising the decomposition temperatures. For instance, the initial temperatures for the reactions $Si_3N_4 + 3C \rightarrow 3SiC + 2N_2\uparrow$ and $Si_3N_4 \rightarrow 3Si + 2N_2\uparrow$ increase from 1484 to 1700 °C and from 1841 to 2034 °C, respectively, with internal nitrogen partial pressure increasing from 1 to 10 bar [71].

However, the initial crystallization temperature for MA-SiBCN powder is somewhat lower than that (1700 °C) of PDCs-SiBCN powder. The Si_3N_4 or SiC_xN_y phases likely form together with SiC and BN(C) phases in the crystalline PDCs-SiBCN powder depending on chemical composition and structures of precursors and pyrolysis process. The difference between the two "types" of powders can be most likely attributed to the strength of chemical bonds in the as-prepared powders. Compared to 3-D networks consisting of strong chemical bonds in PDCs-SiBCN powder, MA most likely results in relatively weak chemical bonds or even dangling bonds at defects within and on the surfaces of amorphous particles, though mechanochemical effects occur during MA, eventually causing relatively lower structural stability at higher temperatures.

Note that the design of chemical composition in MA-SiBCN system is less flexible than that for the precursor route. Numerous problems still exist that need to be addressed in the future development of high-performance MA-SiBCN powder materials. For example, nitrogen cannot be introduced independently during MA. It remains a challenge in developing new, superior MA derived SiBCN ceramics and their matrix composites by inorganic route based on MA and subsequent sintering techniques. Furthermore, attempts should be made to use MA to develop more types of amorphous Si-based materials such as binary (SiC, Si_3N_4), ternary (SiCN, SiBN, SiBC, etc.) and multinary (SiBCN-Me) powder systems.

References

1. Li DX, Yang ZH, Jia DC, Wang SJ, Duan XM, Liang B, Zhu QS, Zhou Y. Structure evolution, amorphization and nucleation studies of carbon-lean to-rich SiBCN powder blends prepared by mechanical alloying. RSC Adv. 2016;6(53):48255–71.
2. Yang ZH, Jia DC, Duan XM, Zhou Y. Microstructure and thermal stabilities in various atmospheres of $SiB_{0.5}C_{1.5}N_{0.5}$ nano-sized powders fabricated by mechanical alloying technique. J Non-Cryst Solids. 2010;356(6):326–33.
3. Apperley DC, Harris RK, Marshall GL, Thompson DP. Nuclear-magnetic-resonance studies of silicon-carbide polytypes. J Am Ceram Soc. 1991;74:777–82.
4. Narisawa M, Iseki T, Katase Y, Okamura K, Oka K, Dohmaru T. Ceramization of reflux-treated polymethylsilane precursors to silicon carbide. J Am Ceram Soc. 2003;86:227–33.
5. Carduner KR, Shinozaki SS, Rokosz MJ, Peters CR, Whalen TJ. Characterization of beta-silicon carbide by Si-29 solid-state NMR, transmission electron-microscopy, and powder X-ray-diffraction. J Am Ceram Soc. 1990;73:2281–6.
6. Zhang YF, Tang YH, Lee CS, Bello I, Lee ST. Nanocrystalline C-BN synthesized by mechanical alloying. Diamond Relat Mater. 1999;8:610–3.
7. Deng ZW, Souda R. XPS studies on silicon carbonitride films prepared by sequential implantation of nitrogen and carbon into silicon. Diamond Relat Mater. 2002;11:1676–82.
8. Xiao X, Li Y, Song L, Peng X, Hu X. Structural analysis and microstructural observation of SiC_xN_y films prepared by reactive sputtering of SiC in N_2 and Ar. Appl Surf Sci. 2000;156:155–60.
9. Meng H, Yu X, Yu J, Wang C. Effect of Dc negative-bias and silicon introduction on performance of Si–B–N composite film by RF-PECD technique. Appl Surf Sci. 2005;245:45–50.
10. Xiong YH, Xiong CS, Wei SQ, Yang HW, Mai YT, Xu W, Yang S, Dai GH, Song SJ, Xiong J, Ren ZM, Zhang J, Pi HL, Xia ZC, Yuan SL. Study on the bonding state for carbon–boron nitrogen with different ball milling time. Appl Surf Sci. 2006;253:2515–21.
11. Maquin B, Derré A, Labrugère C, Trinquecoste M, Chadeyron P, Delhaès P. Submicronic powders containing carbon, boron and nitrogen: their preparation by chemical vapor deposition and their characterization. Carbon. 2000;38:145–56.
12. Schlögl R, Boehm HP. Influence of crystalline perfection and surface species on the X-ray photoelectron spectra of natural and synthetic graphites. Carbon. 1983;21(4):345–58.
13. Jama C, Dessaux O, Goudmand P, Soro JM, Rats D, Stebut JV. Structural and micro-mechanical studies of CN_x films deposited on silicon substrates in a remote nitrogen plasma. Surf Coat Technol. 1999;59:116–9.
14. Oh YS, Cho WS, Kim CS, Lim DS, Cheong DS. XPS investigation of Si_3N_4/SiC nanocomposites prepared using a commercial polymer. J Am Ceram Soc. 1999;82:1076–8.
15. Morjan I, Conde O, Oliveira M, Vasiliu F. Structural characterization of $C_xB_yN_z$ ($x= 0.1$ to $x = 0.2$) layers obtained by laser-driven synthesis. Thin Solid Films. 1999;340:95–105.
16. Zhang GJ, Yang JF, Ohji T. Thermogravimetry, differential thermal analysis, and mass spectrometry study of the silicon nitride–boron carbide–carbon reaction system for the synthesis of silicon carbide–boron nitride composites. J Am Ceram Soc. 2002;85:2256–60.
17. Clegg WJ. Role of carbon in the sintering of boron-doped silicon carbide. J Am Ceram Soc. 2000;83:1039–43.
18. Yang ZH, Jia DC, Zhou Y, Shi PY, Song CB, Lin L. Oxidation resistance of hot-pressed SiC–BN composites. Ceram Int. 2008;34:317–21.
19. Weinmann M, Kamphowe TW, Schuhmacher J, Muller K, Aldinger F. Design of polymeric Si–B–C–N ceramic precursors for application in fiber-reinforced composite materials. Chem Mater. 2000;12:2112–22.
20. Zern A, Mayer J, Janakiraman N, Weinmann M, Bill J. Quantitative EFTEM study of precursor-derived Si–B–C–N ceramics. J Eur Ceram Soc. 2002;22:1621–9.
21. Jansen M, Jüngerman H. A new class of promising ceramics based on amorphous inorganic networks. Curr Opin Solid State Mater Sci. 1997;2:150–7.

22. Butchereit E, Nickel KG, Müller A. Precursor-derived Si–B–C–N ceramics: oxidation kinetics. J Am Ceram Soc. 2001;84:2184–8.
23. Li D, Yang Z, Jia D, Cai DL, Wang SJ, Chen QQ, Zhou Y, Yu DL, Tian YJ. Boron-dependent microstructural evolution, thermal stability, and crystallization of mechanical alloying derived SiBCN. J Am Ceram Soc. 2018;101:1–17.
24. Liu KW, Mucklich F. Thermal stability of nano-RuAl produced by mechanical alloying. Acta Mater. 2001;49:395–403.
25. Tavakoli AH, Gerstel P, Golczewski JA, Bill J. Kinetic effect of boron on the thermal stability of Si–(B–)C–N polymer-derived ceramics. Acta Mater. 2010;58:6002–11.
26. Shaw LL, Yang ZG, Ren RM. Synthesis of nanostructured Si_3N_4/SiC composite powders through high energy reaction milling. Mater Sci Eng A. 1998;244:113–26.
27. Yang ZG, Shaw LL. Synthesis of nanocrystalline SiC at ambient temperature through high energy reaction milling. Nanostruct Mater. 1996;7:873–86.
28. Schmidt H, Borchardt G, Weber S, Scherrer H, Baumann H, Muller A, Bill J. Comparison of ^{30}Si diffusion in amorphous Si–C–N and Si–B–C–N precursor-derived ceramics. J Non-Cryst Solids. 2002;298:232–40.
29. Mueller A, Gerstel P, Weinmann M, Bill J, Aldinger F. Correlation of boron content and high temperature stability in Si–B–C–N ceramics. J Eur Ceram Soc. 2000;20:2655–9.
30. Kumar R, Cai Y, Gerstel P, Rixecker G, Aldinger F. Processing, crystallization and characterization of polymer derived nano-crystalline Si–B–C–N ceramics. J Mater Sci. 2006;41:7088–95.
31. Zhang PF, Jia DC, Yang ZH, Duan XM, Zhou Y. Progress of a novel non-oxide Si–B–C–N ceramic and its matrix composites. J Adv Ceram. 2012;1(3):157–78.
32. Li DX, Yang ZH, Jia DC, Hu CC, Liang B, Zhou Y. Preparation, microstructures, mechanical properties and oxidation resistance of SiBCN/ZrB_2-ZrN ceramics by reactive hot pressing. J Eur Ceram Soc. 2015;35(16):4399–410.
33. Lu FG, Qiu LX, Ding ZH, Guo XY, Lu Y, Zhao XD. High-pressure synthesis and properties characterization of ZrN–ZrB_2 nanocomposites. Chin J High Pressure Phys. 2011;25:104–10.
34. Fahrenholtz WG, Hilmas GE, Talmy IG, Zaykoski JA. Refractory diborides of zirconium and hafnium. J Am Ceram Soc. 2007;90:1347–64.
35. Agaogullari D, Gokce H, Duman I, Ovecoglu ML. Characterization investigations of ZrB_2/ZrC ceramic powders synthesized by mechanical alloying of elemental Zr, B and C blends. J Eur Ceram Soc. 2011;32:1447–55.
36. Tsuchida T, Kawaguchi M, Kodaira K. Synthesis of ZrC and ZrN in air from mechanically activated Zr–C powder mixtures. Solid State Ionics. 1997;101:149–54.
37. Ye D, Jia DC, Yang ZH, Sun ZL, Zhang PF. Microstructure and thermal stability of amorphous SiBCNAl powders fabricated by mechanical alloying. J Alloys Comp. 2010;506(1):88–92.
38. Besling WFA, Goossens A, Meester B, Schoonman J. Laser-induced chemical vapor deposition of nanostructured silicon carbonitride thin films. J Appl Phys. 1998;83:544–53.
39. Malengreau F, Hautier V, Vermeersch M, Sporken R, Caudano R. Chemical interactions at the interface between aluminum nitride and iron-oxide determined by XPS. Surf Sci. 1995;330:75–85.
40. Do T, McIntyre NS, Harshman RA, Lundy ME, Splinter SJ. Application of parallel factor analysis and X-ray photoelectron spectroscopy to the initial stages in oxidation of aluminium. I. The Al 2p photoelectron line. Surf Interf Anal. 1999;27:618–28.
41. Xie XQ, Yang ZG, Ren R, Shaw LL. Solid state Si-29 magic angle spinning NMR: investigation of bond formation and crystallinity of silicon and graphite powder mixtures during high energy milling. Mater Sci Eng A Struct Mater Propert Microstruct Process. 1998;255:39–48.
42. Berger F, Weinmann M, Aldinger F, Müller K. Solid-State NMR Studies of the preparation of Si–Al–C–N ceramics from aluminum-modified polysilazanes and polysilylcarbodiimides. Chem Mater. 2004;16:919–29.
43. Toyoda R, Kitaoka S, Sugahara Y. Modification of perhydropolysilazane with aluminum hydride: preparation of poly(aluminasilazane)s and their conversion into Si–Al–N–C ceramics. J Eur Ceram Soc. 2008;28:271–7.

44. Li XD, Edirisinghe MJ. A new aluminum coordination site in Si–C–Al–N–(O) ceramics. J Am Ceram Soc. 2003;86:2212–4.
45. Ye D, Jia DC, Yang ZH, Duan XM, Zhou Y. Structural and microstructural characterization of $SiB_{0.5}C_{1.5}N_{0.8}Al_{0.3}$ powders prepared by mechanical alloying using aluminum nitride as aluminum source. Ceram Int. 2011;37:2937–40.
46. Liao X Q. Microstructure and properties of the (TiB_2+TiC)/SiBCN composites. Harbin: Harbin Institute of Technology, 2014.
47. Miao Y. Preparation, oxidation resistance and ablation mechanisms of ZrB_2/SiBCN composites. Harbin: Harbin Institute of Technology, 2017.
48. Ao D F. The thermal shock and ablative properties of HfB_2/SiBCN multiphase ceramic. Harbin: Harbin Institute of Technology, 2018.
49. Zhao Y. Microstructure and properties of the ZrC/SiBCN multiphase ceramics by hot pressing. Harbin: Harbin Institute of Technology, 2016.
50. Wang BZ, Yang ZH, Li DX, Cai DL, Niu B, Guan JY, Jia DC, Zhou Y. Effects of TaC addition on microstructure and mechanical properties of SiBCN composite ceramics. Ceram Int. 2019;45:22138–47.
51. Wang BZ, Li DX, Yang ZH, Jia DC, Guan JY, Peng H, Cai DL, He PG, Duan XM, Zhou Y, Zhang T, Gao CG. Microstructural evolution and mechanical properties of in situ nano Ta_4HfC_5 reinforced SiBCN composite ceramics. J Adv Ceram. 2020;9(6):739–48.
52. Chen HT, Tang B, Zhong CW, Yuan Y, Tan YD, Zhang SR. The dielectric constant and quality factor calculation of the microwave dielectric ceramic solid solutions. Ceram Int. 2017;43:7383–6.
53. Zaara K, Chemingui M, Optasanu V, Khitouni M. Solid solution evolution during mechanical alloying in Cu–Nb–Al compounds. Int J Miners Metall Mater. 2019;26:1129–39.
54. Peng H, Li DX, Yang ZH, Duan WJ, Jia DC, Zhou Y. Effects of Ti on the microstructural evolution and mechanical property of the SiBCN–Ti composite ceramics. Materials. 2023;16:3560.
55. Liao N, Jia DC, Yang ZH, Zhou Y, Li YW. Strengthening and toughening effects of MWCNTs on Si_2BC_3N ceramics sintered by SPS technique. Mater Sci Eng A. 2018;725:364–74.
56. Liao N, Jia DC, Yang ZH, Zhou Y. Enhanced mechanical properties and thermal shock resistance of Si_2BC_3N ceramics with SiC coated MWCNTs. J Adv Ceram. 2019;8(1):121–32.
57. Li D X. Spark plasma sintering graphene/SiBCN ceramic and high temperature properties. Harbin: Harbin Institute of Technology, 2014.
58. Gao Y, Mera G, Nguyen H, Morita K, Kleebe H-J, Riedel R. Processing route dramatically influencing the nanostructure of carbon-rich SiCN and SiBCN polymer-derived ceramics. Part I: low temperature thermal transformation. J Eur Ceram Soc. 2012;32(9):1857–66.
59. Muller A, Zern A, Gerstel P, Bill J, Aldinger F. Boron-modified poly(propenylsilazane)-derived Si–B–C–N ceramics: preparation and high temperature properties. J Eur Ceram Soc. 2002;22:1631–43.
60. Franke R, Bender S, Jungermann H, Kroschel M, Jansen M. The determination of structural units in amorphous Si–B–N–C ceramics by means of Si, B, N and C K-XANES spectroscopy. J Electr Spectr Relat Phenom. 1999;101–103:641–5.
61. Jeschke G, Kroschel M, Jansen M. A magnetic resonance study on the structure of amorphous networks in the Si–B–N(–C) system. J Non-Crystall Solids. 1999;260:216–27.
62. Heinemann D, Assenmacher W, Mader W, Kroschel M, Jansen M. Structural characterization of amorphous ceramics in the system Si–B–N–(C) by means of transmission electron microscopy methods. J Mater Res. 1999;14:3746–53.
63. Bill J, Aldinger F. Precursor-derived covalent ceramics. Adv Mater. 1995;7:775–87.
64. Tavakoli AH, Gerstel P, Golczewski JA, Bill J. Kinetic effect of boron on the crystallization of Si_3N_4 in Si–B–C–N polymer-derived ceramics. J Mater Res. 2011;26:600–8.
65. Sarkar S, Gan Z, An L, Zhai L. Structural evolution of polymer-derived amorphous SiBCN ceramics at high temperature. J Phys Chem C. 2011;115:24993–5000.
66. Kroll P. Modelling polymer-derived ceramics. J Eur Ceram Soc. 2005;25:163–74.

67. Schuhmacher J, Berger F, Weinmann M, Bill J, Aldinger F, Muller K. Solid-state NMR and FT IR studies of the preparation of Si–B–C–N ceramics from boron-modified polysilazanes. Appl Organometall Chem. 2001;15:809–19.
68. Haug J, Lamparter P, Weinmann M, Aldinger F. Diffraction study on the atomic structure and phase separation of amorphous ceramics in the Si–(B)–C–N system. 1. Si–C–N ceramics. Chem Mater. 2004;16:72–82.
69. Haug J, Lamparter P, Weinmann M, Aldinger F. Diffraction study on the atomic structure and phase separation of amorphous ceramics in the Si–(B)–C–N system. 2. Si–C–N ceramics. Chem Mater. 2004;16:83–92.
70. Riedel R, Kienzle A, Dressler W, Ruwisch L, Bill J, Aldinger F. A silicoboron carbonitride ceramic stable to 2000 °C. Nature. 1996;382:796–8.
71. Weinmann M, Schuhmacher J, Kummer H, Prinz S, Peng JQ, Seifert HJ, Christ M, Muller K, Bill J, Aldinger F. Synthesis and thermal behavior of novel Si–B–C–N ceramic precursors. Chem Mater. 2000;12:623–32.

Chapter 4
Densification, Microstructure and Crystallization Mechanisms of MA-SiBCN Bulk Materials

Sintering, the oldest but probably an everlasting technique to produce density and microstructure-controlled materials and components from metallic and/or ceramic powders by applying, in general, thermal energy. The MA-SiBCN powder is difficult to be sintered due to the low coefficients of atomic diffusion. Consequently, their full densification is usually achieved by adding aids and/or sintering using high mechanical pressure and high temperature. Sintering of MA-SiBCN generally occurs when packed amorphous nanoparticles are heated to a temperature where there is sufficient atomic motion to grow bonding particles. Pores, particle sizes, second phases and a host of processing variables have all an important impact on many of the final properties of MA-SiBCN bulk materials.

The densification of MA-SiBCN ceramics and their CMCs free of any additives is most often achieved by hot-pressing (HP), spark plasma sintering (SPS), hot isostatic pressing (HIP) or high-pressure sintering (HPS). At temperatures sufficiently high, the crystallization of amorphous MA-SiBCN cannot be avoided; meanwhile, ceramic densification occurs during sintering. Material densification leads to a high packing density of particles beneficial to the mass and energy transfer thus promoting crystallization. Crystallization induced elemental segregation, phase separation, nucleation and grain growth hold the key to densification behavior. Microstructure coarsening is also inherent to the above sintering process, most evident as grain growth of SiC and BN(C), but it is common for porosity reduction to occur as density increases. During coarsening, the SiC and BN(C) grain structure converges to a self-similar character witnessed in both the grain shape distribution and grain size distribution.

D. Li et al., *SiBCN Ceramics and Composites Prepared by Mechanical Alloying*,
https://doi.org/10.1007/978-981-97-3029-2_4

4.1 Densification and Microstructure of Nanocrystalline MA-SiBCN Monoliths

4.1.1 Densification Behavior for Hot-Pressed Monoliths

Careful investigation reveals that the volume shrinkage of the hot-pressed MA-Si_2BC_3N powder includes three stages, suggesting different mechanisms for densification [1, 2]. When temperature is lower than ~1450 °C, the pressing head displacement increases rapidly with a great shrinkage occurring to the ceramic body. Since the loading of the axial pressure starts at 1200 °C and finishes at about 1400–1500 °C, the volume shrinkage in current stage is mainly ascribe to the closer packing of ceramic particles resulting from the applied axial pressure (Fig. 4.1).

Between 1450 and 1800 °C, the pressing head shows a slightly decreased displacement with the increase of sintering temperature, most likely caused by the thermal expansion of graphite mould, pressure head and ceramic body. Note that the shrinkage of the ceramic body is negligible, thus sintering is probably in its infancy. In this stage, atomic diffusion coupled with metallurgical bonding are limited in some narrow regions at grain boundaries. Consequently, very small sintering necks form among neighboring particles, and little has changed to the outline of ceramic particles.

Once sintering temperature exceeds ~1830 °C, the pressing head displacement increases rapidly, and the shrinkage rate of the ceramic body researches its maximum in a short time, giving $d(\Delta L/L_0)/dt \approx 0.03$. In this stage, large volume shrinkage and hence significant densification occurs to the ceramic body, and sintering is most likely in its intermediate stage. The axial pressure-assistant high-temperature gives enough energy for the working of most sintering mechanisms, such as atomic diffusion on particle surface or within particles, sliding, creeping and deformation of ceramic particles, et al.

MA-Si_2BC_3N ceramics show good amorphous nature after annealed at 1800 °C. Further annealing at 1900 °C, obvious diffraction peaks of α/β-SiC appear but

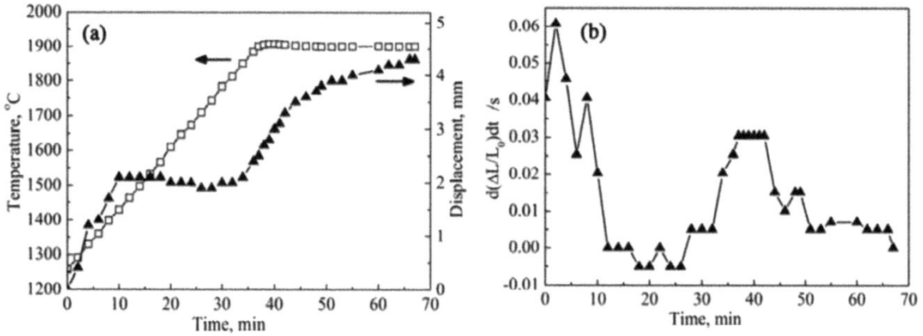

Fig. 4.1 The sintering curves of the MA-Si_2BC_3N ceramics prepared by hot-pressing sintered at 1900 °C/80 MPa/30 min/1 bar N_2: **a** Surface temperature of the die versus indenter displacement; **b** indenter displacement rate versus sintering time [1]

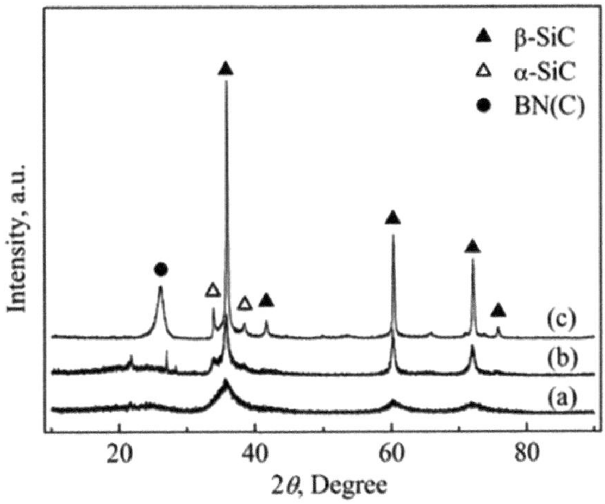

Fig. 4.2 XRD patterns of the hot-pressed MA-Si_2BC_3N ceramics (80 MPa): Sintering at **a** 1800 °C for 0 min; **b** 1900 °C for 0 min; **c** 1900 °C for 10 min [1]

still reserve some amorphous phases; however, the peak intensities for α/β-SiC and BN(C) both increase significantly when hot-pressed at 1900 °C dwelling for 10 min (Fig. 4.2).

MA-Si_2BC_3N ceramics begin to crystallize at 1500 °C under HP 80 MPa for 30 min (Fig. 4.3). From TEM bright-field (BF) image, the change in the contrast is not significant, implying the existence of amorphous bodies or the extremely small β-SiC nanocrystallites, without evidence of BN(C). The diffraction rings in SAED pattern are smooth and continuous with stable brightness. The measured interplanar spacing $d = 0.2548$ nm, $d = 0.1541$ nm, and $d = 0.1319$ nm, are in agreement with that of the (111), (220), and (311) planes for β-SiC, respectively. From HRTEM and inverse FFT images, an amorphous body with indeed embedded β-SiC nanocrystallites ~2.6 nm is observed.

Further hot-pressed at 1600 °C, MA-Si_2BC_3N is composed of SiC nanocrystallites, BN(C) nuclei, and considerable proportion of amorphous phase (Fig. 4.4). The bright field image shows that thc SiC crystallites have grown up with an average size of ~14.9 nm. SAED pattern displays three continuous diffraction rings for the (111), (220) and (311) planes in β-SiC. HRTEM and inverse FFT images suggest the SiC nanocrystallites consisted of several atomic planes, and BN(C) has a primary turbostratic structure. The SiC nanocrystallites are not perfect, and contain many defects or disordered atoms. The atomic planes in BN(C) are squiggly, and the size of the crystal nucleus is about several nanometers. Besides the crystallites, amorphous body still takes a significant proportion in the ceramics. The crystal planes in the two adjacent SiC nanocrystallites have the same orientation, and the disordered atoms show the signs to join them as one grain. Namely, the SiC crystallites may grow up via the integration of the adjacent crystals.

HP sintered at 1700 °C, the crystal grains show larger sizes and the amorphous body accounts for a smaller proportion (Fig. 4.5). The bright-field image and SAED pattern both suggest those larger grains are BN(C) and SiC ~26.4 ± 8.0 nm. Besides,

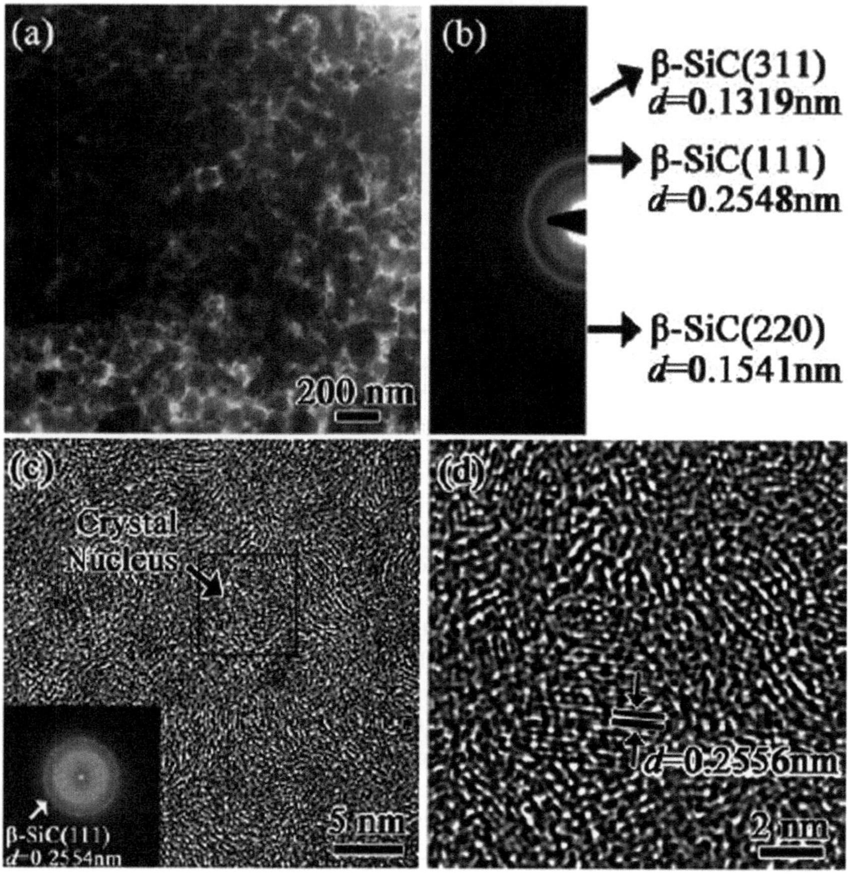

Fig. 4.3 Microstructure of the MA-Si_2BC_3N ceramics hot-pressed at 1500 °C/80 MPa/30 min: **a** Bright-field image; **b** SAED pattern; **c** HRTEM and FFT images; **d** inverse FFT image of the area marked by the black box shown in (**c**) [2]

SAED pattern also shows a continuous diffraction ring with the interplanar spacing of about $d = 0.3517$ nm, which approaches to that of the (0002) planes in graphite or h-BN and is assigned to the graphite-like BN(C). Again, SiC grains show an imperfect atomic arrangement, and contain lots of stacking faults in the crystal region and an amorphous area in the central part. The amorphous region contains only Si and C atoms.

SiC grains may develop in situ from the amorphous Si–C zones. In boundary regions, the crystal planes grow along the [111] direction and toward the amorphous Si–C zone. During the growing process, stacking faults or twins usually occur in the SiC grains. The turbostratic BN(C) is characterized by the layer structure similar to graphite or h-BN, but most of the layers rotated by random angles around the *c*-axis

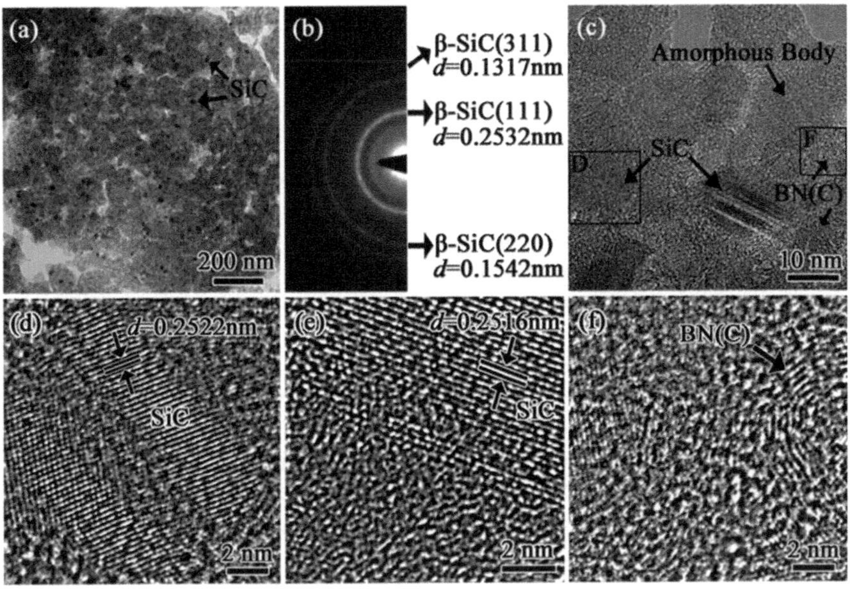

Fig. 4.4 Microstructure of MA-Si_2BC_3N ceramics prepared by hot-pressing sintering at 1600 °C/ 80 MPa/30 min: **a** Bright field image; **b** SAED pattern; **c** HRTEM image; **d–f** corresponding inverse FFT images of the marked areas shown in (**c**) [2]

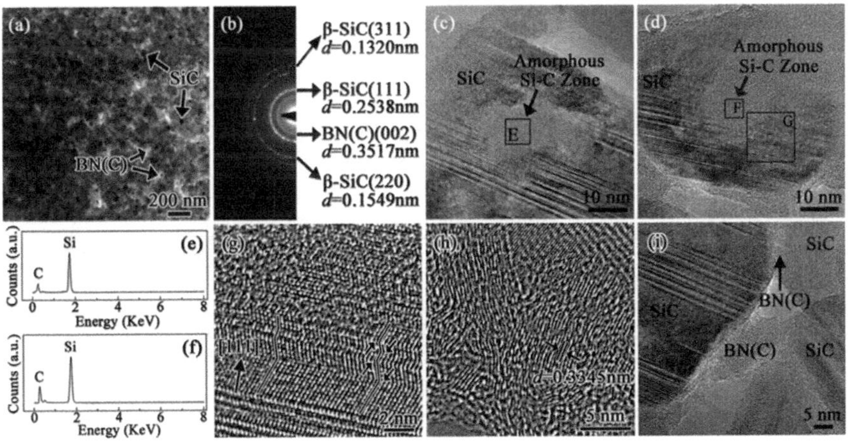

Fig. 4.5 Microstructure of MA-Si_2BC_3N hot-pressed at 1700 °C: **a** Bright-field image; **b** SAED pattern; **c**, **d** HRTEM images; **e**, **f** EDX spectra of the marked areas E and F shown in (**c**) and (**d**), respectively; **g** inverse FFT image of the marked area G shown in (**d**); **h** inverse FFT image of another area showing the turbostratic structure of BN(C); **i** interfacial structure between SiC and BN(C) [2]

or the axis perpendicular to *c*. The SiC grains are separated by the turbostratic BN(C), possible responsible for retarding crystallization and grain growth of SiC.

For the ceramics hot-pressed at 1800 °C, the grains grow faster with high degree of crystallinity (Fig. 4.6). SiC grains have an average size of ~42.4 ± 13.1 nm. The SAED pattern shows the bright but discontinuous diffraction rings for SiC, providing further evidence for larger grains and higher crystallinity. However, the diffraction ring assigned to the (0002) planes of BN(C) is still continuous with a stable brightness, indicating a rather small grain size or a low crystallinity of BN(C). HRTEM images display the co-existence of the SiC nanocrystals, turbostratic BN(C) and amorphous phases. Atomic arrangement in the SiC grains is still not perfect with some structural defects.

Further HP sintered at 1900 °C, the SiC grains have an average grain size of about 79.3 ± 13.6 nm, while BN(C) with small size show no fixed shape (Fig. 4.7). The diffraction rings in SAED pattern assigned to SiC are discontinuous implying the larger grain sizes. The ring assigned to BN(C) is still continuous but has varied brightness, thus turbostratic BN(C) may behaver as oriented growth. Probably, most of the (0002) planes in the turbostratic BN(C) align approximately perpendicular to the loading direction. SiC grains still contain a large amount of stacking faults and the boundary regions are clean and limited to a narrow range. SiC grains are surrounded and separated by turbostratic BN(C). Adjacent to the crystal SiC and BN(C) are those of amorphous phases.

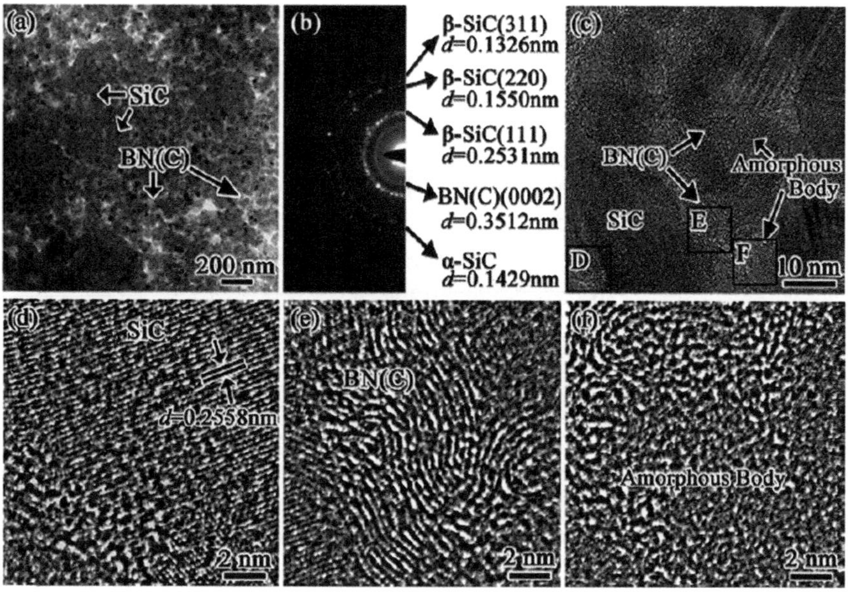

Fig. 4.6 Microstructure of MA-Si_2BC_3N hot-pressed at 1800 °C: **a** Bright-field image; **b** SAED pattern; **c** HRTEM image; **d–f** inverse FFT images of the marked areas D, E, and F shown in (**c**), respectively [2]

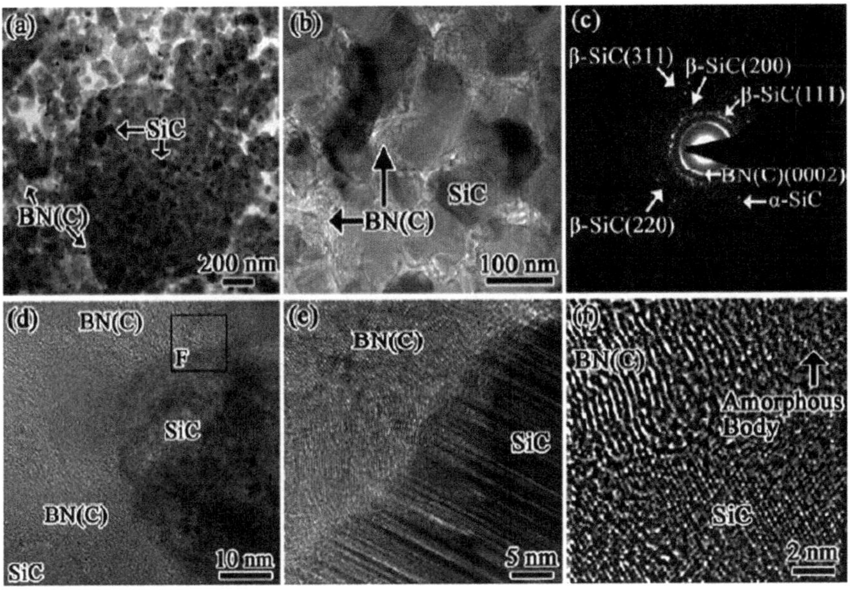

Fig. 4.7 Microstructure of MA-Si_2BC_3N ceramics hot-pressed at 1900 °C. **a, b** Bright-field images; **c** SAED pattern; **d, e** HRTEM images; **f** inverse FFT image of the marked area F shown in (**e**) [2]

Differing from the large quantity of β-SiC and BN(C), α-SiC always has a low content in the as-prepared ceramics. Actually, the α-SiC peaks in the XRD spectra approximately overlap with the beta ones and so do the alpha rings in the SAED patterns. The broad ring, with the interplanar spacing $d \approx 0.25$ nm, results from the diffraction of the low index planes from both α/β-SiC. Specifically, this broad ring corresponds to the interplanar spacings located between $d = 0.240$ nm and $d = 0.265$ nm, covering $d(111)$ in β-SiC; $d(100)$, $d(101)$, and $d(002)$ in 2H-SiC; $d(100)$, $d(101)$, d (102), and $d(004)$ in 4H-SiC; $d(101)$, $d(102)$, and $d(103)$ in α-SiC, etc. In addition, the diffraction ring, with the interplanar spacing $d \approx 0.1540$–0.1550 nm, coincides with that of the (220) planes in β-SiC and the (110) planes in hexagonal α-SiC. The diffraction ring, with $d \approx 0.1317$–0.1326 nm, covers that of the (311) plans in β-SiC, the (112) planes in 2H-SiC, the (114) planes in 4H-SiC, and the (116) planes in α-SiC, etc. Besides the overlapped diffraction rings, other weak signals also appear in SAED pattern for the existence of α-SiC, as shown in Figs. 4.6 and 4.7. Owing to the low content of α-SiC, the bright diffraction rings in the SAED pattern are only labeled as β-SiC.

Coincidentally, SiC grains show an approximately linear growth when sintering temperature is lower than or equal to 1800 °C (Fig. 4.8). The data are fitted with the linear function $S = 0.1309t - 194.41$, where S stands for the SiC grain size (nm) and t is the sintering temperature (°C). When $S = 0$, the temperature is calculated as $t_0 = 1485$ °C. It appears that, if the HP sintering temperature is lower than 1485 °C, there may be no crystals in the ceramic matrix, and a well-defined amorphous

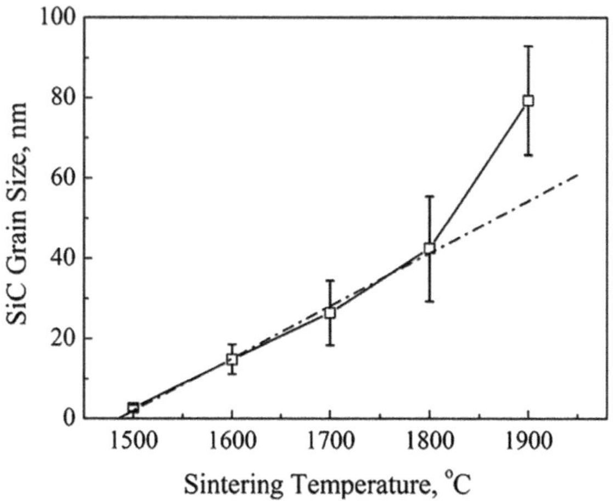

Fig. 4.8 SiC grain size as a function of HP sintering temperature of the MA-Si_2BC_3N ceramics (Other sintering conditions: 80 MPa/30 min) [2]

structure can be retained. As the time factor and the nucleation fundamentals have not been taken into consideration, the inference may be correct only in the particular case, in which the annealing time at the low temperatures is not long enough for overcoming the crystallization kinetic constraints. Once the MA-Si_2BC_3N ceramics are HP sintered at 1900 °C, the grain size increases sharply implying a rapid grain coarsening or ceramic densification.

4.1.2 Microstructure of Spark Plasma Sintered Monoliths

The spark plasma sintering (SPS) technique, also referred to as field-assisted sintering technique [4], or pulsed electric current sintering is a very effective way to consolidate metastable MA-SiBCN materials that resist conventional HP. Heat is produced mainly by joule heating at the contact points of the particles. In addition, the direct current promotes surface cleaning and enhances solid-state sintering compared to conventional sintering techniques [5]. SPS permits very rapid heating and cooling rates thus reducing sintering times, lowering sintering temperatures and leading to fine-grained microstructure, accelerated densification as well as improved mechanical properties. Coincidently, crystallization and excessive grain growth are inhibited due to the apparent low sintering temperatures and short dwell times compared to conventional heating, essential to sinter partial amorphous or nanostructured MA-SiBCN materials.

After SPS at 1800 °C under 25 or 40 MPa [6], the amorphous MA-Si_2BC_3N powder crystallizes to β-SiC and BN(C), essentially devoid of the α-SiC. A trace of ZrO_2 contamination is detected, due to the wear of the milling vessel and balls (Fig. 4.9). The diffraction peak intensities for β-SiC and BN(C) of the 40 MPa sintered

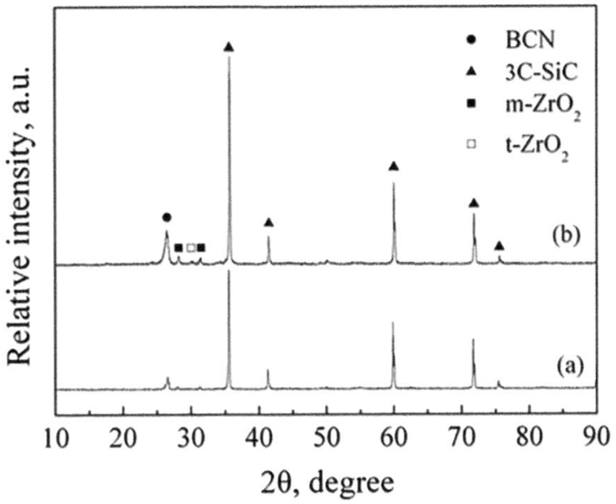

Fig. 4.9 XRD patterns of MA-Si_2BC_3N ceramics prepared by SPS at 1800 °C under different pressure: **a** 25 MPa; **b** 40 MPa [6]

ceramics, are higher than that of the ceramics sintered at 25 MPa. This is due to the increasing pressure promotes the crystallization of the amorphous powder.

For ceramics with different Si/C ratio (MA-$SiBC_2N$, MA-Si_2BC_3N and MA-Si_3BC_4N) densified at 1800 °C/40 MPa/5 min [7], in all cases, they all show sharp diffraction peaks of β-SiC, BN(C) and m/t-ZrO_2, suggesting sufficient amorphous-to-crystalline transformation in spite of the short sintering duration of SPS process (Fig. 4.10). Rough surface is observed for the MA-Si_2BC_3N ceramics SPS consolidated under a pressure of 25 MPa. The grain boundaries between SiC particles (light phase) and BN(C) particles cannot be clearly distinguished resulting from the insufficient crystallization. Further sintered under a higher pressure of 40 MPa, finer and homogeneous microstructure is achieved (Fig. 4.11). The average SiC grain sizes for MA-$SiBC_2N$, MA-Si_2BC_3N and MA-Si_3BC_4N ceramics are similar with all grains below 1 μm. Microstructure of the as-sintered three bulk ceramics is similar, and nanosized BN(C) is mainly distributed around SiC grains (Fig. 4.12).

After SPS, a partial aggregation of nanocrystals appears in MA-$SiBC_2N$ ceramic matrix (Fig. 4.13). A small number of BN(C) crystals (black area) distribute in this aggregation area with a grain size of approximately 100 nm [8]. Actually, this aggregation area also contains few β-SiC ~400 nm, which is in a near-equated shape. Some structural defects such as stacking faults and twinning, are observed inside the SiC grains. The BN(C) is mostly scattered around the SiC grains. The microstructure of SPS MA-Si_2BC_3N ceramics also presents phase composition of BN(C) and SiC (Fig. 4.14). The grain size of SiC is approximately 200–800 nm, whereas BN(C) grain size is much smaller, with most grains being less than 100 nm. The aggregation of nanocrystals is also found.

Interestingly, the amorphous SiC region with an area approximately ~10 nm^2 and amorphous aggregation areas > 100 nm are distributed in the MA-Si_2BC_3N ceramics (Fig. 4.15). The BN(C) also contains considerable layer defects; besides, some BN(C) is severely warped and deformed. The amorphous areas in SPS densified

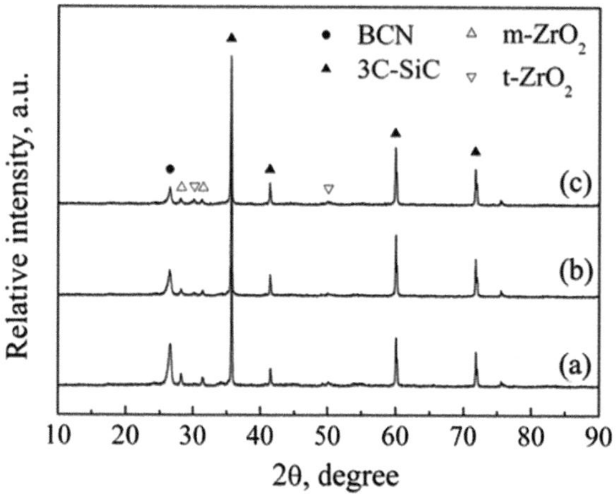

Fig. 4.10 XRD patterns of the as-sintered ceramics with different Si/C ratio prepared by SPS at 1800 °C/40 MPa/5 min: **a** MA-$SiBC_2N$; **b** MA-Si_2BC_3N; **c** MA-Si_3BC_4N [7]

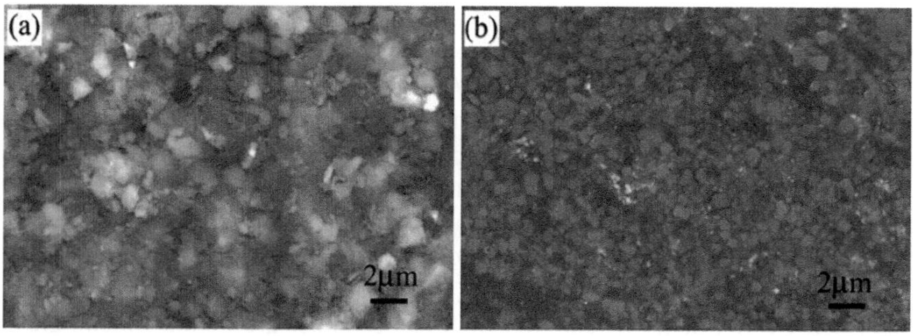

Fig. 4.11 SEM surface morphologies of the MA-Si_2BC_3N ceramics prepared by SPS at 1800 °C under different pressure: **a** 25 MPa; **b** 40 MPa [6]

MA-Si_2BC_3N are lagger than that of HP counterparts. Additionally, more crystal defects are presented in SiC grains and the (0002) planes of the BN(C) are twisted to a higher extent. The remaining considerable amorphous phases may be inherited from the as-milled amorphous powder due to the rapid sintering. Compared with the HP counterparts, the SPS densified ceramics show a more uniform element distribution in amorphous and nanocrystalline aggregation areas.

The SPS densified MA-Si_3BC_4N bulk ceramics show similar microstructure as compared with the formers (Fig. 4.16). Some agglomerating areas with fine grains are also found. In nanocrystals aggregation areas, the BN(C) and SiC phases show larger sizes and SiC content increases with increasing Si/C ratio.

Fig. 4.12 SEM surface morphologies of as-sintered ceramics with different Si/C ratio prepared by SPS at 1800 °C/40 MPa/5 min: **a** MA-$SiBC_2N$; **b** MA-Si_2BC_3N; **c** MA-Si_3BC_4N [7]

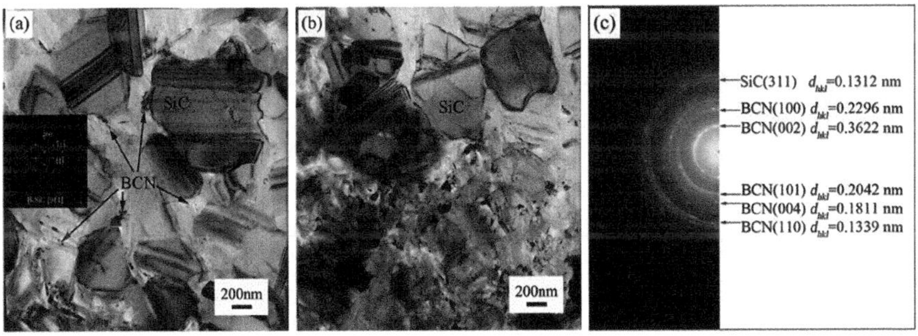

Fig. 4.13 TEM microstructure of MA-$SiBC_2N$ bulk ceramics prepared by SPS at 1800 °C/40 MPa/5 min: **a**, **b** Bright-field images; **c** SAED pattern [8]

4.1.3 Pressure-Dependent Densification of Hot-Isostatic Pressed Monoliths

The hot-isostatic pressing (HIP) technique [9] is distinguished from conventional unidirectional pressing such as HP and SPS. An inert gas acts as the pressure transmission medium as the powder is consolidated to develop a monolith by HIP. The gas pressure is achieved partly by a mechanical compressor and partly by heating the

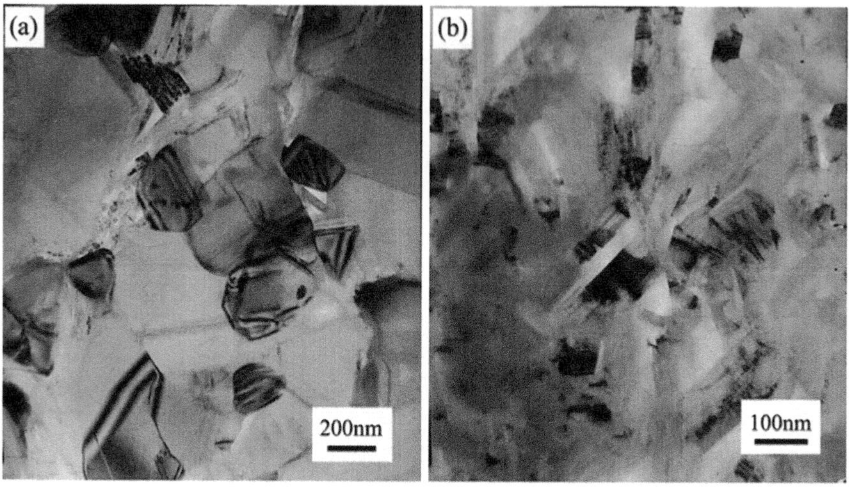

Fig. 4.14 TEM bright field images of MA-Si_2BC_3N bulk ceramics prepared by SPS at 1800 °C/40 MPa/5 min: **a** Low magnification; **b** high magnification [6]

gas in a closed space. Consequently, the pressure is the same and acts in all directions perpendicular to the object surfaces.

HIP cannot be used directly to consolidate MA-SiBCN powder. The powder must first be placed into an envelope with the same shape as predesigned. The envelope must be made from a material (such as metal, alloy or glass) that can deform in a plastic manner at temperatures and pressures high enough during HIP. As the densification occurs, both the mold and the powder compact shrink to the same extent in all direction due to the isostatic pressure. Additionally, HIP is also used as an effective post treatment to eliminate residual closed porosity in a cold pressed, sintered, or cast parts [10].

The MA-Si_2BC_3N ceramics consolidated by HIP sintering, also show the same phase composition of α/β-SiC and BN(C), but present a higher degree of crystallinity (Fig. 4.17). Especially, increasing sintering pressure is significantly beneficial to the increasing crystallinity of BN(C). Compared with the MA-Si_2BC_3N bulk ceramics hot-pressed at 1900 °C/80 MPa/30 min, the density of the HIP bulk ceramics increases rapidly from 88.7 to 97.2% showing pressure-dependent densification.

For MA-Si_2BC_3N ceramics hot-pressed at 1700–1800 °C under 40 MPa, the ceramic particles have a near-spherical outline with the sizes similar to that of the primary particles in the as-milled amorphous powder. Especially HP sintered at 1700 °C, the trace of sintering among the adjacent particles is inapparent. Further HP sintered at 1800 °C, most ceramic particles are sintered together but still remain some micropores. For the HIP sintered MA-Si_2BC_3N bulk ceramics, the interparticle neck formation and growth may dominate the sintering process. The effective sintering of the HIP sintered ceramics may be responsible for their higher relative densities and well-developed microstructure (Fig. 4.18).

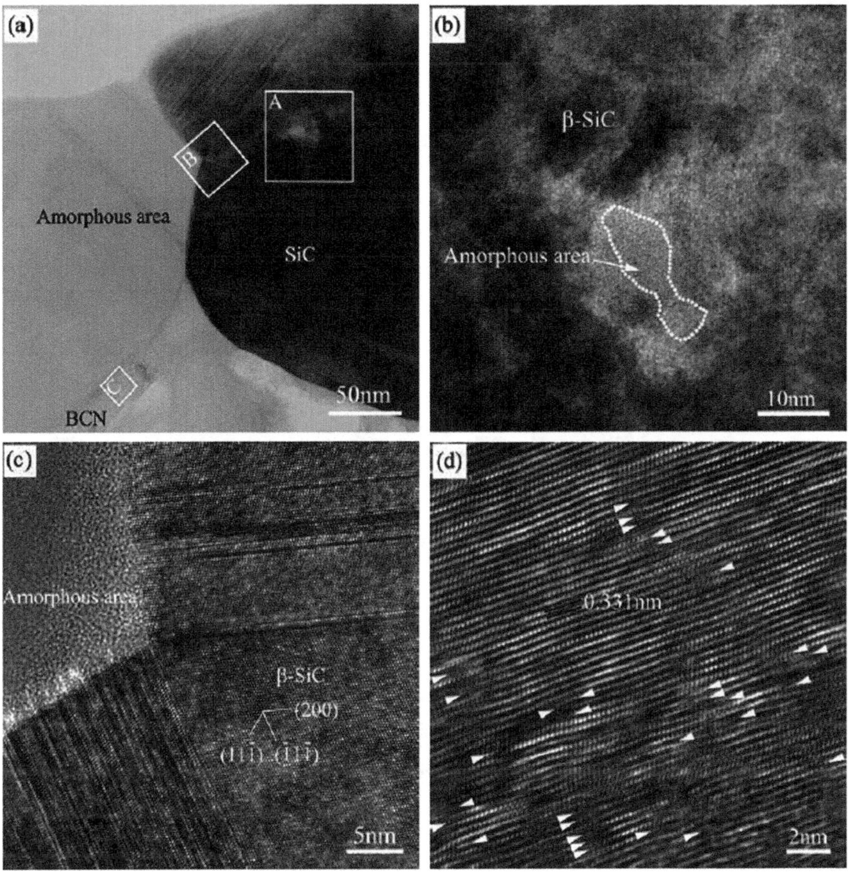

Fig. 4.15 The detailed microstructure of SiC, BN(C) and amorphous phase in MA-Si_2BC_3N ceramics prepared by SPS at 1800 °C/40 MPa/5 min: **a** Bright-field image; **b**–**d** HRTEM images corresponding to the A–C regions [8]

4.2 Crystallization Thermodynamics and Kinetics of Amorphous MA-SiBCN Monoliths

High pressure sintering (HPS) [11, 12] (here defined as $\geq$ 1 GPa) can induce microstructural changes at atomic levels changing physical and chemical properties, increasing reaction rates thus reducing the reaction times and temperatures, and improving bulk densities superior to those achieved by conventional sintering. It can also deform materials leading to changes in surface states and interfacial structures, the degree of crystallinity and phase transition. Consequently, unusual physical chemistry phenomena occur that induce formation of novel SiBCN materials with special microstructures and properties. High pressure sintering provides fully dense MA-Si_2BC_3N materials even at relatively low temperatures. In most instances, sintering

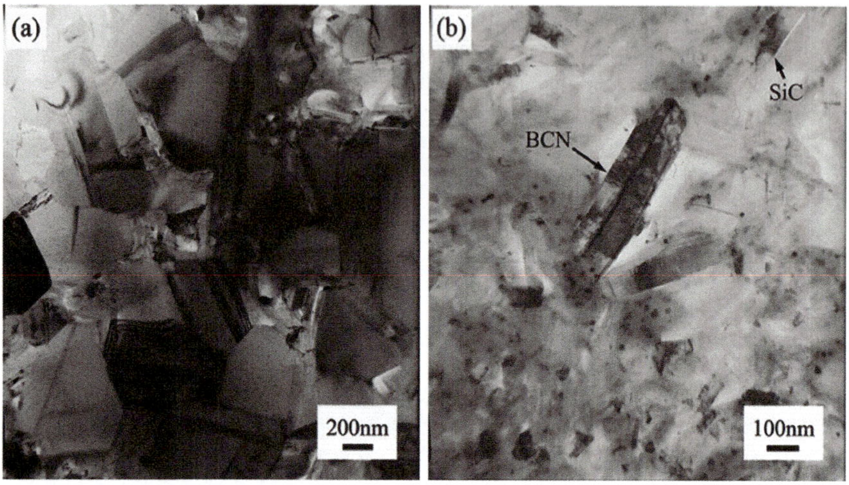

Fig. 4.16 TEM bright-field images of MA-Si_3BC_4N ceramics prepared by SPS at 1800 °C/40 MPa/5 min: **a** Low magnification; **b** high magnification [7]

Fig. 4.17 XRD patterns of MA-Si_2BC_3N bulk ceramics prepared by HIP or HP techniques, respectively: ① HP 1700 °C/40 MPa; ② HP 1800 °C/40 MPa; ③ HIP 1700 °C/190 MPa; ④ HIP 1800 °C/190 MPa

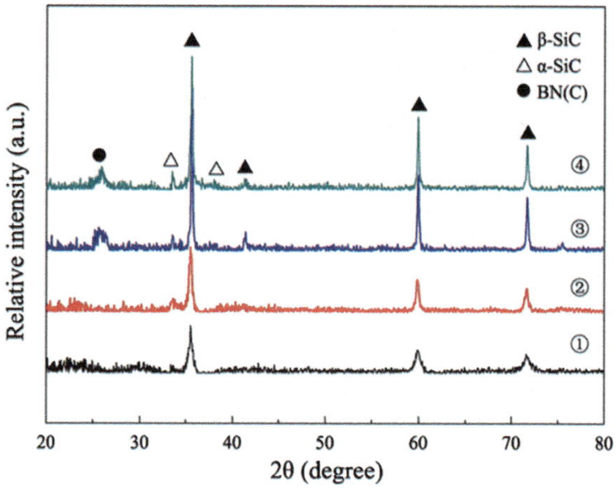

temperatures vary from hundreds of degrees to 3000 °C. The generation of pressures ≤ 8 GPa coupled with temperatures up to 2000 °C is generally achieving using the cubic multi-anvil high pressure apparatus.

4.2.1 Temperature-Dependent Crystallization

XRD patterns show that, similar to mechanically alloyed MA-Si_2BC_3N powder, the samples obtained at 1000 and 1100 °C under 5 GPa have considerably broad

Fig. 4.18 SEM fracture morphologies of MA-Si_2BC_3N bulk ceramics prepared by HIP or HP techniques: **a** HP 1700 °C/40 MPa: **b** HP 1800 °C/40 MPa: **c** HIP 1700 °C/190 MPa; **d** HIP 1800 °C/190 MPa

peaks indicating amorphous microstructure of MA-Si_2BC_3N bulk ceramics [13]. With increasing temperature, the diffraction peaks gradually sharpen and increase in intensity. The β-SiC and BN(C) phases appear at 1600 °C. However, the diffraction peaks are still broad implying nanosized crystals and/or incomplete crystallization (Fig. 4.19).

Slight element agglomeration occurs in as-sintered MA-Si_2BC_3N ceramics. Only a clear diffraction halo appears in the SAED pattern for the ceramics heated at 1000 °C. Here the microstructure consists of so-called dark and bright areas, marked by circles and irregular frames, respectively. The bright areas arise due to ceramic particle bridging, and the dark areas correspond to the ceramic particles themselves containing more Si and C atoms than bright zone. This suggests that the element clustering leading to crystallization requires an incubation period at 1000 °C and 5 GPa (Fig. 4.20).

At 1050 °C, both SAED pattern and HRTEM image do show little if any element agglomeration. Note that a nanosized puckered structure appears in the ceramics, which corresponds closely to the bright area, that is, the bridging area. At 1100 °C, the

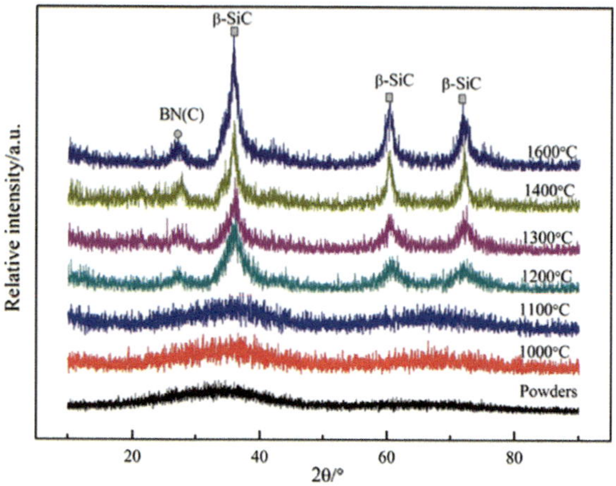

Fig. 4.19 XRD patterns of MA-Si_2BC_3N powder and high pressure sintered (5 GPa/30 min) MA-Si_2BC_3N bulk ceramics [13]

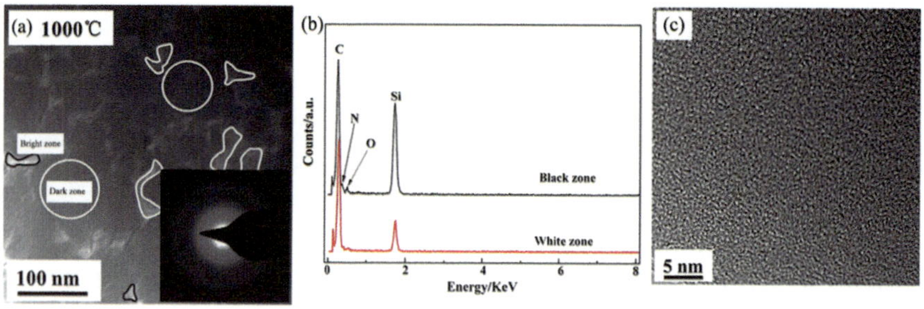

Fig. 4.20 TEM analysis of MA-Si_2BC_3N bulk ceramics prepared by high-pressure sintering at 1000 °C/5 GPa/30 min: **a** Bright field image and corresponding SAED pattern; **b** EDS energy spectra; **c** HRTEM image [13]

diffraction rings are very faint and broad. Both HRTEM and IFFT images show little atom agglomeration at puckered areas. Thus, the MA-Si_2BC_3N ceramics obtained at 1100 °C and 5 GPa is still completely amorphous and a phase segregation is incubating (Fig. 4.21).

No significant changes are observed when sintering temperature increases to 1150 °C. However, sintered at 1200 °C, the broad diffraction rings of (0002) BN(C) and (111) β-SiC lattice planes begin to appear. Moreover, two different nanostructures are observed clearly, although the phase cannot be established. The segregated areas have widths of ≈3 nm and indicate that the phase segregation ends at ≈1200 °C. Thus, amorphous MA-Si_2BC_3N densifies at 1200 °C/5 GPa coincidentally forming amorphous BN(C) and β-SiC marking the beginning of nucleation.

Increasing the temperature to 1250 °C causes the appearance of BN(C) and β-SiC nuclei in the amorphous MA-Si_2BC_3N. The BN(C) phase with widths of 3–5 nm is

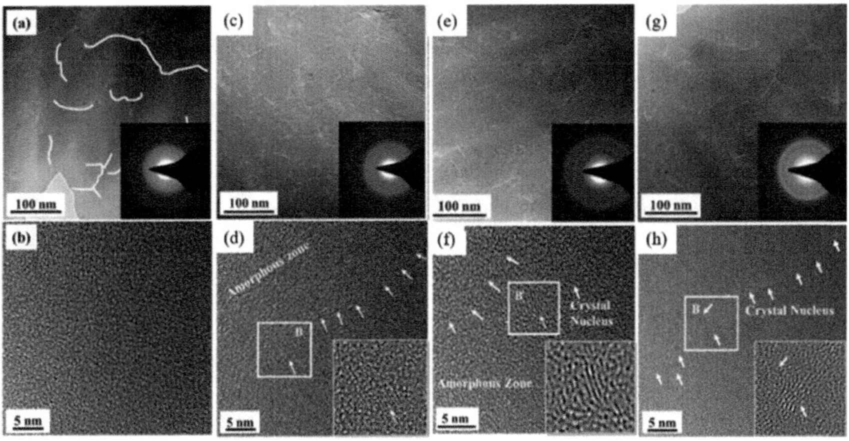

Fig. 4.21 TEM analysis of MA-Si_2BC_3N ceramics prepared by 5 GPa/30 min sintered at different temperature: **a**, **b** 1050 °C; **c**, **d** 1100 °C; **e**, **f** 1150 °C; **g**, **h** 1200 °C [13]

severely distorted and appears to be turbostratic BN(C) akin to turbostratic BN (t-BN) or t-C. The β-SiC nuclei of ≈3 nm appears around t-BN(C) nuclei and develops incompletely so that no clear lattice planes can be observed. At 1300 °C under 5 GPa, MA-Si_2BC_3N bulk ceramics show a high degree of crystallinity. BN(C) and β-SiC nanocrystals (≈5 nm) are now observed. Both grain types develop imperfectly containing some disordered atoms. Thus, at temperatures higher than 1250 °C, BN(C) and β-SiC nanocrystals nucleate and grow (Fig. 4.22).

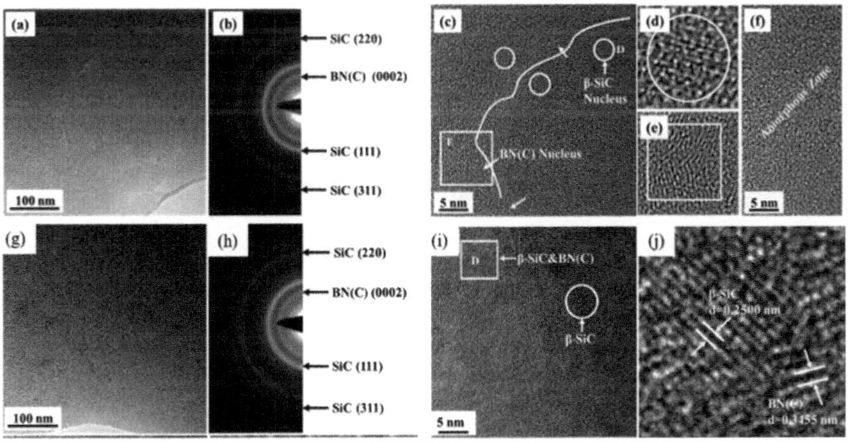

Fig. 4.22 TEM analysis of MA-Si_2BC_3N ceramics prepared by 5 GPa/30 min sintered at different temperature: **a**–**f** 1250 °C; **g**–**j** 1300 °C [13]

The microstructure of MA-Si_2BC_3N sintered at 1300 °C/3 GPa is also studied by EFTEM, as the puckered structures have an acceptable universality during crystallization of amorphous MA-Si_2BC_3N. In an EFTEM image, the brighter area contains more atoms. The puckered structures are poor in Si atoms, and rich in both B and N atoms. Carbon atoms have a uniform distribution in the ceramic, but an agglomeration in part of puckered structures. This indicates that the puckered structure areas are mainly rich in BN(C) and poor in SiC (Fig. 4.23).

At 1400 °C, nanoscale precipitates (sub-10 nm) appear and puckered structures disappear in MA-Si_2BC_3N ceramic matrix. The diffraction rings are consistent with BN(C) and β-SiC crystals. The uneven intensities for the β-SiC diffraction rings indicate increases in size. Some almost perfectly developed b-SiC grains of 5–10 nm enclosed by nanosized *t*-BN(C) crystals. Therefore, the nanoscale precipitates should

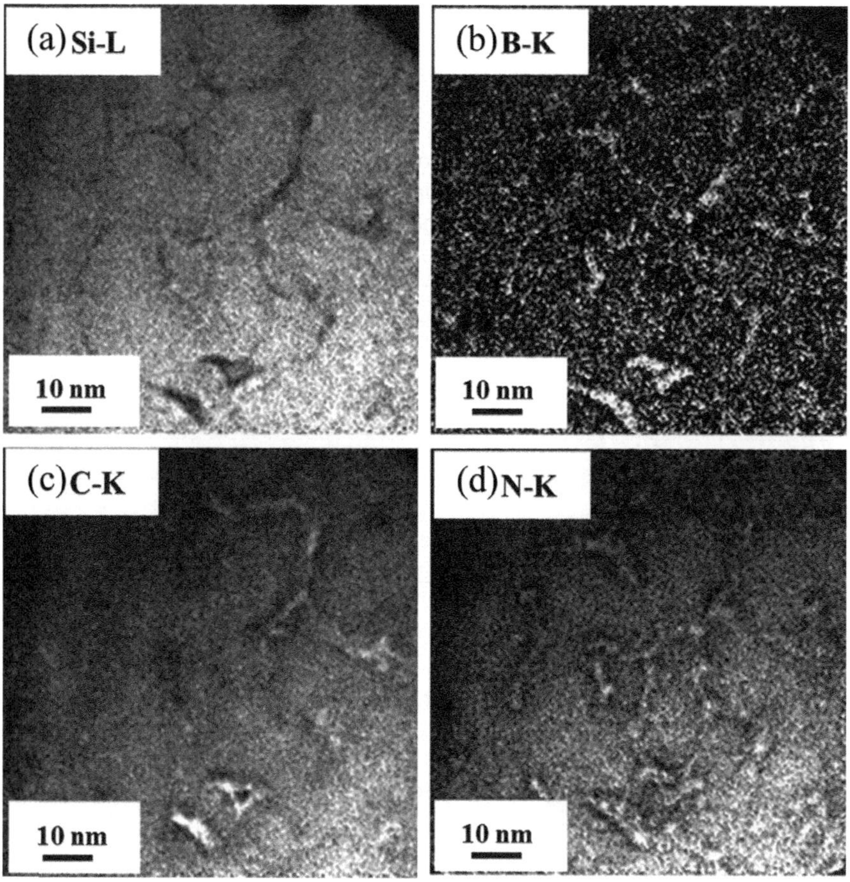

Fig. 4.23 EFTEM images of the MA-Si_2BC_3N bulk ceramics sintered at 1300 °C/3 GPa/30 min: **a** Si-L: **b** B-K; **c** C-K: **d** N-K [13]

be BN(C) and β-SiC crystals. In addition, stacking faults form in a few of β-SiC nanocrystals (Fig. 4.24).

At constant pressure, crystallization of amorphous ceramics depends mainly on temperature, which can be interpreted based on crystallization kinetics: nucleation and subsequent grain growth. In classical theory, the relationship between temperature T, and nucleation rate I, is given by [14]

$$I = \frac{nD}{a_0}\exp\left(-\frac{\Delta G^*}{k_B T}\right) \tag{4.1}$$

where the n is the number of embryos per unit volume, D is the atomic diffusion coefficient, a_0 is the interatomic distance, ΔG^* is the critical free energy required to form a nucleus, i.e., the thermodynamic potential barrier of nucleation, and k_B is

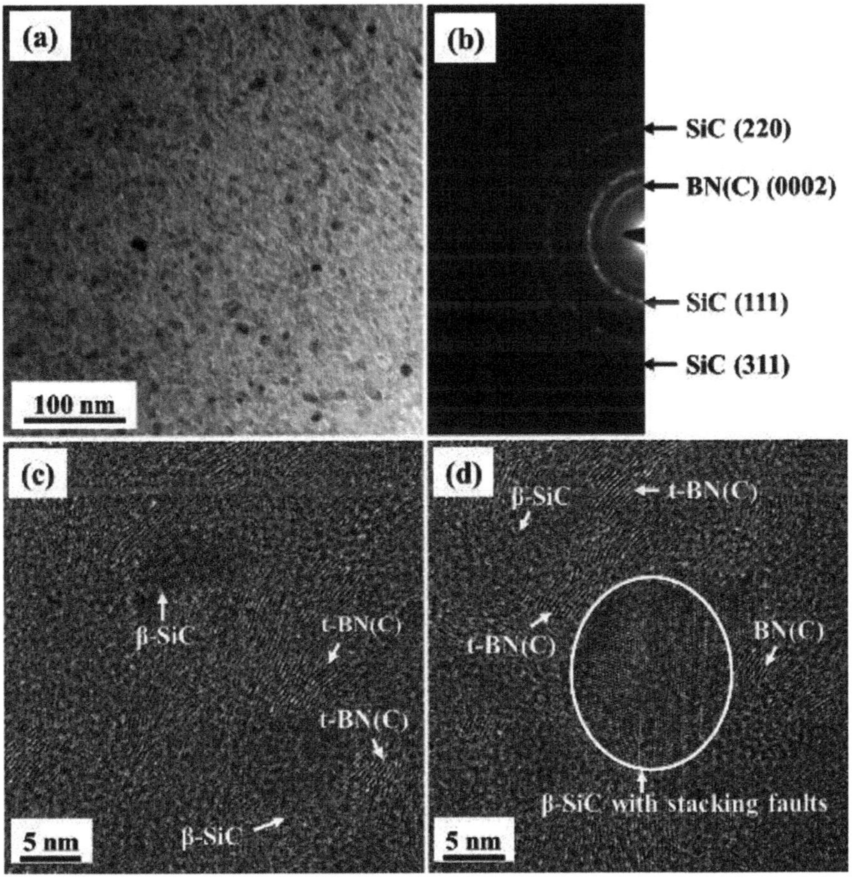

Fig. 4.24 TEM analysis of MA-Si_2BC_3N bulk ceramics sintered at 1400 °C/5 GPa/30 min: **a** Bright field image; **b** SAED pattern; **c**, **d** HRTEM images [13]

Boltzmann's constant. As is well known the nucleation rate increases with increasing temperatures.

Likewise, temperature T, also influences crystal growth rate u, which can be expressed as [15]

$$u = r_o \nu_o \exp\left(-\frac{Q_g}{RT}\right) \tag{4.2}$$

where the r_0 is the atomic diameter, ν_0 is the atomic transition frequency, Q_g is the activation energy for crystal growth, and R is the ideal gas constant. Crystal growth rate increases with the increase in temperature. The two equations above both confirm raising the holding temperature encourages grain nucleation and growth.

4.2.2 Pressure-Dependent Crystallization

Only a diffuse diffraction peak is detected in XRD patterns for MA-Si_2BC_3N bulk ceramics after HPS at 1–3 GPa/1200 °C/30 min, suggesting the bulk ceramics retain a good amorphous structure [16]. Further increasing pressure to 5 GPa, peaks for β-SiC and BN(C) emerge with a broadening of the diffraction peaks and a larger FWHM (Fig. 4.25).

In cases the sintering pressure is $\leq$ 3 GPa, no nanocrystallites or ordered atomic clusters are found in the ceramic matrix (Fig. 4.26). The SAED pattern is consisted of a diffraction halo and several broadening diffraction rings. The HRTEM images show a disordered arrangement of atoms, typically of amorphous nature. Also no precipitates appear on the ceramic matrix when the sintering pressure increases to 3 GPa. However, the SAED pattern shows the broadened diffraction rings of the (111)

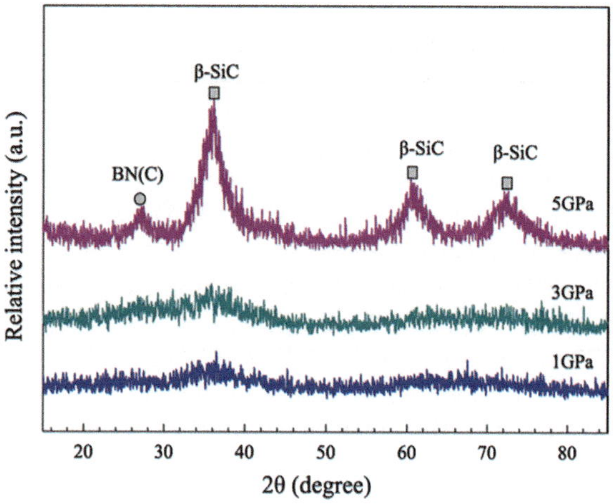

Fig. 4.25 XRD patterns of the MA-Si_2BC_3N ceramics sintered at 1–5 GPa/1200 °C/30 min [16]

planes of SiC and the (0002) planes of the BN(C). HRTEM images further show some ordered atomic arrangements (several to dozens of nanometers), implying the partial crystallization of amorphous phases and the beginning of nucleation (Fig. 4.27).

High pressure can promote crystallization leading to a low-temperature initiation of crystallization. Thus PDCs-$Si_6B_{1.1}C_{10}N_{3.4}$ remains amorphous up to 1700 °C during annealing in nitrogen, whereas high-pressure sintered MA-Si_2BC_3N monolithic ceramics have relatively low starting crystallization temperature of only 1200 °C. Hot-pressed MA-Si_2BC_3N starts to crystallize at ~1485 °C under a uniaxial pressure of 80 MPa.

Given the crystallization kinetic theory, the nucleation rate I, is given by [17]

$$I = I_0 \exp\left(-\frac{\Delta G^* + \Delta G^m}{k_B T}\right) \tag{4.3}$$

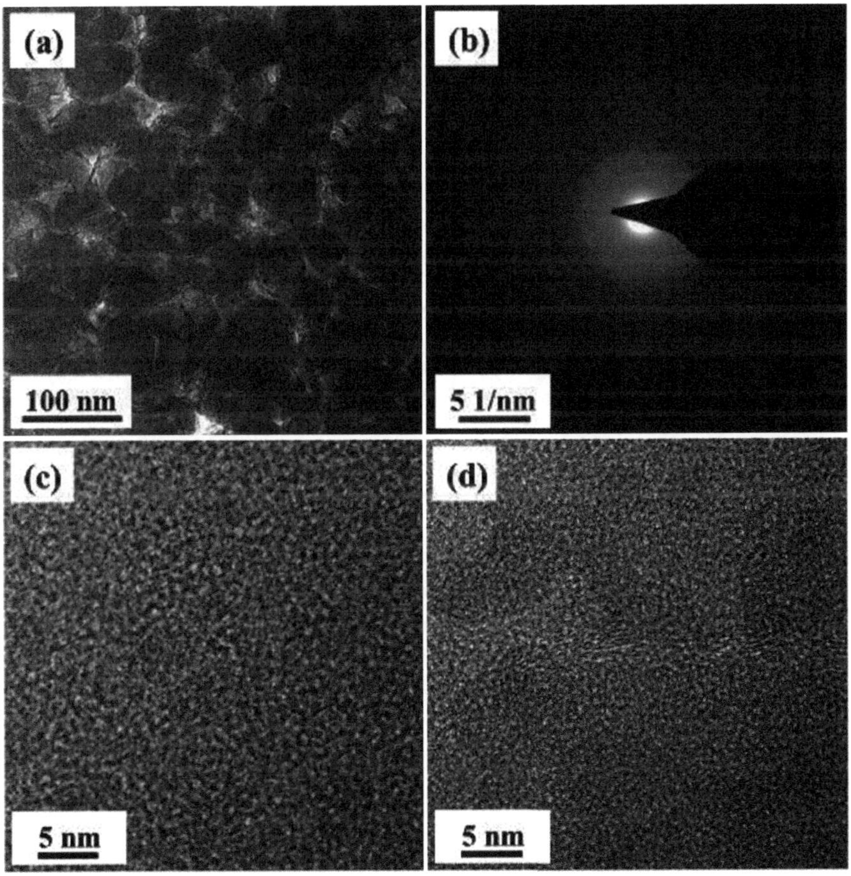

Fig. 4.26 TEM analysis of MA-Si_2BC_3N bulk ceramics prepared by 1200 °C/1 GPa/30 min: **a** Bright-field image; **b** SAED pattern: **c**, **d** HRTEM image [16]

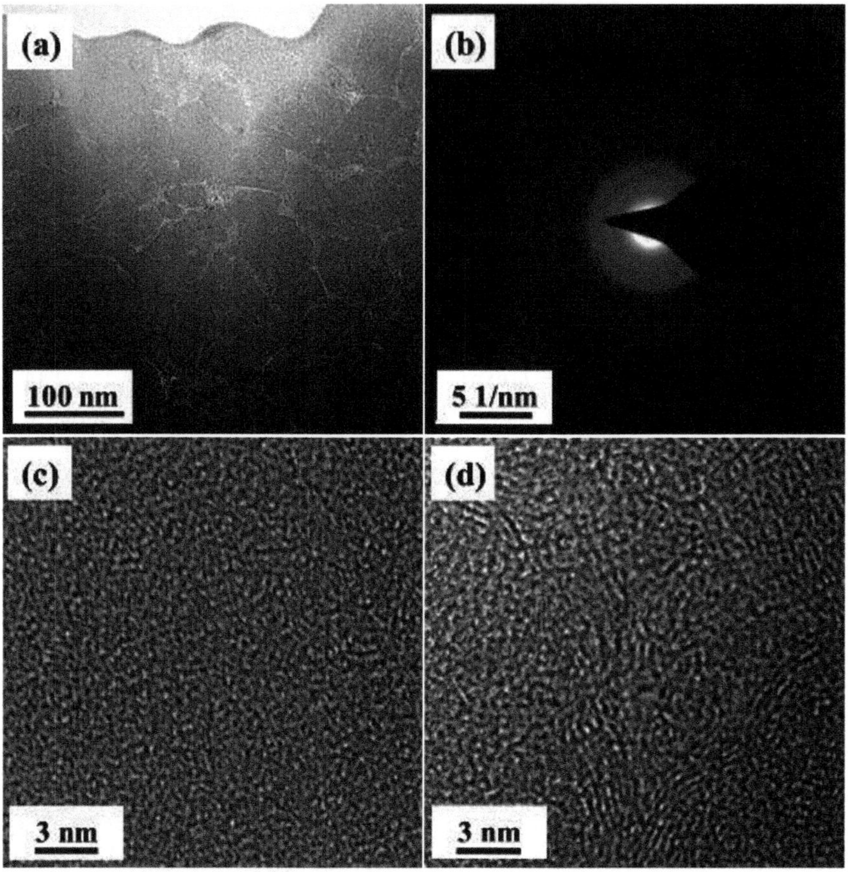

Fig. 4.27 TEM analysis of MA-Si_2BC_3N bulk ceramics prepared by 1200 °C/3 GPa/30 min: **a** Bright field image; **b** SAED pattern: **c**, **d** HRTEM image [16]

where I_0 is a constant, ΔG_m is the activation energy for atomic diffusion and k_B is the Boltzmann's constant. The sum of ΔG^* and ΔG^m decreases with increasing pressure because ΔG^* decreases sharply while ΔG^m increases smoothly. As a result, the nucleation rate I, increases with increasing pressures.

Crystal growth rates are usually dependent on atomic diffusion. The effect of pressure p, on atomic diffusion in solids can be written as [18]:

$$\left(\frac{\partial \ln D}{\partial p}\right)_T = -\frac{V^*}{k_B T} \tag{4.4}$$

where D is the atomic diffusion coefficient, V^* is the activation volume and normally a positive value for the atomic diffusion in amorphous solids. From Eq. (4.4), increases in pressure decrease crystal growth rates.

High pressure favors nucleation but disfavors subsequent growth. High pressure is, therefore, conducive to forming fine and homogeneous microstructures and results in desirable mechanical properties and novel thermophysical characteristics. This is why some nanosized β-SiC and *t*-BN(C) crystals appear in the matrix of MA-Si_2BC_3N monoliths at 1400 °C under 5 GPa and why MA-Si_2BC_3N tends to crystallize at high pressures. In addition to high pressure, the capsule-like structure also has a positive influence on grain refinement. The grain growth of β-SiC and *t*-BN(C) may depend on long-range atomic diffusion. However, the mutual encapsulation or segregation leads to a mutual restriction of atomic diffusion and thus limits β-SiC and *t*-BN(C) grain growth.

Note that high-temperature promotes both nucleation and growth, while high pressure promotes nucleation but constrains growth. The amorphous MA-Si_2BC_3N obtained at 1100 °C under 5 GPa can be attributed to high pressure helpful to reducing the sintering temperature and thus limiting crystallization.

4.2.3 Nucleation Thermodynamics and Kinetic Mechanisms

The following reactions are most likely occurring during the crystallization of amorphous MA-Si_2BC_3N. In the calculated temperature range, the standard Gibbs free energy changes for all the reactions are negative. Expectedly, the β-SiC tends to precipitate preferentially from the amorphous matrix, followed by SiB_6, SiB_3 and B_4C (Fig. 4.28).

$$\mathrm{Si(s) + C(s) \rightarrow SiC(s)} \tag{4.5}$$

$$\mathrm{Si(s) + 6B(s) \rightarrow SiB_6(s)} \tag{4.6}$$

$$\mathrm{Si(s) + 3B(s) \rightarrow SiB_3(s)} \tag{4.7}$$

$$\mathrm{C(s) + 4B(s) \rightarrow B_4C(s)} \tag{4.8}$$

The question why BN(C) (instead of SiC) preferentially nucleates at the particle bridging areas causing SiC to nucleates inside particles is critical to clarifying the nucleation mechanism [19]. At pressures < 36.5 MPa, the enthalpy differences for β-SiC are always lower than those of h-BN and graphite indicating the favorable formation of β-SiC compared to h-BN and graphite, approximately equal to BN(C) (Fig. 4.29). But at 1–7 GPa, the enthalpy differences for h-BN and graphite are always lower than that of β-SiC, and increase more slowly with increasing pressures. This suggests the formation of h-BN and graphite should be more favorable under high pressure than β-SiC. The t-BN(C) in crystalline MA-Si_2BC_3N is mostly likely to consist of heterogeneously distributed t-carbon, t-BN, and B-doped t-carbon layers.

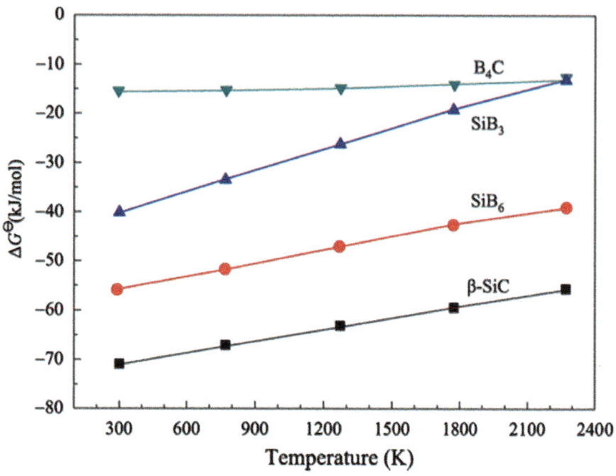

Fig. 4.28 The standard Gibbs free energy curves of the amorphous MA-Si_2BC_3N ceramics [16]

We speculate therefore that the enthalpy difference for t-BN(C) should fall somewhere between these two extremes for h-BN and graphite. As a result, the formation of t-BN(C) is more favorable under high pressure than β-SiC and thus nucleates preferentially prior to β-SiC during crystallization of MA-Si_2BC_3N.

High mechanical stresses produced by high pressure can create some defects at particle surfaces. The defects lead to high energy offering driving force of nucleation and thus can act as nucleation sites. Therefore, the crystallization of amorphous MA-Si_2BC_3N subjected to high pressure is most likely induced by heterogeneous nucleation reducing Gibbs free energy of the whole system.

The nucleation of t-BN(C) and β-SiC in MA-Si_2BC_3N can be regarded as a kind of solid phase transformation involving the changes of Gibbs free energy, interfacial energy and elastic strain energy. In heterogeneous nucleation, the defects such as dislocations and grain boundaries can act as a kind of unstable factor. Nucleating at

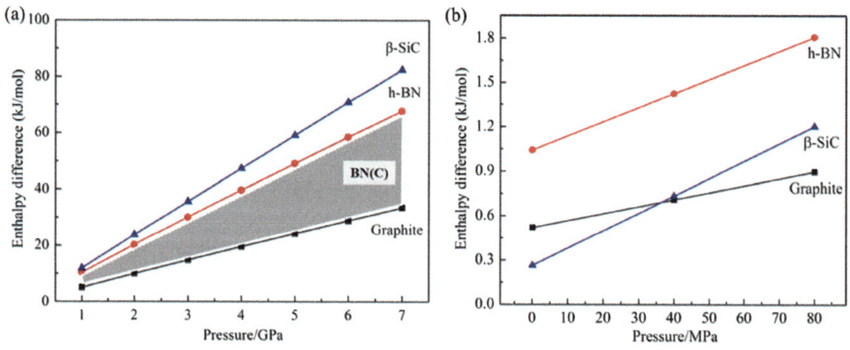

Fig. 4.29 The calculated enthalpies for β-SiC, h-BN and graphite as a function of pressure: **a** Formation of h-BN and graphite is more favorable under high pressure than β-SiC, that is, BN(C) nucleates preferentially prior to β-SiC at pressures ≥ 1 GPa; **b** nucleation precedence of β-SiC over BN(C) at 0.1 MPa ≤ P ≤ 80 MPa [20]

these sites can reduce the Gibbs free energy and release free energy causing decrease in nucleation energy and thus driving nucleation. The energy changes can be given as

$$\Delta G_N = \Delta G + \varepsilon + \gamma - \Delta G_I \tag{4.9}$$

where ΔG_N is the nucleation energy, ΔG is the difference between free energy of the new-born phase and the disappeared parent phase, ε is the elastic strain energy caused by the new phase, γ is the additional interfacial energy stored probably in material system, ΔG_I is the decreased interfacial energy caused by the disappeared original interface. The larger the ΔG_I, the smaller ΔG_N. Therefore, heterogeneous nucleation can decrease the nucleation energy (equivalent to nucleation barrier) via decreasing the area of interface.

The MA-Si_2BC_3N particle bridging area can be seen as a type of interface with some defects acting as unstable factors because of the considerations as follows: (1) Plastic deformation and particle rearrangement due to strong stresses break the chemical bonds on particle surfaces generating dangling broken bonds that then form new bonds; (2) High pressure creates some defects at particle surfaces so that particles with stress disturbed amorphous structures have high internal energy and thus are highly susceptible to further reactions, equivalent to a reduced nucleation energy barrier; (3) The nanosized MA-Si_2BC_3N powders offer an inherent high-specific surface area and surface activity so that the particle surface provides additional active reaction sites for nucleation. Given the first principles investigations, at temperatures high enough, B, N, and partial C atoms (or B–N, B–N–C and B–C–N units) begin to diffuse and concentrate in the particle bridging areas. Meanwhile, Si and C atoms (or Si–C unit) left in particle interiors begin to diffuse and rearrange. The diffusion of atoms leads to phase segregation followed by nucleation. Once the structure of aggregations becomes stable, nucleation will finish, followed by the growth of nuclei with the continuous diffusion and aggregation.

Based on the aforementioned results, nucleation in MA-Si_2BC_3N monoliths at 5 GPa can be summarized as follows (Fig. 4.30). The mechanically alloyed MA-Si_2BC_3N powder shows relatively high stability of amorphous to some extent. Low temperatures cannot provide enough driving forces for atomic diffusion favorable for crystallization so that the sintered ceramics remain amorphous below 1100 °C (Stage I). At 1100–1200 °C, phase segregation occurs caused by atomic diffusion (Stage II). The amorphous BN(C) and SiC tend to concentrate in the particle bridging areas and particle interiors, respectively. At 1200–1300 °C, nucleation of t-BN(C) and β-SiC occurs. The 3–5 nm-sized BN(C) nucleates preferentially in the ceramic particle bridging areas, while 3 nm-sized β-SiC nuclei form inside particles (Stage III). Above 1300 °C, significant quantities of t-BN(C) and β-SiC nuclei form in the amorphous matrix. At the same time, the grain growth occurs and plays a dominant role (Stage IV). At 1400 °C, MA-Si_2BC_3N is composed of < 10 nm β-SiC grains and turbostratic BN(C) grains of 3–5 nm in width and a few remained amorphous phase. The continuously increasing temperatures lead to the development of larger SiC and BN(C) gains.

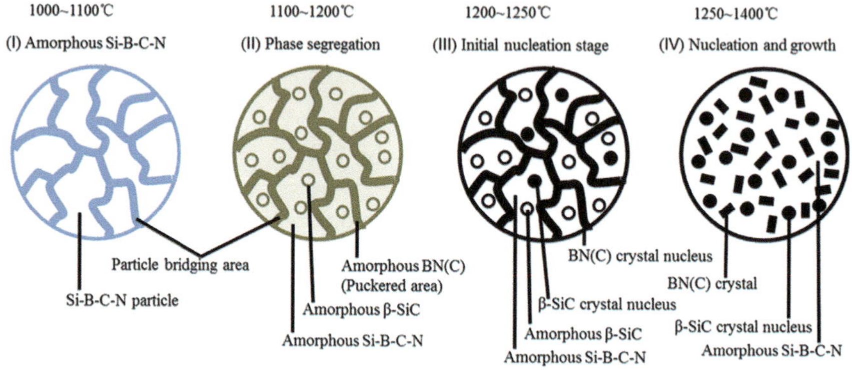

Fig. 4.30 Schematic illustration of nanostructure development in MA-Si_2BC_3N monoliths during HPS at 1000–1400 °C and 5 GPa [13]

4.2.4 *Densification Behavior Under High-Pressure Sintering*

The density increases sharply with temperatures increasing from RT to 1000 °C, but only slightly for 1100–1600 °C. The monolithic density of MA-Si_2BC_3N compacted at RT under 5 GPa is 2.35 g/cm^3, which is just 10% lower than 2.60 g/cm^3 for counterparts hot-pressed at 1900 °C/80 MPa. The bulk density of the monoliths sintered at 1100 and 1600 °C is ≈2.80 g/cm^3, 8% higher than hot-pressed ceramics. These results indicate that high pressure can effectively promote densification of MA-Si_2BC_3N. Highly dense amorphous MA-Si_2BC_3N monoliths without sintering aids are processed successfully at 1100 °C and 5 GPa (Fig. 4.31).

For MA-Si_2BC_3N monoliths consolidated at 1000 °C/5 GPa, micropores are detected on the sample surfaces. As sintering temperature further increases, the

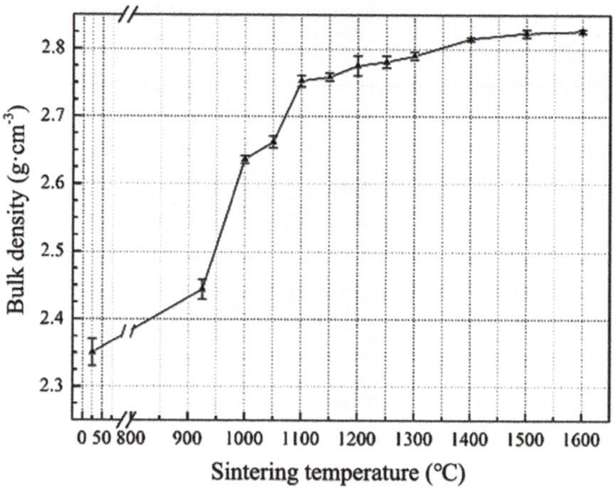

Fig. 4.31 Bulk density of MA-Si_2BC_3N monoliths prepared at RT ~1600 °C/5 GPa/30 min [16]

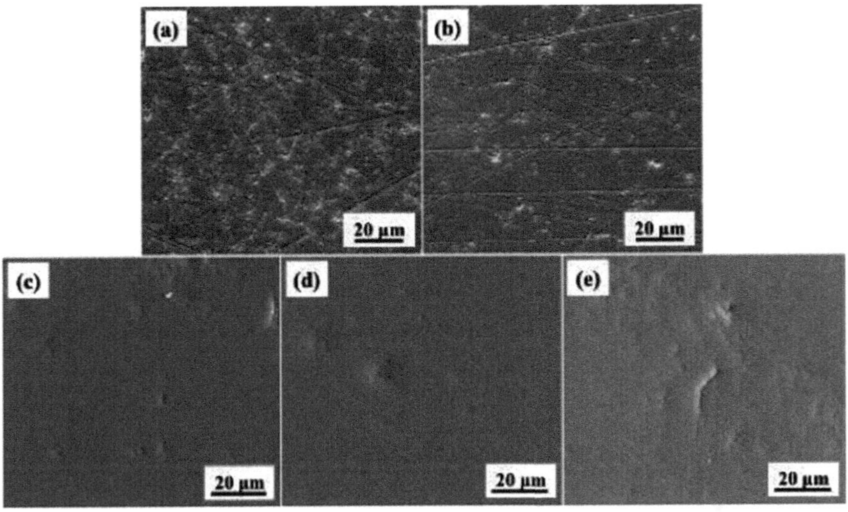

Fig. 4.32 SEM surface morphologies of MA-Si_2BC_3N monoliths (5 GPa/30 min) sintered at different temperature: **a** 1000 °C; **b** 1100 °C; **c** 1200 °C; **d** 1400 °C: **e** 1600 °C [16]

ceramics become denser. Especially, a smooth surface free of any micropores and microcracks is found for the ceramics sintered at 1200–1600 °C (Fig. 4.32).

There are no particular morphologies seen on the fracture surfaces for the MA-Si_2BC_3N bulk ceramics sintered at 1100 °C and 5 GPa (Fig. 4.33). However, finer and denser nanosized near-spherical morphology is obtained at 1600 °C versus RT. The surface pits are caused by grain pull-out or separation during fracture. The nanostructure shown in 1600 °C sintered ceramics are likely crystallized β-SiC and BN(C), whereas the ones sintered at 1100 °C are still amorphous nanoparticles with mechanical interlocks caused by the high pressure induced surface roughness on particle surfaces.

High pressure is widely accepted as a positive factor in the densification of bulk materials. High mechanical stress derived from high pressure breaks agglomerates and permits rapid particle rearrangement and even plastic deformation resulting in closed-packed structures with a reduced porosity. This thereby enhances atomic diffusion and mass transfer at interfaces, and thus greatly promotes densification during sintering. Consequently, using high pressure sintering, dense MA-Si_2BC_3N bulk ceramics can be prepared at high pressure (up to 5 GPa) but relatively low sintering temperatures (about 700–800 °C lower than the hot-pressing sintering). Coincidently, high pressure sintering results in a use of low sintering temperatures, which hinders effectively atomic diffusion limiting crystallization and thus ensures a completely amorphous structure.

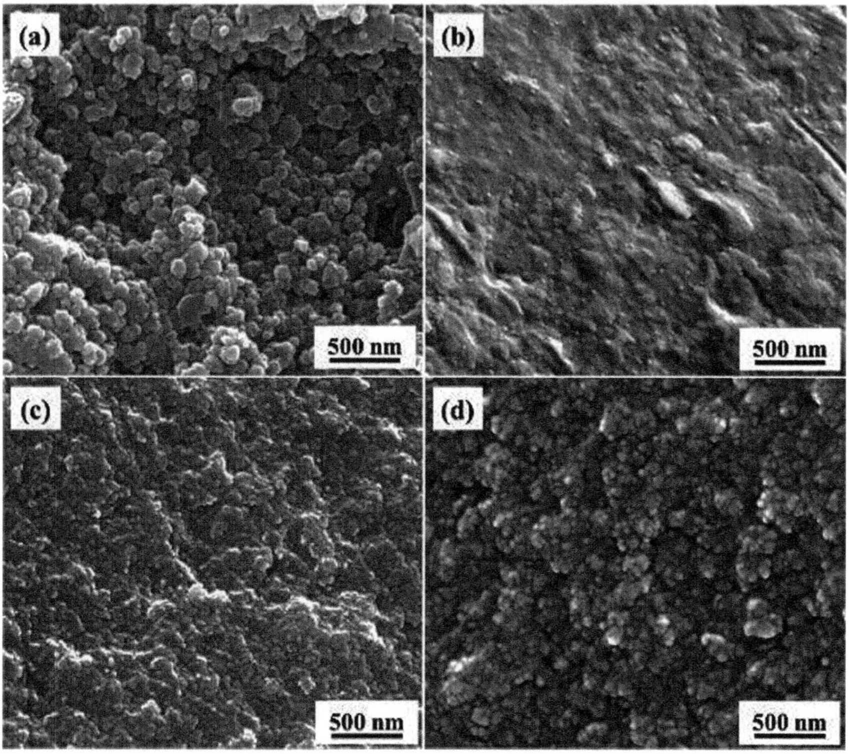

Fig. 4.33 SEM fracture morphologies of MA-Si_2BC_3N monoliths (5 GPa/30 min) sintered at different temperature: **a** RT; **b** 1100 °C; **c** 1400 °C, **d** 1600 °C [16]

4.2.5 Chemical Bonds Evolution

For the as-prepared MA-Si_2BC_3N powder and MA-Si_2BC_3N monoliths, no distinct changes whether the types of chemical bonds or the contents can be detected from Raman and XPS spectra [20]. The broad peak at 1200–1800 cm^{-1} is detected from Raman spectra for the as-milled amorphous powder and as-sintered monoliths at 1100 °C/5 GPa (Fig. 4.34). The peaks arising at ca. 1370, 1528 and 1588 cm^{-1} can be attributed to D band of h-BN and graphite, amorphous carbon, and G band of graphite, respectively. The peaks at 2740 and 2938 cm^{-1} are coincident with the secondary coherence peaks of graphite. The sharp increase in signal intensity above 1800 cm^{-1} is most likely due to the interference from fluorescent impurities in as-milled powder. It is indicative of an amorphous state, that is, the B, C and N atoms exhibit long-range disorder. At 1100 °C, peak located at 1200–1800 cm^{-1} remains broad. However, some fraction of carbon atoms coalesces in the materials sintered at 1100 °C which is characterized by a small peak at 1528 cm^{-1}. With the increasing temperatures to 1600 °C, carbon and h-BN form indicating the occurrence of crystallization.

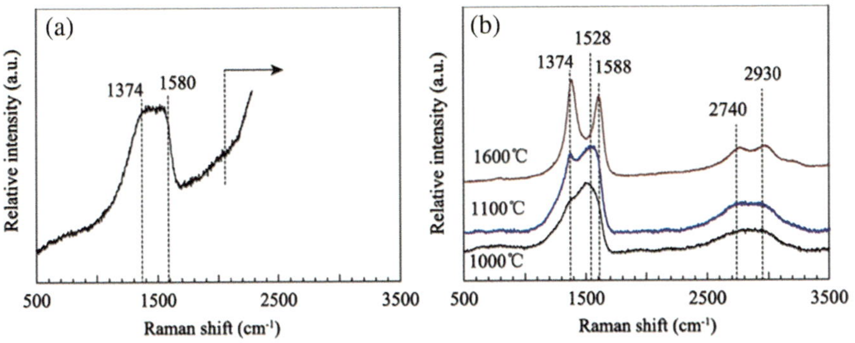

Fig. 4.34 Raman spectra of the MA-Si_2BC_3N materials: **a** As-milled amorphous powder; **b** monoliths sintered at 1000–1600 °C and 5 GPa [20]

XPS spectra suggest both of the MA-Si_2BC_3N ceramics sintered at 1100 or 1600 °C consist of Si–C, C–C, C–B, C–N (sp^2/sp^3), N–B (sp^2/sp^3), C–B–N bonds, etc. (Fig. 4.35). However, the increasing temperatures lead to the collapse of the amorphous network structure with the reduction in C–N (sp^3) and C–B–N bonds in ceramics sintered at 1600 °C (Table 4.1).

Generally, the β-SiC NMR peak locates at −17 ppm, while α-SiC signal peaks are positioned at −15, −20, and −25 ppm, respectively. After sintered at 1100 °C under 5 GPa, a very broad resonance signal centered at around −107.43 ppm is observed, attributed to SiO_4 units in four coordinate forms. The sharp resonance signals at around −11.37 ppm suggests the formation of SiC_4 units, close to that of α-SiC. The very broad peaks centered at 0–50 ppm can be assigned to the tri-coordinated B

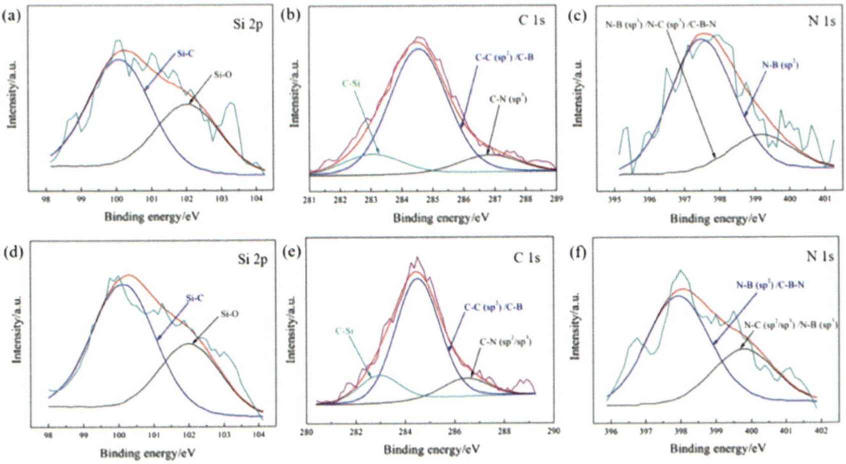

Fig. 4.35 XPS spectra of the MA-Si_2BC_3N monoliths (5 GPa/30 min) sintered at different temperature: **a–c** 1100 °C; **d–f** 1600 °C [20]

Table 4.1 XPS results of the MA-Si_2BC_3N monoliths prepared by high-pressure sintering [20]

Elemental lines	Chemical bonds	1100 °C/5 GPa/30 min		1600 °C/5 GPa/30 min	
		Binding energy (eV)	Content (%)	Binding energy (eV)	Content (%)
Si 2p	Si–C	100.10	14.20	100.14	13.63
	Si–O	102.04	9.41	102.02	7.62
B 1s	B–N (sp^2)–B–C	190.40	5.06	190.32	5.94
	B–C–N	192.93	4.15	192.11	3.44
C 1s	C–Si	283.14	7.37	282.59	8.56
	C–C (sp^2)/C–B	284.62	44.01	284.37	44.91
	C–N (sp^3)	286.80	6.21	286.00	7.52
N 1s	N–B (sp^2)	397.50	7.63	397.92	5.64
	N–C (sp^3)/N–B (sp^3)/C–B–N	399.19	1.96	399.78	2.75

atoms with BC_xN_{3-x} ($0 \leq x \leq 3$) units including BN_3 planar triangles in turbostratic-BN (t-BN, 22 ppm) and/or in h-BN (27 ppm), and BN_2C units (35–40 ppm). The summit of peaks locates at ~27.45 ppm. For MA-Si_2BC_3N ceramics sintered at 1600 °C under 5 GPa, the SiC_4 units shift from −11.37 to −20.21 ppm, assigned to α-SiC. Coincidentally, the summit of ^{11}B peaks shifts from −27.45 to 11.65 ppm. It suggests to some extent that partial amorphous BN domains are gradually separated from (turbostratic) BN(C), then aggregated and rearranged to form relatively ordered h-BN at 1600 °C (Fig. 4.36).

Taken together, the results from NMR spectra indicate that the degree of order of atomic cluster increases with increases in heating temperatures. Coincidently, silicon atoms are mainly fourfold coordinated by carbon atoms in forms of SiC_4 units at high temperatures. Partial units transfer from amorphous, to β-SiC, then to α-SiC. High sintering temperatures also lead to partial phase-segregation of evenly distributed BN(C) to relatively ordered h-BN and graphite.

4.3 Phase Composition and Microstructural Evolution of MA-SiBCN Composites

4.3.1 Effect of Metal Particles

(1) Zr particles introduced by MA

For the HP sintered one-step milled Zr containing powders, obviously crystallized peaks are observed and the main phases are those of ZrB_2, ZrN, SiC, BN(C), and few m-ZrO_2 and ZrO_x [21, 22]. The diffraction peaks at 33.97°, 39.42°, 56.75° and

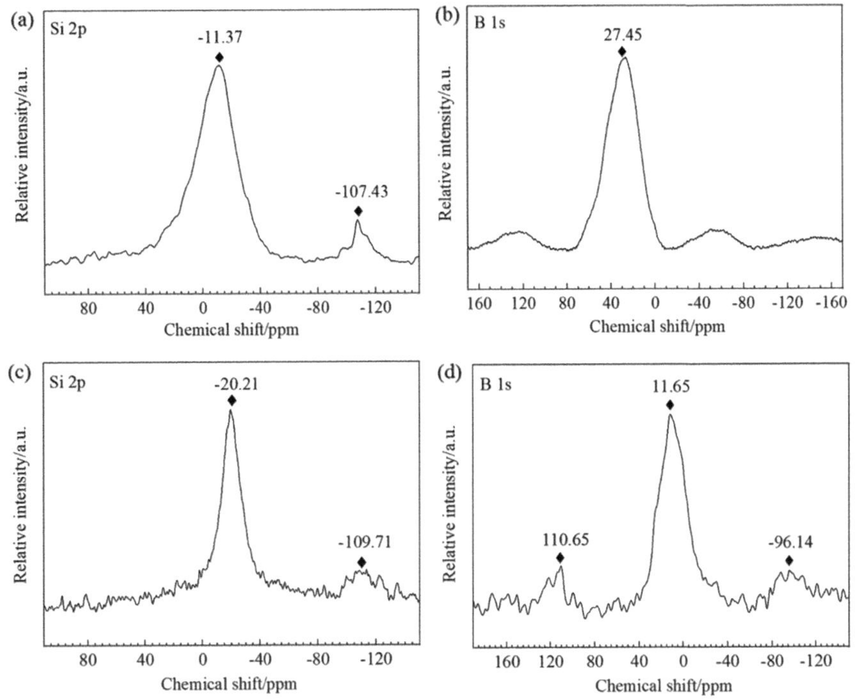

Fig. 4.36 The ^{29}Si and ^{11}B NMR spectra of MA-Si_2BC_3N monoliths (5 GPa/30 min) sintered at different temperature: **a**, **b** 1100 °C; **c**, **d** 1600 °C [16]

71.42° correspond to (111), (200), (220) and (311) lattice planes of ZrN, while peaks of 25.8°, 32.96° and 42.10° assign to (001), (100) and (101) lattice planes of ZrB_2. Besides, peaks at 28.31° and 31.49° correspond to ($\bar{1}$11) and (111) lattice planes of the m-ZrO_2. With a higher content of Zr–B mixtures, the diffraction peaks of ZrB_2 and ZrN increase along with the decreasing intensity of SiC (Fig. 4.37).

With the increasing sintering temperature, the bulk density of MA-$Si_2B_5C_3NZr_2$ bulk ceramics steadily increases (Fig. 4.38). MA-$Si_2B_5C_3NZr_2$ ceramics consolidated at 1800–1900 °C present some macro-porosity and cracks; however, a relatively dense structure can be obtained after rising the sintering temperature to 2000 °C (Fig. 4.39). For the two-step milling MA-$Si_2B_2C_3NZr_{0.5}$ power (c-Si, graphite and h-BN powders in a mole ratio of 2:3:1 are milled for 20 h, and then the pre-milled Zr–B powders (5 h) are added and milled continued for another 5 h), they are hot-pressed at 1900 °C, and no obvious pores and cracks are found on surface and fracture morphologies. The impact of ball milling method on microstructure of the two bulk ceramics remains small and shows a similar microstructure; besides, the grains of ZrB_2–ZrN are homogenously distributed with average particle sizes less than 1 μm for both ceramics (Fig. 4.40).

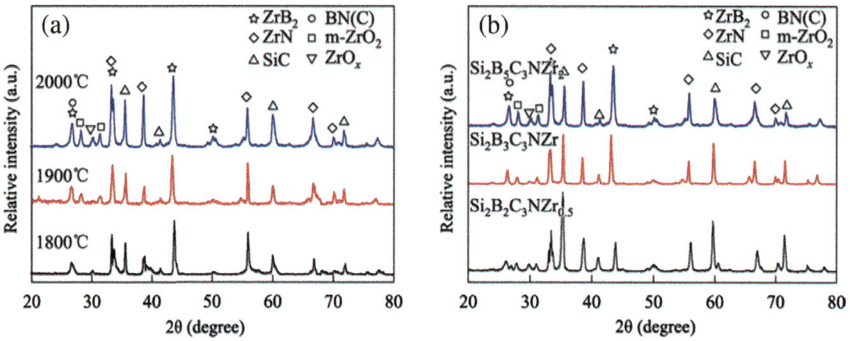

Fig. 4.37 XRD patterns of MA-SiBCNZr ceramics prepared by hot-pressing sintering: **a** MA-$Si_2B_5C_3NZr_2$ sintered at 1800–2000 °C; **b** MA-$Si_2B_2C_3NZr_{0.5}$, MA-$Si_2B_3C_3NZr$ and MA-$Si_2B_5C_3NZr_2$ sintered at 1900 °C [22]

Fig. 4.38 The bulk density of MA-$Si_2B_5C_3NZr_2$ ceramics prepared by hot-pressing sintering at different temperature [21]

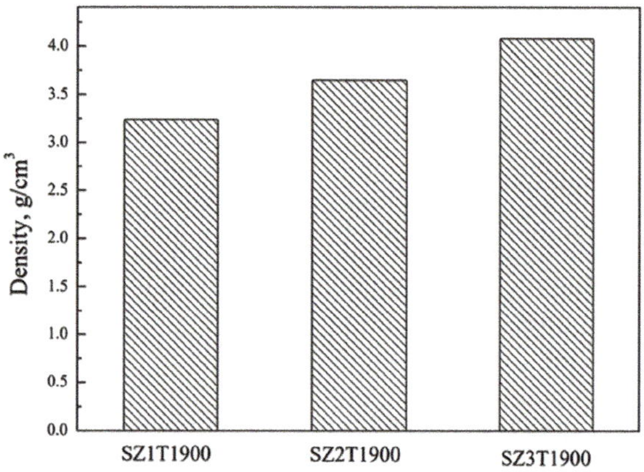

(2) Al particles introduced by MA

The amorphous MA-$Si_2BC_3NAl_{0.6}$ powder (using Al particles as the Al source) are hot-pressed at 1800–1950 °C/40 MPa or 1900 °C/40–50 MPa [23]. Accordingly, the as-sintered MA-$Si_2BC_3NAl_{0.6}$ bulk ceramics all exhibit obvious diffraction peaks of β-SiC, BN(C), AlN and AlON. The diffraction peak intensities of the crystallized phases increase with the increasing sintering temperature or pressure (Fig. 4.41).

Due to the existence of some amorphous phases, the theoretical density of the as-sintered MA-$Si_2BC_3NAl_{0.6}$ bulk ceramics cannot be estimated. The bulk density of the MA-$Si_2BC_3NAl_{0.6}$ sintered at 1800 °C is only 2.52 g/cm^3; however, the bulk density of the ceramics sintered at 1900 °C reaches to 2.80 g/cm^3. Densification of the MA-$Si_2BC_3NAl_{0.6}$ ceramics can also be induced by increasing sintering pressure (Fig. 4.42).

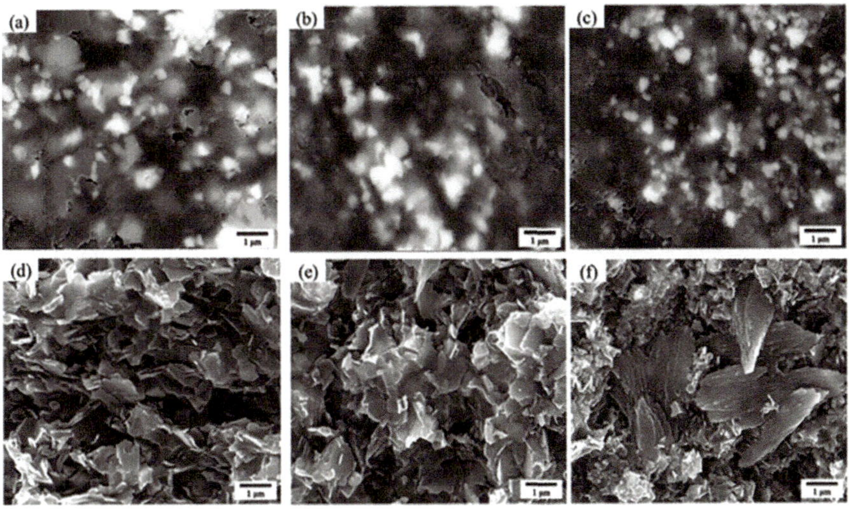

Fig. 4.39 SEM surface and fracture morphologies of MA-$Si_2B_5C_3NZr_2$ ceramics hot-pressed at different temperature: **a**, **d** 1800 °C; **b**, **e** 1900 °C; **c**, **f** 2000 °C [22]

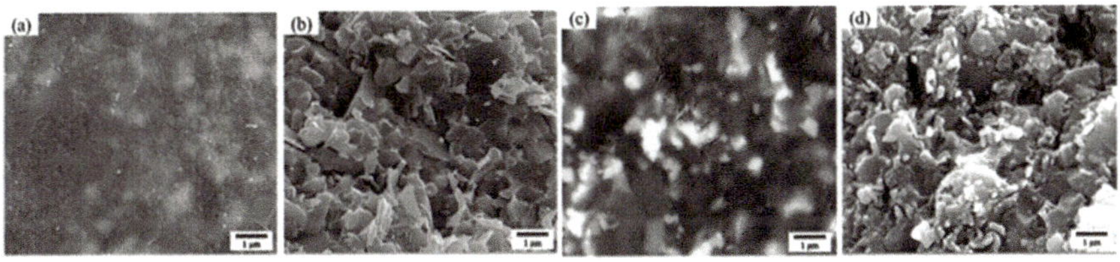

Fig. 4.40 SEM surface and fracture morphologies of hot-pressed MA-$Si_2B_2C_3NZr_{0.5}$ ceramics (from two-step milling powder): Sintered at **a** **b** 1900 °C; **c** **d** 2000 °C [22]

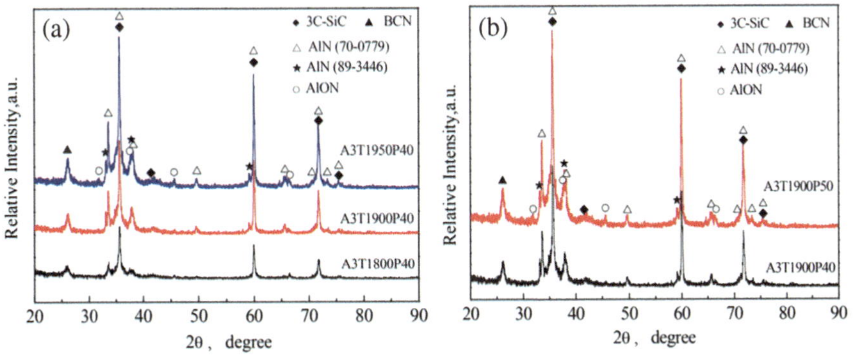

Fig. 4.41 XRD patterns of MA-$Si_2BC_3NAl_{0.6}$ bulk ceramics prepared by hot-pressing sintering: **a** 1800–1950 °C/40 MPa/30 min; **b** 1900 °C/40–50 MPa/30 min [23]

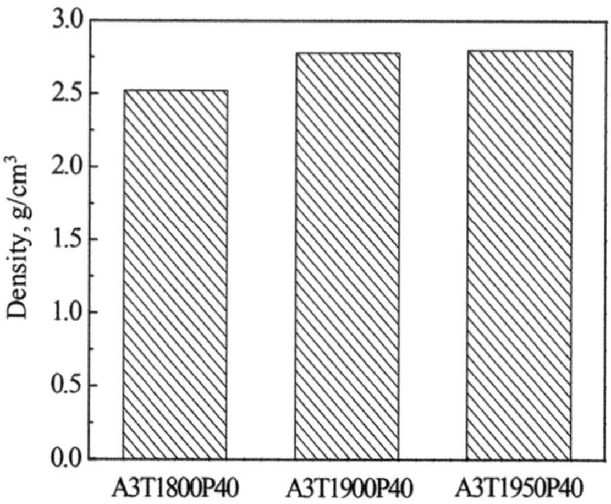

Fig. 4.42 The bulk density of the MA-$Si_2BC_3NAl_{0.6}$ bulk ceramics prepared by hot-pressing sintering at 1800–1950 °C/40 MPa/ 30 min [23]

For MA-$Si_2BC_3NAl_{0.6}$ ceramics sintered under 1 bar Ar, some micropores ~5 μm are observed on the monolithic surfaces; however, ceramics sintered under 1 bar N_2 are free of micropores and microcracks with fine and homogeneous microstructure [24]. The fracture morphologies for the two ceramics are also different. Lamellar crystallites are observed for ceramics sintered under N_2, possibly resulting from nitrogen promoting crystallization of lamellar BN(C) (Fig. 4.43). For MA-$Si_2BC_3NAl_{0.6}$ ceramics sintered at higher temperatures, dense surface structure is also found on the sample surfaces (Fig. 4.44).

The SiC and lamellar BN(C) phases are both observed for the two ceramics sintered under different atmosphere. A thin layer of amorphous phase is found near to the SiC grains. Such amorphous structure may remain from the as-milled amorphous powder. Some agglomerating areas composed of very small grains are observed. The grain sizes of the nanocrystals for the ceramics sintered under Ar are smaller than their counterparts. The SAED pattern indicates the agglomerating areas are mainly BN(C) and AlN (Fig. 4.45).

The amorphous MA-$Si_2BC_3NAl_{0.6}$ (using Al particles as the Al source) and MA-$Si_2BC_3N_{1.6}Al_{0.6}$ (using AlN particles as the Al source) powders with the same Al content are hot-pressed at 1900 °C/50 MPa/30 min. XRD results confirm the as-sintered MA-$Si_2BC_3N_{1.6}Al_{0.6}$ monoliths contain only three phases, β-SiC, BN(C) and AlN (PDF#70-0779). Using Al as the Al source, the AlN (PDF#70-0779 and PDF#89-3446), AlON, SiC and BN(C) form in the final hot-pressed $Si_2BC_3N_{1.6}Al_{0.6}$ ceramics (Fig. 4.46). AlN can most likely acts as sintering aids thus promoting ceramic densification.

The SEM morphologies of the two MA-$Si_2BC_3NAl_{0.6}$ and MA-$Si_2BC_3N_{1.6}Al_{0.6}$ ceramics show similar surface structure (Fig. 4.47). However, MA-$Si_2BC_3N_{1.6}Al_{0.6}$ ceramics seem more denser, possibly resulting from the promoting sintering of AlON phase (Fig. 4.48). The resulting MA-$Si_2BC_3N_{1.6}Al_{0.6}$ ceramics have a bulk density of 2.65 g/cm^3, lower than that of MA-$Si_2BC_3NAl_{0.6}$ ceramics. The Al addition promotes the grain growth to about 100–200 nm within the MA-$Si_2BC_3NAl_{0.6}$ matrix. The

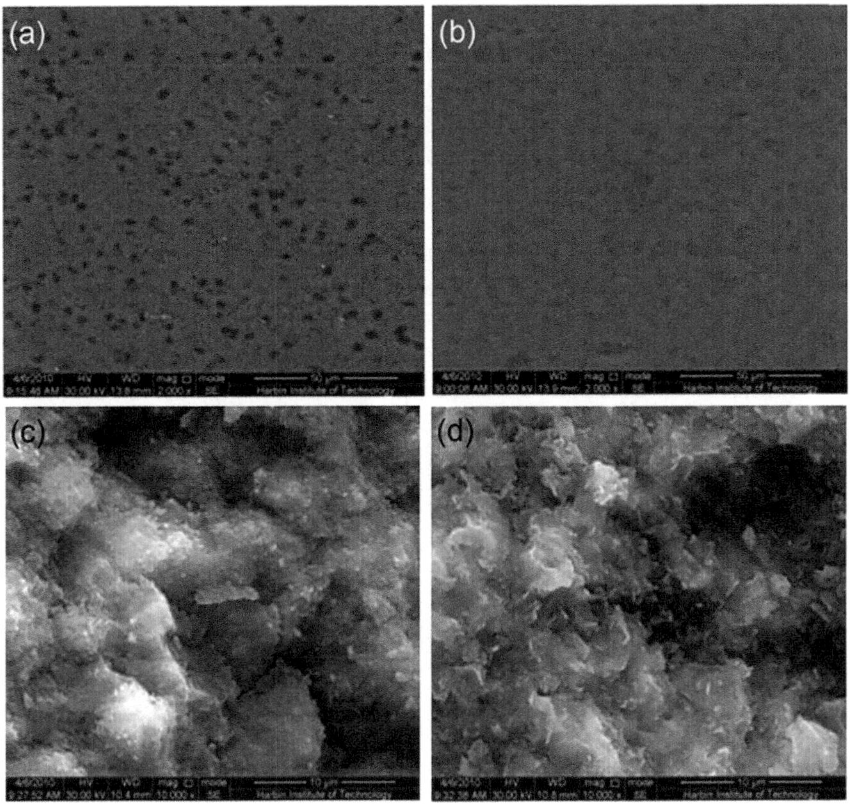

Fig. 4.43 SEM surface and fracture morphologies of MA-$Si_2BC_3NAl_{0.6}$ ceramics prepared by hot-pressing sintering at 1800 °C/30 MPa/30 min under different atmosphere (0.1 MPa): **a**, **c** under Ar; **c**, **d** sintered under N_2 [24]

SAED pattern is a sequence of polycrystalline diffraction spots and HRTEM images clearly suggest the stacking faults and twining in β-SiC crystals (Fig. 4.49).

The AlN introduction inhibits the grain growth. For MA-$Si_2BC_3N_{1.6}Al_{0.6}$ ceramics, an average grain size of ~50 nm is measured; besides, some structural defects are also found in β-SiC grains (Fig. 4.50). In this case, BN(C) phase contains more crystal defects inside the grains, including a large number of disordered atomic arrangements. The interfaces between SiC and AlN are clear and clean without trac of oxides (Fig. 4.51).

(3) Zr–Al particles introduced by MA

For all the investigated MA-Si_2BC_3N ceramics with or without Zr–Al incorporation, main phases of BN(C), α-SiC and β-SiC are identified [25]. Obviously, the relative peak intensities of β-SiC increase with the addition of Zr–Al, suggesting additive-induced grain growth of β-SiC. Nevertheless, higher Zr–Al additive shows no further

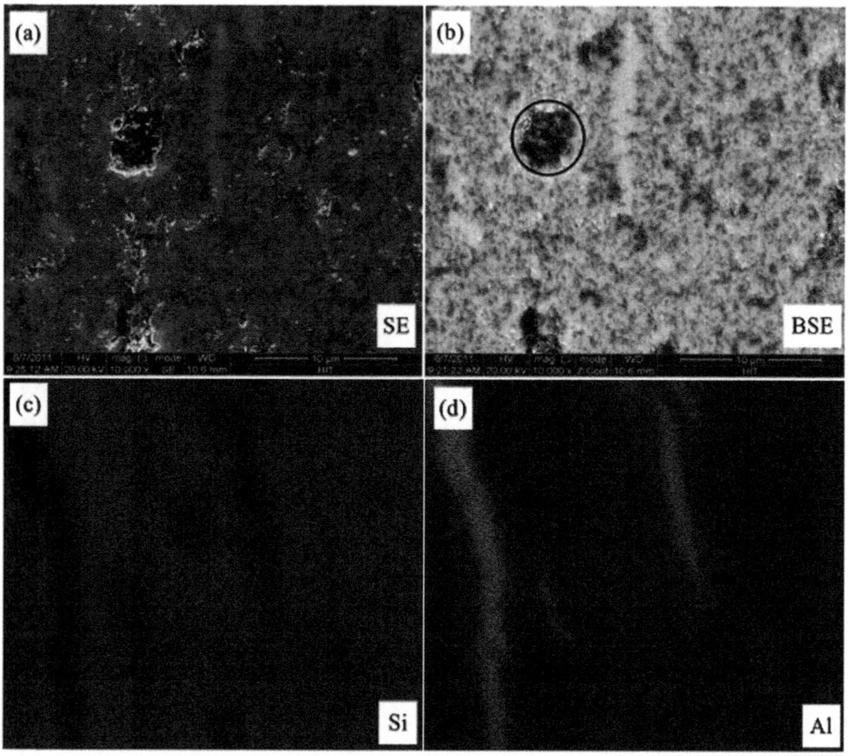

Fig. 4.44 SEM surface morphologies and element distribution of MA-$Si_2BC_3NAl_{0.6}$ ceramics hot-pressed at 1950 °C/40 MPa/30 min: **a** Secondary electron morphology; **b** backscattered electron morphology: **c** Si mapping; **d** Al mapping (micropores are caused by the mechanical force during sample preparation process) [23]

stimulation on the formation of β-SiC. Coincidentally, no Al-containing phases have been detected regardless of the additive content (Fig. 4.52).

The Zr–Al incorporated ceramics present well-developed structures. Specifically, larger BN(C) plates and SiC particles are embedded with each other. With only 1 mol.% Zr–Al additive, the fracture surfaces are much denser and rougher. Further increasing additive content, dense fracture surfaces are also observed and no apparent difference is detected. The observation of the rougher surfaces in the additives containing ceramics should be ascribed to the "pulling-out" of well-developed BN(C) plates and the debonding between BN(C) plates and SiC particles (Fig. 4.53).

Taken ceramics with 5 mol.% Zr–Al additive for example, the grain size of SiC is mainly around 200–500 nm, much larger than that of the pristine ceramics. A relative ambiguous interface separates ZrO_2 and SiC grains, implying ZrO_2 induced the grain growth of SiC. Element distribution mappings confirm the existence of well-dispersed SiC and BN(C). Some ZrO_2 and Al_2O_3 are observed adhere to SiC grains,

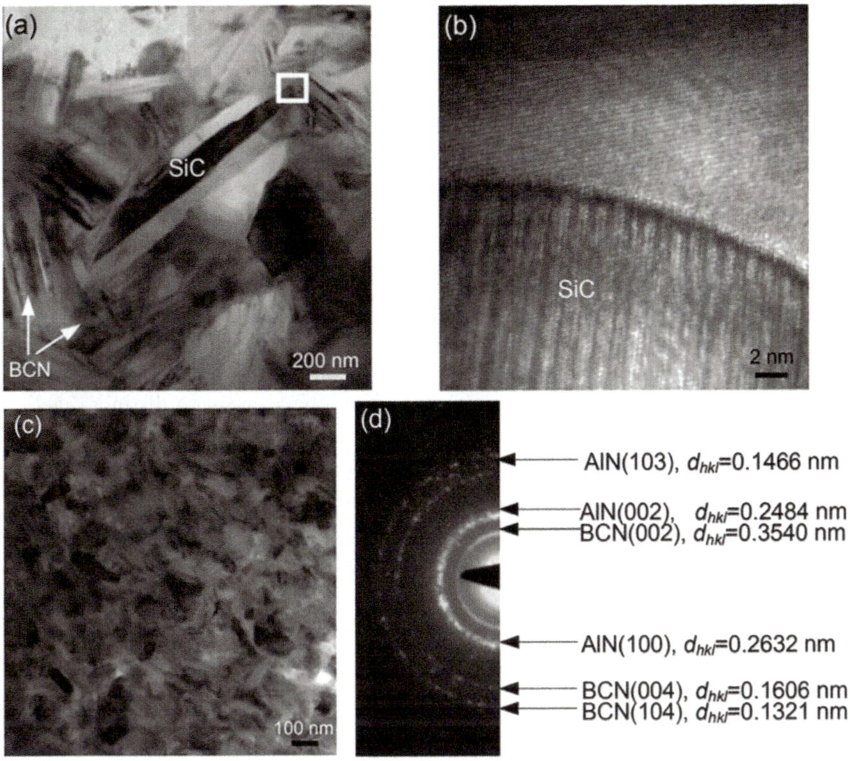

Fig. 4.45 TEM analysis of MA-$Si_2BC_3NAl_{0.6}$ bulk ceramics sintered at 1800 °C and 30 MPa under Ar **a**, **b** or N_2 **c**, **d**: **a** grain growth of SiC and BN(C); **b** HRTEM image of the area surrounded by the white box in (**a**); **c** nano-crystal agglomerating areas; **d** SAED pattern [24]

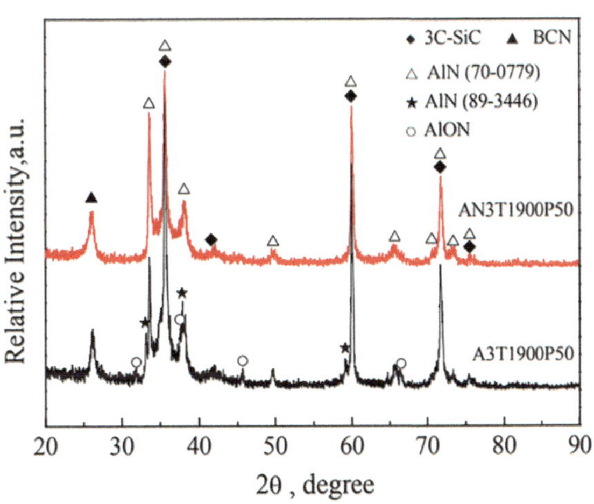

Fig. 4.46 XRD patterns of the bulk ceramics hot-pressed at 1900 °C/ 50 MPa/30 min: **a** MA-$Si_2BC_3NAl_{0.6}$ (using Al particles as Al source); **b** MA-$Si_2BC_3N_{1.6}Al_{0.6}$ (using AlN particle as Al source) [23]

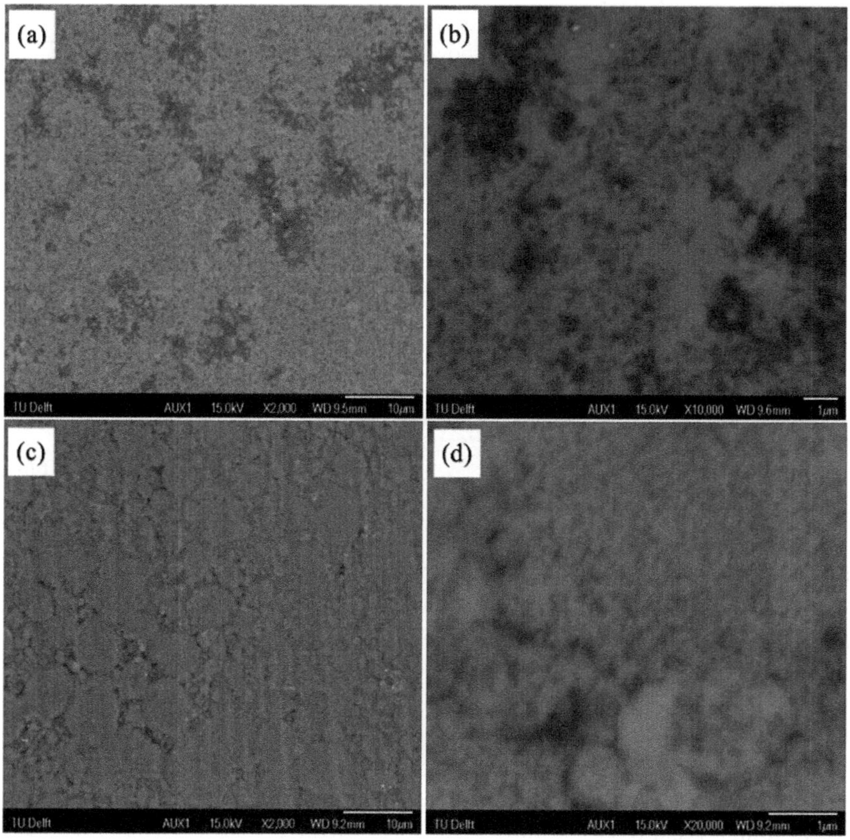

Fig. 4.47 Comparison of the SEM surface morphologies of the as-sintered bulk ceramics: **a, b** MA-$Si_2BC_3NAl_{0.6}$; **c, d** MA-$Si_2BC_3N_{1.6}Al_{0.6}$ [23]

suggesting the oxides coming from oxidation of Zr–Al additive that can stimulate the development of BN(C) and SiC (Fig. 4.54).

The bulk density and relative density of Zr–Al containing ceramics are improved obviously as compared with the pristine ceramics. For example, bulk ceramics with only 1 mol.% Zr–Al show density of ~2.70 g/cm^3, with 2 mol.% Zr–Al obtained density of ~2.71 g/cm^3, and with 5 mol.% Zr–Al obtained density of ~2.75 g/cm^3, much higher than that of monolithic counterparts. Note that no significant densification is further achieved with increasing Zr–Al additive.

(4) Mo particles introduced by MA

The following reactions may occur after introducing metal Mo into the MA-Si_2BC_3N ceramic matrix [26, 27]. At 25–2000 °C, the probable products are those of MoC, Mo_2C, Mo_3Si and $MoSi_2$. The formation of Mo_2N is thermodynamically infeasible under equilibrium circumstance (Fig. 4.55).

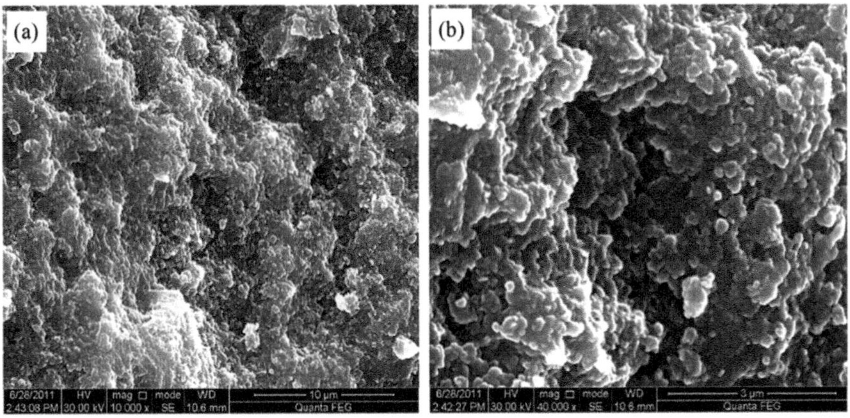

Fig. 4.48 SEM fracture morphologies of the MA-$Si_2BC_3N_{1.6}Al_{0.6}$ ceramics hot-pressed 1900 °C/50 MPa/30 min: **a** Low resolution; **b** high resolution [23]

$$2Mo(s) + C(s) \rightarrow Mo_2C(s) \tag{4.10}$$

$$Mo(s) + C(s) \rightarrow MoC(s) \tag{4.11}$$

$$3Mo(s) + Si(s) \rightarrow Mo_3Si(s) \tag{4.12}$$

$$Mo(s) + 2Si(s) \rightarrow MoSi_2(s) \tag{4.13}$$

$$2Mo(s) + BN(S) \rightarrow Mo_2N(s) + B(s) \tag{4.14}$$

$$8Mo(s) + Si_3N_4(s) \rightarrow 4Mo_2N(s) + 3Si(s) \tag{4.15}$$

$$Mo(s) + SiC(s) \rightarrow MoC(s) + Si(s) \tag{4.16}$$

$$2Mo(s) + SiC(s) \rightarrow Mo_2C(s) + Si(s) \tag{4.17}$$

The products change regularly with increasing MA-Si_2BC_3N content. With 10 or 20 vol.% MA-Si_2BC_3N, there is still unreacted Mo. With the increasing MA-Si_2BC_3N, substantial Mo is consumed and finally disappears when 30 vol.% MA-Si_2BC_3N added. With 10–30 vol.% MA-Si_2BC_3N addition, reactions between Mo and MA-Si_2BC_3N lead to MoC, Mo_2C, Mo_3Si, Mo_5Si_3 or $MoSi_2$. When 40–60 vol.% MA-Si_2BC_3N is added, the main products are $Mo_{4.8}Si_3C_{0.6}$, MoC and Mo_2C. With 70–90 vol.% MA-Si_2BC_3N, MoC and Mo_2C disappear, and the products are those of $Mo_{4.8}Si_3C_{0.6}$ (Fig. 4.56). Generally, the reaction products among Mo and C

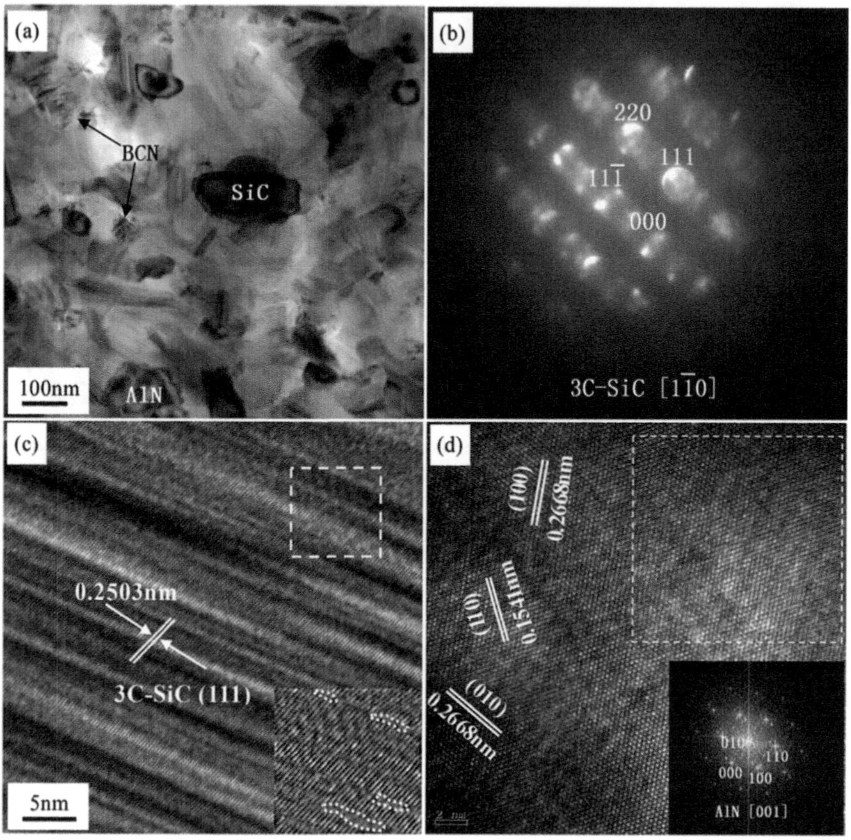

Fig. 4.49 TEM analysis of MA-$Si_2BC_3NAl_{0.6}$ ceramics prepared by hot-pressed at 1900 °C: **a** Bright-field image: **b** SAED pattern: **c** HRETM image of β-SiC; **d** HRETM image of AlN [23]

are mainly Mo_2C and MoC. With few MA-Si_2BC_3N addition, there tends to form Mo_3Si, while Mo_5Si_3 and $MoSi_2$ form with increasing MA-Si_2BC_3N content. The following Mo–Si containing products change in turn with increasing MA-Si_2BC_3N content: $Mo_3Si \rightarrow Mo_3Si + Mo_5Si_3 + MoSi_2 \rightarrow Mo_{4.8}Si_3C_{0.6} + Mo_5Si_3 + MoSi_2 \rightarrow Mo_{4.8}Si_3C_{0.6}$.

The phase composition of the as-sintered ceramics with or without MgO–Al_2O_3–SiO_2 (MAS) additive does not change obviously, but differs in the FWHM of each diffraction peaks. The ceramics with 90 vol.% MA-Si_2BC_3N are composed of $Mo_{4.8}Si_3C_{0.6}$, SiC, and BN(C), while no traces of Al_2O_3 and/or SiO_2 are found for ceramics with MAS additive. The 10 wt.% MAS additive strongly promotes the crystallization and therefore densification of MA-Si_2BC_3N ceramics during HP sintering (Fig. 4.57).

On the surfaces of the ceramics with 10 vol.% MA-Si_2BC_3N, a small number of black areas are distributed in the large gray areas, indicating some MA-Si_2BC_3N

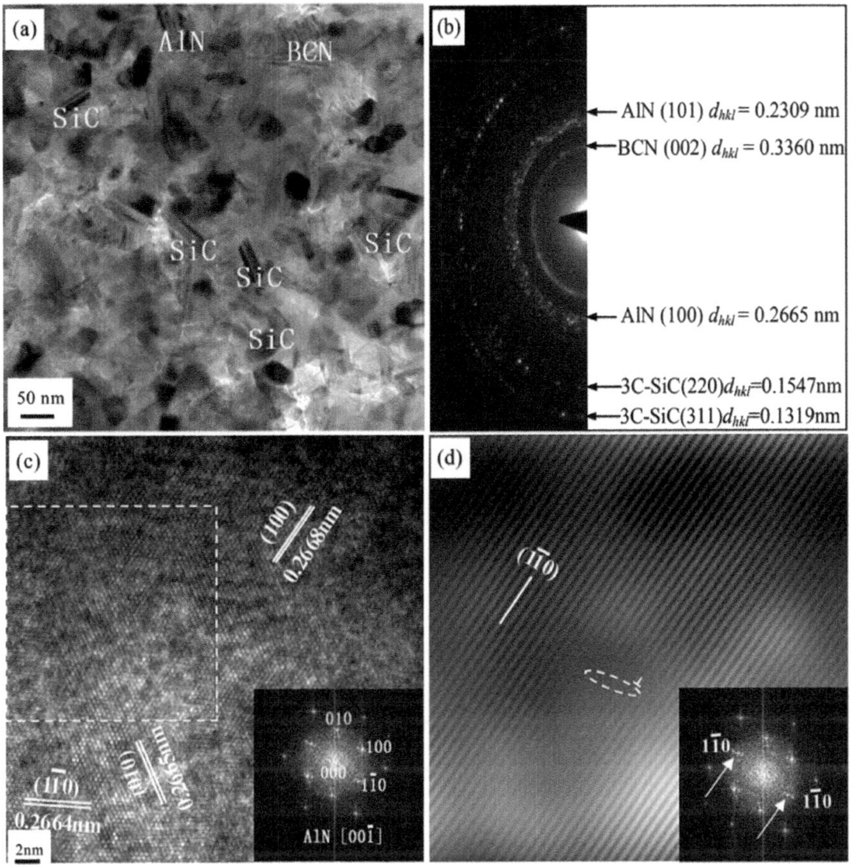

Fig. 4.50 TEM analysis of MA-$Si_2BC_3N_{1.6}Al_{0.6}$ ceramics hot-pressed at 1900 °C: **a** Bright-field image; **b** SAED pattern: **c** HRTEM image of AlN; **d** inverse FFT image of box in (**c**) [23]

does not react with Mo at this time. With the increasing MA-Si_2BC_3N content, the black areas gradually increase. When 40 vol.% MA-Si_2BC_3N added, the black areas gradually converge to form a cloud-like structure. With 70 vol.% MA-Si_2BC_3N, the products are mainly $Mo_{4.8}Si_3C_{0.6}$ and SiC, accompanied with the decreasing gray areas and the increasing black areas (Fig. 4.58).

With 10 wt.% MAS addition, the exfoliated particles and flakes disappear due to the fluidity of the glassy phase formed by the additive (Fig. 4.59). The bulk density of Mo is 10.22 g/cm^3, that of Mo_2C is 9.1 g/cm^3, that of MoC is 9.2 g/cm^3 and that of Mo–Si compounds is 6–8 g/cm^3. With increasing MA-Si_2BC_3N content, the bulk density of the composite ceramics shows a downward trend except with the 50 vol.% MA-Si_2BC_3N addition. The 10 wt.% MAS addition greatly promotes the densification of Mo–Si_2BC_3N composite ceramics, obtaining bulk density of

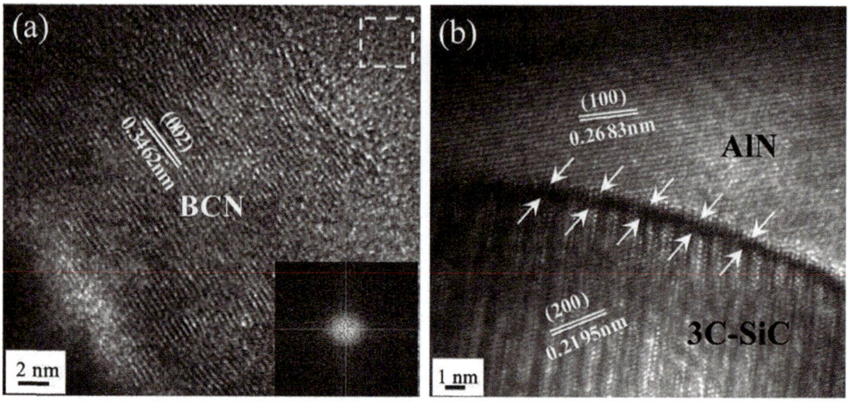

Fig. 4.51 HRTEM images of the MA-$Si_2BC_3N_{1.6}Al_{0.6}$ ceramics hot-pressed at 1900 °C/50 MPa/30 min: **a** BN(C) structure; **b** interfacial structure between AlN and SiC [23]

Fig. 4.52 XRD patterns of MA-Si_2BC_3N bulk ceramics with different Zr–Al additive (Zr–Al mole ratio of 1:1) [25]

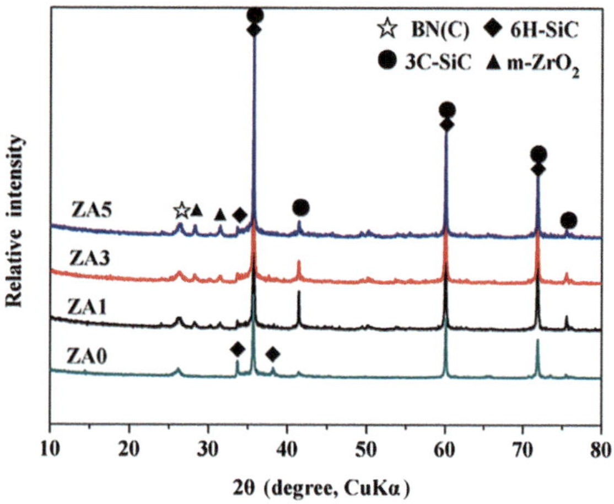

2.44 g/cm^3 higher than that ~1.91 g/cm^3 of pure MA-Si_2BC_3N monoliths. The load–displacement curves of the Mo–Si_2BC_3N ceramics suggest a better strength of the composites with lower MA-Si_2BC_3N addition (Fig. 4.60).

(5) Al_4SiC_4 particles introduced by MA

The amorphous Al_4SiC_4 powder can be prepared by MA technique (Fig. 4.61). After annealing at 1200 °C, the broad and blunt peaks of Al_4SiC_4 are detected [28]. Further annealing at 1400 °C, narrower peaks assigned to Al_4SiC_4 are also identified. However, no Al_4SiC_4 phases are determined when heated to 1600–1800 °C, instead of the strong and sharp peaks of AlN and SiCN, as well as a small amount of graphite. For ceramics sintered by SPS at 1900 °C/40 MPa/5 min, the main phases of BN(C)

Fig. 4.53 SEM fracture morphologies of the bulk ceramics with or without Zr–Al (mole ratio of 1:1) additive: **a**, **b** Pure MA-Si_2BC_3N; **c**, **d** + 1 mol.% Zr–Al; **e** + 3 mol.% Zr–Al; **f** + 5 mol.% Zr–Al [25]

and α/β-SiC are identified. Obviously, the relative peak intensities of BN(C) increase with the addition of Al_4SiC_4 additive (Fig. 4.62).

The grain sizes of SiC (100–150 nm) and BN(C) (up to 2 μm) are apparently larger than those observed in the pristine MA-Si_2BC_3N ceramics. Thus, Al_4SiC_4 addition enhances the development of ceramic matrix. Coincidentally, mainly BN(C), Al_4SiC_4 and Al_4C_3/AlN are found in agglomerating areas. In addition to the BN(C), the formation of Al_4C_3/AlN can be attributed to the partial decomposition of Al_4SiC_4 at this sintering temperature. Simultaneously, the Al can also react with sintering atmosphere to form some AlN (Fig. 4.63).

(6) Cu–Ti particles introduced by pressure infiltration

After melting at 1100–1115 °C, liquid Cu forms spherical droplets on surfaces of porous MA-Si_2BC_3N and their contact angle is stabilized at 112°–114° (Fig. 4.64). Pure Cu is non-wettable on the surfaces of porous MA-Si_2BC_3N ceramics [29]. After adding 10 at.% Ti, the wettability between $Cu_{90}Ti_{10}$ and porous MA-Si_2BC_3N is improved significantly finally forming hemispherical droplets at 1040 °C. For temperature higher than 1050 °C, the height of $Cu_{90}Ti_{10}$ droplets gradually decreases and the final contact angle is levelling out at ~44.3°. The melting point of Cu can be further reduced by increasing Ti content. The $Cu_{80}Ti_{20}$ heated at 920 °C leads to the hemispherical droplets that spread rapidly on the surfaces of porous MA-Si_2BC_3N. At 1000 °C, the contact angle between $Cu_{90}Ti_{10}$ and porous MA-Si_2BC_3N stabilize at ~17.27°, and it further decreases to ~15.60° at 1100 °C.

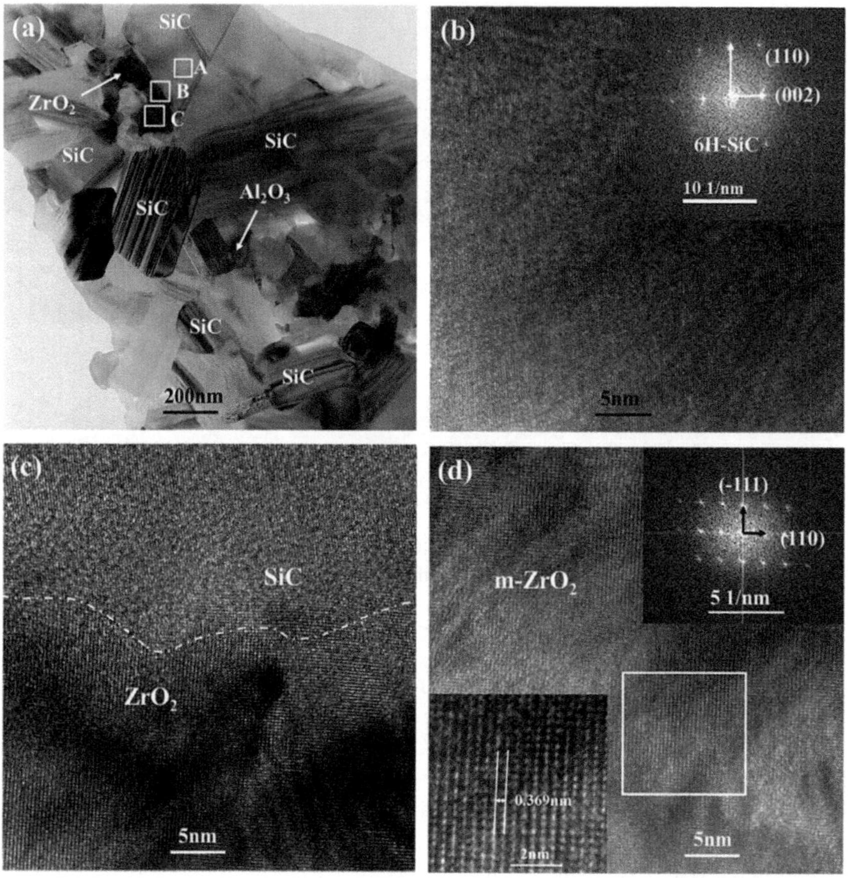

Fig. 4.54 TEM analysis of MA-Si_2BC_3N ceramics with 5 mol% Zr–Al additive: **a** Bright-field image; **b–d** HRTEM images of corresponding regions "A–C" in (**a**) [25]

Prior to the infiltration of metal Cu–Ti, the black and rough surface of the porous MA-Si_2BC_3N structure is observed. After infiltration, the surfaces of the MA-Si_2BC_3N become smooth and denser with distinctive metallic luster (Fig. 4.65). The liquid $Cu_{90}Ti_{10}$ spreads well on the porous ceramic surfaces, but the maximum penetration depth only stops at 612.17 μm. The larger the pore size, the greater the $Cu_{90}Ti_{10}$ penetrating depth is measured. The porous MA-Si_2BC_3N ceramic matrix is rich in Si, while the metal layer is mainly composed of Cu. The Ti mainly distributes at the interfaces of metal and MA-Si_2BC_3N, and a diffusion layer 2–4 μm is observed. The oxide film on the surfaces of the liquid metal may hinder the interfacial reactions thus affecting the effective infiltration (Fig. 4.66).

As compared to $Cu_{90}Ti_{10}$, the penetration depth of $Cu_{80}Ti_{20}$ into the porous MA-Si_2BC_3N matrix increase greatly, reaching to an optimized value of 1631.70 μm. After some $Cu_{80}Ti_{20}$ infiltration, Ti elements are concentrated at the interfaces of

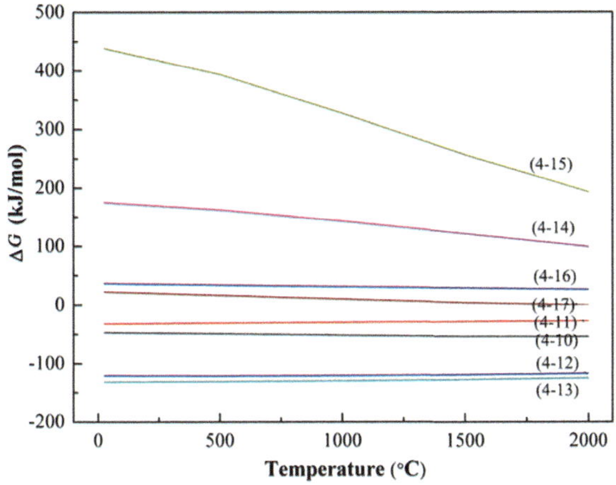

Fig. 4.55 The Gibbs free energy change curve for the possible chemical reactions between metal Mo and MA-Si_2BC_3N in the temperature range of 25–2000 °C [26]

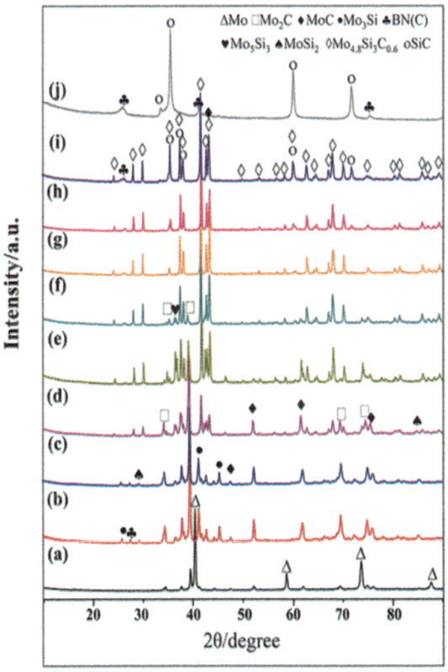

Contents of SiBCN/vol%	Phase composition
10	Mo, Mo_2C, MoC, Mo_3Si, BN(C)
20	Mo, Mo_2C, MoC, Mo_3Si, Mo_5Si_3, $MoSi_2$, BN(C)
30	Mo_2C, MoC, Mo_3Si, Mo_5Si_3, $MoSi_2$, BN(C)
40	$Mo_{4.8}Si_3C_{0.6}$, Mo_2C, MoC, Mo_5Si_3, $MoSi_2$, BN(C)
50	$Mo_{4.8}Si_3C_{0.6}$, Mo_2C, MoC, BN(C)
60	$Mo_{4.8}Si_3C_{0.6}$, Mo_2C, MoC, BN(C)
70	$Mo_{4.8}Si_3C_{0.6}$, SiC, BN(C)
80	$Mo_{4.8}Si_3C_{0.6}$, SiC, BN(C)
90	$Mo_{4.8}Si_3C_{0.6}$, SiC, BN(C)
100	SiC, BN(C)

Fig. 4.56 XRD patterns of the Mo–Si_2BC_3N composite ceramics hot-pressed at 1900 °C/60 MPa/30 min [26]

metal/Si_2BC_3N near the ceramic layer. The sufficient amount of Ti acts as the active element that reacts with the ceramic matrix, and therefore forms a bilayer structure of "ceramic–metal–ceramic" at the interfaces between metal Cu and MA-Si_2BC_3N. Such a reaction layer with desirable thickness and bonding strength can effectively promote the wetting between Cu and the pore wall of porous MA-Si_2BC_3N, thereby increasing the penetration depth of $Cu_{80}Ti_{20}$ (Fig. 4.67).

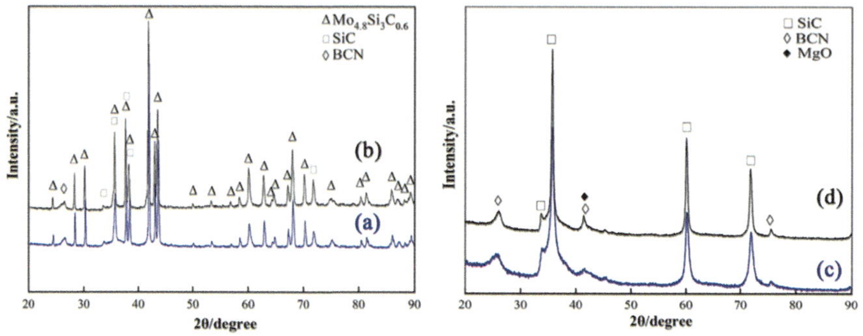

Fig. 4.57 XRD patterns of hot-pressed Mo–Si_2BC_3N composite ceramics with or without 10 wt.% MAS additive: **a** 90 vol.% MA-Si_2BC_3N; **b** 90 vol.% MA-Si_2BC_3N + MAS; **c** 100 vol.% MA-Si_2BC_3N; **d** 100 vol.% MA-Si_2BC_3N + MAS [26]

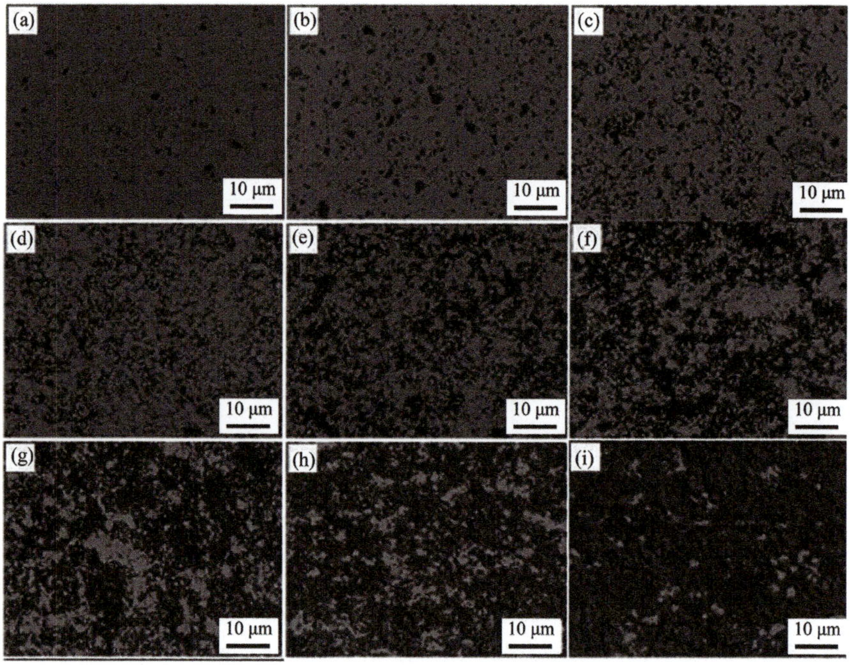

Fig. 4.58 SEM surface morphologies of the Mo–Si_2BC_3N composite ceramics: **a** + 10 vol.% MA-Si_2BC_3N; **b** + 20 vol.% MA-Si_2BC_3N; **c** + 30 vol.% MA-Si_2BC_3N; **d** + 40 vol.% MA-Si_2BC_3N; **e** + 50 vol.% MA-Si_2BC_3N. **f** + 60 vol.% MA-Si_2BC_3N; **g** + 70 vol.% MA-Si_2BC_3N; **h** + 80 vol.% MA-Si_2BC_3N; **i** + 90 vol.% MA-Si_2BC_3N [26]

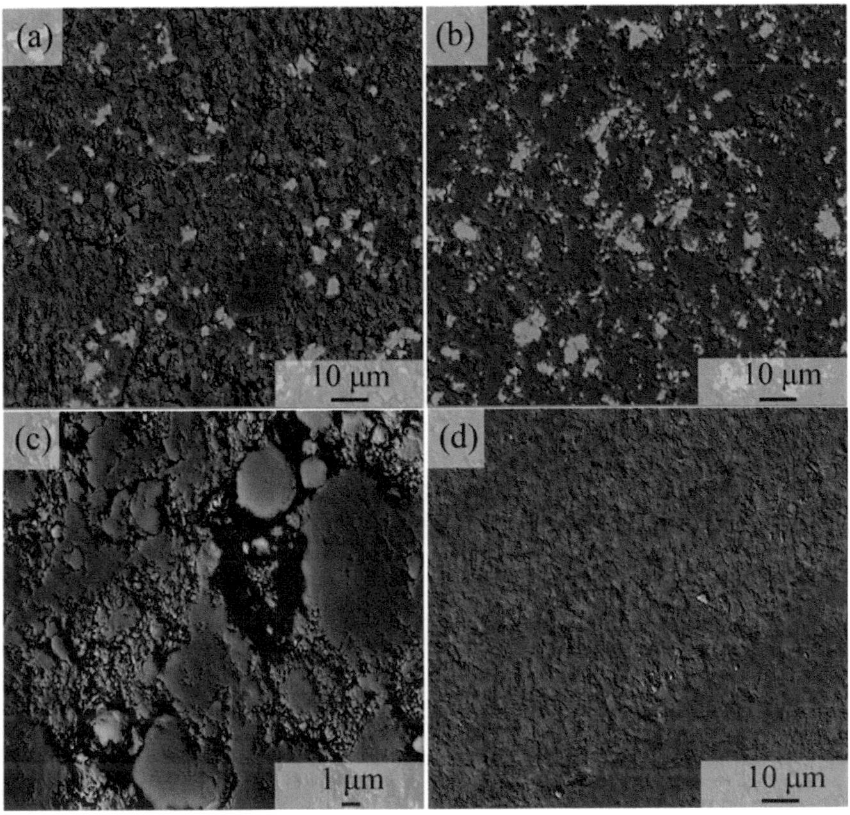

Fig. 4.59 SEM images of the hot-pressed Mo–Si_2BC_3N composite ceramics with or without 10 wt.% MAS additive: **a** 90 vol.% MA-Si_2BC_3N; **b** 90 vol.% MA-Si_2BC_3N + MAS; **c** 100 vol.% MA-Si_2BC_3N; **d** 100 vol.% MA-Si_2BC_3N + MAS [26]

Due to the complex reactions occurring, the phase composition of as-infiltrated $Cu_{80}Ti_{20}/Si_2BC_3N$ composite ceramics is complicated, which are primarily composed of SiC, BN(C), Cu_3Ti_2, TiC(N), Cu_5Si, TiB_2, Cu_3Si_2, and few C, Cu and Ti (Fig. 4.68). The possible chemical reactions are as follows:

$$\mathrm{Ti(s) + SiC(s) \rightarrow TiC(s) + Si(s)} \tag{4.18}$$

$$\mathrm{Si(s) + 5Cu(s) \rightarrow Cu_5Si(s)} \tag{4.19}$$

$$\mathrm{5Cu(s) + SiC(s) \rightarrow Cu_5Si(s) + C} \tag{4.20}$$

$$\mathrm{3Ti(s) + 2BN(C)(s) \rightarrow TiB_2(s) + 2TiN(s) + 2C(s)} \tag{4.21}$$

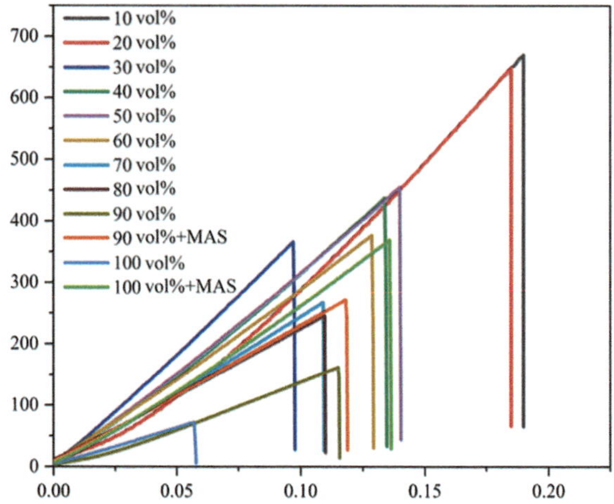

Fig. 4.60 Load–displacement curves of the Mo–Si_2BC_3N composite ceramics with different content of MA-Si_2BC_3N and 10 wt.% MAS additive [26]

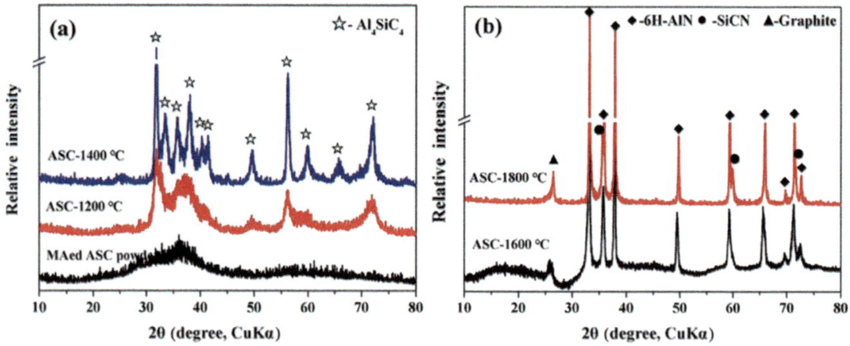

Fig. 4.61 XRD patterns of as-milled Al_4SiC_4 before and after annealed at different temperature: **a** RT ~1400 °C; **b** 1600–1800 °C [28]

The critical infiltration processes by which the active metal Ti enhances the wettability of Cu and porous MA-Si_2BC_3N are controlled by Eqs. (4.18) and (4.19). The formed TiC is very stable and the resulting Si can react with the liquid Cu to generate Cu_5Si compounds, finally forming an interfacial reaction layer composed of TiC and Cu_5Si with a moderate bonding strength. Besides, reactions between Ti and BN(C) phase can produce some TiB_2 and TiN, which can avoid the over-reaction of metal Ti and SiC.

With deceasing infiltrated $Cu_{80}Ti_{20}$ content, both the measured densities and theoretical densities of the three composites decreases. However, the measured density decreases more quickly than the theoretical density. For the composites with 30 vol.% MA-Si_2BC_3N, the measured density is higher than its theoretical density; however,

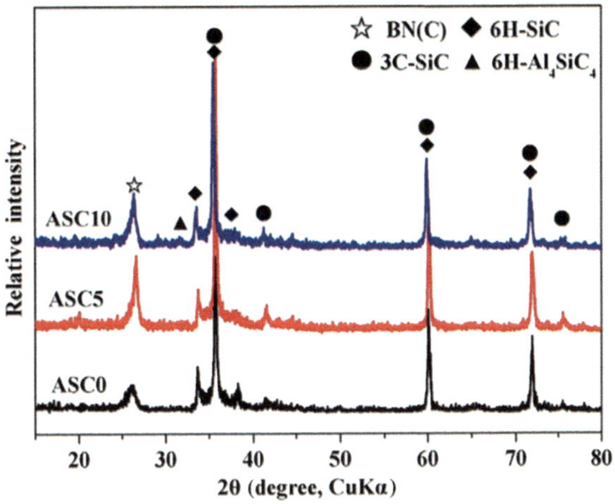

Fig. 4.62 XRD patterns of MA-Si_2BC_3N ceramics with various Al_4SiC_4 additive consolidated at SPS 1900 °C/40 MPa/5 min [28]

the measured density of the composite ceramics with 40 vol.% MA-Si_2BC_3N is lower than its theoretical density. The $Cu_{80}Ti_{20}$/Si_2BC_3N with 35 vol.% MA-Si_2BC_3N shows similar values of measured density and theoretical density (Fig. 4.69 and Table 4.2).

Perpendicular to the freezing direction (FD), the yellow strip of infiltrated $Cu_{80}Ti_{20}$ and black region of MA-Si_2BC_3N matrix are found for composites with 30 vol.% MA-Si_2BC_3N. Besides, an interfacial reaction layer between the metal and ceramic matrix is also confirmed. Parallel to the FD, the metal layers and ceramic layers are alternately parallel arranged for composites with 35 vol.% MA-Si_2BC_3N. Perpendicular to the FD, these layers are also approximately parallel arrangement. For composites with 40 vol.% MA-Si_2BC_3N content, the metal layers parallel to the FD gradually become discontinuous due to the higher solid content. The reaction products of the metal and ceramic matrix block the infiltrated channels and thus retard the liquid metal from penetrating deeply (Fig. 4.70). The reaction layer between metal $Cu_{80}Ti_{20}$ and MA-Si_2BC_3N is in thickness of 10–15 μm for 30 vol.% MA-Si_2BC_3N containing composites, followed by the composites with 35 vol.% MA-Si_2BC_3N (5–8 μm) and composites with 40 vol.% MA-Si_2BC_3N (1–4 μm).

At the interfaces of MA-Si_2BC_3N and $Cu_{80}Ti_{20}$, the bright regions are the infiltrated $Cu_{80}Ti_{20}$, and the dark zones are those of ceramic matrix. Since Cu and Ti metals have both penetrated into the porous MA-Si_2BC_3N matrix, a Ti–rich reaction layer containing B, C and N elements positions between Cu and the ceramic matrix. (Fig. 4.71). In the inner reaction layer, the dark regions are the Cu and Si containing compounds with some SiC. The bright ones present two distinct morphologies: the irregular structure with a small size of 90–120 nm and a rod-shaped structure with a larger size of 200–350 nm in width. The rod-shaped grains are those of TiB_xN_y, while the irregular structure is composed of mainly TiC_xN_y (Fig. 4.72). The formed interfacial structure between the porous MA-Si_2BC_3N and $Cu_{80}Ti_{20}$ is governed by the

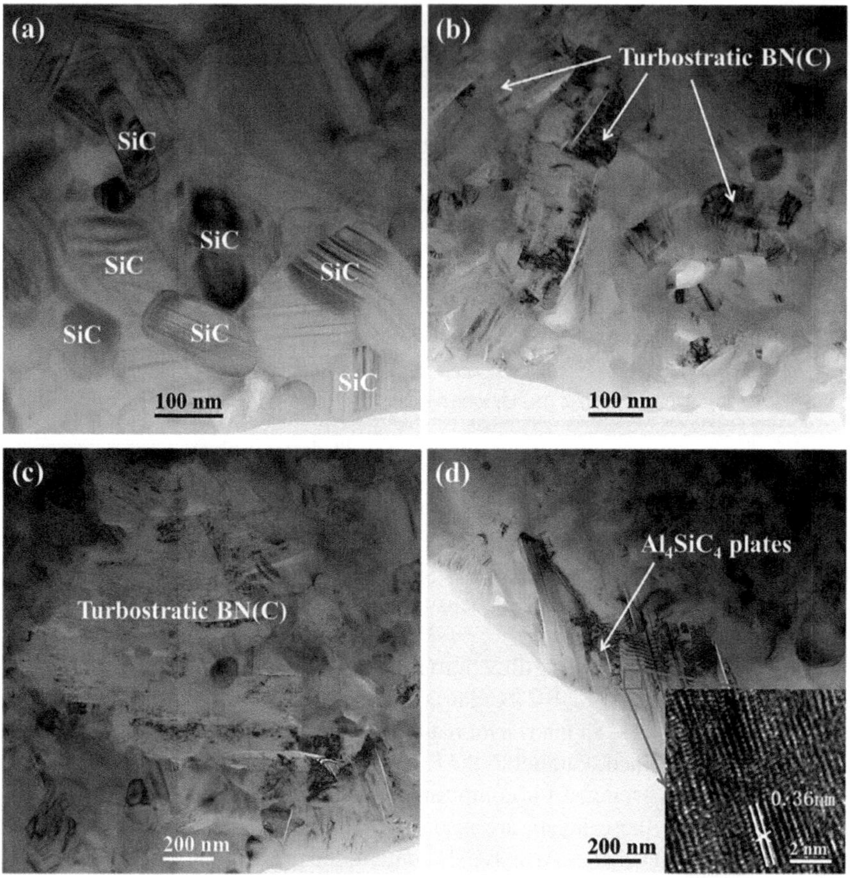

Fig. 4.63 TEM micrographs of MA-Si_2BC_3N with 5 wt.% Al_4SiC_4 addition: **a** Overview; **b**, **c** growth of turbostratic BN(C); **d** Al_4SiC_4 structure [28]

diffusion-based chemical reactions, in which metals Cu and Si diffuse and therefore react with ceramic matrix (Fig. 4.73).

4.3.2 Effect of Oxide Sintering Additives

(1) ZrO_2 Particles Introduced by MA

Full densification of the MA-Si_2BC_3N powder free from any additives is difficult, since under an axial pressure of 80 MPa, significant sintering occurs just when the sintering temperature is higher than 1830 °C [30]. As expected, the addition of AlN or ZrO_2 indeed promotes the sintering and therefore densification of the MA-Si_2BC_3N (Fig. 4.74).

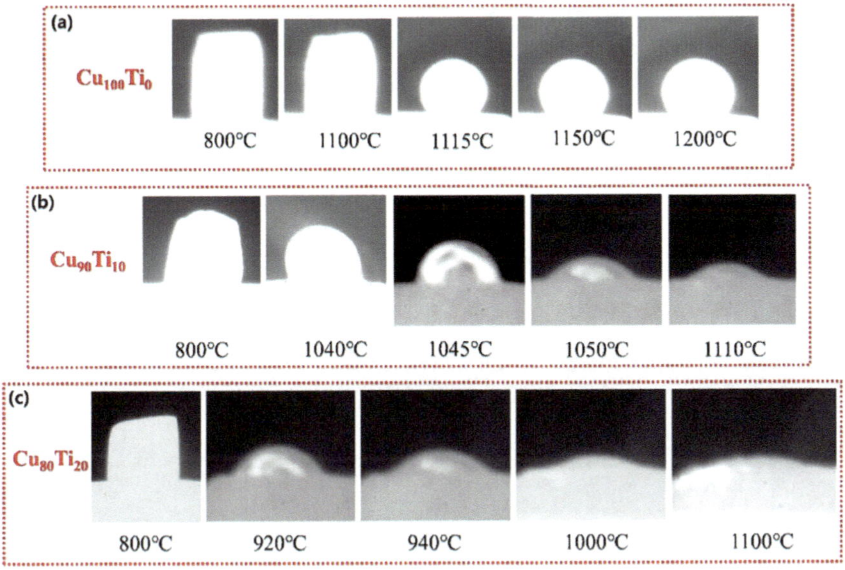

Fig. 4.64 The optical photographs showing the wetting behavior between Cu–Ti and porous MA-Si_2BC_3N ceramics: Infiltrated with **a** pure Cu, **b** $Cu_{90}Ti_{10}$ and **c** $Cu_{80}Ti_{20}$ [29]

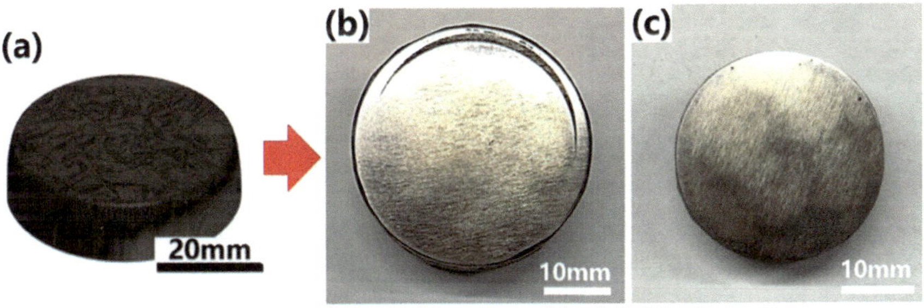

Fig. 4.65 Macroscopic photographs of the (Cu–Ti)/Si_2BC_3N composites: **a** Before infiltration; **b** infiltration of $Cu_{90}Ti_{10}$; **c** infiltration of $Cu_{80}Ti_{20}$ [29]

The densification process of the MA-Si_2BC_3N powder containing AlN or ZrO_2 (5 mol.%) also consists of three stages: the closer packing of ceramic particles ascribed to the applied axial pressure, the initial sintering stage and the intermediate sintering stage. However, with the addition of AlN or ZrO_2, significant densification occurs to the ceramic body at about ~1760 °C and ~1720 °C, respectively, much lower than 1830 °C for the pure MA-Si_2BC_3N ceramics. The shrinkage rate of the ceramic body with 5 mol.% AlN or ZrO_2 reaches its maximum 3 and 5 min early, obtaining $d(\Delta L/L_0)/dt \approx 0.035$ and $d(\Delta L/L_0)/dt \approx 0.045$, respectively.

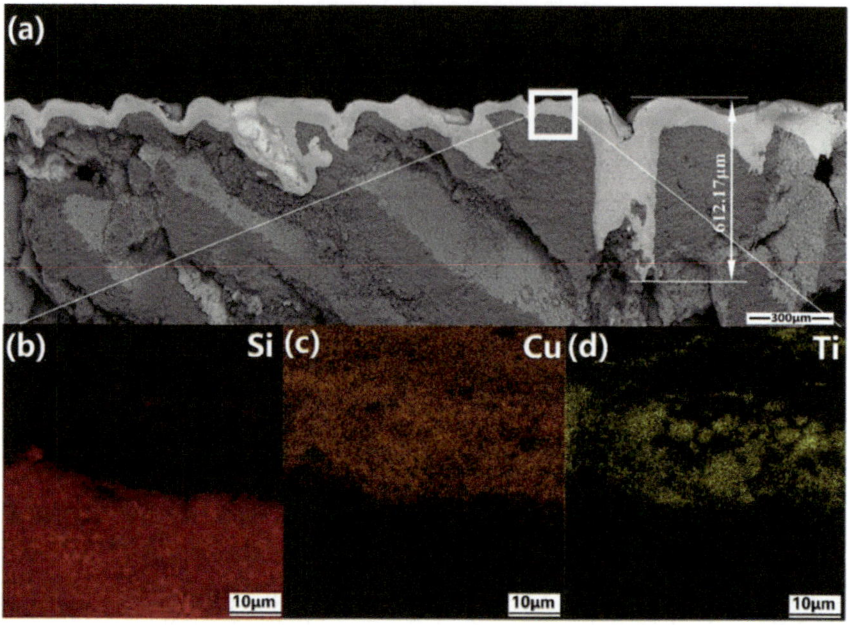

Fig. 4.66 Effect of $Cu_{90}Ti_{10}$ on the infiltration behavior of porous MA-Si_2BC_3N ceramics: **a** SEM cross-sectional photograph: **b–d** corresponding elemental distribution [29]

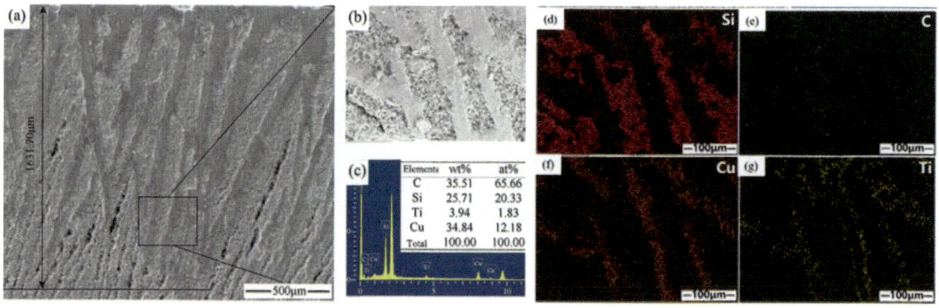

Fig. 4.67 Effect of $Cu_{80}Ti_{20}$ on the infiltration behavior of porous MA-Si_2BC_3N ceramics: **a**, **b** SEM cross-sectional photograph; **c** element distribution of the white box in (**a**); **d–g** corresponding element distribution [29]

Some small holes exist in the hot-pressed additive-free ceramics, while ceramics with 5 mol.% AlN or ZrO_2 show relatively higher density and large particle deformation. This is in agreement with the sintering curves. Hot-pressed at 1900 °C and under a pressure of 80 MPa for 30 min, the bulk density and relative density of the as-prepared additive-free bulk ceramics are 2.52 g/cm^3 and 88.7%, respectively. However, the addition of 5 mol.% AlN or ZrO_2 significantly increases the relative density to 96.9 and 97.7% (Fig. 4.75).

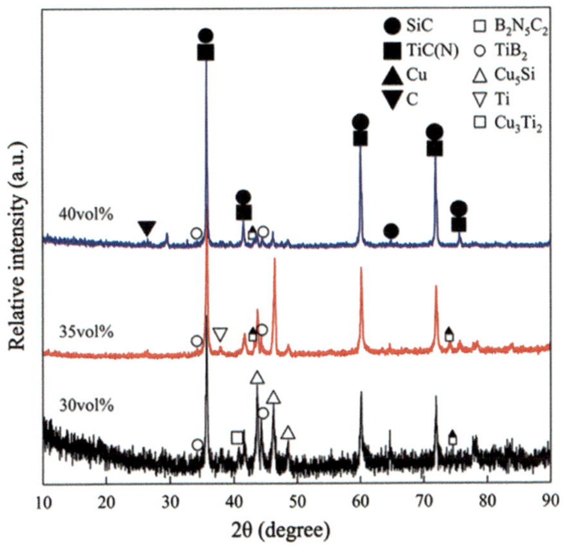

Fig. 4.68 XRD patterns of the as-infiltrated $Cu_{80}Ti_{20}$/Si_2BC_3N composite ceramics with different MA-Si_2BC_3N content [29]

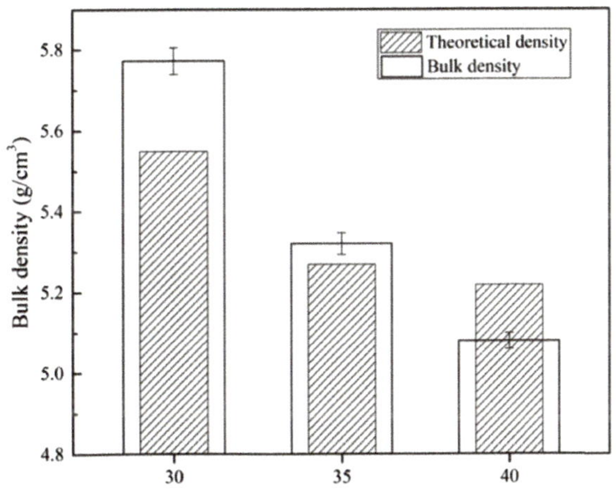

Fig. 4.69 Comparison of the measured bulk densities and theoretical densities of the $Cu_{80}Ti_{20}$/Si_2BC_3N composite ceramics with different MA-Si_2BC_3N content [29]

Table 4.2 The bulk density, theoretical density and relative density of the $Cu_{80}Ti_{20}$/Si_2BC_3N composite ceramics with different MA-Si_2BC_3N content [29]

MA-Si_2BC_3N content (vol.%)	$Cu_{80}Ti_{20}$ content (vol.%)	Theoretical density (g/cm^3)	Bulk density (g/cm^3)	Relative density (%)
30	51.49	5.55	5.77	104.0
35	49.76	5.27	5.32	100.9
40	48.70	5.22	5.08	97.3

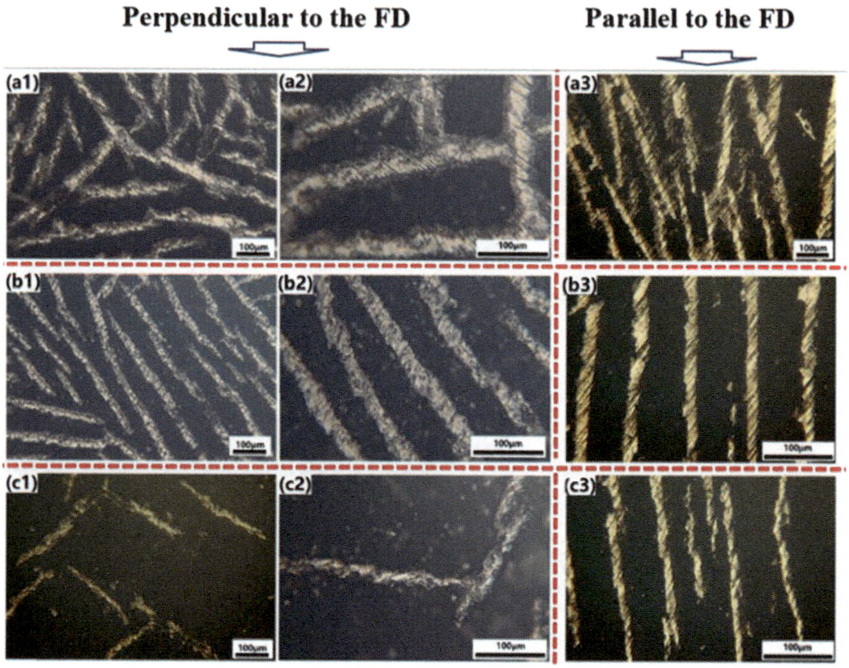

Fig. 4.70 Metallographic photographs of the various $Cu_{80}Ti_{20}/Si_2BC_3N$ composites: **a** 30 vol.% MA-Si_2BC_3N; **b** 35 vol.% MA-Si_2BC_3N; **c** 40 vol.% MA-Si_2BC_3N (FD: freezing direction) [29]

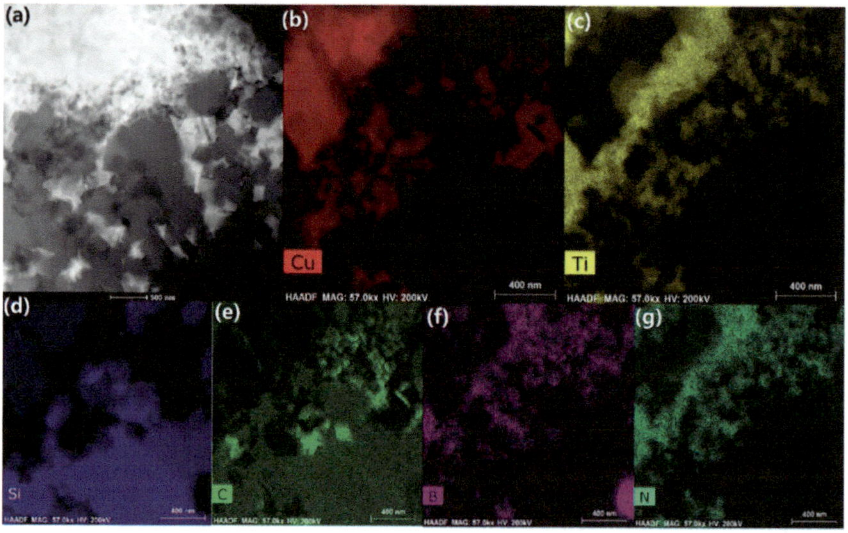

Fig. 4.71 Microstructure and element distribution of the reaction interface of the $Cu_{80}Ti_{20}/Si_2BC_2N$ composite ceramics with 35 vol.% MA-Si_2BC_3N content: **a** STEM image: **b–g** corresponding element mappings [29]

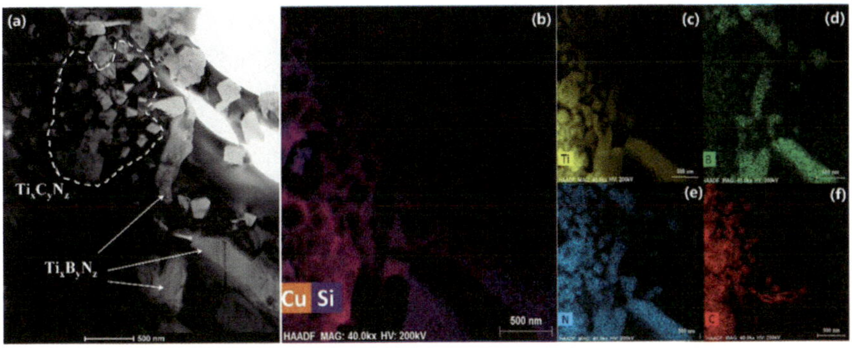

Fig. 4.72 Microstructure and element distribution of the inner reaction region of the $Cu_{80}Ti_{20}$/Si_2BC_3N composites with 35vol.% MA-Si_2BC_3N: **a** STEM image; **b**–**f** corresponding element distribution [29]

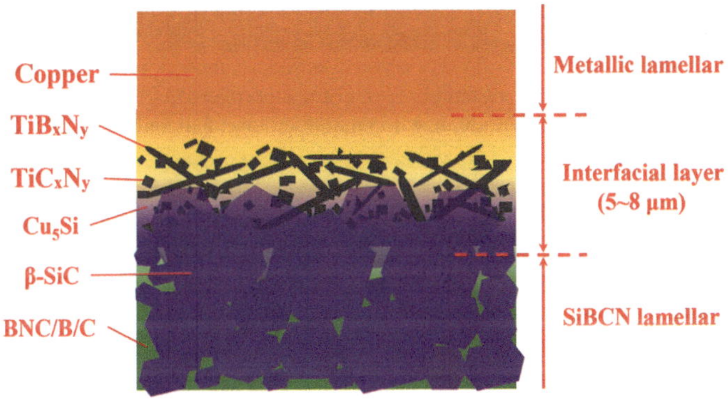

Fig. 4.73 Schematic showing the interfacial structure and phase distribution of the $Cu_{80}Ti_{20}$/Si_2BC_3N composite ceramics with 35 vol.% MA-Si_2BC_3N content [29]

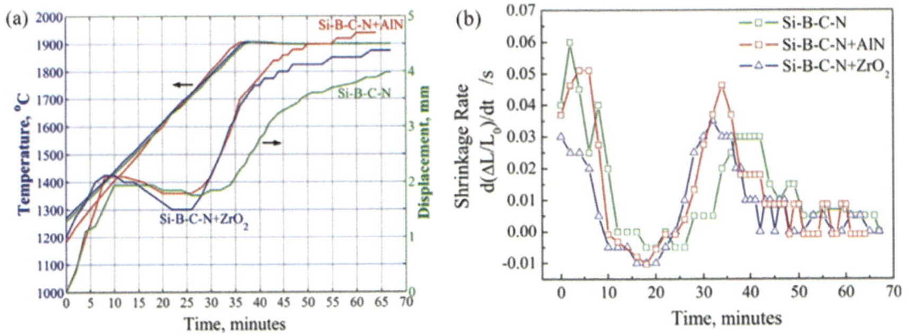

Fig. 4.74 The sintering curves of the amorphous MA-Si_2BC_3N powders with 5 mol% ZrO_2 or AlN additive: **a** The temperature and pressing head displacement variation versus sintering time; **b** the shrinkage rate versus sintering time [30]

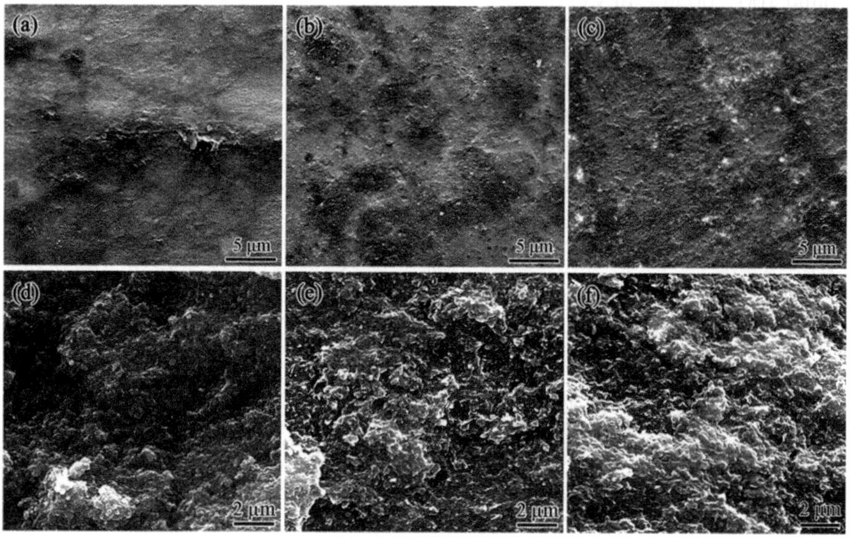

Fig. 4.75 SEM surface and fracture morphologies of hot-pressed MA-Si_2BC_3N ceramics at 1900 °C/80 MPa/30 min: **a**, **d** MA-Si_2BC_3N; **b**, **e** + 5 mol% ZrO_2; **c**, **f** + 5 mol% A1N [30]

All the as-sintered bulk ceramics present a similar phase structure consisting of β-SiC, α-SiC and BN(C) (Fig. 4.76). For pure MA-Si_2BC_3N, the diffraction peaks at $2\theta = 35.42°$ and $2\theta = 25.63°$ are assigned to the (111) planes of β-SiC and the (0002) planes of BN(C), respectively. With the addition of AlN or ZrO_2, the diffraction peaks of the β-SiC (111) planes are increased to $2\theta = 35.63°$ or $2\theta = 35.56°$, respectively. Additionally, the diffraction peaks of the BN(C) (0002) planes are increased to $2\theta = 26.25°$ or $2\theta = 26.05°$, corresponding to a smaller interplanar spacing of $d = 0.33939$ nm or $d = 0.34195$ nm, which is closer to the d(0002) of h-BN or graphite. The addition of 5 mol.% AlN or ZrO_2 greatly increases the crystallinity of both α/β-SiC and BN(C).

For ceramics without any additive, SiC has an average grain size of ~96 nm and BN(C) grain is only several dozens of nanometers. BN(C) distributes homogeneously and most likely has a turbostratic structure. The SiC grain displays alternately dark and bright contrast, indicating severe lattice distortion. With the addition of 5 mol.% AlN or ZrO_2, the average grain size of SiC increases to ~200 and ~300 nm, respectively, and BN(C) is also larger than ~100 nm (Fig. 4.77). The addition of 5 mol.% AlN or ZrO_2 leads to the agglomeration and crystallization of ceramic matrix.

Some SiC grains also contain N and Al, implying the limited AlN solid solution in SiC. AlN has a wurtzite structure, belonging to the space group of *P63mc*, just as the same of α-SiC. Thus, at high sintering temperatures, AlN can form solid solution with α-SiC promoting short-range atomic diffusion and hence densification of the ceramic matrix (Fig. 4.78). Based on Z (atomic number) contrast, the bright regions represent ZrO_2, and the gray areas refer to lighter elements, namely Si, B, C or N. ZrO_2 is heterogeneously distributed in the ceramic matrix. In some parts, ZrO_2 grains

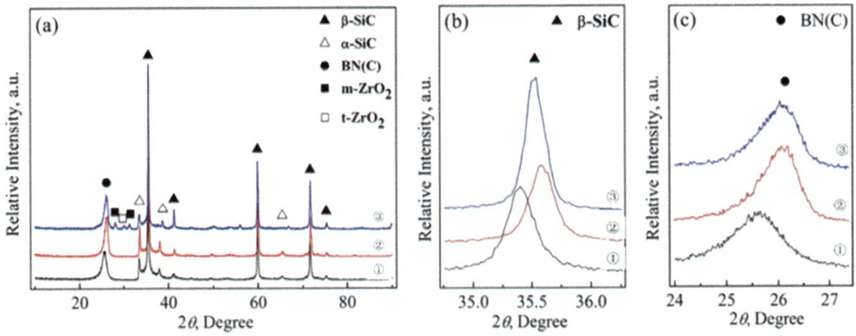

Fig. 4.76 XRD patterns of various MA-Si_2BC_3N bulk ceramics hot-pressed at 1900 °C/8 MPa/30 min: **a** 2θ of 10–90°; **b** 2θ of 34–37°; **c** 2θ of 24–28°. ① Pure MA-Si_2BC_3N; ② + 5 mol.% AlN; ③ + 5 mol.% ZrO_2 [30]

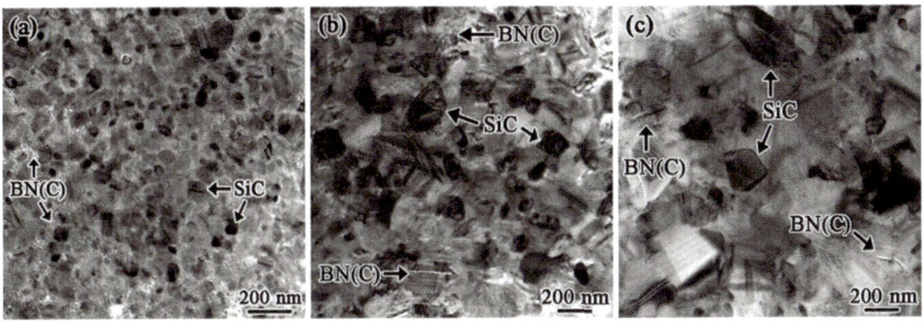

Fig. 4.77 Microstructure comparison of the as-sintered MA-Si_2BC_3N ceramics hot-pressed at 1900 °C/80 MPa/30 min: **a** Pure MA-Si_2BC_3N; **b** + 5 mol.% ZrO_2; **c** + 5 mol.% AlN [30]

aggregate and have an average grain size of about 200–300 nm; in other areas, the slender m-ZrO_2 grains are clearly identified (Fig. 4.79).

(2) MgO–ZrO_2–SiO_2 Particles Introduced by MA

For the low content of sintering aids MgO–ZrO_2–SiO_2 (MZS) and MgO–$ZrSiO_4$ (MZ), their crystallization peaks are undetectable (Fig. 4.80). With 10 wt.% MZS or MZ aids, the bulk densities of MA-Si_2BC_3N ceramics are ~2.69 or ~2.78 g/cm^3, respectively. This seems to indicate a better effect of MZ on ceramic densification [31].

The addition of the two sintering aids can both effectively lower the porosity of the ceramic matrix and contribute to the densification of ceramic body owing to the liquid phase sintering behavior. For example, the porosity in MZS containing ceramics is significantly reduced, along with the residual pores becoming smaller and isolated. The ceramic particles seem to be flattened, indicating the elimination of the pores and the densification of the matrix with the help of MZS addition. Some residual pores are still remained in MZ containing ceramics (Fig. 4.81). Nano-sized

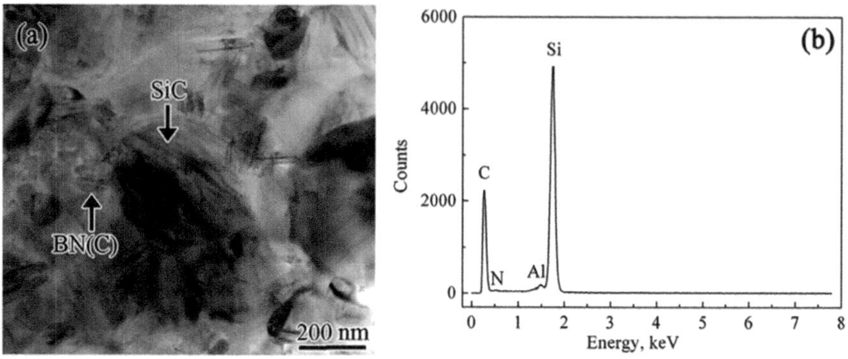

Fig. 4.78 TEM analysis of the hot-pressed MA-Si_2BC_3N bulk ceramics with 5 mol.% AlN addition: **a** Bright-field image; **b** EDS result of SiC grain in (**a**) [30]

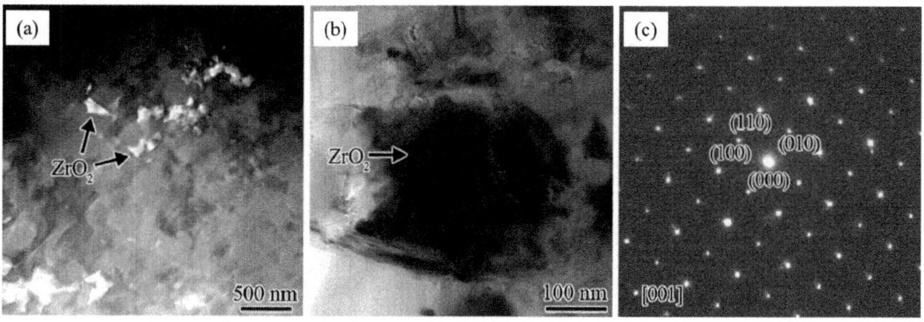

Fig. 4.79 TEM analysis of the hot-pressed MA-Si_2BC_3N bulk ceramics with 5 mol.% ZrO_2 addition: **a** HAADF images: **b** bright-field image; **d** SAED pattern of ZrO_2 [30]

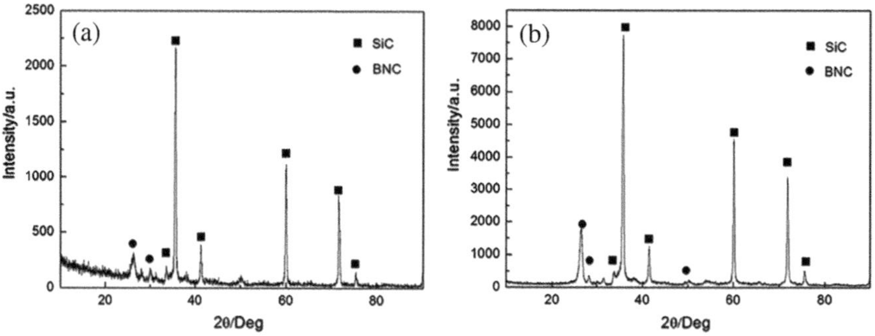

Fig. 4.80 XRD patterns of hot-pressed MA-Si_2BC_3N ceramics with sintering aids: **a** + 10 wt.% MZS; **b** + 10 wt.% MZ [31]

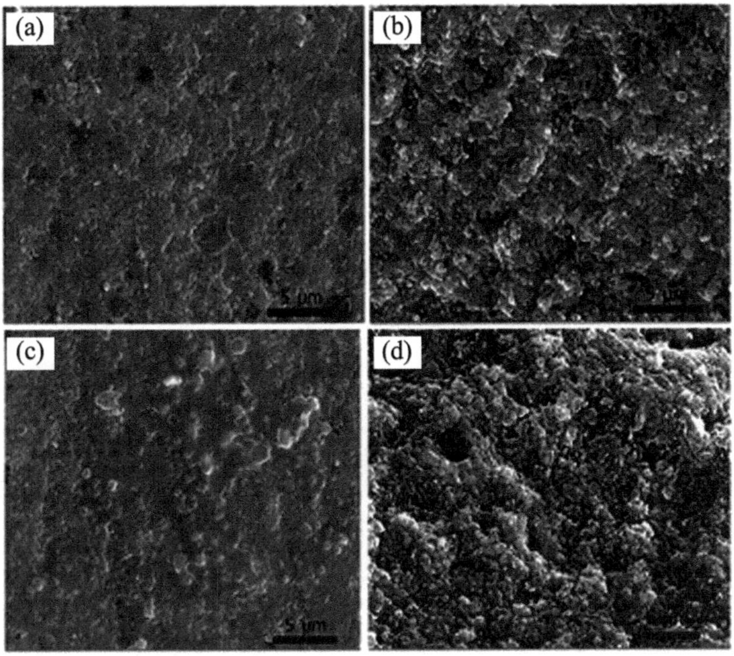

Fig. 4.81 SEM surface and fracture morphologies of the hot-pressed MA-Si_2BC_3N monoliths with different sintering aids: **a**, **b** + 10 wt.% MZS; **c**, **d** + 10 wt.% MZ [31]

BN(C) and SiC are uniformly distributed in the matrix without abnormal growth. The SiC grain sizes in additives-containing ceramics vary from 100 to 400 nm, while the BN(C) grain sizes are much smaller than SiC. The addition of MZS and MZ greatly promotes the grain growth of BN(C) and SiC.

(3) MgO–Al_2O_3–SiO_2 Particles Introduced by Wet Mixing

After hot-pressed at 1800 °C/60 MPa/60 min, the MA-Si_2BC_3N monoliths show MgO–Al_2O_3–SiO_2 (MAS) dependent phase composition [32]. The as-sintered ceramics with only 3–6 wt.% MAS are composed of SiC and BN(C), while 10–20 wt.% MAS addition favors the Al_2SiO_5 formation. The bulk density of the MA-Si_2BC_3N ceramics first increases and then decreases with the increasing MAS content, while the porosity behaves opposite result. With 6 wt.% MAS addition, MA-Si_2BC_3N monoliths exhibit the highest degree of densification with density of 2.59 g/cm^3 (Fig. 4.82).

The 6 wt.% MAS addition does not significantly promote the grain development of the ceramic matrix, with grain sizes of SiC and BN(C) less than 100 nm. The SiC and BN(C) also encapsulate each other to form a "nanocapsule-like" microstructure. No low melting point oxides have been observed at the grain boundaries. The traces of oxides or their reacted products are undetectable owing to their low content or possible in a solution state in matrix (Fig. 4.83).

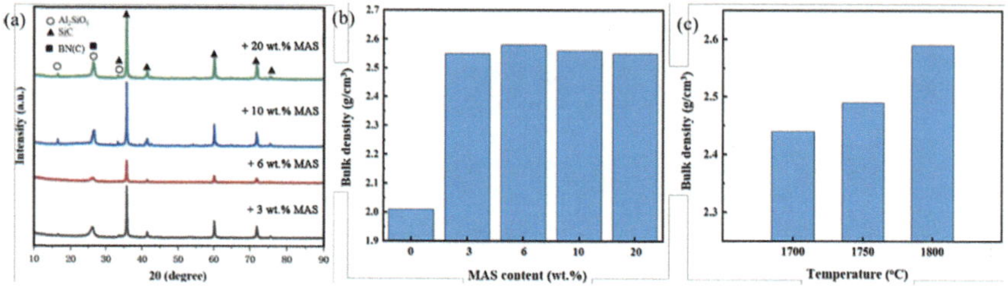

Fig. 4.82 Phase composition and density of the hot-pressed MA-Si_2BC_3N with various MAS addition: **a** XRD patterns: **b**, **c** bulk density [32]

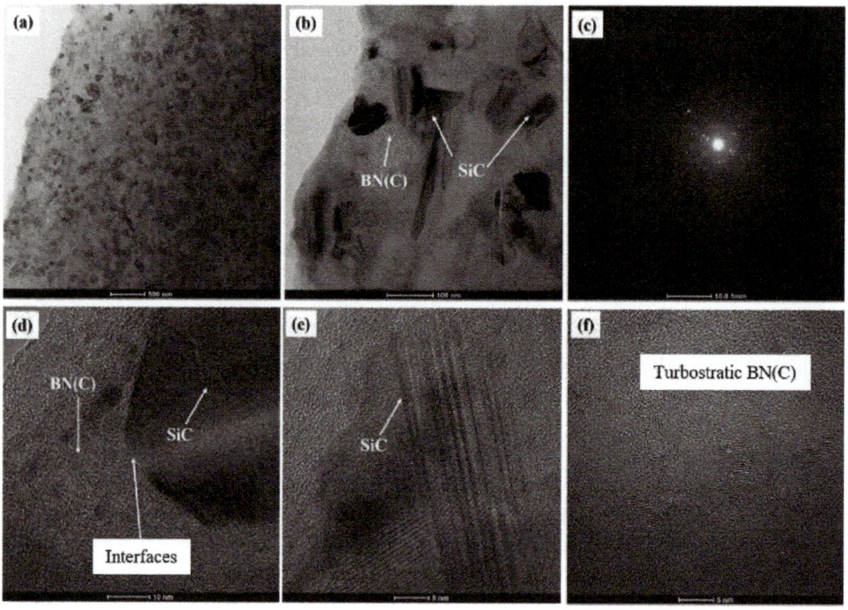

Fig. 4.83 TEM analysis of the hot-pressed MA-Si_2BC_3N with 6 wt.% MAS addition: **a**, **b** Bright field image; **c** SAED pattern; **d**–**f** HRTEM images [32]

Liquid-phase assisted densification occurs at 1800 °C for the hot-pressed MA-Si_2BC_3N bulk ceramics with 6 wt.% MAS addition, though some pores are also found on the fracture surface. As indicated, the ceramic matrix becomes denser and pull-out of BN(C) platelets works with increasing MAS content. Evidently, liquid-phase sintering and formation of BN(C) platelets are promoted by adding MAS (Fig. 4.84).

Fig. 4.84 Fracture morphologies of the hot-pressed MA-Si_2BC_3N with various MAS addition: **a** + 3 wt.% MAS; **b** + 6 wt.% MAS; **c** + 10 wt.% MAS; **d** + 20 wt.% MAS; **f** + 30 wt.% MAS; **e** + 40 wt.% MAS [32]

4.3.3 Effect of Ultrahigh Temperature Ceramic Components

(1) ZrC Particles Introduced by Sol–gel

The raw materials of zirconium propanol, acetylacetone, furfuryl alcohol and hydrochloride with a mole ratio of 1:1:1.8:1 are used to prepare ZrC sol–gel. After hot-pressed at 1900 °C/60 MPa/60 min, crystalline peaks corresponded to α/β-SiC, BN(C) and ZrC have been detected [33]. The residual ZrO_2 is not found or below the limits of detection. Obviously, the content of *in-situ* formed ZrC increases with increasing ZrC content (Fig. 4.85).

After sintering, the *in-situ* formed α/β-SiC, BN(C) and ZrC show homogeneous grain sizes of < 200 nm. The ZrC and SiC grains are found to be equilateral shape, but most BN(C) grains present layered structure. All grains distribute stochastically and disorderly, and the ceramic matrix has no specific orientation. Remarkably, the grain size of ZrC is finer than those of SiC and BN(C). The three-phase interfacial structure of ZrC, SiC and BN(C) is clean without any low melting second-phases. Besides, the initial turbostratic structure of BN(C) have changed to an ordered structure that stacking along its (0002) base sheets (Fig. 4.86). The Si and N have the opposite distribution while C is almost overlapped with Si and Zr. It thus suggests that C tends to react with Zr and Si rather than N. Though excessive furfuryl alcohol decomposed to provide some carbon source, partial C offered by MA-Si_2BC_3N matrix can also react with Zr to produce ordered structure of BN(C).

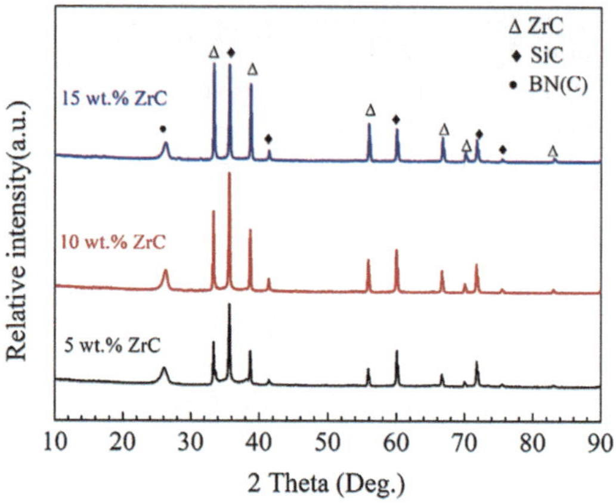

Fig. 4.85 XRD patterns of the various ZrC/Si_2BC_3N monoliths hot-pressed at 1900 °C/60 MPa/60 min [33]

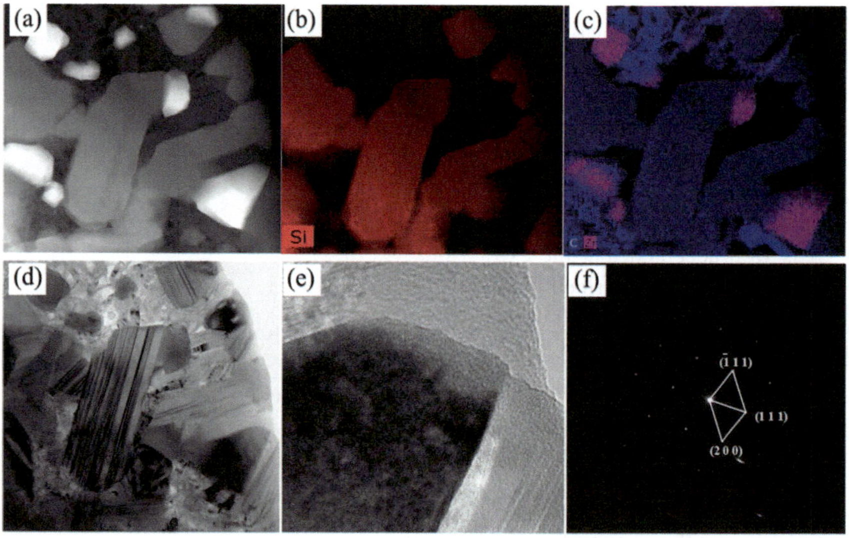

Fig. 4.86 TEM observation of the 15 wt.% ZrC containing MA-Si_2BC_3N monoliths: **a** HADDF image: **b** Si mapping: **c** ZrC mapping: **d** bright-field image: **e** HRTEM image of region "1" in (**a**); **f** SAED pattern of region "2" in (**a**) [33]

The as-sintered ZrC/Si_2BC_3N monoliths exhibit a nearly porosity-free surface. It is expected that the *in-situ* formed ZrC grains can disperse uniformly between SiC and/or BN(C), though ZrC is very difficult to be sintered. In fact, defects such as voids and pores between these nanocrystals are not fulfilled by finer ZrC (Fig. 4.87). The fracture morphologies reveal some particles with approximately spherical characteristics. Generally, the facture surfaces are irregular and small rectangle grains of ZrC can be distinguished occasionally. This means that many sintering mechanisms such as high-temperature creep and sliding rearrangement of ceramic particles have taken modest effects. Consequently, the fracture surfaces are mostly caused by intergranular fracture, suggesting a moderate interface bonding between the grain boundaries; however, the transgranular fracture for ZrC grains is also found. Note that full densification of ZrC/Si_2BC_3N is not yet finished at current hot-pressing sintering condition. Many deep pits are observed on the fracture surfaces, implying the debonding of boundaries and the desquamation of grains are the main factors contributing to the toughening though its overall effects are still limited. The bulk density of the as-sintered ZrC/Si_2BC_3N monoliths increases with increasing ZrC content, but the relative density decreases slightly (Fig. 4.88).

(2) Ta_4HfC_5 Particles Introduced by MA

After hot-pressed at 1900 °C/60 MPa/60 min under 1 bar Ar or vacuum, the Ta_4HfC_5/Si_2BC_3N composite ceramics mainly consist of BN(C), α/β-SiC and Ta_4HfC_5. The diffraction peak intensity of the Ta_4HfC_5 is positively correlated with the Ta_4HfC_5 content. With the increase of Ta_4HfC_5 content, the diffraction peaks of α-SiC decrease while β-SiC increases. This thus means that the *in-situ* Ta_4HfC_5 affects the β → α transition of SiC [34]. Two Raman peaks in the range of 60–260 cm^{-1} are gradually enhanced with the increasing content of the *in-situ* Ta_4HfC_5. The emergence of Raman peak at 1374 cm^{-1} results from the D-side peak edge of graphite and the scattering peak of h-BN. However, the generation of Raman peak at 1588 cm^{-1} should be assigned to the G-side peak of graphite (Fig. 4.89).

Fig. 4.87 SEM surface and fracture morphologies of the hot-pressed ZrC/Si_2BC_3N monoliths: **a**, **d** + 5 wt.% ZrC; **b**, **e** + 10 wt.% ZrC; **c**, **f** + 15 wt.% ZrC [33]

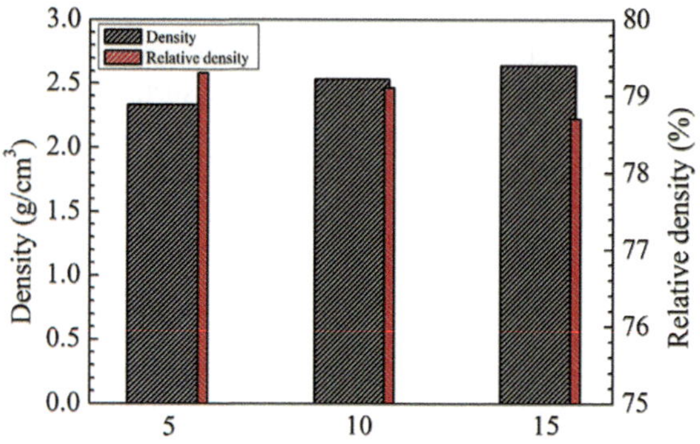

Fig. 4.88 Bulk density and relative density of the as-sintered ZrC/Si_2BC_3N monoliths [33]

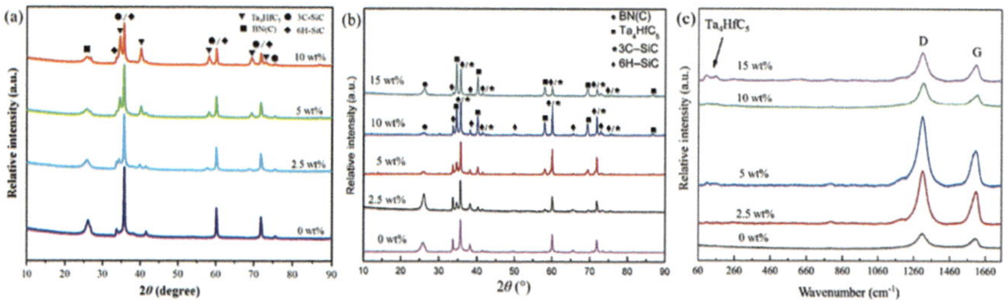

Fig. 4.89 Phase composition of the as-sintered Ta_4HfC_5/Si_2BC_3N composite ceramics: **a** Sintered under Ar; **b** sintered under N_2; **c** Raman spectra [34]

After hot-pressed at 1900 oC under vacuum, the as-sintered Ta_4HfC_5/Si_2BC_3N composite ceramics show dense surface structure without trace of micropores and microcracks (Fig. 4.90). With 2.5 wt.% Ta_4HfC_5 addition, most of the Ta_4HfC_5 (~10 nm) are distributed in BN(C) region in the form of nanocrystals. The Ta_4HfC_5 grains are thereby separated by BN(C). This structure is beneficial to the grain refinement of Ta_4HfC_5 and constrains short-range diffusion of the atoms. With 10 wt.% Ta_4HfC_5 addition, a part of nanosized Ta_4HfC_5 ~10 nm uniformly distributes within the BN(C) region, and another nano Ta_4HfC_5 > 10 nm is randomly dispersed in the ceramic matrix. (Fig. 4.91).

(3) TaC Particles Introduced by MA

As expected, the diffraction peaks assigned to TaC tend to broadening and the diffraction intensities decrease obviously after MA due to the grain refinement induced by mechanochemical action. In all cases, the broad diffraction peaks assigned to TaC are detected. Thus, MA derived TaC crystals do not react with other components during milling process. After HP sintered at 1900 °C/60 MPa/60 min, the crystalized nanocrystals are those of BN(C), SiC and TaC [35]. Again, introduction of nano TaC does not react with other components during sintering. However, the diffraction peak

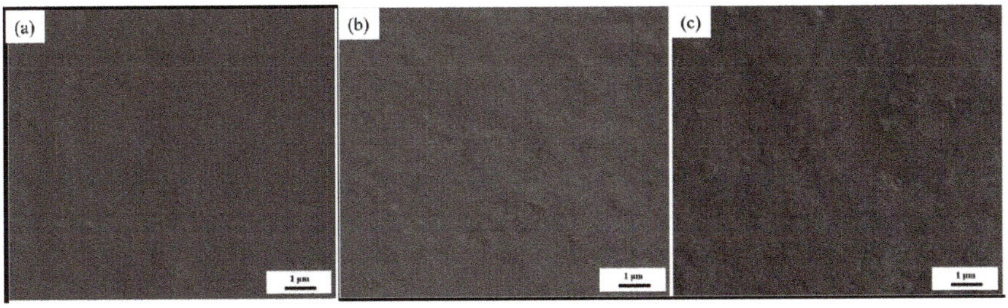

Fig. 4.90 Surface morphologies of the as-sintered Ta_4HfC_5/Si_2BC_3N composite ceramics: **a** + 5 wt.% Ta_4HfC_5; **b** + 10 wt.% Ta_4HfC_5; **c** + 15 wt.% Ta_4HfC_5 [34]

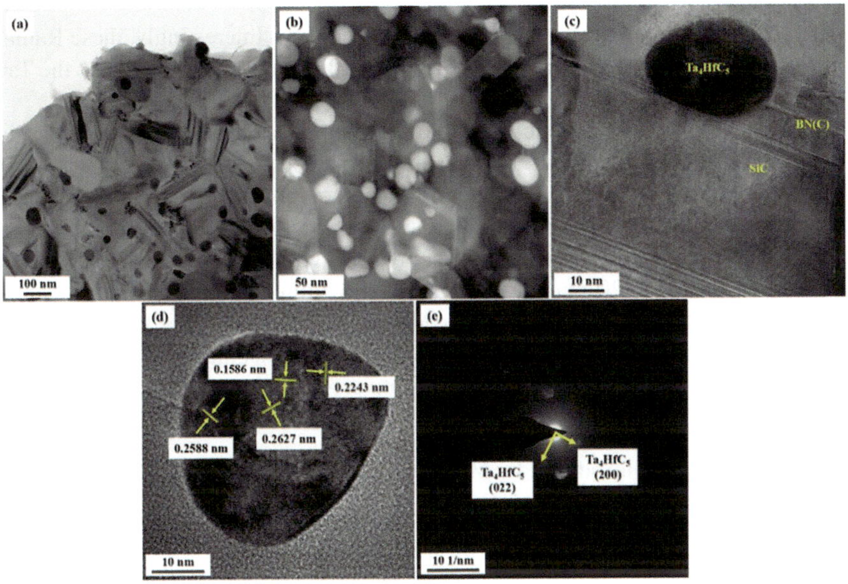

Fig. 4.91 TEM analysis of the 10 wt.% Ta_4HfC_5 containing MA-Si_2BC_3N composite ceramics: **a** Bright-field image; **b** HAADF image; **c**, **d** HRTEM image; **e** SAED pattern [34]

position of TaC shifts to a low-angle compared with the standard pattern (JCPDS No. 89-3831), while that SiC matches well with the standard pattern (JCPDS No. 89-2640). This is mainly due to the mismatch of thermal expansion coefficients between TaC (6.3×0^{-6} °C^{-1}) and MA-Si_2BC_3N ceramic matrix [(3.0–5.0) $\times 10^{-6}$ °C^{-1}]. In this case, the nano TaC is subjected to tensile stress more susceptible to lattice distortion, leading to a tiny change of crystal plane spacing (Fig. 4.92).

Raman peak at 1374 cm^{-1} is a superposition resulting from the D-side peak edge of graphite and the scattering peak of h-BN. The emergence of 1588 cm^{-1} should be assigned to the G-side peak of graphite. It seems that the carbon is graphitized and tends to be ordered after HP sintering. The intensity of Raman peak corresponding

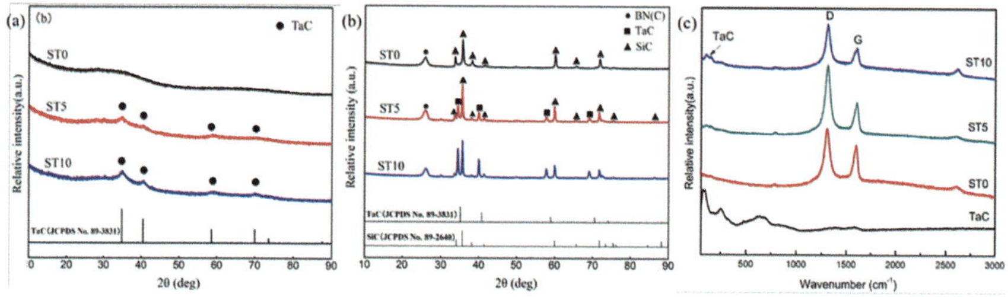

Fig. 4.92 Phase composition of the TaC/Si_2BC_3N composite ceramics hot-pressed at 1900 °C/60 MPa/60 min: **a** XRD patterns of powder mixture; **b** XRD patterns of as-sintered ceramics; **c** Raman spectra [35]

to TaC increases gradually with increasing TaC content. Interestingly, these Raman peaks of TaC become broader as compared to the TaC power, mainly due to the TaC refinement induced by high-energy ball milling.

The fracture surfaces of the TaC/Si_2BC_3N composite ceramics mainly consist of stacking nano and micron particles free of micropores showing a relatively high degree of densification. Particles in these composite ceramics are finer and denser resulting from the uniform distribution of and therefore grains refinement (Fig. 4.93). TaC distributes in two distinct regions with different grain sizes. Some TaC grains about 10–20 nm disperse randomly within the ceramic matrix. Others with smaller size of 3–5 nm are imbedded in turbostratic BN(C). Nano TaC is wrapped by turbostratic BN(C) and SiC nanocrystals so that the individual nanocrystals are isolated from each other thereby effectively hindering abnormal grain growth (Fig. 4.94).

(4) TiCN Particles Introduced by MA

Obviously, the amorphous phases have crystallized into nanostructures after hot-pressing sintering, including the main phases of BN(C), SiC, and TiCN, and the minority of Si_2N_2O and TiB_2 [36]. Naturally, the diffraction peaks of the TiCN phases gradually enhanced with the increase in Ti content (Fig. 4.95). The *in-situ* formed Si_2N_2O must facilitate the sintering densification during HP sintering, but reasons for the Si_2N_2O formation remain to be answered.

SiC are in sizes of ~70 nm while that of TiCN are below 30 nm, but the sizes of Si_2N_2O are over 100 nm. The BN(C) uniformly distributes between the nanosized SiC and TiCN responsible for the slow growth of nano SiC and TiCN. The clean grain boundaries of BN(C)/Si_2N_2O and TiCN/BN(C), contain about four atomic layers with a thickness of ~2 nm (Fig. 4.96).

In all cases, the as-sintered TiCN/Si_2BC_3N bulk ceramics are dense without traces of microcracks and/or micropores. Substantial BN(C) are in the form of layered structure, in which distributed with some nanosized SiC and TiCN. The increasing TiCN content promotes the growth of BN(C) platelets (Fig. 4.97).

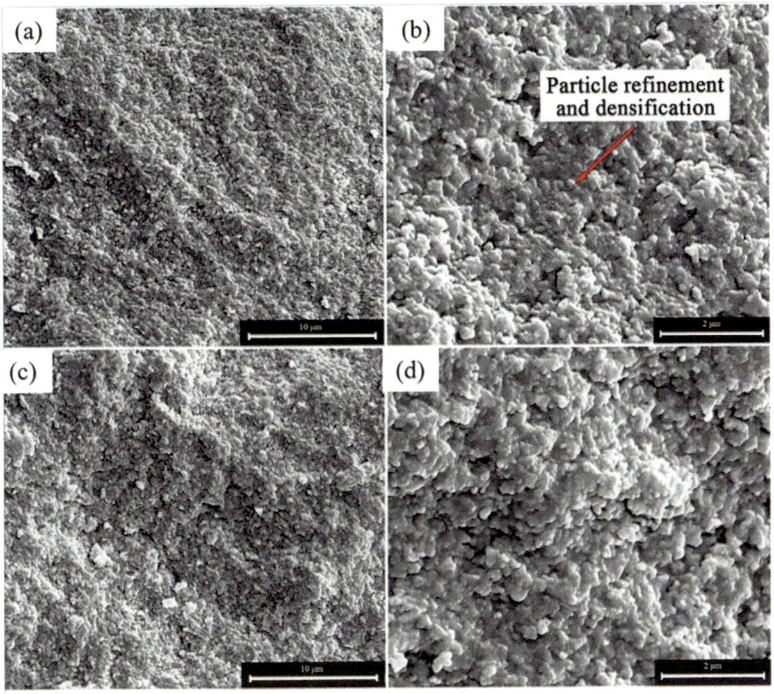

Fig. 4.93 Fracture morphologies of the TaC/Si_2BC_3N composite ceramics hot-pressed at 1900 °C/60 MPa/60 min: **a**, **b** + 5 wt.% TaC; **c**, **d** + 10 wt.% TaC [35]

(5) TiB_2–TiC Particles Introduced by MA

The introduction of TiB has reacted with MA-Si_2BC_3N ceramic matrix to generate new phases of TiC and TiB_2 [37]. However, introduction of TiC and/or TiB_2 can maintain their structure stability within the ceramic matrix after sintering. Thus hot-pressed (TiC_2–TiB_2)/Si_2BC_3N composite ceramics with various Ti-compounds additions are mainly composed of α-SiC, β-SiC, BN(C), TiB_2 and TiC. With increasing sintering temperature and/or pressure, high degrees of crystallinity for the SiC and BN(C) are observed (Fig. 4.98).

The theoretical densities of TiB, TiB_2, TiC and MA-Si_2BC_3N are calculated as 4.51, 4.52, 4.93 and 2.84 g/cm^3, respectively. Thus, the calculated theoretical densities of the composite ceramics with additions of 10 vol.% TiB_2, 10 vol.% TiC, 10 vol.% (TiB_2–TiC) (TiB_2/TiC mole ratio of 2:1), 10 vol.% (TiC–TiB_2) (TiC/TiB_2 mole ratio 2:1), are 2.98, 3.02, 2.99 and 3.01 g/cm^3. Obviously, the additions of 10 vol.% Ti-compounds effectively densify the ceramic matrix (Table 4.3).

After hot-pressed at 1900 °C/40 MPa, merely TiB addition containing composite ceramics show a flat and dense surface structure. The MA-Si_2BC_3N ceramic matrix

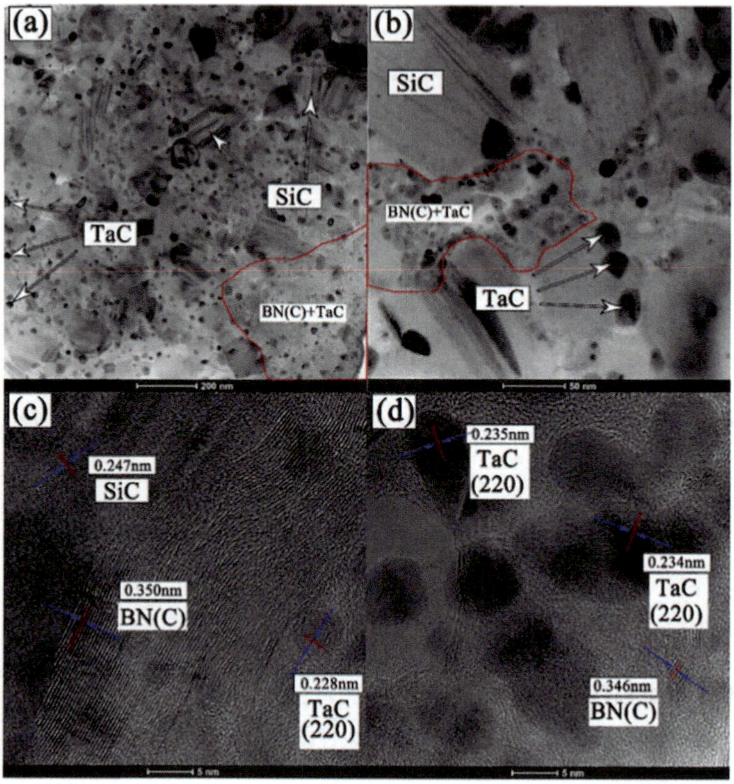

Fig. 4.94 TEM analysis of the 10 wt.% TaC containing MA-Si_2BC_3N bulk ceramics: **a**, **b** Bright-field images; **c**, **d** HRTEM images [35]

Fig. 4.95 XRD patterns of the hot-pressed TiCN/Si_2BC_3N bulk ceramics [36]

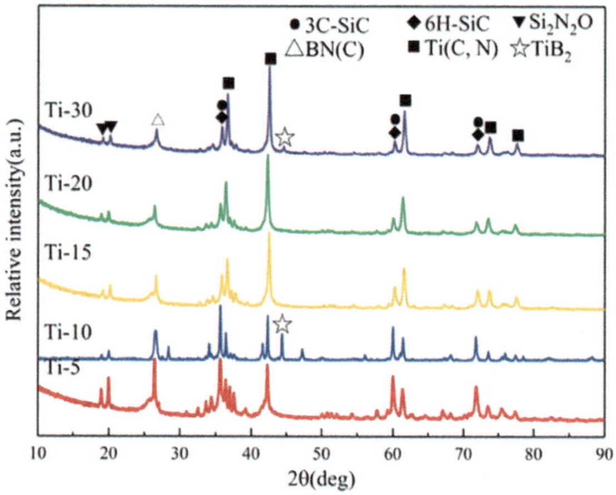

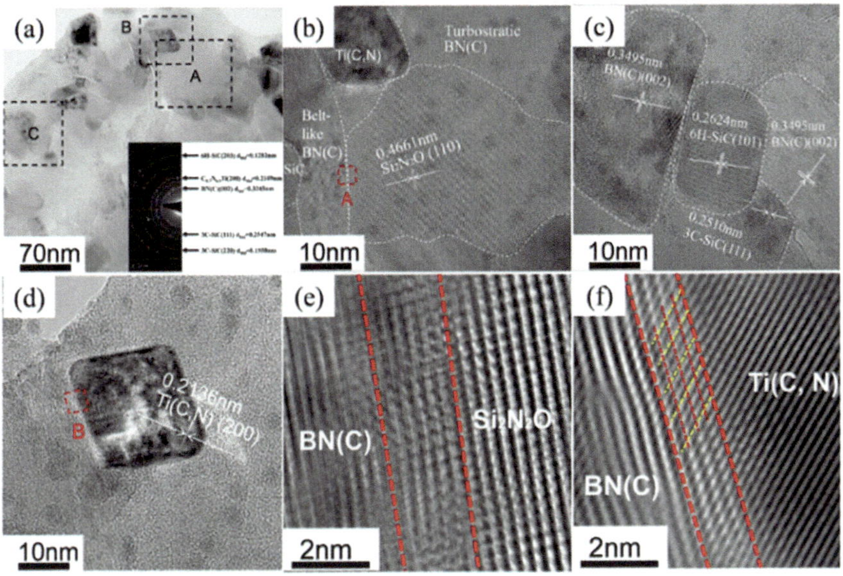

Fig. 4.96 TEM analysis of the hot-pressed TiCN/Si_2BC_3N bulk ceramics: **a** Bright-field image and SAED pattern; **b–d** HRTEM images of the regions "A-C" in (**a**); **e** inverse FFT image of region "A" in (**b**); **f** inverse FFT image of region "B" in (**d**) [36]

Fig. 4.97 Fracture morphologies of the hot-pressed TiCN/Si_2BC_3N bulk ceramics: **a** + 5 wt.% TiCN; **b** + 15 wt.% TiCN; **c** + 30 wt.% TiCN [36]

shows the dark grey contrast, while the brighter ones are those of TiC, TiB_2 or TiB_2–TiC (Fig. 4.99). Coincidentally, a reaction layer forms between the second-phases and ceramic matrix, and the reinforcements are in a micron or submicron scale. The transgranular fracture seems happen to the coarse grains of reinforcements, and intergranular fracture to the rest of grains (Fig. 4.100).

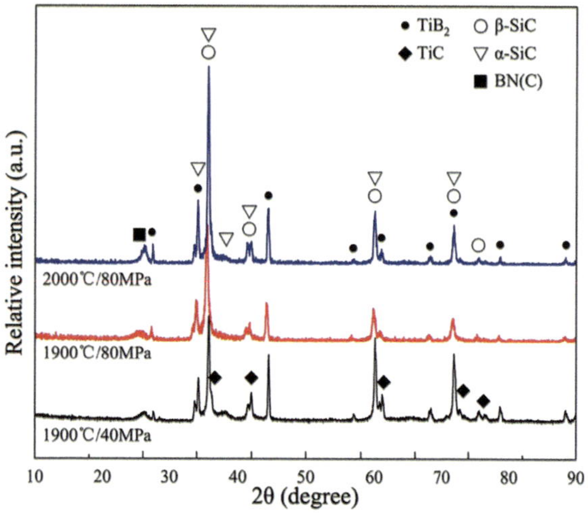

Fig. 4.98 XRD patterns of various (TiB_2–TiC)/Si_2BC_3N composite ceramics hot-pressed at 1900–2000 °C/40–80 MPa/30 min ([a]mole ratio of TiB_2/TiC is 1:2; [b]mole ratio of TiB_2/TiC is 2:1) [38]

Table 4.3 Bulk density and relative density of (TiB_2–TiC)/Si_2BC_3N composite ceramics hot-pressed at 1900 °C/40 MPa/30 min [37]

Ceramic composition	Bulk density (g/cm^3)	Relative density (%)
MA-Si_2BC_3N	2.03	71.5
+ 10 vol.% TiB	2.80	94.1
+ 10 vol.% TiC	2.66	88.1
+ 10 vol.% TiB_2	2.62	88.0
+ 10 vol.% (TiB_2–TiC)[b]	2.66	88.4
+ 10 vol.% (TiB_2–TiC)[a]	2.31	77.3

[a]TiB_2/TiC mole ratio of 1:2; [b] TiB_2/TiC mole ratio of 2:1

(6) ZrB_2 particles introduced by sol–gel or MA

The zirconium n-propoxide was used as the Zr source via the sol–gel method [38, 39]. After SPS densified at 2000 °C/40 MPa/5 min, the resulting ZrB_2/Si_2BC_3N composite ceramics are mainly composed of SiC, ZrB_2 and BN(C). The added ZrO_2 have completely reacted with BN(C) to form some ZrB_2 (Fig. 4.101). The bulk density of the as-sintered bulk ceramics is ~3.24 g/cm^3 as determined by the Archimedes method, much higher than that ~2.74 g/cm^3 of pure MA-Si_2BC_3N.

After SPS densification, obvious grain growth of SiC ≈1–3 μm and ZrB_2 ≈1–1.5 μm has appeared as compared to the pure ceramics. BN(C) phases exhibit two predominate structures: one is the turbostratic structure, another can be regarded as the disordered atomic arrangement (Fig. 4.102). Unexpectedly, partial residual ZrO_2 have been observed on the grain surfaces of ZrB_2; besides, smaller C particles are also precipitated on the ZrB_2 surfaces. A clear element enrichment can be observed between the grain boundaries of SiC and BN(C).

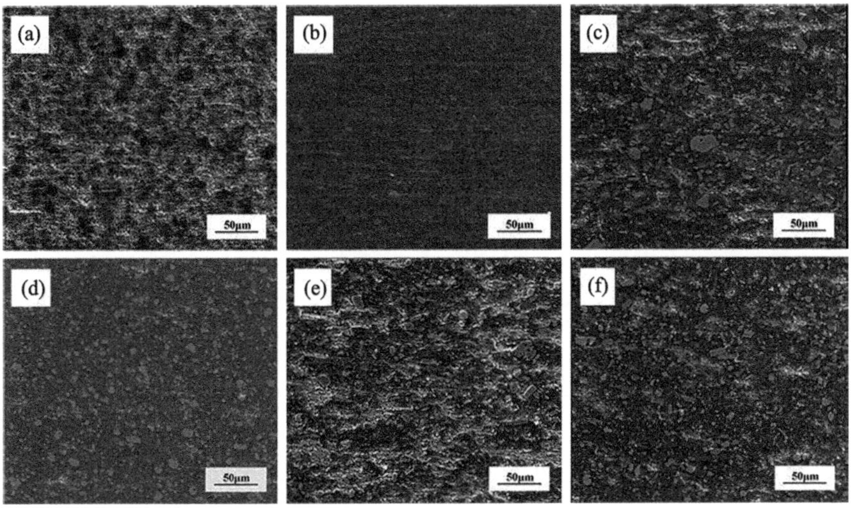

Fig. 4.99 SEM surface morphologies of the as-sintered (TiB_2–TiC)/Si_2BC_3N composite ceramics: **a** MA-Si_2BC_3N; **b** + 10 vol.% TiB; **c** + 10 vol.% TiB_2; **d** + 10 vol.% TiC; **e** + 10 vol.% (TiB_2–TiC), TiB_2/TiC mole ratio of 1:2; **f** + 10 vol.% (TiB_2–TiC), TiB_2/TiC mole ratio of 1:2 [37]

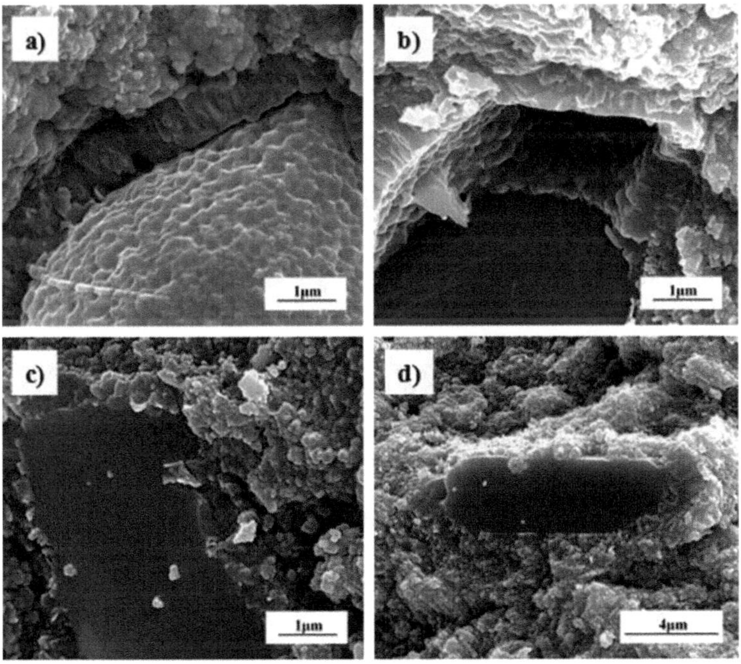

Fig. 4.100 SEM fracture morphologies of the as-sintered (TiB_2–TiC)/Si_2BC_3N composite ceramics: **a**, **b** + 10 vol.% TiC; **c** + 10 vol.% (TiB_2–TiC), TiB_2/TiC mole ratio of 2:1; (d) + 10 vol.% (TiB_2–TiC), TiB_2/TiC mole ratio of 1:2 [37]

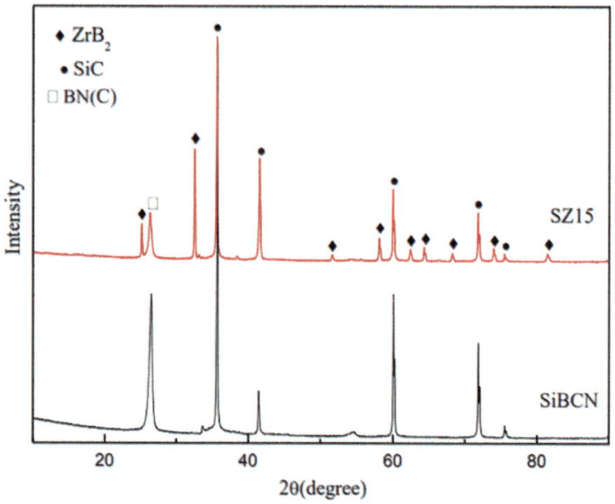

Fig. 4.101 XRD patterns of the a-sintered ZrB_2/Si_2BC_3N composite ceramics prepared by SPS at 2000 °C/40 MPa/5 min [38]

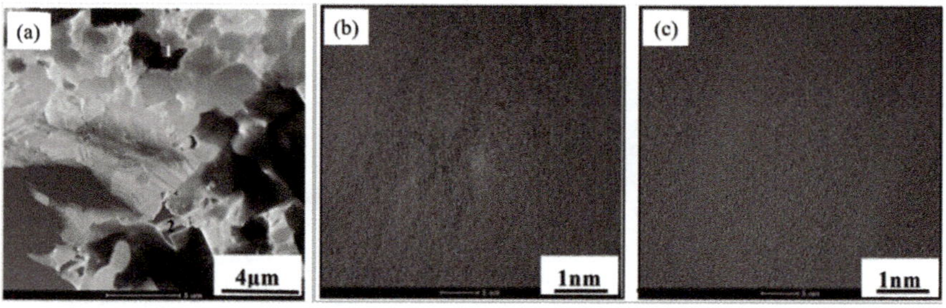

Fig. 4.102 TEM analysis of SPS densified ZrB_2/Si_2BC_3N composite ceramics: **a** Bright-field image, **b** HRTEM image of zone "1" in (**a**); **c** HRTEM image of zone "2" in (**a**) [39]

The *in-situ* formed ZrB_2 can favor the densification the MA-Si_2BC_3N ceramic matrix; however, for the 20 wt.% ZrB_2 containing composite ceramics, an evident aggregation of ZrB_2 has been found. The grain sizes of the *in-situ* formed ZrB_2 are in sizes of 1–1.5 um and lamellar BN(C) structure benefits the fracture toughness of ceramics (Fig. 4.103).

Zirconium *n*-propoxide, boric acid and furfuryl alcohol ($C_5H_6O_2$) are added as the precursors of zirconia, boron oxide and carbon so as to *in-situ* forming ZrB_2 in a MA-Si_2BC_3N matrix [40]. After SPS at 1900 °C/40 MPa/5 min, the *in-situ* ZrB_2 is found but residual ZrO_2 are also detected. Sintered at 2000 °C, the residual ZrO_2 has all changed to ZrB_2 without any trace of oxides. The peak intensities for the ZrB_2 increases with increasing content of *in-situ* formed ZrB_2 (Fig. 4.104).

The SiC grain sizes for 15 wt.% ZrB_2 containing MA-Si_2BC_3N composite ceramics are about 1–1.2 μm, but the sizes of ZrB_2 grains are 300–600 nm. The BN(C) grains are smaller with no fixed shape, and are mainly distributed between

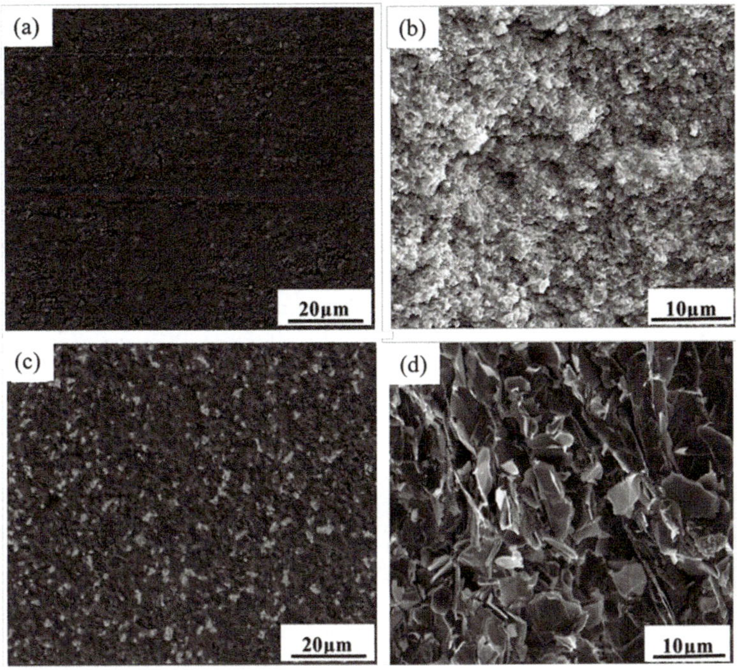

Fig. 4.103 SEM surface and fracture morphologies of various ZrB_2/Si_2BC_3N composite ceramics: **a**, **b** Pure MA-Si_2BC_3N; **c**, **d** + 15 wt.% ZrB_2 [39]

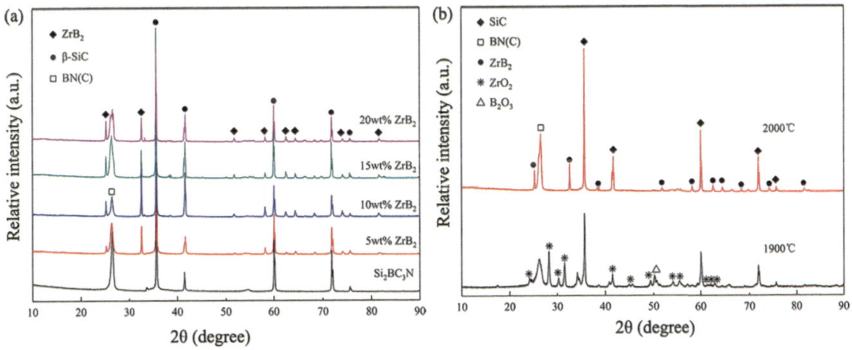

Fig. 4.104 XRD patterns of ZrB_2/Si_2BC_3N composite ceramics: **a** Sintered at 2000 °C/40 MPa/5 min; **b** sintered at 1900–2000 °C/40 MPa/5 min [40]

SiC and ZrB_2 (Fig. 4.105). After polishing, the surfaces of the MA-Si_2BC_3N bulk ceramics still exhibit some micropores. These surface defects steadily diminish as the *in-situ* formed ZrB_2 content increases (Fig. 4.106).

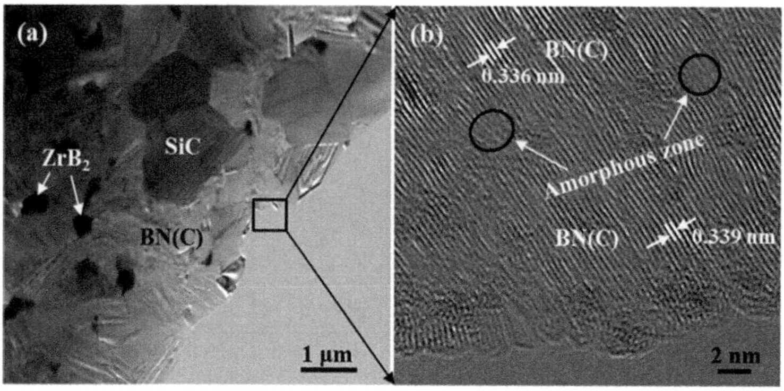

Fig. 4.105 TEM analysis of 15 wt.% ZrB_2 incorporating MA-Si_2BC_3N bulk ceramics prepared by SPS at 2000 °C/40 MPa/5 min: **a** Bright-field image; **b** HRETM image of BN(C) [38]

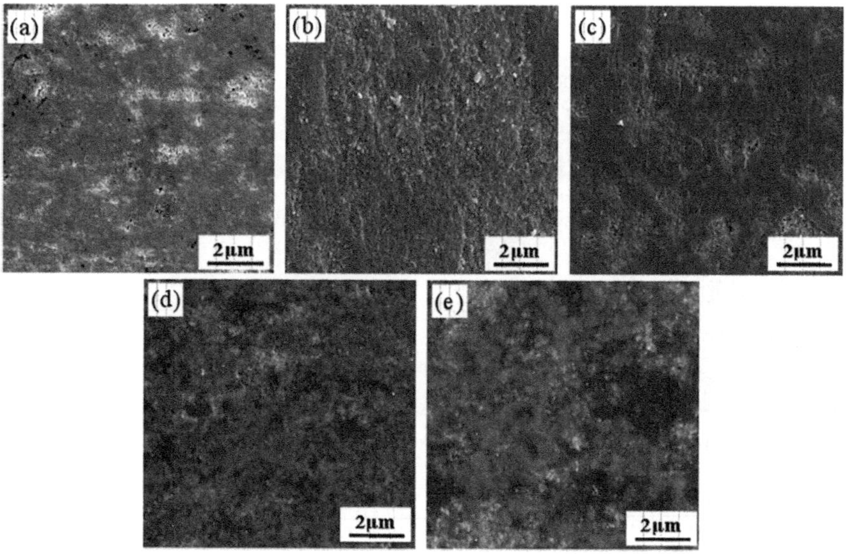

Fig. 4.106 SEM surface morphologies of various ZrB_2/Si_2BC_3N composite ceramics prepared by SPS at 2000 °C/40 MPa/5 min: **a** Pure MA-Si_2BC_3N; **b** + 5 wt.% ZrB_2; **c** + 10 wt.% ZrB_2; **d** + 15 wt.% ZrB_2; **e** + 20 wt.% ZrB_2 [38]

The MA derived nanosized ZrB_2 is also used as reinforcements. As exhibited, and the intensity of the characteristic diffraction peaks for micron ZrB_2 decreases significantly accompanied with the broadened FWHM [41]. The as-milled ZrB_2 powder are primarily agglomerates of nanoparticles. After SPS at 2000 °C/40 MPa/5 min, the new phase of ZrO_2 is detected in ZrB_2/Si_2BC_3N composite ceramics (Fig. 4.107). The detection of monoclinic ZrO_2 should be attributed to the preferential oxidation of ZrB_2 during SPS. With the increasing nano ZrB_2, higher relative intensities of

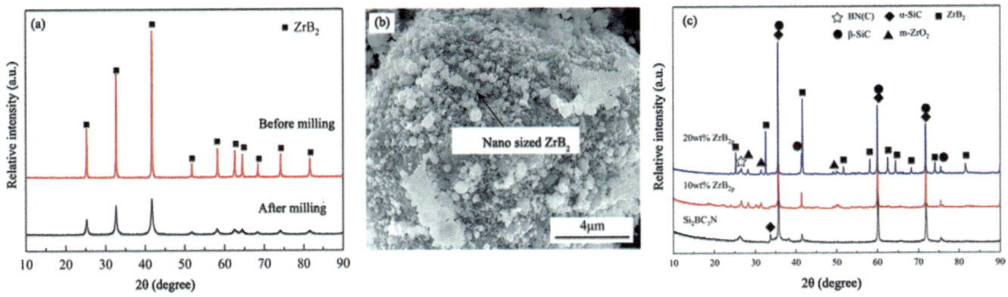

Fig. 4.107 MA derived nano ZrB_2 powder and SPS densified ZrB_2/Si_2BC_3N composite ceramics at 2000 °C/40 MPa/5 min: **a**, **b** Nano ZrB_2 powder; **b** ZrB_2/Si_2BC_3N [41]

the ZrB_2 peaks are observed. The bulk density and relative density of the as-sintered ZrB_2/Si_2BC_3N composite ceramics increase with the introduction of 10–20 wt.% ZrB_2 nanoparticles (Table 4.4).

The nano/submicron particles stack together in pristine MA-Si_2BC_3N ceramics. In comparison, the ZrB_2 incorporated bulk ceramics present well-developed structures. With the increase of ZrB_2 content, larger BN platelets are observed. The microstructures are familiar with those observed in sol–gel derived ZrB_2 modified MA-Si_2BC_3N. The well-developed structures can be argued to the formation of ZrO_2, which stimulated the solid diffusion (Fig. 4.108).

The grain sizes of SiC and BN(C) are mainly around 0.5–1 and 1–3 μm, much larger than those observed in pristine MA-Si_2BC_3N. In addition, the grain size of ZrB_2 also increases from tens of nanometer to 0.5–2 μm. The observed larger SiC, BN and ZrB_2 grains confirms the stimulation effect of ZrB_2 on grain growth during SPS sintering. Interestingly, a clear interface separates the SiC and SiC_xN_y grain, and some ZrB_2 are also adjacent to the SiC_xN_y (Fig. 4.109).

Careful investigations on the interfacial structures of the adjacent phases, partial ZrB_2 grains are enwrapped by a crystallized graphite. Of course, there is also a thin graphite layer distributed between the ZrB_2 and ZrO_2 grains. Actually, a large number of independent BN and C structures are observed in 20 wt.% ZrB_2 incorporating ceramics. It is possible that the independent BN and C phases arise from the phase separation of BN(C) structure at high-temperatures. This process also relies on the presence of nano-ZrO_2, which lowers the potential barrier for the reaction to proceed.

Table 4.4 The bulk density and relative density of various ZrB_2/Si_2BC_3N composite ceramics prepared by SPS at 2000 °C/40 MPa/5 min [41]

Ceramic composition	Bulk density (g/cm^3)	Relative density (%)
MA-Si_2BC_3N①	2.58	91.1
MA-Si_2BC_3N	2.42	85.6
+ 10 wt.% ZrB_2	2.78	93.2
+ 20 wt.% ZrB_2	2.99	93.4

①Hot-pressing sintering at 1900 °C/60 MPa/30 min

Fig. 4.108 SEM fracture morphologies of various ZrB_2/Si_2BC_3N composite ceramics: **a**, **b** Pure MA-Si_2BC_3N; **c**, **d** + 10 wt.% ZrB_2: **e**, **f** + 20 wt.% ZrB_2 [41]

The element distribution map clearly confirms that a graphite layer and scattered ZrO_2 grains surround ZrB_2, and SiC and BN phases are uniformly distributed around the ZrO_2 grains. Besides, some new-born ZrC have etched on the surfaces of ZrB_2 grains.

(7) HfB_2 particles introduced by MA

The densification process of the hot-pressed HfB_2/Si_2BC_3N composite ceramics can be divided into three stages [42, 43]: (i) In range of 500–1200 °C, the displacement of the pressing head decreases continuously, resulting from the expansion of powders and graphite die. Applied axial load from 1200 to 1600 °C, the pressing head rapidly changes with intimate powder packing and reducing free volume; (ii) At 1600–1815 °C, the displacement of pressing head continuously increases, but the volume shrinkage remains almost unchanged; (iii) At above 1815 °C, the pressing head displacement rises to the mountain with a $d(L/L_0)/dt \approx 0.01$–0.03. Therefore, the effective sintering densification of HfB_2/Si_2BC_3N starts at about 1815 °C (Fig. 4.110).

After hot-pressing sintering, the main crystalline peaks of HfC, HfB_2, α/β-SiC, HfO_2 and some BN(C) are observed. Interestingly, the formation of HfC and HfB_2 occurs preferentially with only a trace of HfO_2 detected. The diffraction peak intensities for SiC crystals decrease as the HfC and HfB_2 content increases (Fig. 4.111). The theoretical density of HfB_2 and HfC are 10.5 and 12.7 g/cm^3, higher than the theoretical density of MA-Si_2BC_3N ~2.84 g/cm^3. The open porosity of the composite ceramics decreases with increasing HfB_2, and the 30 vol.% HfB_2 incorporating bulk ceramics retain the lowest porosity of 1.59% (Table 4.5).

The denser ceramics are obtained after the introduction of HfB_2 and *in-situ* formation of HfC and HfO_2. On polished surfaces, the grey areas are those of

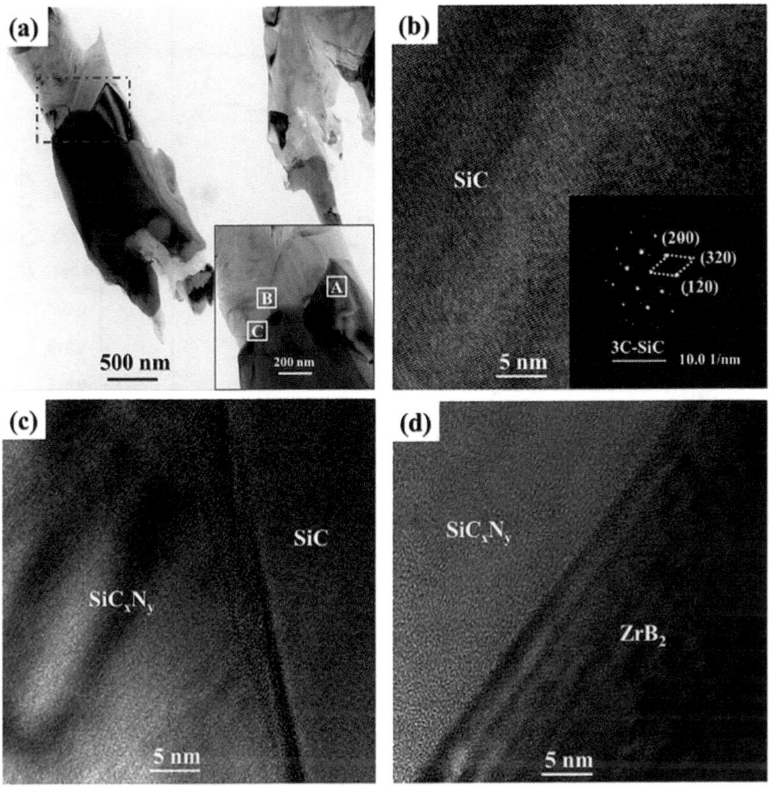

Fig. 4.109 TEM analysis of 20 wt.% ZrB_2 incorporating MA-Si_2BC_3N bulk ceramics: **a** Bright-field image; **b–d** HRTEM images of regions "A–C" in (**a**) [41]

the HfB_2 and HfC, while the white areas indicate the MA-Si_2BC_3N agglomeration. From fracture morphologies, HfB_2 and HfC containing ceramics show well-developed BN(C), HfB_2, HfC and SiC. Specifically, SiC, HfB_2 and HfC phases show isometric grain morphologies, while BN(C) exhibits a laminated structure. The well-developed microstructure for HfB_2/SiBCN composite ceramics should be argued to the formation of HfO_2 that favors densification (Fig. 4.112).

(8) LaB_6 particles introduced by MA

The LaB_6 powder is introduced by two MA ways [16, 44]: (i) one-step ball milling, namely LaB_6, graphite, h-BN and c-Si powder mixtures are high-energy ball-milled simultaneously for 30 h; (ii) two-step ball milling, the pre-milled amorphous MA-Si_2BC_3N powder and LaB_6 are milled together for only 24 h. For one-step milling combined with SPS technique, the displacement increases sharply at about 1160 °C causing a considerable increase in displacement change rate. This is most likely due to initial densification of powder mixtures. Another contribution likely comes

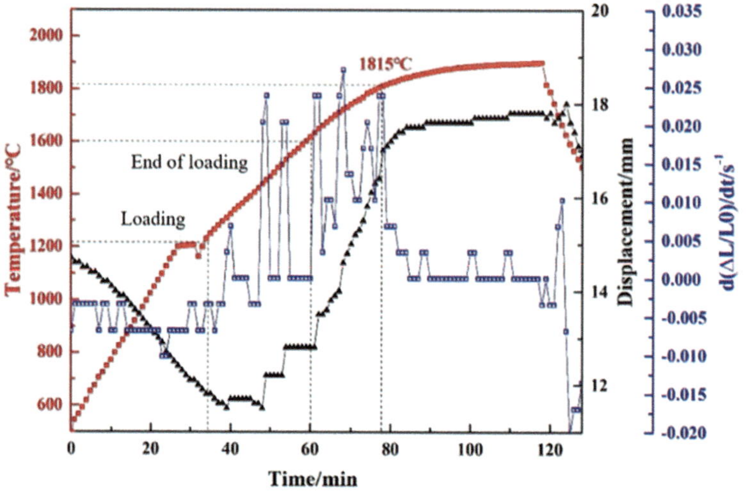

Fig. 4.110 The sintering curves of the hot-pressed HfB_2/Si_2BC_3N composite ceramics [42]

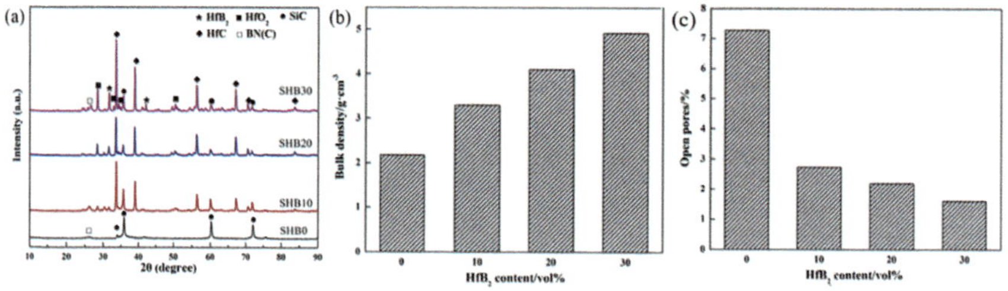

Fig. 4.111 Phase composition and densities of various HfB_2/Si_2BC_3N composite ceramics hot-pressed at 1900 °C/60 MPa/30 min: **a** XRD patterns; **b** bulk density; **c** open porosity [42]

Table 4.5 The apparent density, bulk density and open porosity of the various HfB_2/Si_2BC_3N composite ceramics hot-pressed at 1900 °C/60 MPa/30 min [42]

Ceramic composition	Apparent density (g/cm^3)	Bulk density (g/cm^3)	Open porosity (%)
MA-Si_2BC_3N	2.22	2.17	7.26
+ 10 wt.% HfB_2	3.37	3.29	2.73
+ 20 wt.% HfB_2	4.17	4.10	2.19
+ 30 wt.% HfB_2	5.05	4.91	1.59

from material composition changes at this temperature. A slight increase in displacement is observed at ~1700 °C due to continuous densification and/or composition changes. Little displacement change is observed above 1800 °C signaling the end of densification (Fig. 4.113).

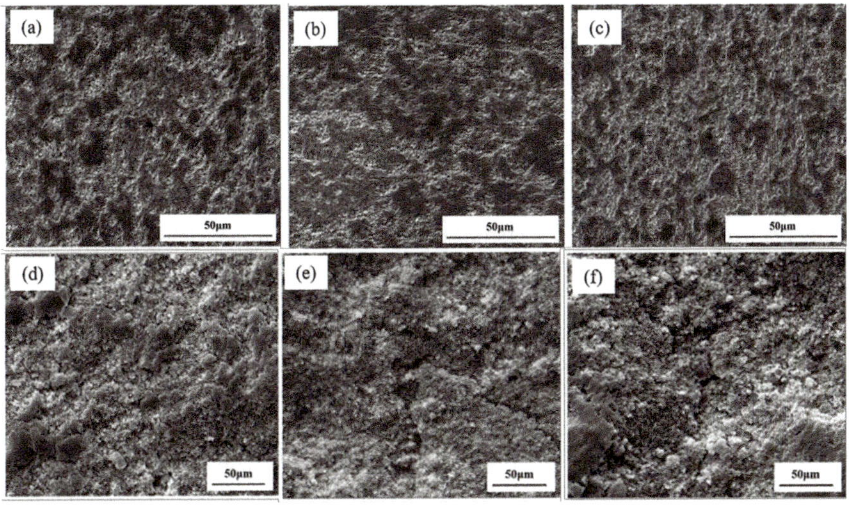

Fig. 4.112 SEM surface and fracture morphologies of various HfB_2/Si_2BC_3N composite ceramics: **a**, **d** + 10 vol.% HfB_2; **b**, **e** + 20 vol.% HfB_2: **c**, **f** + 30 vol.% HfB_2 [42]

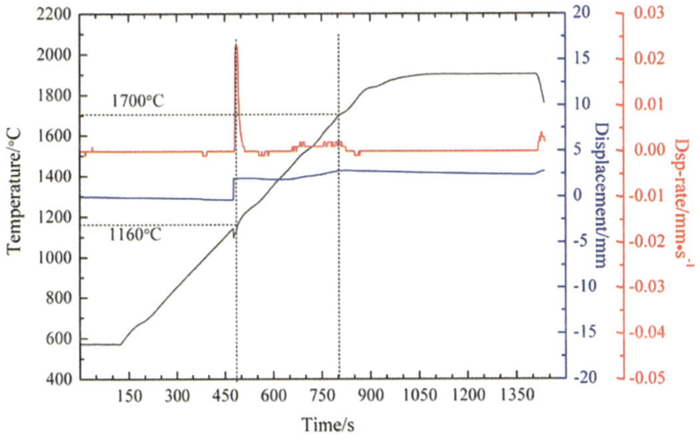

Fig. 4.113 Sintering curves of the LaB_6/Si_2BC_3N bulk ceramics prepared by a combination of one-step ball milling and SPS technique [44]

These two displacement events during SPS likely arise first sharp increase at 1160 °C due dominantly to the densification versus composition changes induced by crystallization. The as-prepared powders by mechanical alloying are porous due to the loose packing and agglomeration of nano-sized particles. In part the process involves particle rearrangement and even plastic deformation responsible for the initial densification characterized by the sharp increase in displacement. Then the second slight increase at 1700 °C is due to formation of La–B–C–N compounds and further densification. Coincidently, the nucleation and growth of SiC and BN(C) occur in this stage (1300–1900 °C). Therefore, the grain boundary sliding and atomic diffusion contribute to the continuous densification.

As compared to the hot-pressed PDCs-SiBCN with SiC addition, the rapid densification starts at 1660 °C [45]. Below 1660 °C, only highly packed particles are observed in all the ceramic samples. The densities show a minor increase because of solid state particles rearrangement. Above 1660 °C, the densities increase rapidly due to the grain boundary sliding. Grain-boundary diffusion enhanced by B and C elements from PDCs-SiBCN is also responsible for the further densification.

The LaB_6/Si_2BC_3N composite powder heated to 1000 °C at 50 MPa shows only LaB_6 peaks the same as the original mechanically alloyed powder. At 1300 °C, the LaB_6 peaks disappear as it reacts with other components. Coincidently, the peaks of α/β-SiC and BN(C) arise due to the crystallization of amorphous phases. On heating to 1800 °C, the BN(C) peak at 26.5° 2θ increases in intensity and α-SiC at 32° 2θ disappears. Moreover, several peaks for undefined phases appear at 22–30° 2θ implying the formation of $La_5(B_4N_9)$ (JCPDS#89-1789) and/or $La_{15}B_{14}C_{19}$ (JCPDS#86-1803) possibly due to a solid-solution of carbon and/or nitrogen in LaB_6 at high temperatures. When the temperature rises up to 1900 °C, the BN(C) peak decreases in intensity relative to SiC attributed to the turbostratic structure of BN(C) together with the higher intrinsic growth rate of SiC than BN(C) (Fig. 4.114).

The bulk density of 5 wt.% LaB_6 containing MA-Si_2BC_3N bulk ceramics increases by up to 40% on heating from 1000 to 1300 °C, then by 17% (1300–1800 °C), and finally remains almost unchanged (1800–1900 °C). The LaB_6/Si_2BC_3N composite ceramics sintered at 1000–1300 °C provide very low densities with spherical particle fracture morphologies. The well-developed lamellar BN(C) inducing grain pull-out, crack bridging and crack deflection are found (Fig. 4.115).

In the previous reports [46–48], the elongated α-SiC grains form by phase transformation at high-temperatures or the solution-reprecipitation process involving the introduction of α-SiC seeds and rare earth oxides, while the growth of elongated Si_3N_4 grains is attributed to the preferential segregation of rare earth cations to and

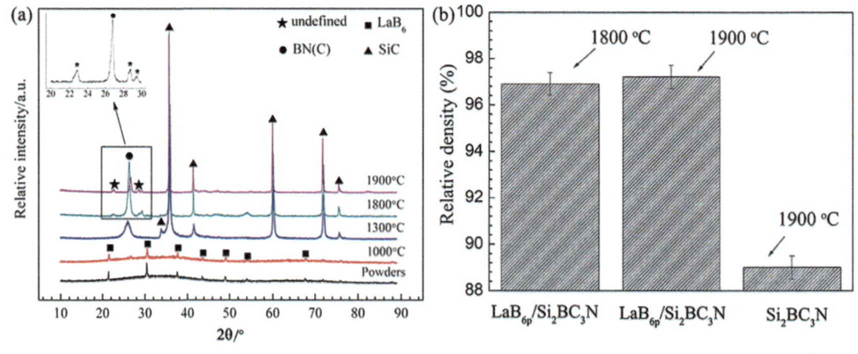

Fig. 4.114 Phase composition and density of the LaB_6/Si_2BC_3N bulk ceramics prepared by a combination of one-step ball milling and SPS technique: **a** XRD patterns; **b** relative density [44]

Fig. 4.115 SEM fracture morphologies of LaB_6/Si_2BC_3N ceramics prepared by a combination of one-step ball milling and SPS: **a** 1000 °C; **b** 1300 °C; **c** 1500 °C; **d** 1900 °C [44]

the adsorption on the prism planes of hexagonal grains. It thus can be inferred that LaB_6 drives the formation of BN(C) flakes since some compounds of $La_5B_2C_6$ as well as BN(C) phases are detected in La/BCN materials [49]. As expected, thin BN(C) flakes are formed and arranged randomly in La/BCN *vs.* BCN materials (both consolidated by SPS at 1900 °C/50 MPa). The element mappings indicate the homogenous distribution of La over the interlayers of BN(C) except for a few agglomerates at boundaries. Therefore, the solid-solution of La occurs in BN(C) and in its interlayers at least, just as it should be shown in Fig. 4.116. The La-containing BN(C) exhibits a preferred orientation along the (0002) plane leading to the lamellar structure. It also provides strong evidences of growth mechanism of lamellar BN(C) induced by La.

For two-step mechanical alloyed LaB_6/Si_2BC_3N powder, their sintering densification process can also be divided into three stages. The major difference is the initial densification temperature of ~1360 °C, higher than the former ~1160 °C. As the sintering temperature increased, the bulk density of the as-sintered ceramics changes gradually: At initial stage of 1250–1450 °C, the bulk density increased from 1.7 to 2.4 g/cm^3. In the second stage (1450–1550 °C), the bulk density increases to 2.75 g/cm^3. For sintering temperature $\geq$ 1550 °C, the bulk density is almost unchanged, indicating the end of densification for the two-step milled ceramics (Fig. 4.117).

Except for the matrix phases, two peaks around $2\theta = 28.5°$ and $2\theta = 47.5°$ of $LaBO_3$ are checked for the two-step milled LaB_6/Si_2BC_3N powder. Since the LaB_6/

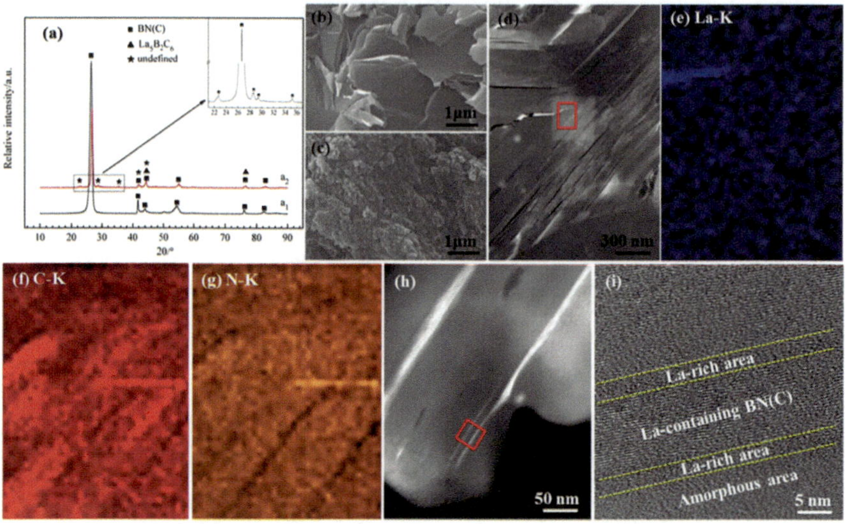

Fig. 4.116 Phase composition and microstructures of the as-sintered La/BNC and BNC ceramics by SPS at 1900 °C/50 MPa: **a** XRD patterns for (a_1) BNC and (a_2) La/BNC; fracture surfaces of La/BNC (**b**) and BNC (**c**); **d** STEM image of La/BNC; Element mappings of La K-edge (**e**), C K-edge (**f**) and N K-edge (**g**) for the marked box zone in (**d**); **h** HAADF image; **i** HRTEM image for the marked box zone in (**h**) [49]

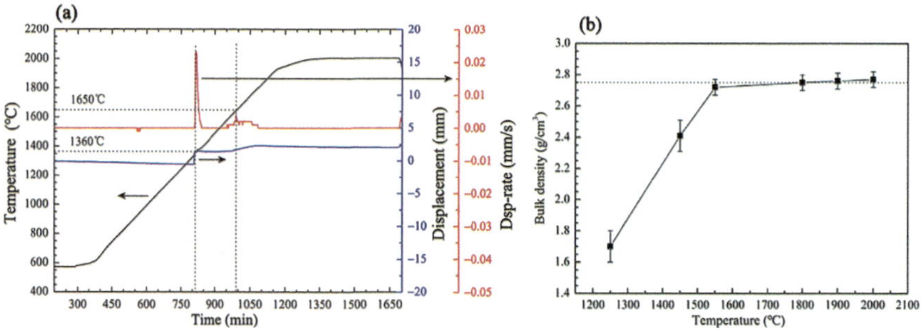

Fig. 4.117 SPS densification behavior and bulk density change of the LaB_6/Si_2BC_3N composite ceramics prepared by a combination of two-step ball milling and SPS technique: **a** Sintering curves; **b** bulk density [16]

Si_2BC_3N powder mixtures are milled under high purity Ar and LaB_6 starts to oxidize > 700 °C, the limited impurities of $LaBO_3$ may originate from the raw LaB_6 powder. After sintering at 1250 °C, the diffraction peak intensities of the LaB_6 are reduced and finally disappears at temperature higher than 1450 °C. The partial solid solution of La into BN(C) is one of the reasons for the disappearance of the LaB_6 diffraction peak. Consequently, the diffraction peak for the BN(C) (0002) shifts from $2\theta = 26.1°$ to $2\theta = 26.8°$ (Fig. 4.118).

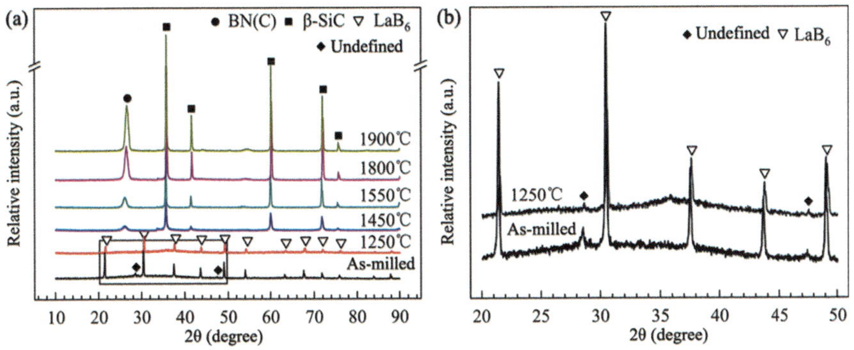

Fig. 4.118 XRD patterns of LaB_6/Si_2BC_3N composite ceramics prepared by a combination of two-step ball milling and SPS technique: **a** Sintered at various temperature; **b** enlarged of the box region in (**a**) [16]

LaB_6/Si_2BC_3N ceramics show gradual densification as sintering temperature raises from 1250 to 2000 °C, accompanied with the growth of lamellar structure of BN(C) flakes. The paths of crack propagation induced by Vickers indentation on MA-Si_2BC_3N ceramic surface are relatively straight. Conversely, LaB_6/Si_2BC_3N ceramics exhibit a zigzag crack path mainly propagating along the grain boundaries. Toughening mechanisms of pull-out of flakes, crack bridging and crack deflection are observed as a consequence of the increase in fracture toughness of the LaB_6/Si_2BC_3N. The lamellar structure of BN(C) flakes most likely plays a dominant role in toughening LaB_6/Si_2BC_3N ceramics (Figs. 4.119 and 4.120).

4.3.4 Effect of 1-D or 2-D Reinforcements

The intrinsic brittleness of SiBCN based materials is a great obstacle limiting their widespread applications under high-temperature ablation and thermal shock conditions. Carbon fiber (C_f) and silicon carbide fiber (SiC_f) are often used to reinforce SiBCN ceramic matrix composites (CMCs) to overcome such drawbacks (brittleness and low reliability). In this regard, the continuous carbon fiber reinforced PDCs-SiBCN CMCs processed by precursor impregnation and pyrolysis (PIP) provide tough composites. Fiber pull-out during fracture contributes to improved mechanical properties and such composites generally show non-brittle fracture behavior coupled with improved high temperature flexural strength and excellent creep resistance. In contrast, short fibers are more suitable for the development of MA-SiBCN CMCs by inorganic processing method as they allow a more uniform dispersion in as-milled amorphous powder.

Fig. 4.119 SEM fracture morphologies of the LaB_6/Si_2BC_3N composite ceramics prepared by a combination of two-step ball milling and SPS technique: sintered at **a** 1250 °C; **b** 1450 °C; **c** 1550 °C; **d** 1800 °C; **e** 1900 °C; **f** 2000 °C [16]

Fig. 4.120 SEM image for the indentation induced crack propagation in the as-sintered LaB_6/Si_2BC_3N composite ceramics [16]

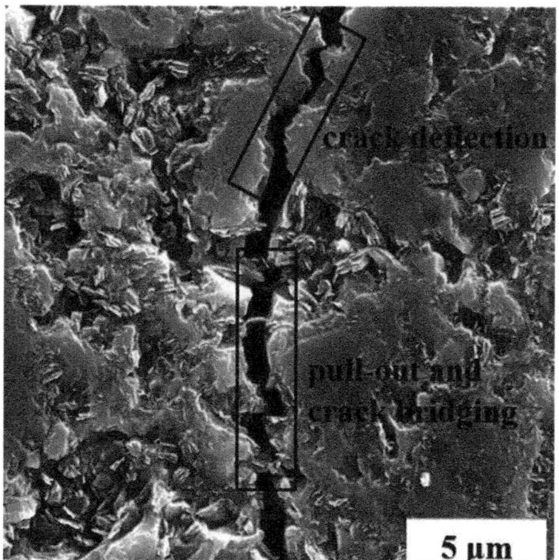

(1) C_f/Si_2BC_3N composites

Without sintering additive, the 20 vol.% C_f containing composites are composed of BN(C), α-SiC and β-SiC [50]. With the addition of 7.5 mol.% ZrO_2, the phase composition of the matrix has changed into BN(C), α/β-SiC, t-ZrO_2 and m-ZrO_2. Additionally, the peak position of BN(C) and SiC moves towards the high-angle direction. These changes imply that the addition of ZrO_2 promotes the formation of β-SiC, and improves the crystallinity of each phase (Fig. 4.121).

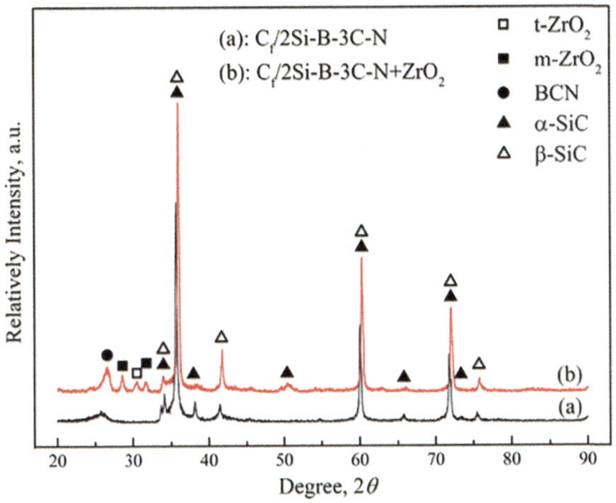

Fig. 4.121 XRD patterns of the C_f/Si_2BC_3N composites hot-pressed at 1800 °C/40 MPa/30 min: **a** + 7.5 mol.% ZrO_2; **b** without ZrO_2 additive [50]

For the additive-free C_f/Si_2BC_3N composites, a transition region exists between the carbon fiber and the ceramic matrix. The heterogeneous nucleation and abnormal grain growth exist in these interfacial regions. That is, the average size of SiC grains close to the carbon fiber is ~249.10 nm, while that of the SiC grains far away from the carbon fiber is just 50.45 nm. During hot-pressing, the carbon atoms on the carbon fiber surface are more active and have a higher concentration than those in the ceramic matrix. When the temperature approaches to the crystallization temperature of the amorphous phases, it gives priority to the nucleation of SiC grains on the carbon fiber surfaces. Relatively, the amorphous phases close to the carbon fiber has a faster atomic diffusion velocity and a lower crystallization temperature. As a result, SiC grains in the interfacial zone have a larger growing rate and a longer growing time, and hence a larger average grain size. The heterogeneous nucleation and abnormal grain growth may change the phase composition of the ceramic matrix (Fig. 4.122).

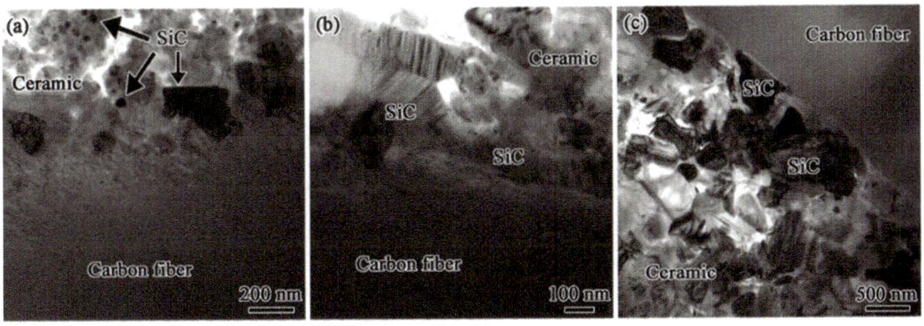

Fig. 4.122 Microstructure of the 20 vol.% C_f/Si_2BC_3N composites prepared by hot-pressing sintering at 1800 °C/40 MPa/30 min: **a, b** Without ZrO_2 additive; **c** + 7.5 mol.% ZrO_2 [50]

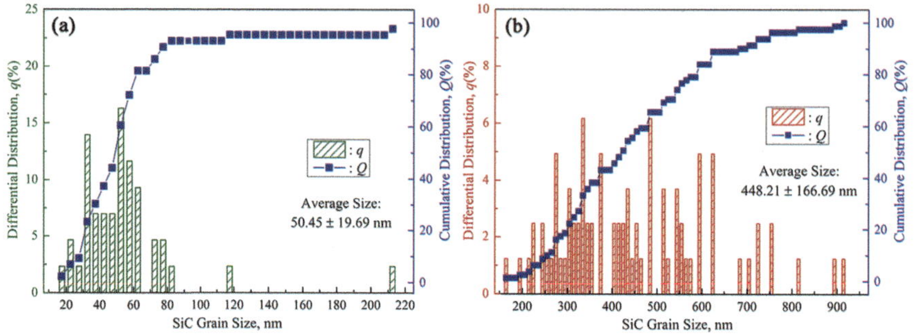

Fig. 4.123 The statistical distribution of the SiC grain sizes in 20 vol.% C_f/Si_2BC_3N composites: **a, b** Without ZrO_2 additive; **c** + 7.5 mol.% ZrO_2 [50]

The interface reactions may strengthen the interfacial bonding, and therefore the carbon fibers are difficult to be pulled out when the composite is fracturing. Thus, the toughening effects by carbon fibers are limited. With the addition of ZrO_2, the SiC grains close to or far away from the carbon fibers almost have the same average size of ~448.21 nm, much larger than the additive-free counterparts (Fig. 4.123). No abnormal grain growth exists in the ceramic matrix or in the interfacial region, and the interface is well defined between the fibers and ceramic matrix. The statistical data, also indicate SiC grains in the additive-free composites have a narrow size distribution, and the average size has a small standard deviation of ~19.69 nm. While with 7.5 mol.% ZrO_2 addition, the SiC grains are dispersedly distributed between 100 and 1000 nm, and the standard deviation of the average grain size is ~166.69 nm.

Though the 20 vol.% C_f/Si_2BC_3N composites are hot-pressed at 1800 °C and under a pressure of 40 MPa, they still contain a large number of micropores and holes. Most of the aggregates or particles still show near spherical shape. Expectedly, the ceramic particles are difficult to be deformed at this sintering condition. Additionally, the pull-out parts of the fibers are adhered with some ceramic particles, implying severe interface reactions occurring between fibers and ceramic matrix. Both the inferior sinterability of the amorphous powder and the aggregation of carbon fibers result in the large porosity of the as-prepared composites (Fig. 4.124).

(2) C_f/Si_2BC_3N-Zr composites

For fiber-free Si_2BC_3N-Zr ceramics, new phases of ZrB_2 and ZrO_2 form besides the original matrix phases of BN(C) and α/β-SiC [51]. The diffraction intensity of the ZrB_2 becomes stronger as the sintering temperature raises. In this case, ZrN is not detected. However, for C_f/Si_2BC_3N–Zr composites, they are comprised of new phase of ZrB_2, ZrN and ZrO_2. Of special interest is the formation of ZrN phase, which is not found during mechanical alloying process, but finally forms during hot-pressing sintering. The "carbon fibers driving ZrN formation" is the likely explanation, since some residual Zr may react with BN coating to form ZrN (Fig. 4.125).

The equiaxed particles are those of β-SiC grains with some stacking faults (Fig. 4.126). The ellipsoid particles in the gray region are suggested to be ZrB_2

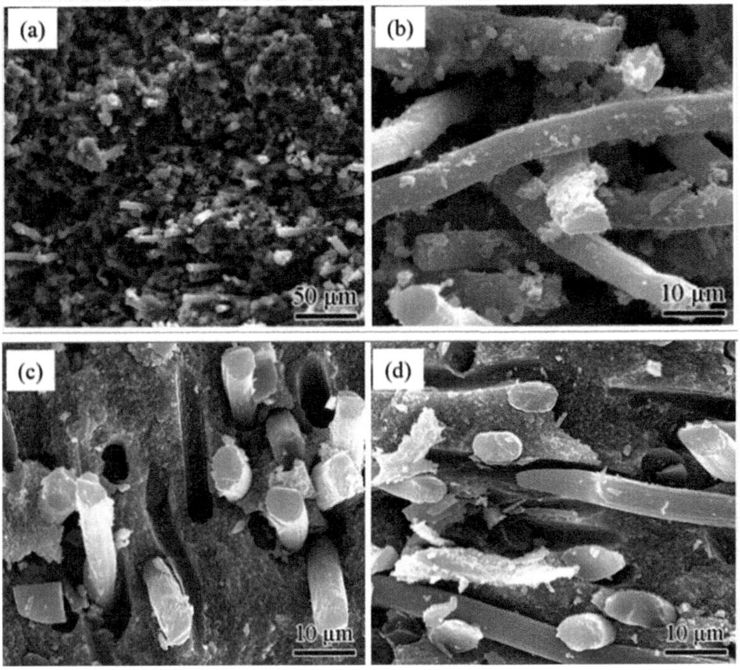

Fig. 4.124 SEM fracture morphologies of the 20 vol.% C_f/Si_2BC_3N composites hot-pressed at 1800 °C/40 MPa/30 min: **a**, **b** Without ZrO_2 additive; **c** + 7.5 mol.% ZrO_2 [50]

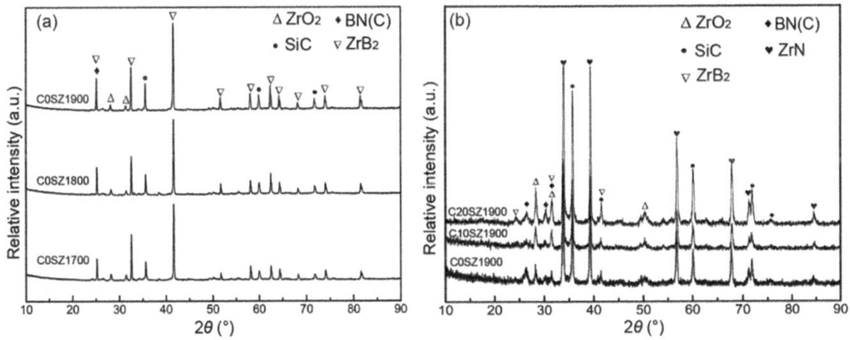

Fig. 4.125 XRD patterns of the Si_2BC_3N–Zr ceramics and C_f/Si_2BC_3N–Zr composites: **a** Si_2BC_3N–Zr sintered at different temperature; **b** h-BN coated C_f/Si_2BC_3N–Zr CMCs [51]

grains with a size distribution ranging from 10 to 100 nm. The white region between SiC and ZrB_2 grains, can be assigned to BN or graphite. No visible cracks and pores are observed on the polished surfaces for the fiber-free Si_2BC_3N-Zr ceramics (Fig. 4.127). For fiber containing composites, some neighboring fibers are agglomerated to form clusters. Such stacking or agglomeration is detrimental to the material

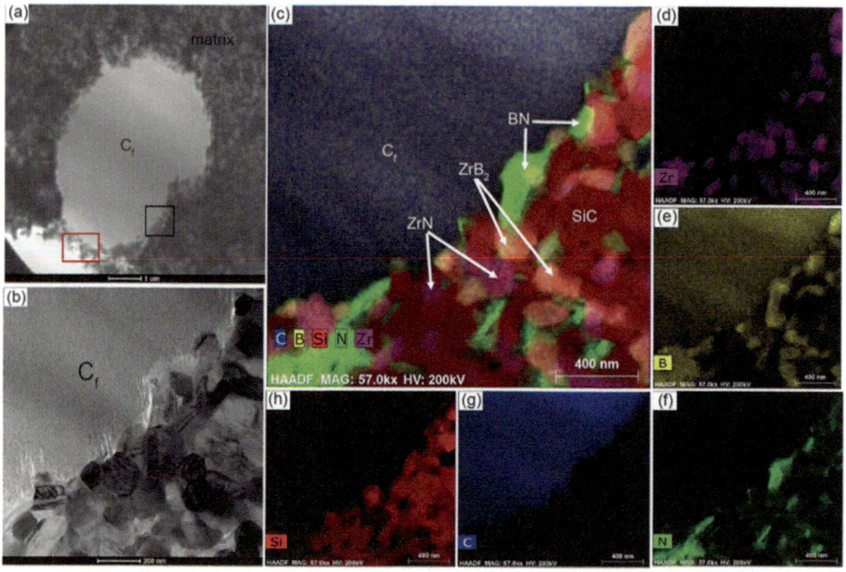

Fig. 4.126 TEM analysis of the 20 vol.% C_f/Si_2BC_3N–Zr composites (fiber coated with BN): **a** Low-magnification; **b** high-magnification; **c** Si, B, C, N, and Zr mapping; **d** Zr mapping; **e** B mapping; **f** N mapping; **g** C mapping; **h** Si mapping [51]

densification. However, carbon fibers are distributed along different directions, i.e., parallel to, at an angle to and perpendicular to the fracture planes.

(3) $(C_f–ZrB_2)/Si_2BC_3N$ composites

MA-Si_2BC_3N composites containing 10 vol.% C_f and 15 wt.% ZrB_2 are SPS densified at 2000 °C/40 MPa/5 min, and they all demonstrate the presence of ZrB_2, BN(C) and SiC [52]. Absence of ZrO_2 suggests their full reactions with B_2O_3 and C at this sintering condition. Significantly, the grain sizes of ZrB_2 and SiC are both over 100 nm, larger than those in pure ceramics; besides, a weak h-BN interface between ZrB_2 and C_f remains free of impurities and debris (Fig. 4.128).

For $(C_f–ZrB_2)/Si_2BC_3N$ composites without h-BN coating, micropores and defects are found on the sample surfaces. The breakage of C_f originates from these surface defects. The degradation of fibers should be attributed to chemical reactions between ZrO_2 and C_f during SPS process. Once h-BN is coated on the fibers, the resulting composites show a smooth surface. The pull-out of fibers is confirmed, as judged by the holes left on the fractured surface. These holes are perpendicular to the fracture interface toughening composites. The surfaces of the pulled-out fibers are smooth and dense, with few ceramic particles attached (Fig. 4.129).

(4) C_f/Si_2BC_3N composites with PyC coating

The bulk density of 20 vol.% C_f/Si_2BC_3N composites with pyrolytic carbon (PyC) coating increases with the increasing sintering temperature, which is only 1.92 g/cm^3

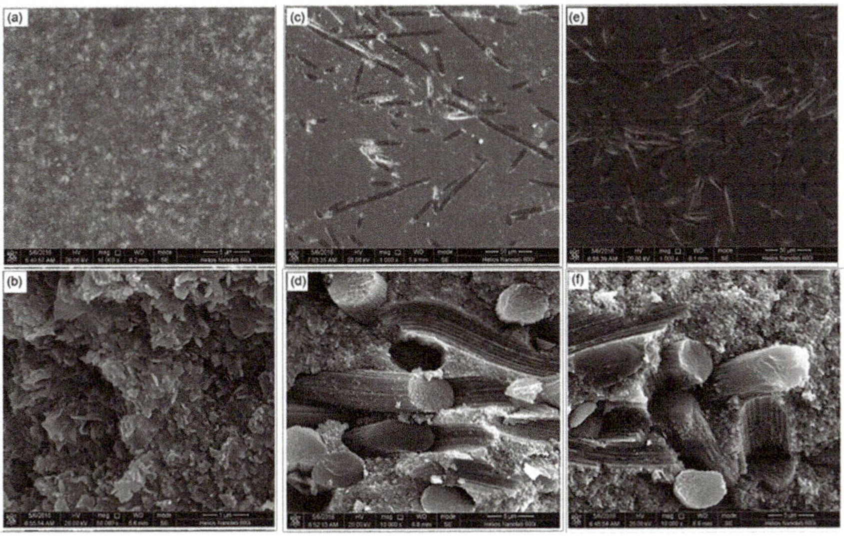

Fig. 4.127 Surface and fracture morphologies of the various C_f/Si_2BC_3N–Zr composites sintered at 1900 °C/60 MPa/30 min: **a**, **b** Pure Si_2BC_3N–Zr; **c**, **d** + 10 vol.% C_f; **e**, **f** + 20 vol.% C_f [51]

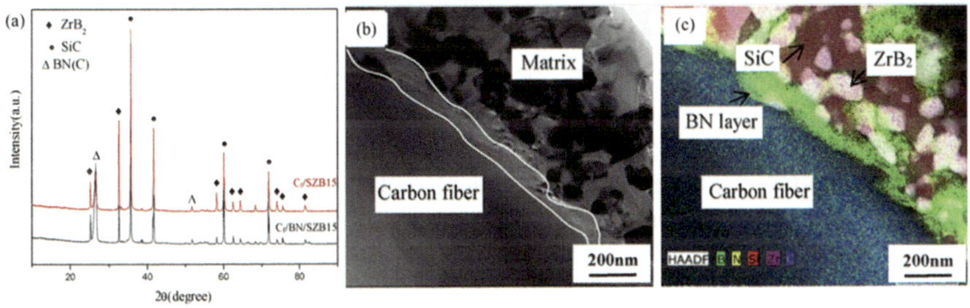

Fig. 4.128 Structure characterization of the (C_f–ZrB_2)/Si_2BC_3N composites SPS densified at 2000 °C/40 MPa/5 min: **a** XRD patterns; **b**, **c** STEM image and element distribution map [52]

at 1800 °C but increases to 2.14 g/cm^3 at 1900 °C, and reaches 2.18 g/cm^3 at 2000 °C. Consequently, their relative densities are calculated as 71, 81, and 83%, respectively. With the same sintering process, the relative density of pure MA-Si_2BC_3N ceramics can reach to 74, 88, and 93%, respectively. Obviously, the addition of carbon fibers strongly hinders densification, mainly due to bridging among fibers [53]. Most fibers are distributed in the plane perpendicular to the pressing direction (horizontal plane). In fiber-dense area, different degrees of matrix shedding occur exposing the fiber end, because fiber bridging in the fiber-dense area causes difficulty of sintering densification and decrease of matrix strength (Fig. 4.130).

The axial thermal expansion coefficient of carbon fiber -0.7×10^{-6} °C^{-1} is much lower than that (3–5) $\times$ 10^{-6} °C^{-1} of ceramic matrix. Due to the thermal

Fig. 4.129 SEM surface and fracture morphologies of the (C_f–ZrB_2)/Si_2BC_3N composites: **a–c** Unmodified C_f; **d–f** C_f coated with h-BN [52]

expansion mismatch, thermal stress between the fibers and the matrix exceeds the tensile strength of the matrix, resulting in a large number of micro-cracks in the matrix and promoting the shedding of the matrix. With the increasing sintering temperature, the area of matrix shedding decreases gradually, and only a small amount of matrix shedding occurs on the surface for the composites sintered at 2000 °C. The illustration in the upper right corner is the surface morphology parallel to horizontal plane. Due to less fiber bridging in the horizontal plane, the surface is flatter with fewer matrix shedding. With increasing sintering temperature, matrix shedding in horizontal plane also further decreases, and there is almost no matrix shedding on the surfaces for composites sintered at 2000 °C.

The SiC nanocrystals are uniformly distributed in ceramic matrix, with mean grain sizes of 30–50 nm, but a small number of SiC grains are abnormally grown ~100 nm. Although SiC nanocrystals have a high crystallization degree, there are high-density stacking faults and twins in SiC nanocrystals. The BN(C) is mostly distributed in the edge of the SiC grains. The segregation regions of Si and C atoms in residual amorphous phase may become the crystal nucleus of SiC (Fig. 4.131).

The fracture surfaces of the as-sintered composites are uneven, indicating obvious crack-deflection near fiber bundles. A large number of pulled out fibers and the residual pits suggest debonding and pull-out are the main toughening mechanisms during fracture. As sintering temperature increases, fiber pull-out length gradually decreases due to increase in interfacial bonding strength. Specifically, stress concentration at interfaces cannot be alleviated by fiber debonding and at fiber ends; stress

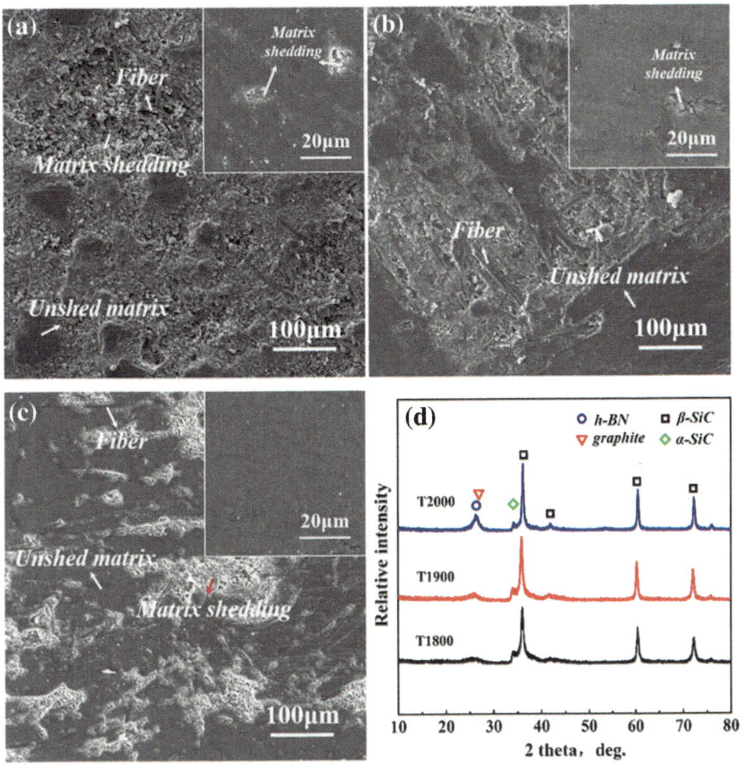

Fig. 4.130 SEM surface morphologies and XRD patterns of 20 vol.% C_f/Si_2BC_3N composites with PyC coating hot-pressed at different temperature: Sintered at **a** 1800 °C; **b** 1900 °C; **c** 2000 °C; **d** XRD patterns [53]

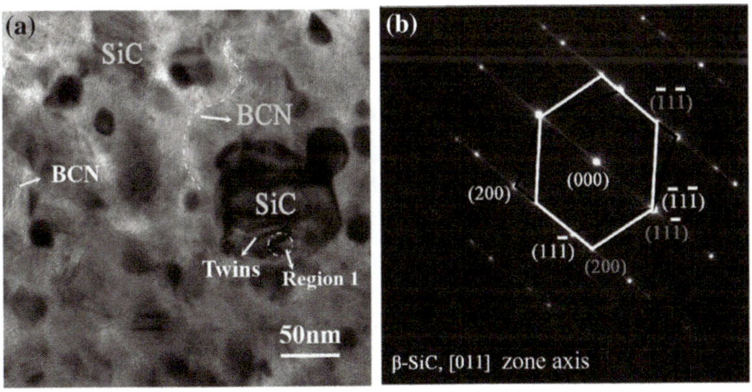

Fig. 4.131 TEM investigation on 20 vol.% C_f/Si_2BC_3N composites hot-pressed at 1900 °C: **a** Bright-field image; **b** SAED pattern [53]

quickly reaches the fiber strength, resulting in a closer fracture position to the fiber end and shorter fiber pull-out length. For C_f/Si_2BC_3N composites without PyC coating, the carbon fiber becomes blurred with obvious distortion. Compared with the PyC coated ones, the pull-out length of carbon fibers is shorter, and many ceramic particles are adhered on the fiber surfaces, implying strong interfacial bonding. The fractured fibers of all the PyC coated C_f/Si_2BC_3N composites retain the original morphology without signs of corrosion and deformation damage (Fig. 4.132).

(5) C_f/Si_2BC_3N-MAS composites with PyC coating

The as-sintered C_f/Si_2BC_3N-MAS composites are composed of α/β-SiC and BN(C). The diffraction peaks for MAS are not detected due to their low content or below the limits of detection. The resulting PyC coating bonds well with the carbon fiber surface, when 50 g/L of phenolic resin impregnation concentration is used and pyrolysis temperature of 800 °C is adopted [32]. With 6 wt.% MAS addition, the bulk

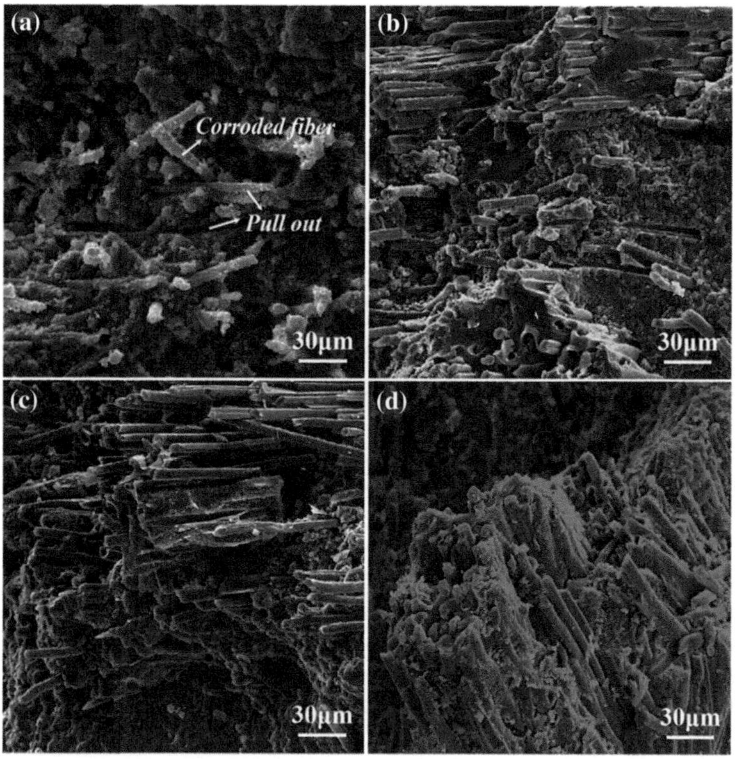

Fig. 4.132 SEM fracture morphologies of 20 vol.% C_f/Si_2BC_3N composites hot-pressed at different temperature: **a** Without PyC coating, sintered at 1800 °C; With PyC coating sintered at **b** 1800 °C; **c** 1900 °C; **d** 2000 °C [53]

densities of the C_f/Si_2BC_3N-MAS composites decrease gradually with increasing fiber content; however, the porosity of the CMCs show irregular change with low porosity below 3.2% (Fig. 4.133). The effective pull-out of fibers is observed for all the resulting composites; however, agglomeration of fibers is frequently encountered with increasing fiber content. The addition of 6.0 wt.% MAS greatly promotes the densification of the C_f/Si_2BC_3N matrix (Fig. 4.134).

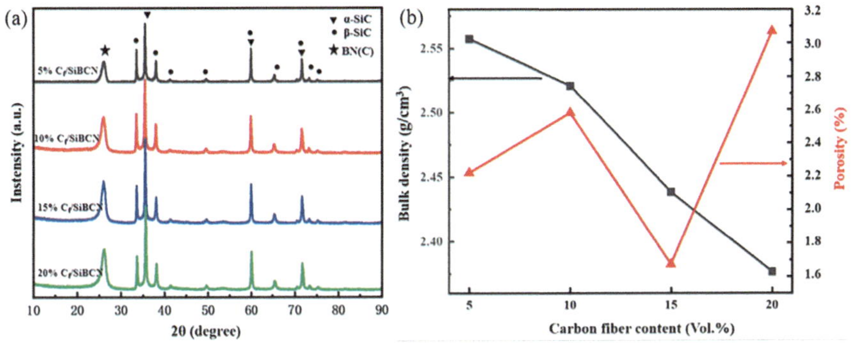

Fig. 4.133 Phase composition and density of C_f/Si_2BC_3N-MAS composites with PyC coating hot-pressed at 1800 °C/60 MPa: **a** XRD patterns; **b** bulk density and porosity [32]

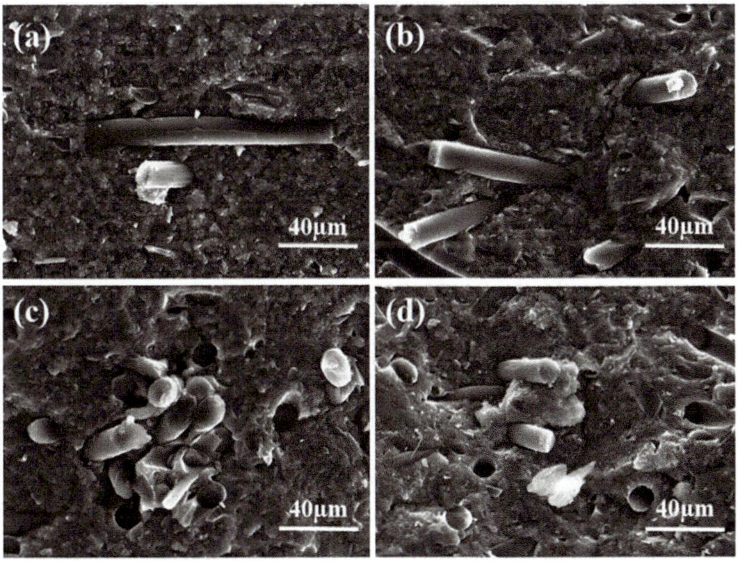

Fig. 4.134 Fracture morphologies of hot-pressed C_f/Si_2BC_3N-MAS composites with PyC coating: **a** + 5 vol.% C_f; **b** + 10 vol.% C_f; **c** + 15 vol.% C_f; **d** + 20 vol.% C_f [32]

(6) SiC_f/Si_2BC_3N composites

The SLFC1 SiC fibers are used as the reinforcements. After hot-pressed at 1800–1900 °C/60 MPa, the 20 vol.% SiC_f/Si_2BC_3N composites mainly consist of nano-sized SiC ~100 nm and BN(C) [54]. The SiC fibers are severely crystallized and the size of the resulting SiC grains is also ~100 nm. The interfaces between ceramic matrix and fibers are hard to be observed by TEM (Fig. 4.135).

For all the as-sintered 20 vol.% SiC_f/Si_2BC_3N composites, fiber pull-out is rarely observed on the fractured surfaces. In rare cases, the length of fiber pull-out is too short so the SiC_f toughening effect cannot be insured. The fibers have sheared off at the fracture surfaces suggesting the strong interfacial bonding of fiber/matrix and/or the strength degradation of fibers (Fig. 4.136).

The layer-fractured phenomenon of the SiC fibers can be generally found on the fracture surfaces. Protrusions or pits exist in the center of SiC fibers and the thickness of fractured layer is ~2 μm, basically equivalent to the SiC sheath. On the fractured surfaces of the composites prepared at 1900 °C/60 MPa/60 min, similar phenomenon is also observed, but the reaction layer thickness increases and few protrusions or pits are left in the center area of fibers (Fig. 4.137).

(7) SiC_f/Si_2BC_3N composites with amorphous graphite coating

To avoid the strong interfacial reactions, the amorphous graphite coated KD-II SiC_f are introduced into the MA-Si_2BC_3N ceramic matrix. After hot-pressed at 1900 °C, a distinct reaction interface about 30–50 nm composed of nano-SiC forms, whose size is similar to the amorphous graphite coating. In some cases, abnormal growth of SiC grains is found at the interfaces. Besides, some micropores form at the interfacial regions. Some free carbon accumulates near the interfaces. The SiC grain sizes in ceramic matrix are in a range of 100–200 nm while the grain sizes inside the fiber matrix are of 30–50 nm (Fig. 4.138).

The reaction interface structure of the KD-II SiC_f reinforced MA-Si_2BC_3N composites (without amorphous graphite coating) is similar to the former [54, 55], but differs in the size of reaction layer (Fig. 4.139). Abnormal growth of SiC grains is also found and the thickness of SiC reaction layer is about 100–120 nm. Some

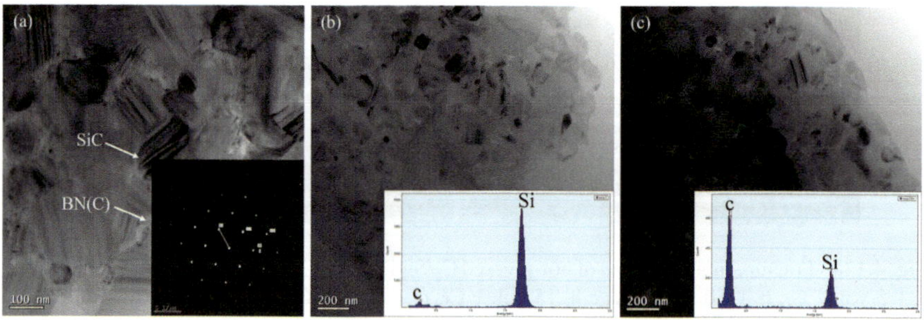

Fig. 4.135 TEM analysis of 20 vol.% SiC_f/Si_2BC_3N composites: **a** Hot-pressed at 1900 °C/60 MPa/60 min; **b** hot-pressed at 1800 °C/60 MPa/60 min, inner SiC_f and **c** ceramic matrix [54]

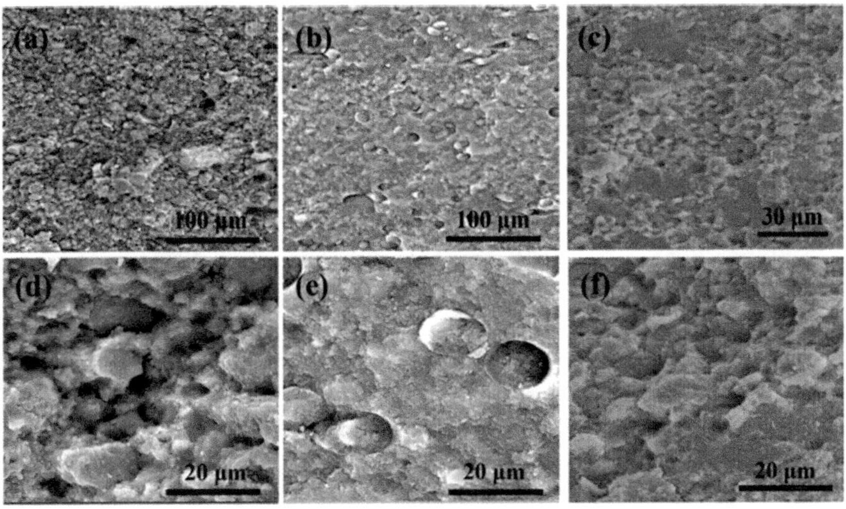

Fig. 4.136 SEM fracture morphologies of hot-pressed 20 vol.% SiC_f/Si_2BC_3N composites: Sintered at **a**, **d** 1800 °C/60 MPa/60 min; **b**, **e** 1900 °C/60 MPa/30 min; **c**, **f** 1900 °C/60 MPa/60 min [54]

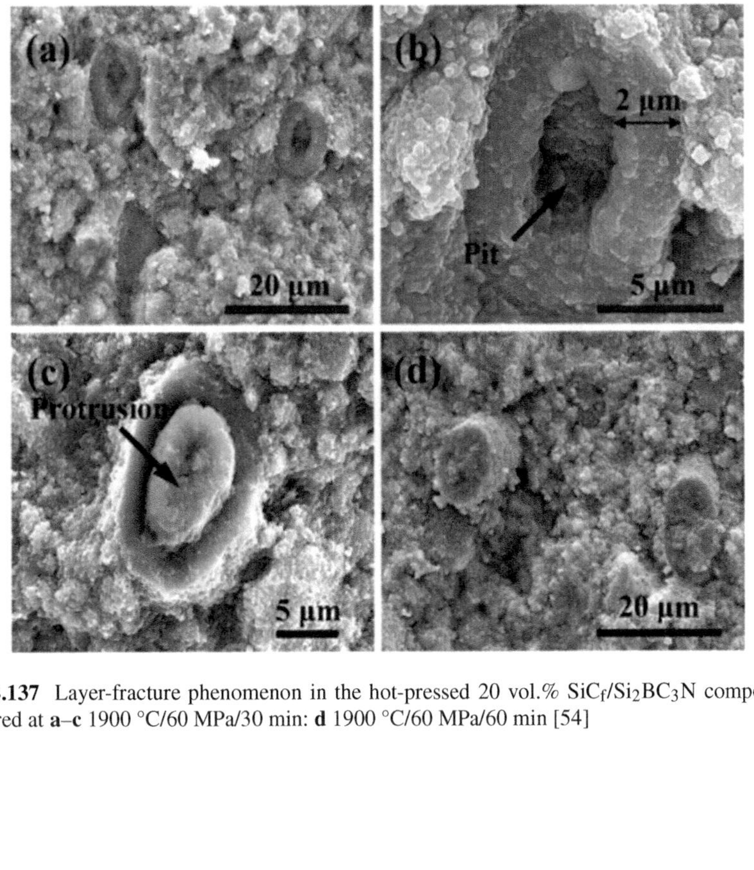

Fig. 4.137 Layer-fracture phenomenon in the hot-pressed 20 vol.% SiC_f/Si_2BC_3N composites: Sintered at **a**–**c** 1900 °C/60 MPa/30 min: **d** 1900 °C/60 MPa/60 min [54]

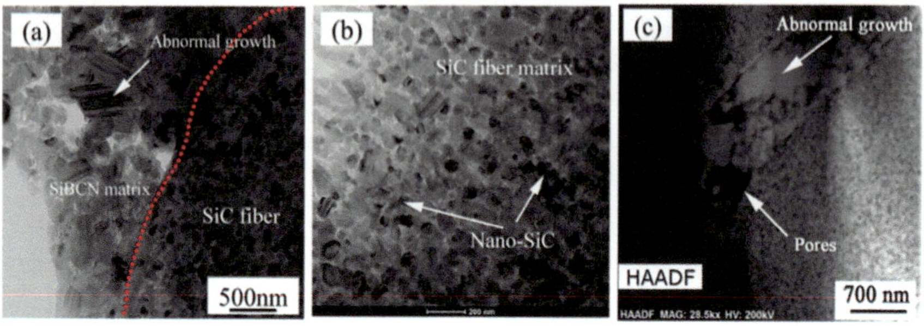

Fig. 4.138 TEM microstructure of the 10 vol.% SiC_f (KD-II) reinforced MA-Si_2BC_3N CMCs (fibers coated with amorphous graphite): **a** Fiber/ceramic matrix interfaces; **b** inner fiber matrix; **c** abnormal grain growth of SiC and pore distribution at the interfaces [55]

oxygens have reacted with Si to form SiO_2 at the interfaces. Though the thickness of reaction interfaces is larger than that of the former, it is suggested that pre-treatment to remove the impurities and oxide layer on SiC fiber surface prior to utility is enough without *in-situ* amorphous graphite coating to optimize the strength of interfacial bonding. The crystallization of SiC fibers and subsequently forming interfacial structure during HP sintering should be responsible for the poorer performance of the 10 vol.% SiC_f/Si_2BC_3N composites (Fig. 4.140).

The as-sintered composites all present smooth fracture morphologies with very limited pullout length of SiC_f (Fig. 4.141). Thus, surface modification of SiC fibers at a low temperature avoiding crystallization is necessary or perhaps chooses much higher thermal-stability SiC fibers or changes the nature of reinforcements.

(8) SiC_f/Si_2BC_3N composites with composite coating

After HP at 1900 °C/60 MPa/30 min, the 5 vol.% SiC_f containing MA-Si_2BC_3N composites with different coating modifications all show obvious crystallization and the main phases are β-SiC, α-SiC and BN(C) [57, 58]. In all cases, the coating

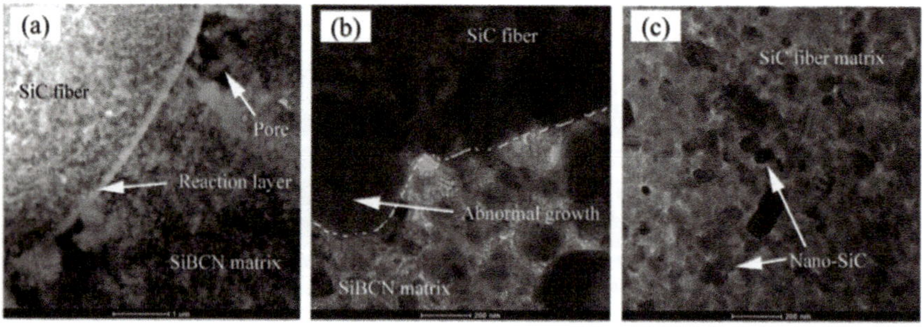

Fig. 4.139 TEM microstructure of the 10 vol.% SiC_f (KD-II) reinforced MA-Si_2BC_3N CMCs (with unmodified fibers): **a** Fiber/ceramic matrix interfaces; **b** abnormal growth of SiC at the interfaces; **c** inner fiber matrix [55]

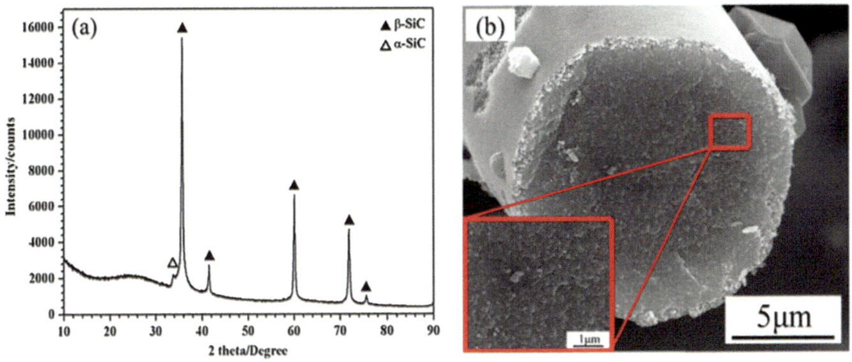

Fig. 4.140 Phase composition and microstructure showing the influence of sintering temperature leading to fiber crystallization: **a** XRD pattern; **b** SEM morphologies of SiC fibers heat-treated at 1900 °C for 30 min under 1 bar N_2 (Particles on fiber surfaces are those of silica and inserted magnified picture indicates the fiber matrix composed of nano-grains) [55]

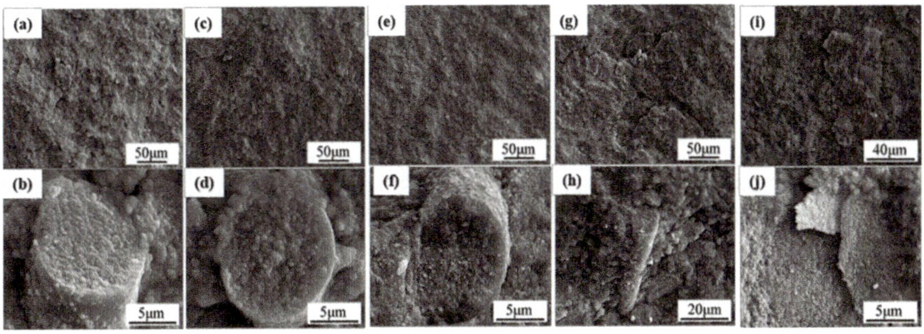

Fig. 4.141 The fracture morphologies of the various SiC_f (KD-II) reinforced MA-Si_2BC_3N CMCs with **a**–**h** or without **i**, **j** amorphous graphite coating: **a**, **b** + 5 vol.% SiC_f; **c**, **d** + 10 vol.% SiC_f; **e**, **f** + 15 vol.% SiC_f; **g**, **h** + 20 vol.% SiC_f; **i**, **j** + 10 vol.% SiC_f [56]

composition are not detected (Fig. 4.142). The polished composites show relatively dense surface structure and SiC_f are preferred to distribute perpendicular to the HP direction. In few cases, some fibers are scattered parallel to the HP direction (Fig. 4.143).

The composite coatings, such as C/BN, C/ZrB_2, ZrB_2/BN and C/ZrB_2/BN are prepared on the KD-II SiC_f surfaces. The amorphous graphite coating (C) has partly changed to ordered structure during the C/BN, C/ZrB_2 and C/ZrB_2/BN composite coatings processing. For C/BN composite coating, the fiber surfaces are relatively smooth, basically achieving uniform coating. The C/ZrB_2 coating is essentially consistent though a few particles attached to the coating surfaces; however, the C/ZrB_2 coated fibers tend to overlap together after ultrasonic dispersion. Note that part of the SiC_f fractures during C/ZrB_2 composite coating processing. Unlike the

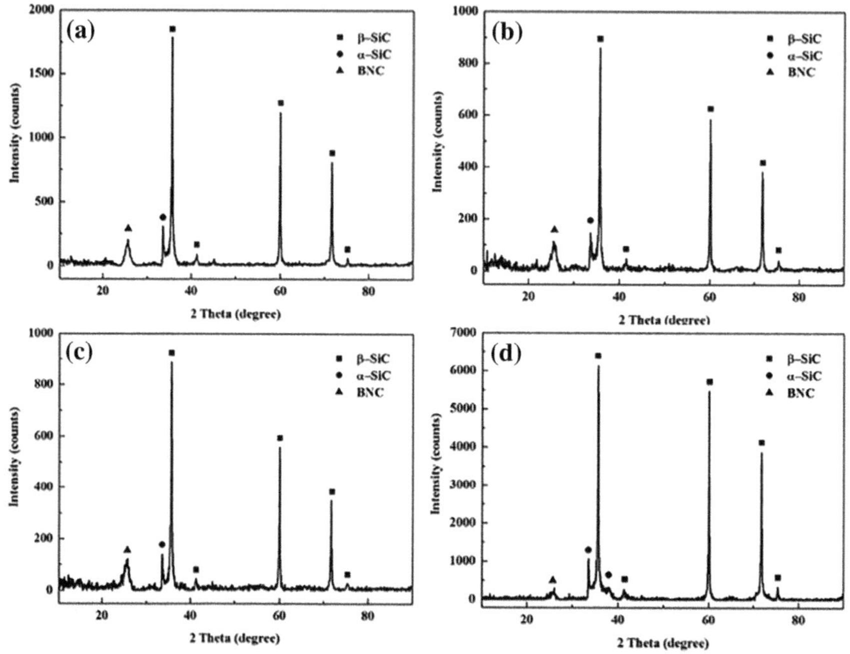

Fig. 4.142 XRD patterns of various 5 vol.% SiC_f (KD-II) reinforced MA-Si_2BC_3N CMCs hot-pressed at 1900 °C/60 MPa/30 min: **a** Unmodified fibers: fibers coated with **b** amorphous graphite: **c** ZrB_2; **d** h-BN [57]

double-layer coating, the three-layer composite coating C/ZrB_2/BN shows inhomogeneous distribution of each component with substantial nanoparticles anchored on fiber surfaces (Fig. 4.144).

After introduction of 5 vol.% KD-II SiC_f coated with various composite coating, the amorphous MA-Si_2BC_3N matrix has crystallized to β-SiC, α-SiC and BN(C) after HP sintering (Fig. 4.145). The as-sintered composites are relatively dense with relatively low porosity. Most of the SiC_f in the composites are anisotropic and perpendicular to the HP direction, induced by the axial pressure (Fig. 4.146). The C/ZrB_2 coated SiC_f/Si_2BC_3N composites show a crack propagation leading to the peel-off of fibers. In rare cases, debonding and pulling out of short SiC_f are found for all the investigated composites. Generally, formation of strong interfacial bonding between SiC_f and MA-Si_2BC_3N matrix or limited strength retention of SiC_f after HP sintering irrespective of what coating used, are both observed (Fig. 4.147).

Contrary to expectations, the relative densities of as-coated 5 vol.% SiC_f/Si_2BC_3N composites are lower than that of its counterparts (Fig. 4.148). Especially, the relative density of ZrB_2/BN coated SiC_f/Si_2BC_3N composites decreases greatly. The unmodified composites have the highest relative density of ~84.2% (Table 4.6). The undesirable relative density of the as-coated composites should be partially attributed

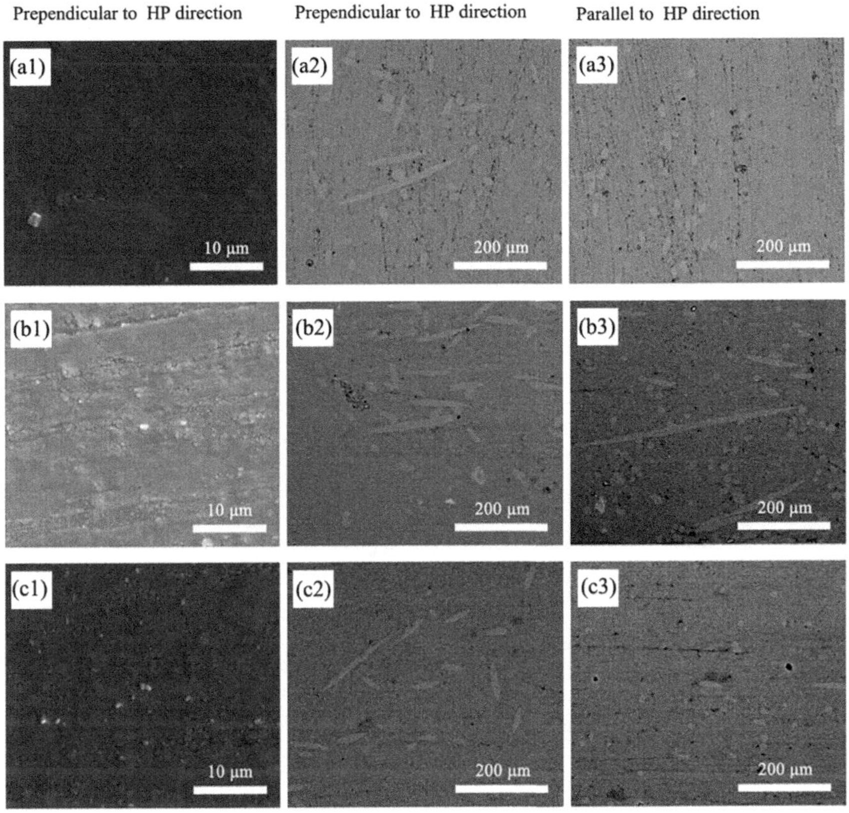

Fig. 4.143 SEM surface morphologies of various 5 vol.% SiC_f (KD-II) reinforced MA-Si_2BC_3N CMCs: Fibers coated with **a** amorphous graphite; **b** ZrB_2; **c** h-BN [57]

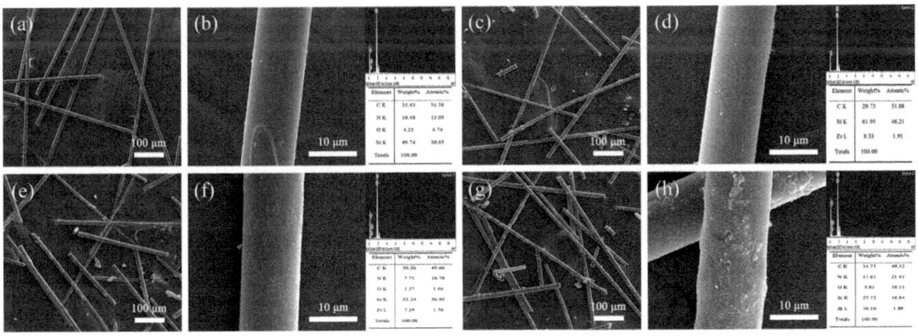

Fig. 4.144 SEM surface morphologies of the various 5 vol.% SiC_f (KD-II) reinforced MA-Si_2BC_3N CMCs: Fibers coated with **a** C/BN; **b** C/ZrB_2; **c** ZrB_2/BN; **d** C/ZrB_2/BN [57]

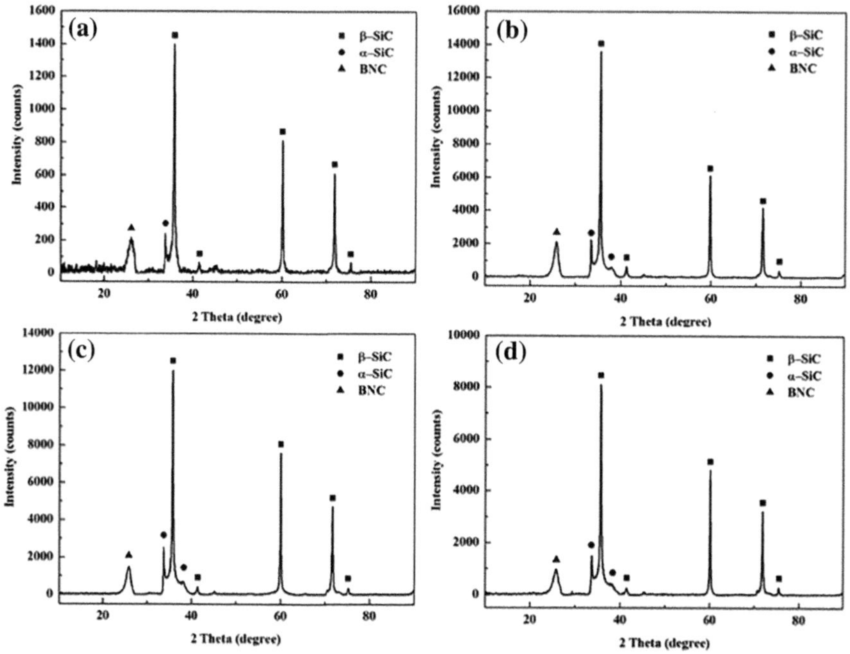

Fig. 4.145 XRD patterns of various 5 vol.% SiC_f (KD-II) reinforced MA-Si_2BC_3N CMCs: **a** Fibers coated with **a** C/BN; **b** C/ZrB_2; **c** ZrB_2/BN; **d** C/ZrB_2/BN [57]

to the relatively low uniformity of coating, bad physicochemical compatibility of coating and ceramic matrix, and poor dispersibility of fibers in ceramic matrix.

(9) SiC_f/Si_2BC_3N-MAS composites

The KD-SA SiC_f with higher thermal stability is used [59]. Without MAS additive, the porosity of the hot-pressed SiC_f/Si_2BC_3N composites is in range of 14.8–19.7%, much higher than those < 3.2% of SiC_f/Si_2BC_3N. Consequently, the bulk density of the composites retains only 2.14–2.20 g/cm^3 (Table 4.7). However, with 10 wt.% MAS addition, the SiC_f/Si_2BC_3N composites are almost fully densified. All the investigated composites with MAS additive show bulk density higher than 2.68 g/cm^3 and porosity lower than 0.6% regardless of h-BN coating use (Table 4.8).

Without MAS additive, the relatively weak bonding interfaces between the fibers and ceramic matrix are guaranteed, thus the h-BN coated 20 vol.% fibers reinforced composites exhibit the pseudo-plastic fracture characteristics. The failure morphologies can be divided into two parts, fist is ragged pull-out of fiber, and second is uneven fracture. After adding 10 wt.% MAS, fracture surfaces of the composite materials are very even and tend to smooth, also suggesting undesirable toughening effects (Fig. 4.149). Without h-BN coating, the as-sintered SiC_f/Si_2BC_3N composites with 10 wt.% MAS addition are also rarely observed fiber extraction and fiber bridging at the fracture surfaces. Thus, though the MAS additive effectively enhances the

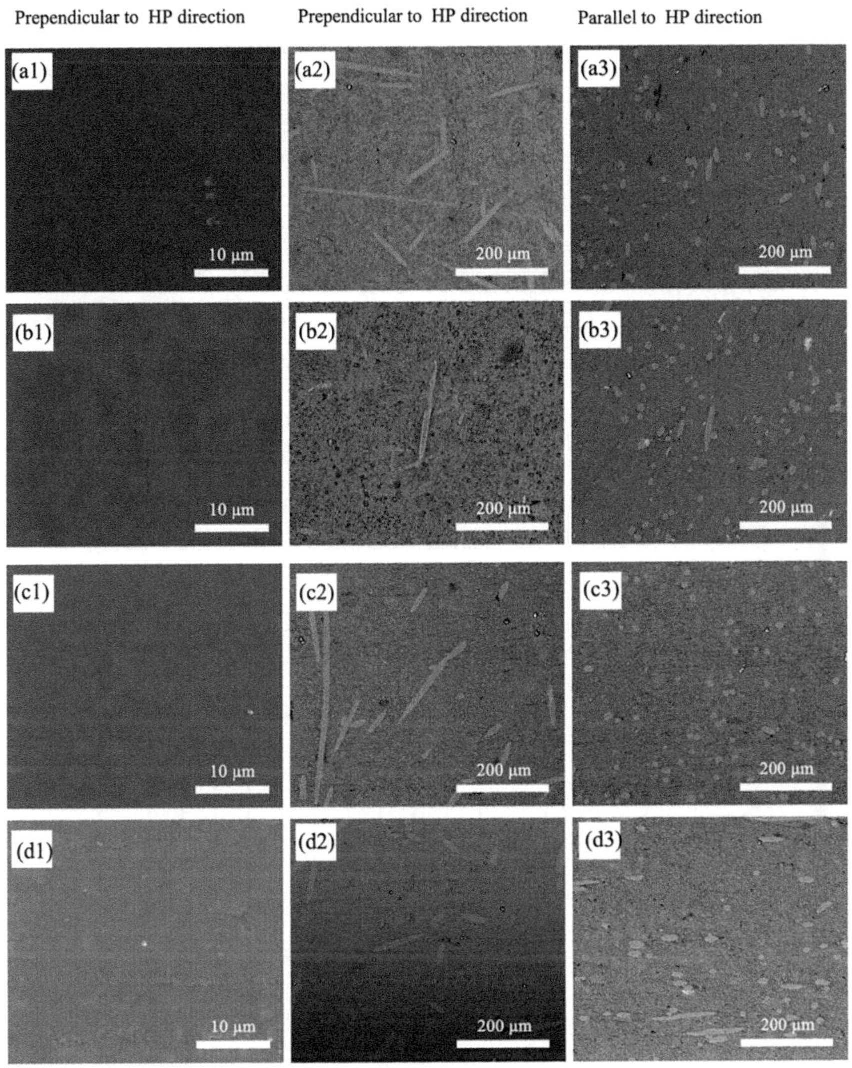

Fig. 4.146 SEM surface morphologies of various 5 vol.% SiC_f (KD-II) reinforced MA-Si_2BC_3N CMCs: Fibers coated with **a** C/BN; **b** C/ZrB_2; **c** ZrB_2/BN; **d** C/ZrB_2/BN [57]

densification of ceramic matrix, they also favor the formation of stronger interfacial bonding between SiC_f and ceramic matrix.

(10) (C_f–SiC_f)/Si_2BC_3N composites

The 10 vol.% SiC_f (SLFC1, 12 μm in average diameter, 1–2 mm in length), and 10 vol.% C_f (TX-3, 7 μm in average diameter, 1–2 mm in length) are both adopted [60]. The h-BN or PyC coating is deposited on (SiC_f–C_f) surfaces by dip-coating

Fig. 4.147 SEM fracture morphologies of various 5 vol.% SiC_f (KD-II) reinforced MA-Si_2BC_3N CMCs: **a** Unmodified fibers; fibers coated with **b** amorphous C; **c** ZrB_2; **d** BN; **e** C/BN; **f** C/ZrB_2; **g** ZrB_2/BN; **h** C/ZrB_2/BN [57]

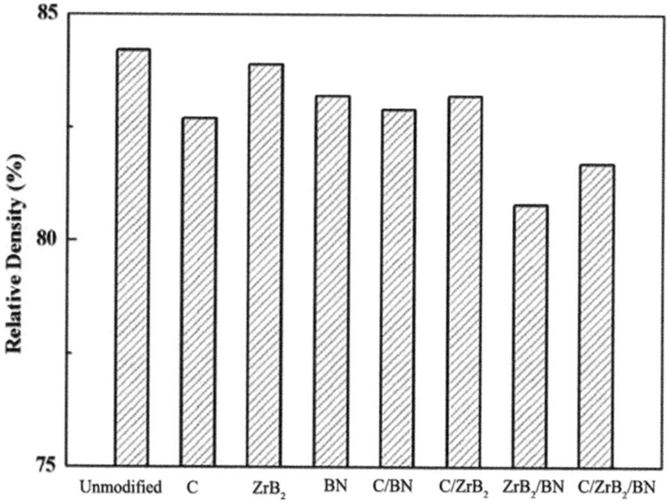

Fig. 4.148 The relative density of various 5 vol.% SiC_f (KD-II) reinforced MA-Si_2BC_3N CMCs hot-pressed at 1900 °C/60 MPa/30 min [57]

method. After hot-pressed at 1900 °C/60 MPa/30 min, the as-sintered three (C_f–SiC_f)/Si_2BC_3N composites contain main phases of β-SiC, α-SiC and BN(C). After coated with h-BN, the relative contents of each phase in (C_f–SiC_f)/Si_2BC_3N composites almost remain unchanged. But the decrease of α-SiC and slight increase of β-SiC contents are found after PyC coating processing, likely as a result of the α → β phase transition during HP sintering (Fig. 4.150).

Table 4.6 The bulk density and relative density of the various 5 vol.% SiC_f (KD-II) reinforced MA-Si_2BC_3N CMCs hot-pressed at 1900 °C/60 MPa/30 min [57]

SiC_f/ Si_2BC_3N	Unmodified	With amorphous C	With ZrB_2	With BN	With C/ BN	With C/ ZrB_2	With ZrB_2/ BN	With C/ ZrB_2/ BN
Bulk density (g/cm^3)	2.39	2.35	2.38	2.36	2.35	2.36	2.29	2.32
Relative density (%)	84.2	82.7	83.9	83.2	82.9	83.2	80.8	81.7

Table 4.7 Bulk density and porosity of various SiC_f/Si_2BC_3N CMCs without MAS addition [59]

Composition*	+ 5 vol.% SiC_f	+ 10 vol.% SiC_f	+ 15 vol.% SiC_f	+ 20 vol.% SiC_f
Bulk density (g/cm^3)	2.20	2.14	2.14	2.17
Porosity (%)	14.8	18.9	19.7	19.1

* The SiC_f is coated with h-BN

Table 4.8 Bulk density and porosity of various SiC_f/Si_2BC_3N CMCs with 10 wt.% MAS addition [59]

Composition	+ 5 vol.% SiC_f	+ 10 vol.% SiC_f	+ 20 vol.% SiC_f	* + 5 vol.% SiC_f	* + 20 vol.% SiC_f
Bulk density (g/cm^3)	2.70	2.76	2.69	2.68*	2.71
Porosity (%)	0.28	0.27	0.42	0.55	0.27

* The SiC_f is not coated with h-BN

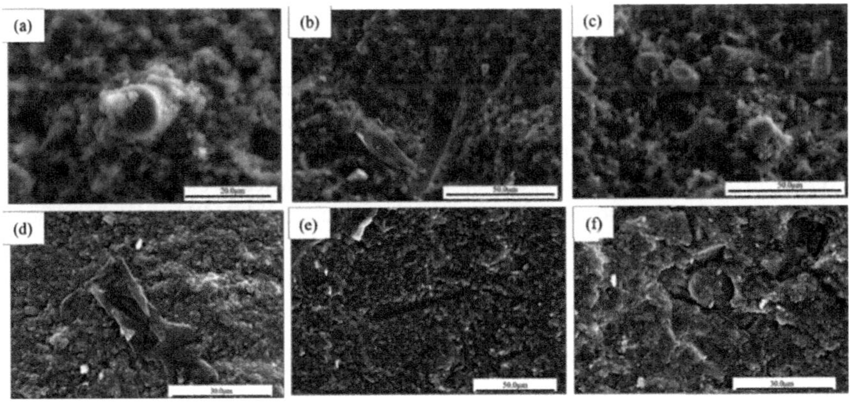

Fig. 4.149 SEM fracture morphologies of h-BN coated SiC_f reinforced MA-Si_2BC_3N CMCs without **a**–**c** or with **d**–**f** MAS addition (hot-pressed condition: 1850 °C/60 MPa/60 min) [59]

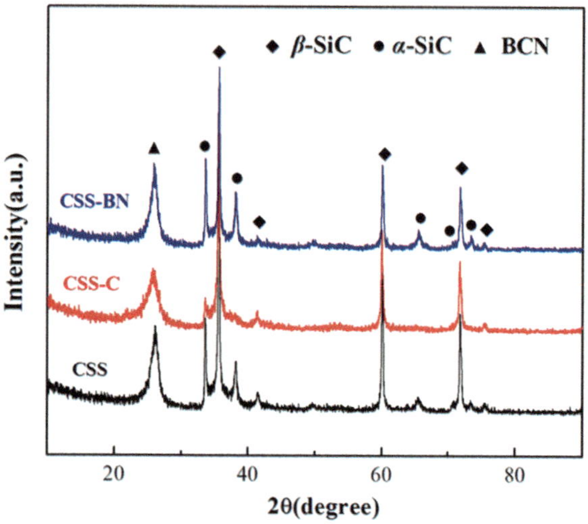

Fig. 4.150 XRD patterns of the various (C_f-SiC_f)/Si_2BC_3N CMCs hot-pressed at 1900 °C/60 MPa/30 min (perpendicular to the hot-pressing direction) [60]

After HP sintering, all the (C_f–SiC_f)/Si_2BC_3N composites present relatively homogenous structure with preferred orientation of short fibers in the plane perpendicular to HP direction, and only a few fibers are found in parallel direction. Some pores and flutes are observed frequently on the surfaces of the composites, mainly distributed around the overlapping area of fibers (Fig. 4.151). With the coating of PyC and h-BN weak interfaces, the bulk density of the (C_f–SiC_f)/Si_2BC_3N composites drops from 2.24 to 2.04 g/cm^3 and 2.18 g/cm^3, and the relative densities are reduced from 80.9 to 73.5 and 78.8%, respectively.

Cracks between fibers and ceramic matrix are observed and the introduction of PyC and h-BN coating seems to increase the pore sizes. The fibers are prone to agglomerate at some zones. Actually, only few C_f realize the limited pull-out and bridging function to toughened composites. Nearly all the SiC_f are degraded or partly reacted with the ceramic matrix (Fig. 4.152). It can be seen that a distinct reaction layer of SiC ~1.5 μm is surrounded on the C_f surfaces after PyC processing; in this case, the interfacial debonding is found. After coated with h-BN, such interfacial reaction among the fibers and matrix can be avoided. Generally, all the as-sintered (C_f-SiC_f)/Si_2BC_3N CMCs have a multimodal distribution of pore channel, which subsequently leads to an identical nonhomogeneous microstructure (Fig. 4.153).

(11) (C_f–SiC_f)/Si_2BC_3N–$ZrSiO_4$ composites

With $ZrSiO_4$ additive, except for the matrix phases of β-SiC, α-SiC and BN(C), the t-ZrO_2, and m-ZrO_2 are observed without any trace of $ZrSiO_4$. Though sintering aid $ZrSiO_4$ is used, the hot-pressed (C_f–SiC_f)/Si_2BC_3N composites (1900 °C/60 MPa/30 min) still show some pores distributed on the sample surfaces (Fig. 4.154). But the porosity reduction and density improvement can both be achieved by adding

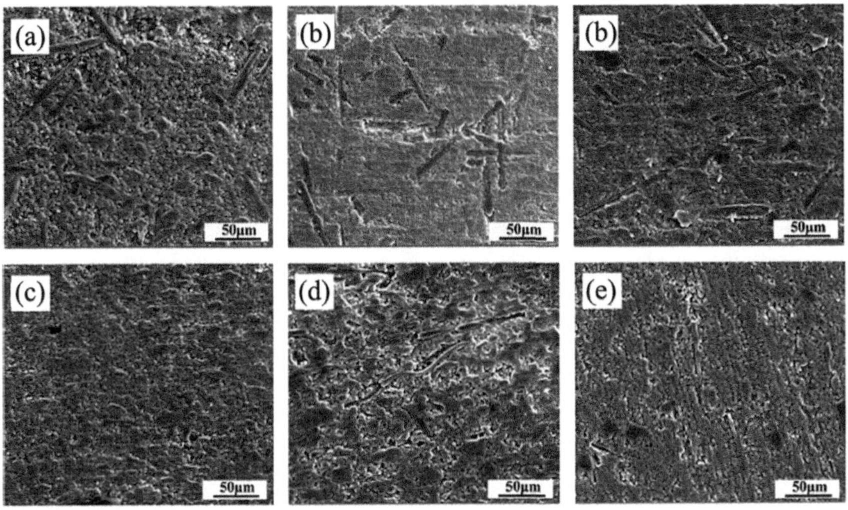

Fig. 4.151 SEM surface morphologies of the $(C_f–SiC_f)/Si_2BC_3N$ CMCs perpendicular (**a**–**c**) or parallel (**d**–**f**) to the HP direction: **a** Unmodified fibers; **b** fibers coated with PyC; **c** fibers coated with h-BN [60]

Fig. 4.152 SEM fracture morphologies of the $(C_f–SiC_f)/Si_2BC_3N$ CMCs perpendicular to the HP direction: **a**, **d** Unmodified fibers; **b**, **e** fibers coated with PyC; **c**, **f** fibers coated with h-BN [60]

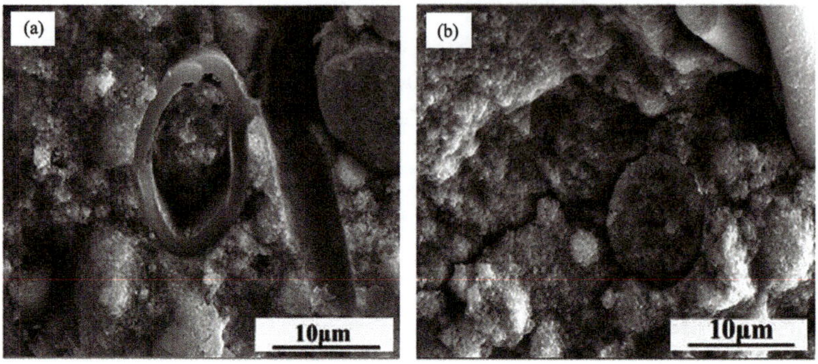

Fig. 4.153 Interfacial structures of the (C_f–SiC_f)/Si_2BC_3N CMCs hot-pressed at 1900 °C/60 MPa/30 min: **a** Fibers coated with PyC; **b** fibers coated with h-BN [60]

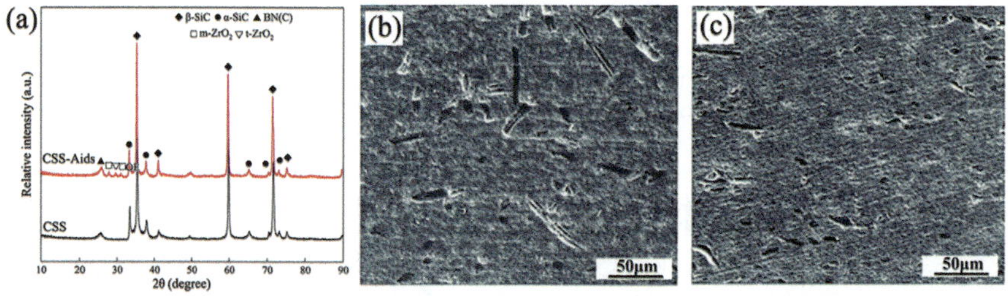

Fig. 4.154 Phase composition and surface morphologies of hot-pressed (C_f–SiC_f)/Si_2BC_3N CMCs with or without $ZrSiO_4$: **a** XRD patterns; **b** perpendicular to HP direction; **c** parallel to HP direction (fibers are uncoated with PyC or h-BN) [60]

$ZrSiO_4$. With the help of $ZrSiO_4$, the bulk density of the composites raises from 2.24 to 2.58 g/cm^3, and the relative density climbs from 80.9 to 93.1%.

In exterior layer, a jagged fracture surface with limited fiber pullout and an obvious de-cohesion between fibers and matrix leaving holes are found. But the pullout length of the fibers is short irrespective of $ZrSiO_4$ addition. These fracture characteristics can be attributed to the existence of inter-bundle pores and the strong interfacial bonding between fibers and matrix (Fig. 4.155).

(12) MWCNTs/Si_2BC_3N composites

Obviously, β-SiC, α-SiC and BN(C) phases are detected in all the as-sintered MWCNTs/Si_2BC_3N CMCs [61, 62]. The relative peak intensity of β-SiC increases while α-SiC decreases gradually with the incorporation of MWCNTs regardless of the SiC coating used. This variation implies that MWCNTs stimulated the formation of β-SiC in ceramic matrix (Fig. 4.156).

The smaller BN(C) platelets and SiC particles are both observed for the as-sintered MWCNTs/Si_2BC_3N CMCs. Increasing unmodified MWCNTs to 3 vol.%, the sizes of the BN(C) platelets are further reduced and spherical particles are observed. These

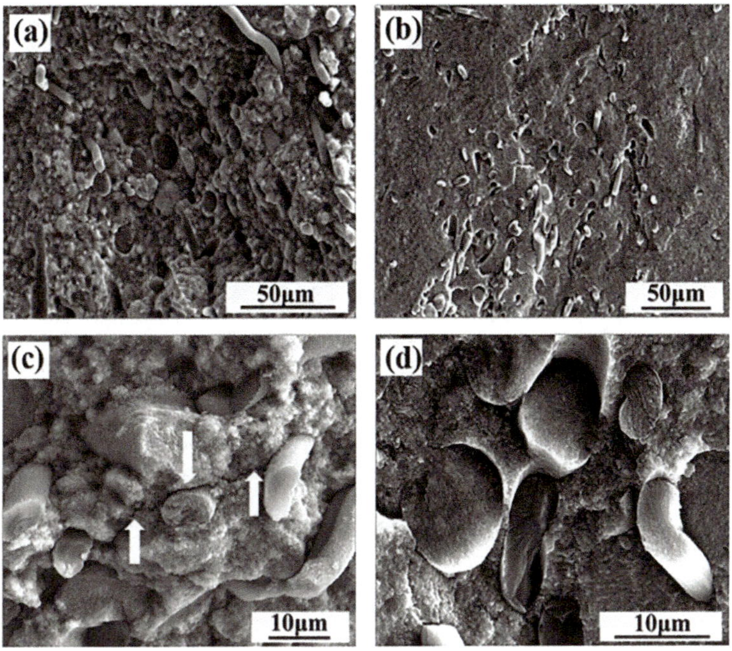

Fig. 4.155 SEM fracture morphologies of the (C_f–SiC_f)/Si_2BC_3N CMCs hot-pressed with $ZrSiO_4$ addition: **a**, **c** Unmodified fibers; **b**, **d** unmodified fibers with $ZrSiO_4$ additive [60]

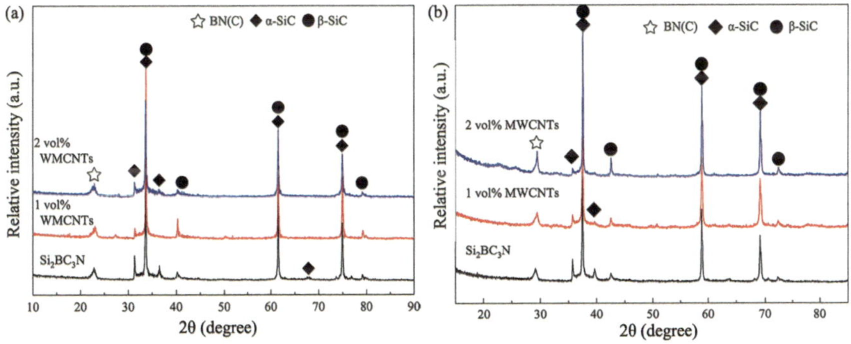

Fig. 4.156 XRD patterns of the MWCNTs/Si_2BC_3N CMCs prepared by SPS at 1900 °C/40 MPa/5 min: **a** MWCNTs without SiC coating; **b** MWCNTs with SiC coating [61, 62]

observations suggest MWCNTs hinders the growth of BN(C) and SiC, which should be argued to the segregation of MWCNTs regardless of the SiC coating (Fig. 4.157).

In comparison with pure MA-Si_2BC_3N, the SiC grain sizes of the 1 vol.% unmodified MWCNTs containing composites are reduced slightly to 50–70 nm; besides, a characteristic interfacial structure between turbostratic BN(C) and SiC is observed.

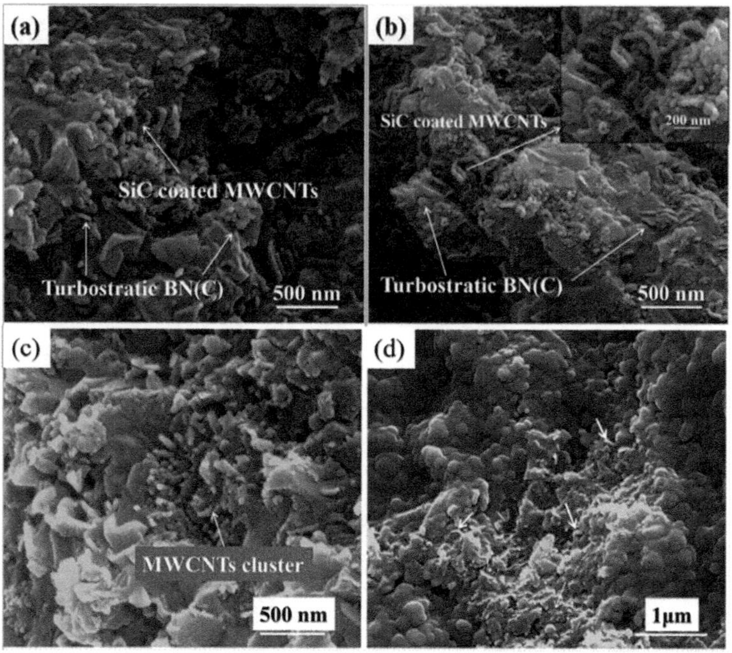

Fig. 4.157 SEM fracture morphologies of the SPS densified MWCNTs/Si_2BC_3N CMCs: **a**, **b** MWCNTs with SiC coating; **c**, **d** MWCNTs without SiC coating [61, 62]

In addition, the MWCNTs ~27.5 nm in diameter is pinning in the matrix between SiC and turbostratic BN(C). With 3 vol.% unmodified MWCNTs addition, the SiC grain sizes further decrease to ~50 nm and considerable MWCNTs are also found to pin the matrix (Fig. 4.158). Similar phase structure and phase distribution are found for the 2 vol.% modified MWCNTs containing composites. Coincidentally, a MWCNTs bridges both SiC grains are found; however, the SiC and BN(C) grains are larger than those observed in pristine MA-Si_2BC_3N ceramics. Thus, the SiC coating on MWCNTs promotes the development of each phase in ceramic matrix.

The composites with 1 vol.% unmodified MWCNTs exhibit higher relative density as compared with the pure ceramics. But the addition of 3 vol.% unmodified MWCNTs lowers the relative density. Obviously, incorporation of 1 vol.% MWCNTs or 1 vol.% SiC coated MWCNTs have no apparent effect on the densification behavior. In comparison, the introduction of 2 vol.% SiC coated MWCNTs improves the relative density of the composites (Table 4.9).

(13) GPLs/Si_2BC_3N composites

After SPS densified at 1800 °C/40 MPa/3 min, the amorphous MA-Si_2BC_3N powder has crystallized to form crystalline α/β-SiC and BN(C), but the corresponding peaks of graphene platelets (GPLs) are not found due to its low content or the overlapping at $2\theta = 24$–26° [63]. The peaks at 35.61°, 41.61°, 60.11°, 72.21° and 75.71° correspond

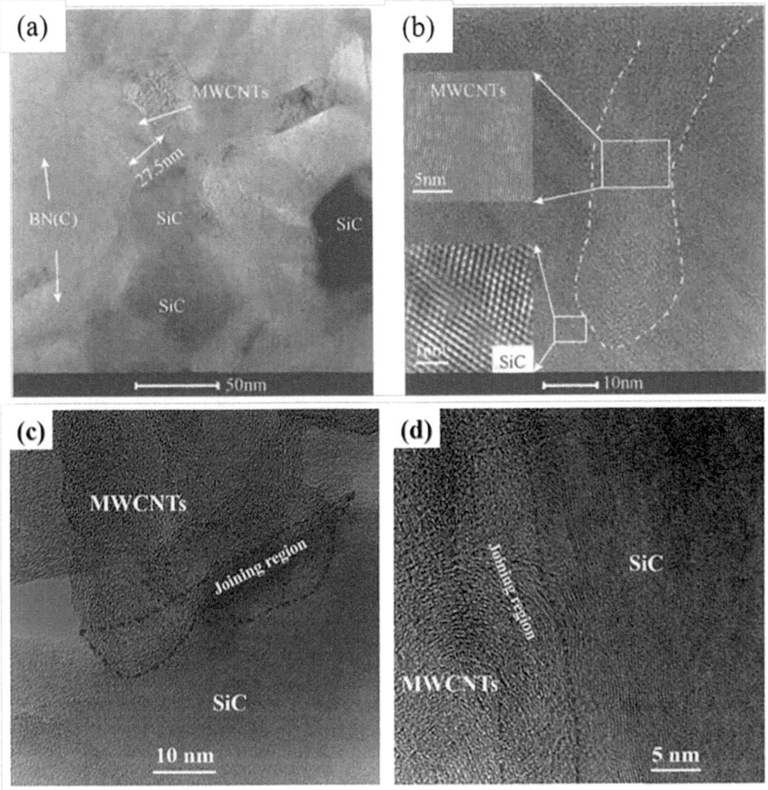

Fig. 4.158 TEM analysis of SPS densified MWCNTs/Si_2BC_3N CMCs: **a** MWCNTs without SiC coating; **b** MWCNTs with SiC coating [61]

Table 4.9 The bulk density and relative density of the MWCNTs toughened MA-Si_2BC_3N CMCs prepared by SPS at 1900 °C/40 MPa/5 min [61, 62]

Composition	Bulk density (g/cm^3)	Relative density (%)
+ 1 vol.% unmodified MWCNTs	2.58	91.2
+ 3 vol.% unmodified MWCNTs	2.41	85.8
+ 1 vol.% modified MWCNTs	2.59	91.7
+ 2 vol.% modified MWCNTs	2.61	93.1

to β-SiC and peaks at 33.71° and 38.41° to α-SiC. Another peak emerging at 35.61 assigns to a mixture of β-SiC and α-SiC (Fig. 4.159).

Nanosized SiC and BN(C) grains without abnormal growth show relatively uniform distribution in ceramic matrix after introducing some GPLs. Similarly, lots of light-and-dark stripes observed in many SiC grains illustrate the existence of substantial stacking faults. The BN(C) with no fixed shape has smaller size scale than SiC,

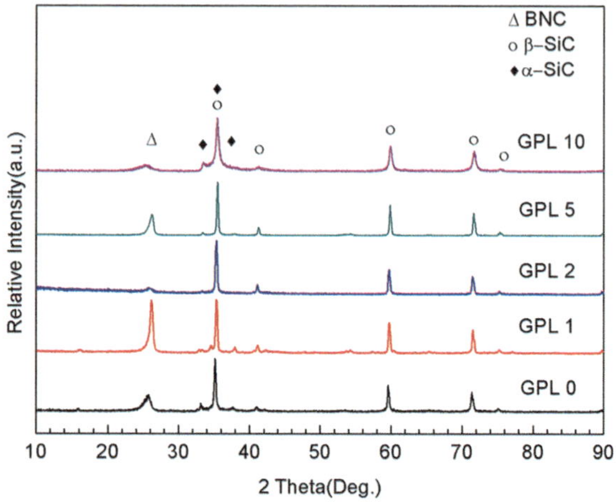

Fig. 4.159 XRD patterns of SPS densified GPLs/Si_2BC_3N CMCs (1800 °C/40 MPa/5 min) [63]

and mainly distributes among SiC grains or their junction regions. There also exist amorphous regions and nanosized BNC aggregation regions. GPLs mainly distribute in amorphous zones or nanosized BN(C) aggregation regions (Fig. 4.160). The introduction of limited GPLs strongly weakens the densification of the MA-Si_2BC_3N ceramic matrix (Fig. 4.161).

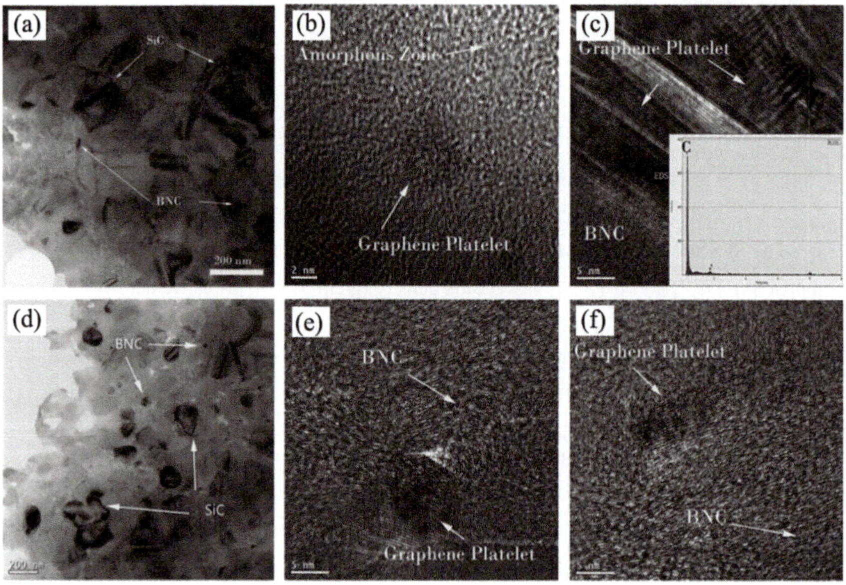

Fig. 4.160 TEM analysis of SPS densified GPLs/Si_2BC_3N CMCs (1800 °C/40 MPa/5 min): **a–c** + 1 vol.% GPLs; **d–f** + 2 vol.% GPLs

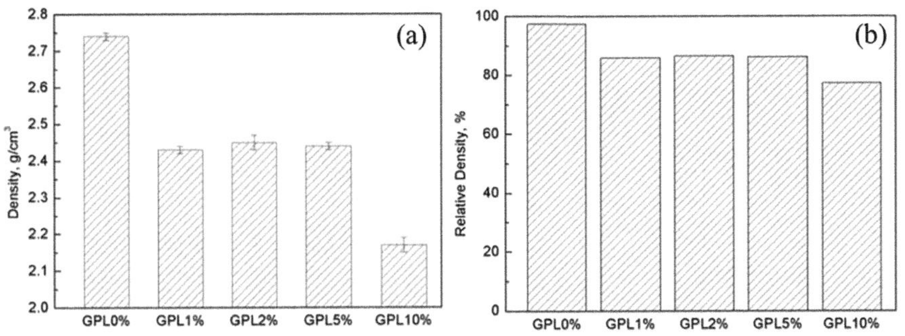

Fig. 4.161 Bulk density and relative density of SPS densified GPLs/Si_2BC_3N CMCs (1800 °C/40 MPa/5 min): **a** Bulk density; **b** relative density

4.4 Comparison of Sintering Methods on Densification of SiBCN Materials

During uniaxial hot pressing (HP) sintering, the as-milled amorphous SiBCN particles show good thermal plasticity and relatively low resistance to deformation. The required molding pressures are therefore much lower than that for cold pressing. The simultaneous use of high pressures and temperatures promotes transfer of mass and energy thus reducing sintering temperatures and dwelling times while retarding grain coarsening compared with pressureless sintering, which provides fully dense products with fine microstructures. The sintering method characteristics applied for the MA-SiBCN based materials are summarized in Table 4.10.

The first is the most commonly used in MA-SiBCN research as high uniaxial HP pressure can provide > 90% or more of theoretical density. The major drawbacks of this method are the limited geometry and the preferred alignment of reinforcements perpendicular to the pressing direction. This imposes limitations on the versatility and applications of materials processed this way. For example, amorphous MA-Si_2BC_3N powder sintered at 1900 °C/80 MPa in N_2 consists of ca. 100 nm-sized β-SiC and t-BN(C) grains sometimes together with residual amorphous phases. They show densities up to 2.52 g/cm^3; however, sintering addition of Al, AlN, Zr, ZrO_2, $ZrSiO_4$, $MgO–Al_2O_3–SiO_2$, $MgO–ZrO_2–SiO_2$, etc. particles further promote the densification of the composite powders, accelerates the crystallization and grain growth, and improves the density of the as-sintered composite ceramics. Significant densification occurs to the MA-Si_2BC_3N powder free from any additives once the temperature is higher than 1830 °C. The addition of AlN or ZrO_2 decreases this critical temperature to ~1760 and ~1730 °C, respectively.

Unfortunately, HP sintering of fiber-composites in an inert or nitrogen atmosphere as well as in gas-pressure sintering (≈0.8 MPa of nitrogen overpressure), densification can be hampered by impingement of fibers to form a skeletal network which can inhibit further shrinkage and thus result in a lower sintered density and higher percentage porosity. Typically for more than 10 vol.% carbon fibers the final density

Table 4.10 Comparison of sintering method on the densification of MA-SiBCN based materials [64]

Parameters	Sintering methods				
	HP*1	HP*2	SPS*3	HIP	HPS*4
Temperature (°C)	≤ 2300	≤ 2000	≤ 2100	≤ 2100	≤ 1800
Pressure	≤ 100 kN	≤ 600 kN	≤ 200 kN	≤ 200 MPa	≤ 8 GPa
Heating rate (°C/min)	≤ 25	≤ 20	≤ 100	≤ 20	≤ 100
Cooling	Furnace-cooling in general				
Atmosphere	Vac./N_2/Ar			Ar	O_2-poor
Sample size (mm)	≤ Φ70 × 45	≤ Hundreds	≤ Φ150	≤ Hundreds	≤ Φ10
Sample shape	Simple	Simple	Simple	Simple/ Irregular	Simple
Applicability	Science and Engineering	Science and Engineering	Science and Engineering	Science and Engineering	Sciences

* Referential furnaces

(1) HIGH MULTI 10,000, Nihon Dempa Kogyo Co., Ltd.

(2) ZKLRY-400, Rui Zhuzhou in High-frequency Equipment Co., Ltd.

(3) Pressure ≤ 200 kN and sample size ≤ Φ30 for 3.20MK-IV, Sumitomo Coal Mining Co., Ltd.; Pressure ≤ 100 kN and sample size ≤ Φ150 for SPS-300-100, shanghai Haoyue Electric Technology Co., Ltd.

(4) Cubic multi-anvil apparatus CS-1B, Guilin Metallurgical Mechanical Plant

is less than 90% of theoretical when processed this way, which leads to a reduction in many mechanical properties such as strength, and changes to some physical properties such as thermal conductivity. Purely from a processing viewpoint, one can envision that HP is very straightforward and produces a good but anisotropic product, though from a manufacturing perspective this process is the least accommodating in terms of shaping.

HP has also been used to densify PDCs-Si(B)CN materials often obtained by the consolidation of already pyrolyzed powders rather than from precursor powders to avoid gas release and pore formation during pyrolysis and microcracks formed therefrom. In one case, the polyborosilazane is chosen as the precursor, pyrolyzed at 1000 °C in inert atmosphere, ground, sieved and eventually mixed with SiC powder to prepare composite ceramics by HP sintering [45]. Results indicate that the amorphous PDCs-SiBCN acts as a sintering aid for SiC ceramics promoting effective densification leading to nearly fully dense samples at 1830 °C and 40 MPa for 90 min. But at temperatures < 1660 °C the density of all the HP densified SiC/SiBCN ceramics shows a minor increase, while above 1660 °C the density increases rapidly due to the grain boundary sliding with a non-Newtonian viscous boundary phase. After grain boundary sliding, the grain-boundary diffusion enhanced by B and C elements is responsible for the further sintering densification.

The MA-Si_2BC_3N monoliths processing using SPS from MA amorphous powder offer comparable density with more retaining amorphous phases than those hot-pressed counterparts. Consolidated at 1800 °C under a pressure of 25 MPa for 3 min, the SiC/BN(C) grain boundaries for the SPS densified MA-Si_2BC_3N ceramics cannot be clearly distinguished. As pressure increased to 40 MPa, the SiC almost completely crystallizes with grain sizes less than 1 μm. The nanosized BN(C) is mainly distributed at boundaries among SiC grains. Relative densities of the resulting MA-Si_2BC_3N bulk ceramics increases from 76.6 to 92.4% with sintering pressure from 25 to 40 MPa. The β-SiC tends to coarsening with increasing Si/C ratio, but no changes are observed for the microstructure characteristics of sub-micrometer sized β-SiC and nano-sized BN(C). The measured bulk densities of the MA-$SiBC_2N$, MA-Si_2BC_3N and MA-Si_3BC_4N, are 2.49, 2.62 and 2.84 g/cm^3 respectively.

The sintering additive-free amorphous PDCs-SiBCN powders are synthesized in nitrogen or ammonia atmosphere to modify their carbon content and subsequently SPS to densify monoliths [65]. The nitrogen-treated PDCs-SiBCN powder sintered at 1500–1900 °C leads to monoliths with a density increasing from 2.4 to 2.6 g/cm^3. The SiC nucleates preferentially around 1700 °C, whereas Si_3N_4 grows at 1800 °C. At this temperature, PDCs-SiBCN monoliths are pure electrical conductor nanocomposites mainly composed of nano Si_3N_4 and nano SiC embedded in a turbostratic BN(C) matrix. Above 1800 °C, Si_3N_4 is not identified. In comparison, the ammonia-treated PDCs-SiBCN powders SPS sintered at 1800 °C represent dense insulator composites mainly composed of Si_3N_4 and BN with an excellent stability in air.

High pressure sintering (HPS) has been used to processing SiC, Si_3N_4, TaC, Al_2O_3, ZrO_2, ZrB_2, BN, etc. binary ceramic systems [64]. In case of quaternary SiBCN, the use of high pressure leads to a lower sintering temperature limiting crystallization and thus ensuring completely amorphous MA-Si_2BC_3N monoliths. Increases in heating temperatures lead to the nucleation and growth of SiC and BN(C) grains, along with partial transformation of [SiC_4] units from amorphous to α/β-SiC and collapse of B–C–N bonds. High pressures also shift the initial crystallization temperatures of amorphous MA-Si_2BC_3N powder from 1400 (1 atm) to 1150 °C (5 GPa). The SiC nucleates prior to BN(C) at 1 atm, while it goes in the opposite order at high pressures ($\geq$1 GPa).

With the increasing temperatures at 5 GPa, MA-Si_2BC_3N undergoes partial phase segregation, followed by initiation of nucleation, and then nucleation and growth. BN(C) nucleates preferentially at bridges between ceramic particles causing SiC to concentrate in particle interiors thus forming "nanocapsule-like" structure. Above 1250 °C, nucleation and growth of turbostratic BN(C) and β-SiC crystallites occur simultaneously, and the sizes of the two crystals are < 10 nm at 1400 °C. Unfortunately, HPS is restricted at present to prepare millimeter-sized samples and thus serves primarily to investigate fundamental scientific issues rather than to prepare components for engineering applications.

The hot-isostatic pressing (HIP) [66–68] is considerably more complex and expensive than hot-pressing but for additive-free monolithic MA-Si_2BC_3N it normally produces a superior product with near-isotropic properties. Expectedly, the HIP sintered MA-Si_2BC_3N monoliths thus offer higher relative densities than those of HP or SPS sintered counterparts. Coincidently, special envelopes are co-requisite [69–72]. After being HIP processed, the MA-Si_2BC_3N monoliths have higher average properties and, significantly, narrower width of their distribution. This is of significant importance for the accurate prediction of fatigue and creep lives and for the design of high-performance SiBCN based components. However, for the fiber-composites, fibers bridging and the formation of fiber "nests" can be more of a problem.

The post-HIP greatly improves the bulk density and mechanical properties of MA-Si_2BC_3N monoliths. Exemplify after post-HIP treatment (1900 °C/190 MPa/1 h), the density of the hot-pressed (1700 °C/50 MPa/30 min) MA-Si_2BC_3N monoliths increases by ca. 35% (from 2.0 to 2.7 g/cm^3). A shrinkage ratio of ca. 20% in the radial direction is measured.

HIP is also applied to the PDCs-SiBCN, PDCs-SiCN, β-SiAlON and SiON multi-component system [73–76]. As-pyrolyzed PDCs-SiBCN ceramics are annealed in nitrogen at 1400 °C for 80 h, or encapsulated in quartz and HIP sintered at 1600 °C and 200 MPa for 4 h leads to increases in density improving the creep resistance at high-temperatures [73]. As-pyrolyzed PDCs-SiCN ceramics have porosities of ~20% and deform up to a strain of ~8% in preliminary uniaxial compression tests. Significantly, the density of the PDCs-SiCN ceramics increases from 1.9 to 2.4 g/cm^3 when HIP sintered at 1600 °C/980 MPa, although a few large pores remained [74]. The SiAlON ceramics encapsulated HIP sintering at 1700 °C and 200 MPa gives essentially fully dense (>99.5% of theoretical density) single-phase β-$Si_{3-x}Al_xO_xN_{4-x}$ with $x \approx 0.25$ to 1.90 [75]. However, there is no need to use HIP at temperatures much above ≈1700 °C for these SiAlON ceramics. For SiON ceramics [76], all ceramics are fully densified to the calculated theoretical density of each final composition, regardless of conversion yield to SiON. A strong dependence for SiON on formation temperature is observed, especially below 1800 °C. For SiON materials with 1 wt.% Y_2O_3 addition, the phase transformation from α-Si_3N_4 to β-Si_3N_4 is facilitated since the temperature needed to obtain a higher residual amount of β-Si_3N_4 than of α-Si_3N_4 is lowered to 1800 °C. The two HIP sinterings at 1800 °C for 2 h and at 1900 °C for 4 h results in higher yield than at 1850 °C for 1 h and 1950 °C for 2 h, respectively.

There is growing need to develop dense and large-scale (dozens or even hundreds of centimeters) SiBCN structural parts to meet performance requirements for multiple applications. However, HP, SPS and HPS technologies are currently ineffective in preparing the sufficiently large products. Thus, HIP densified SiBCN based materials represents a mechanism for achieving such objectives. The basic scientific issues relevant to HIP sintered SiBCN are worthy of in-depth investigation.

References

1. Zhang P F. Hot pressing behavior and high-temperature properties of mechanically alloyed 2Si–B–3C–N ceramic. Harbin: Harbin Institute of Technology, 2013.
2. Zhang PF, Jia DC, Yang ZH, Duan XM, Zhou Y. Crystallization and microstructural evolution process from the mechanically alloyed amorphous Si–B–C–N powder to the hot-pressed nano SiC/BN(C) ceramic. J Mater Sci. 2012;47:7291–304.
3. Zhang PF, Jia DC, Yang ZH, Duan XM, Zhou Y. Microstructural features and properties of the nano-crystalline SiC/BN(C) composite ceramic prepared from the mechanically alloy SiBCN powder. J Alloy Compd. 2012;537:346–56.
4. Tokita M. Trends in advanced SPS spark plasma sintering system and technology. J Soc Powder Technol. 1993;30:790–804.
5. Vanmeensel K, Laptev A, Hennicke J, Vleugels J, Van der Biest O. Modelling of the temperature distribution during field assisted sintering. Acta Mater. 2005;53:4379–88.
6. Yang ZH, Jia DC, Zhou Y, Zhang JX. Processing and characterization of $SiB_{0.5}C_{1.5}N_{0.5}$ produced by mechanical alloying and subsequent spark plasma sintering. Mater Sci Eng A. 2008;488:241–6.
7. Yang ZH, Jia DC, Duan XM, Sun KN, Zhou Y. Effect of Si/C ratio and their content on the microstructure and properties of Si–B–C–N ceramics prepared by spark plasma sintering techniques. Mater Sci Eng A. 2011;528:1944–8.
8. Yang Z H. Microstructure and high-temperature properties of the Si–B–C–N MA-powders and ceramics. Harbin: Harbin Institute of Technology, 2008.
9. Loh NL, Sia KY. An overview of hot isostatic pressing. J Mater Process Technol. 1992;30:45–65.
10. Atkinson HV, Davies S. Fundamental aspects of hot isostatic pressing: an overview. Metall and Mater Trans A. 2000;31:2981–3000.
11. Huang Q, Yu D, Xu B, Hu WT, Ma YM, Wang YB, Zhao ZS, Wen B, He JL, Liu ZY, Tian YJ. Nanotwinned diamond with unprecedented hardness and stability. Nature. 2014;510:250–3.
12. Tian YJ, Xu B, Yu D, Ma YM, Wang YB, Jiang Y, Hu WT, Tang CC, Gao YF, Luo K, Zhao ZS, Wang LM, Wen B, He JL, Liu ZY. Ultrahard nanotwinned cubic boron nitride. Nature. 2013;493:385–8.
13. Liang B, Yang ZH, Chen QQ, Wang SJ, Duan XM, Jia DC, Zhou Y, Luo K, Yu DL, Tian YJ. Crystallization behavior of amorphous Si_2BC_3N ceramic monolith subjected to high-pressure. J Am Ceram Soc. 2015;98:3788–96.
14. Dai DS, Han NQ. Amorphous physics (In Chinese). Beijing: Electronic Industry Press; 1989.
15. Lu K. Phase transformation from an amorphous alloy into nanocrystalline materials. Acta Metall Sin. 1994;30:1–21.
16. Liang B. Crystallization and high temperature oxidation mechanisms of amorphous Si_2BC_3N ceramics by high pressure sintering. Harbin: Harbin Institute of Technology, 2017.
17. Wu W, Sun HY, Li H, Li XH, Liu BT, Zhang XY. Pressure-induced preferential growth of nanocrystals in amorphous $Nd_9Fe_{85}B_6$. Nanotechnology. 2008;19:285603.
18. Faupel F, Frank W, Macht MP, Mehrer H, Naundorf V, Rätzke K, Schober HR, Sharma SK, Teichler H. Diffusion in metallic glasses and supercooled melts. Rev Mod Phys. 2003;75:237–80.
19. Liang B, Jia DC, Liao XQ, Zhu QS, Chen QQ, Yang ZH, Tian F, Li XY, Zhou Y. Microstructural evolution of amorphous Si_2BC_3N nanopowders upon heating at high temperatures: high pressures reverse the nucleation order of SiC and BN(C). J Am Ceram Soc. 2018;101:4321–30.
20. Liang B, Yang ZH, Rao JC, Cai DL, Duan XM, Jia DC, Zhou Y, Yu DL, Tian YJ. Highly dense amorphous Si_2BC_3N monoliths with excellent mechanical properties prepared by high pressure sintering. J Am Ceram Soc. 2015;98:3782–7.
21. Hu C C. Microstructure and Properties of The Si–B–C–N–Zr MA-powders and ceramics. Harbin: Harbin Institute of Technology, 2013.

22. Li DX, Yang ZH, Jia DC, Hu CC, Liang B, Zhou Y. Preparation, microstructures, mechanical properties and oxidation resistance of SiBCN/ZrB_2–ZrN ceramics by reactive hot pressing. J Eur Ceram Soc. 2015;35:4399–410.
23. Ye D. Microstructure and oxidation resistance of the mechanically alloyed Si–B–C–N–Al powders and ceramics. Harbin: Harbin Institute of Technology, 2012.
24. Ye D, Jia DC, Yang ZH, Sun ZL, Zhang PF. Microstructures and mechanical properties of SiBCNAl ceramics produced by mechanical alloying and subsequent hot pressing. J Zhejiang Univ –Sci A. 2010;11:761–5.
25. Liao N, Jia DC, Yang ZH, Zhou Y, Li YW. Enhanced mechanical properties, thermal shock resistance and oxidation resistance of Si_2BC_3N ceramics with Zr–Al addition. Mater Sci Eng A. 2018;725:364–74.
26. Zhang M, Hou JN, Li DX, Liang B, Niu ZB, Zhang Q, Yang ZH, Jia DC, Cai DL, Duan XM, He PG, Zhou Y. Mo-SiBCN metal-ceramic composites with enhanced and tunable thermophysical properties and thermal shock resistance. Ceram Int. 2022;48:5744–51.
27. Hou J N. Microstructural design and thermal shock resistance of Mo-SiBCN gradient composite materials. Harbin: Harbin Institute of Technology, 2016.
28. Liao N, Jia DC, Yang ZH, Zhou Y, Li YW. Mechanical properties and thermal shock resistance of Si_2BC_3N ceramics with ternary Al_4SiC_4 additive. Ceram Int. 2018;44:9009–17.
29. Wang G Y. Preparation and properties of SIBCN(Zr) gradient porous ceramics prepared by ice molding and their bicontinuous composites. Harbin: Harbin Institute of Technology, 2018.
30. Zhang PF, Yang B, Lu Z, Jia DC. Effect of AlN and ZrO_2 on the microstructure and property of the 2Si–B–3C–N ceramic. Ceram Int. 2018;44:3406–11.
31. Li DX, Yang ZH, Mao ZB, Jia DC, Cai DL, Liang B, Duan XM, He PG, Rao JC. Microstructures, mechanical properties and oxidation resistance of SiBCN ceramics with the addition of MgO, ZrO_2 and SiO_2 (MZS) as sintering additives. RSC Adv. 2015;5:52194–205.
32. Zhoutang J E. Microstructure and high temperature thermal shock, ablation properties of the C_f/SiBCN-MAS composites. Harbin: Harbin Institute of Technology, 2023.
33. Yang ZH, Li DX, Zhao Y, Miao Y, Rao JC, Cai DL, Duan XM, He PG, Jia DC, Zhou Y. *In Situ* ZrC/Si–B–C–N monoliths prepared by sol-gel and reactive hot-pressing: processing, microstructure, mechanical properties and oxidation behavior. J Alloy Compd. 2019;811:151687.
34. Wang BZ, Li DX, Yang ZH, Jia DC, Guan JY, Peng H, Cai DL, He PG, Duan XM, Zhou Y, Zhang T, Gao CG. Microstructural evolution and mechanical properties of *in situ* nano Ta_4HfC_5 reinforced SiBCN composite ceramics. J Adv Ceram. 2020;9:739–48.
35. Wang BZ, Yang ZH, Li DX, Cai DL, Niu B, Guan JY, Jia DC, Zhou Y. Effects of TaC addition on microstructure and mechanical properties of SiBCN composite ceramics. Ceram Int. 2019;45:22138–47.
36. Peng H, Li DX, Yang ZH, Duan WJ, Jia DC, Zhou Y. Effects of Ti on the microstructural evolution and mechanical property of the SiBCN-Ti composite ceramics. Materials. 2023;16:3560.
37. Liao X Q. Microstructure and properties of the (TiB_2+TiC)/SiBCN composites. Harbin: Harbin Institute of Technology, 2014.
38. Miao Y. Preparation, oxidation resistance and ablation mechanisms of ZrB_2/SiBCN composites. Harbin: Harbin Institute of Technology, 2017.
39. Miao Y, Yang ZH, Liang B, Li Q, Chen QQ, Jia DC, Cheng YB, Zhou Y. A novel in situ synthesis of SiBCN-Zr composites prepared by a sol-gel process and spark plasma sintering. Dalton Trans. 2016;45:12739–44.
40. Miao Y, Yang ZH, Rao JC, Duan XM, He PG, Jia DC, Cheng YB, Zhou Y. Influence of sol-gel derived ZrB_2 additions on microstructure and mechanical properties of SiBCN composites. Ceram Int. 2017;43:4372–8.
41. Liao N, Jia DC, Yang ZH, Niu B, Zhou Y, Li YW. Enhanced mechanical properties, thermal shock resistance and ablation resistance of Si_2BC_3N ceramics with nano ZrB_2 addition. J Eur Ceram Soc. 2019;39:846–59.

42. Ao D F. The thermal shock and ablative properties of HfB_2/SiBCN multiphase ceramic. Harbin: Harbin Institute of Technology, 2018.
43. Zhu QS, Liao XQ, Ao DF, Liang B, Li DX, Yang ZH, Jia DC, Zhou Y, Zhang T, Gao CG, Huang YJ. Enhanced ablation resistance of HfB_2-HfC/SiBCN ceramics under an oxyacetylene torch environment. Corros Sci. 2021;187:109509.
44. Liang B, Yang ZH, Miao Y, Li Q, Zhu QS, Liao XQ, Jia DC, Zhou Y. Microstructural evolution, mechanical and thermal properties of LaB_6 embedded in Si–B–C–N prepared by spark plasma sintering. Ceram Int. 2017;43:4814–20.
45. Lu B, Zhang Y. Densification behavior and microstructure evolution of hot-pressed SiC–SiBCN ceramics. Ceram Int. 2015;41:8541–51.
46. Kim YW, Mitomo M, Hirotsuru H. Microstructural development of silicon carbide containing large seed grains. J Am Ceram Soc. 1997;80:99–105.
47. Liu Q, Ye F, Gao Y, Liu SC, Yang HX, Zhou ZQ. Development of elongated 6H–SiC grains in reaction-bonded porous SiC ceramics. Scripta Mater. 2014;71:13–6.
48. Becher PF, Painter GS, Shibata N, Waters SB, Lin HT. Effects of rare-earth (RE) intergranular adsorption on the phase transformation, microstructure evolution, and mechanical properties in silicon nitride with RE_2O_3+MgO additives: RE= La, Gd, and Lu. J Am Ceram Soc. 2008;91:2328–36.
49. Liang B, Yang ZH, Miao Y, Zhu QS, Liao XQ, Li Q, Tian Z, Jia DC, Zhou Y. Si–B–C–N monoliths with LaB_6-induced well-developed BN(C) flakes. Mater Lett. 2017;187:36–9.
50. Zhang PF, Jia DC, Yang ZH, Yang B, Wang GX. Research on the interfacial structure and property improvement of the C_{fiber}/2Si–B–3C–N ceramic matrix composite. Mater Charact. 2018;142:59–67.
51. Zhu QS, Wang GY, Liao XQ, Li DX, Yang ZH, Jia DC, Duan XM, He PG, Zhou Y. Effects of Zr and chopped c fiber on microstructure and mechanical properties of SiBCN ceramics. Sci China Technol Sci. 2020;68:1520–30.
52. Miao Y, Zhang FN, Yang ZH, Jia DC, Zhou Y. Incorporation of BN-coated carbon fibers into ZrB_2/SiBCN ceramic composites and their ablation behavior. J Eur Ceram Soc. 2020;40:1078–85.
53. Niu ZB, Wang BZ, Pan LJ, Li DX, Jia DC, Yang ZH, Peng H, Ye SQ, Zhou Y. Mechanical and thermal shock properties of C_f/SiBCN composite: effect of sintering densification and fiber coating. J Am Ceram Soc. 2022;105:4321–35.
54. Wang JY, Yang ZH, Duan XM, Jia DC, Zhou Y. Microstructure and mechanical properties of SiC_f/SiBCN ceramic matrix composites. J Adv Ceram. 2015;4:31–8.
55. Li DX, Wu DX, Yang ZH, Zhou YY, Wang SJ, Duan XM, Jia DC, Zhu QS, Zhou Y. Effects of in situ amorphous graphite coating on ablation resistance of SiC fiber reinforced SiBCN ceramics in an oxyacetylene flame. Corros Sci. 2016;113:31–45.
56. Wu D X. Effect of SiC_f surface amorphous carbon layer on the properties of SiC_f/SiBCN composite materials. Harbin: Harbin Institute of Technology, 2016.
57. Zhou Y Y. Preparation and characterization of composite coating on SiC fiber surface. Harbin: Harbin Institute of Technology, 2016.
58. Zhang M, Li DX, Zhou YY, Yang ZH, Jia DC, Cai DL, Niu ZB. Synthesis of coatings on SiC fibers and their effects on microstructure and mechanical properties of SiC_f/SiBCN COMPOSITES. J Am Ceram Soc. 2021;104:6589–600.
59. Xiao J Z. Microstructure and thermal shock ablative behavior of the SiC_f/SiBCN composites. Harbin: Harbin Institute of Technology, 2023.
60. Li Y T. Mechanical properties, thermal shock and ablation resistance of (C_f–SiC_f)/SiBCN composites. Harbin: Harbin Institute of Technology, 2014.
61. Liao N, Jia DC, Yang ZH, Zhou Y, Li YW. Strengthening and toughening effects of MWCNTs on Si_2BC_3N ceramics sintered by SPS technique. Mater Sci Eng A. 2018;725:364–74.
62. Liao N, Jia DC, Yang ZH, Zhou Y. Enhanced mechanical properties and thermal shock resistance of Si_2BC_3N ceramics with SiC coated MWCNTs. J Adv Ceram. 2019;8:121–32.
63. Li DX, Yang ZH, Jia DC, Wu DX, Zhu QS, Liang B, Wang SJ, Zhou Y. Microstructure, oxidation and thermal shock resistance of graphene reinforced SiBCN ceramics. Ceram Int. 2016;42:4429–44.

64. Jia DC, Liang B, Yang ZH, Zhou Y. Metastable Si–B–C–N ceramics and their matrix composites developed by inorganic route based on mechanical alloying: fabrication, microstructures, properties and their relevant basic scientific issues. Prog Mater Sci. 2018;98:1–67.
65. Bechelany MC, Salameh C, Viard A, Guichaoua L, Rossignol F, Chartier T, Bernard S, Miele P. Preparation of polymer-derived Si–B–C–N monoliths by spark plasma sintering technique. J Eur Ceram Soc. 2015;35:1361–74.
66. Hoffmann MJ, Geyer A, Oberacker R. Potential of the sinter-HIP-technique for the development of high-temperature resistant Si_3N_4-ceramics. J Eur Ceram Soc. 1999;19:2359–66.
67. Shinoda Y, Nagano T, Wakai F. Fabrication of nanograined silicon carbide by ultrahigh-pressure hot isostatic pressing. J Am Ceram Soc. 1999;82:771–3.
68. Dong SM, Jiang DL, Tan SH, Guo JK. Preparation and characterization of nano-structured monolithic SiC and Si_3N_4/SiC composite by hot isostatic pressing. J Mater Sci Lett. 1997;16:1080–3.
69. Dong SM, Jiang DL, Tan SH, Guo JK. Hot isostatic pressing and post-hot isostatic pressing of SiC-β-sialon composites. Mater Lett. 1996;29:259–63.
70. Bhandhubanyong P, Akhadejdamrong T. Forming of silicon nitride by the HIP process. J Mater Process Technol. 1997;63:277–80.
71. Babcsan JK, Maros MB, Arato P. Microstructural features of the post heat-treated Si_3N_4 based ceramics. Silic Indus. 2004;69:199–203.
72. Vassen R, Stover D. Processing and properties of nanophase non-oxide ceramics. Mater Sci Eng A. 2001;301:59–68.
73. Christ M, Thurn G, Weinmann M, Bill J, Aldinger F. High-temperature mechanical properties of Si–B–C–N-precursor-derived amorphous ceramics and the applicability of deformation models developed for metallic glasses. J Am Ceram Soc. 2000;83:3025–32.
74. Ishihara S, Gu H, Bill J, Aldinger F, Wakai F. Densification of precursor-derived Si–C–N ceramics by high-pressure hot isostatic pressing. J Am Ceram Soc. 2002;85:1706–12.
75. Ekström T, Olsson PO. β-sialon ceramics prepared at 1700°C by hot isostatic pressing. J Am Ceram Soc. 1989;72:1722–4.
76. Larker R. Reaction sintering and properties of silicon oxynitride densified by hot isostatic pressing. J Am Ceram Soc. 1992;75:62–6.

Chapter 5
Mechanical Behavior and Strengthening/Toughening Mechanisms of MA-SiBCN Materials

Interest in MA-SiBCN ceramics is driven by their applicability at high temperatures. Many MA-SiBCN materials now can offer a tantalizing combination of properties, including low density, high stiffness, good environmental resistance, and desirable strength at ambient and elevated temperatures. Unfortunately, monolithic versions of MA-SiBCN materials suffer from one serious and obvious shortcoming, namely low toughness. An effective way to overcome the low fracture toughness is to fabricate them as composites. Another route for improving the fracture toughness of MA-SiBCN ceramics is to prepare fiber reinforced ceramic matrix composites (FRCMCs). Coincidentally, ceramic/metal particles, whiskers, and short fibers reinforced MA-SiBCN ceramic composites have apparent toughness and reliability comparable to that of monoliths, though toughness is gained at the expense of some loss in other characteristics.

A further issue in the design and processing of optimal FRCMCs, is the interface which often controls the failure process. The introduction of a weak but not too weak interface is key to superior mechanical and thermophysical performance. Significantly, a homogeneous and uniform distribution of the second phases in MA-SiBCN matrix is essential required to achieve effective load-bearing capacity of reinforcements. Enhanced mechanical and thermophysical properties can be achieved once the homogeneous distribution of reinforcements is in a nano or submicron scale. Whilst there still remain many applications for which MA-SiBCN monoliths are the better choice.

D. Li et al., *SiBCN Ceramics and Composites Prepared by Mechanical Alloying*,
https://doi.org/10.1007/978-981-97-3029-2_5

5.1 Mechanical Behavior and Thermophysical Properties of Nanocrystalline MA-SiBCN Monoliths

5.1.1 Hot-Pressed Monoliths

(1) *Room temperature mechanical properties*

The room temperature (RT) mechanical properties of the hot-pressing (HP) sintered MA-Si_2BC_3N monoliths are closely related to the sintering temperatures [1]. Significantly, the noticeable improvement in density and mechanical property occurs only when the ceramics are sintered at 1900 °C. At temperatures lower than or equal to 1800 °C, these properties are relatively small and have insignificant enhancement with the increase of the sintering temperature. The small relative densities and the slightly varied properties in 1500–1900 °C range suggest sintering is most likely in its initial stage or earlier intermediate stage (Table 5.1).

Sintering pressure strongly influences the density and mechanical property, while the effects of dwelling time and protective atmosphere are finite (Table 5.2). When heated at 1900 °C/80 MPa under 8 bar N_2, the mechanical properties of the as-prepared MA-Si_2BC_3N ceramics are significantly enhanced and their bulk density, flexural strength, Young's modulus, fracture toughness and Vickers hardness obtained at 2.64 g/cm^3, 360.7 MPa, 155.9 GPa, 2.94 MPa m$^{1/2}$ and 6.3 GPa, respectively [2]. When the pressure is decreased to 50 MPa (under the same temperature of 1900 °C in 8 bar N_2), the density, flexural strength, Young's modulus and fracture toughness are rapidly reduced to 2.2 g/cm^3, 89.4 MPa, 33.6 GPa and 1.25 MPa m$^{1/2}$, respectively. In case the temperature is lowered to 1800 °C (sintering pressure is maintained at 80 MPa under 8 bar N_2), these properties are decreased to 2.12 g/cm^3, 66.3 MPa, 30.6 GPa and 1.71 MPa m$^{1/2}$, respectively.

At 1900 °C/80 MPa, as the dwelling time increases from 10 to 30 min, the bulk density and mechanical properties of the MA-Si_2BC_3N ceramics are slightly improved. At this sintering stage, almost all sintering mechanisms has clearly done its stuff, therefore the mechanical properties improvement by extending dwelling time is limited. Hot-pressing at 1900 °C/80 MPa/30 min under 1 bar or 8 bar N_2,

Table 5.1 Bulk density and mechanical properties of the MA-Si_2BC_3N ceramics prepared by hot-pressing at 1500–2000 °C (80 MPa/30 min/1 bar N_2, hearting rate 20 K/min) [1]

Sintering temperature (°C)	Bulk density (g/cm^3)	Flexural strength (MPa)	Young's modulus (GPa)	Facture toughness (MPa m$^{1/2}$)	Vickers' hardness (GPa)
1500	1.80	89.8 ± 8.6	36.5 ± 1.9	1.23 ± 0.17	2.2 ± 0.1
1600	1.96	94.3 ± 6.7	44.6 ± 1.5	1.32 ± 0.14	3.0 ± 0.1
1700	2.17	109.1 ± 7.2	57.3 ± 0.6	1.47 ± 0.06	3.4 ± 0.1
1800	2.25	121.1 ± 6.7	62.7 ± 0.7	1.71 ± 0.42	4.5 ± 0.3
1900	2.52	331.1 ± 40.5	139.0 ± 16.0	2.81 ± 0.89	5.7 ± 0.4

Table 5.2 Bulk density and mechanical properties of the MA-Si_2BC_3N ceramics prepared by different hot-pressing condition [2]

Hot-pressing conditions	Bulk density (g/cm^3)	Flexural strength (MPa)	Young's modulus (GPa)	Facture toughness (MPa $m^{1/2}$)	Vickers' hardness (GPa)
1800 °C/80 MPa/30 min/8 bar N_2	2.12	66.3 ± 7.2	30.6 ± 2.4	1.71 ± 0.42	3.4 ± 0.1
1900 °C/50 MPa/30 min/8 bar N_2	2.20	89.4 ± 9.3	33.6 ± 3.6	1.25 ± 0.31	2.4 ± 0.1
1900 °C/80 MPa/10 min/8 bar N_2	2.53	305.8 ± 31.6	106.8 ± 3.0	2.96 ± 0.14	6.2 ± 0.2
1900 °C/80 MPa/30 min/8 bar N_2	2.64	360.7 ± 19.9	155.9 ± 17.2	2.94 ± 0.59	6.3 ± 0.2
1900 °C/80 MPa/30 min/1 bar N_2	2.52	331.1 ± 40.5	139.0 ± 16.0	2.81 ± 0.89	5.7 ± 0.4

sintering of the MA-Si_2BC_3N ceramics is basically completed with low-porosity, thus their mechanical properties remain almost unchanged.

The mechanical performance of the hot-pressed MA-Si_2BC_3N bulk ceramics is also associated with the powder milling condition and raw material composition [3, 4]. For one-step milling MA-Si_2BC_3N powder using c-Si, h-BN, and graphite as raw materials, the flexural strength of resulting bulk ceramics (CS2H1850) hot-pressed at 1850 °C is obtained at 191.7 MPa, much lower than that ~ 312.9 MPa of ceramics (CS2H1900) sintered at 1900 °C (Fig. 5.1 and Table 5.3).

Using Si_3N_4, B_4C, c-Si, h-BN, amorphous carbon, and graphite as raw materials, the bulk densities of MA-Si_2BC_3N bulk ceramics (CD2H1900 and ID2H1900) prepared by two-step high-energy ball milling combined with hot-pressing are not as high as the former (CS2H1900), but their bend strengths are up to 446.6 MPa and 423.4 MPa, respectively. Besides, the ID2H1900 ceramics exhibit the highest fracture toughness and Vickers hardness, reaching 5.10 MPa $m^{1/2}$ and 5.5 GPa, respectively, which is relevant to the crystal sizes of SiC precipitates. The CS2H1850 ceramics display the lowest Young's modulus, obtained at only 107.9 GPa. Other ceramics of CS2H1900, CD2H1900 and ID2H1900, all show similar Young's modulus of about 135–145 GPa.

Nevertheless, to make clear "how high" or "how low" are the properties obtained for the hot-pressed MA-Si_2BC_3N bulk ceramics, the data reported in the literature of the magnetron sputtered SiBCN films, the PDCs-Si(B)CN bulk ceramics and the hot-pressed SiC bulk ceramics are compared [5–7]. The bending strength and the Young's modulus of the as-prepared MA-Si_2BC_3N ceramics hot-pressed at 1500–1800 °C are much smaller than those of the PDCs-SiCN or the hot-pressed SiC

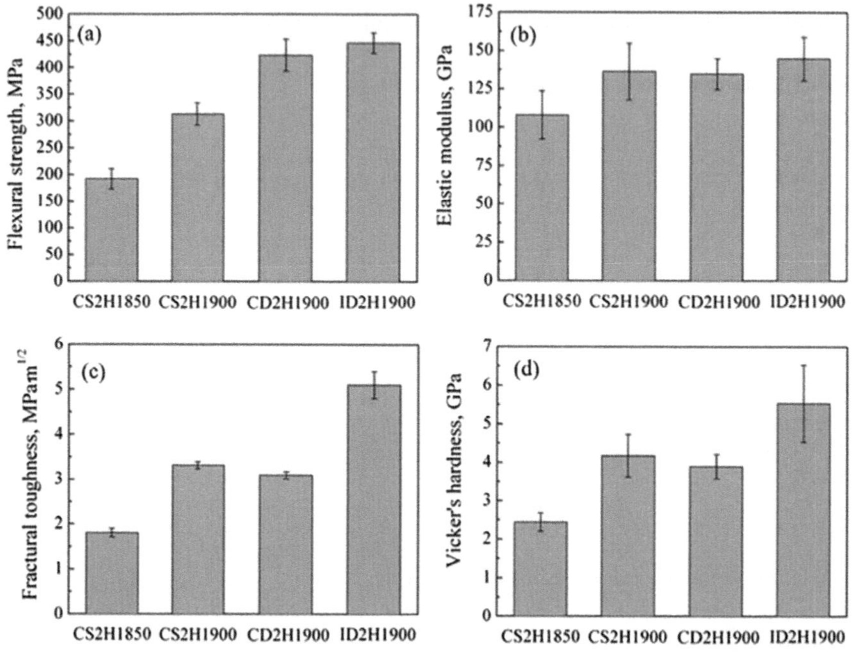

Fig. 5.1 Room temperature mechanical properties of various MA-Si_2BC_3N ceramics hot-pressed at different condition: **a** Flexural strength; **b** Young's modulus; **c** Fracture toughness; **d** Vickers' hardness [4]

Table 5.3 Bulk density and mechanical properties of the MA-Si_2BC_3N bulk ceramics prepared by at 1850–1900 °C under a pressure of 40 MPa for 30 min in 1 bar N_2 [3, 4]

Samples	Bulk density (g/cm^3)	Flexural strength (MPa)	Young's modulus (GPa)	Facture toughness (MPa $m^{1/2}$)	Vickers' hardness (GPa)
[a]CS2H1850	2.59	191.7 ± 20.7	107.9 ± 9.5	1.80 ± 0.07	2.4 ± 0.3
[a]CS2H1900	2.82	312.8 ± 5.2	136.3 ± 17.8	3.31 ± 0.02	4.2 ± 0.5
[b]CD2H1900	2.67	423.4 ± 43.0	134.6 ± 13.5	3.09 ± 0.05	3.9 ± 0.4
[c]ID2H1900	2.76	446.6	145.0	5.10	5.5

[a]One-step milling: The c-Si, h-BN and graphite powders are simultaneously milled for 20 h
[b]Two-step milling: Both Si and partial graphite powders in mole ratio of 1:1 are milled for 15 h, and then the residual graphite and h-BN powders are added for another 5 h of ball-milling
[c]Two-step milling: initial ball milling of Si_3N_4, B_4C and amorphous C for 8 h, followed by adding c-Si and graphite (molar ratio 1:1) further milling for 8 h, and finally adding h-BN, and graphite milled for 4 h

ceramics [6, 7]. In addition, the Young's modulus is also much smaller than those of the magnetron sputtered SiBCN films [8, 9].

For MA-Si_2BC_3N ceramics hot-pressed at 1900 °C/80 MPa/30 min, their bulk densities approach to that of the PDCS-SiBCN bulk ceramics or the magnetron sputtered SiBCN films [6]. The bending strength and the Young's modulus are approximately one-third of those of the hot-pressed SiC ceramics [7]. Although the Young's modulus comes closer to that of the magnetron sputtered SiBCN films, it is still only half of some values reported for the latter [8, 9]. The fracture toughness is slightly lower than that of the hot-pressed or the pressureless-sintered SiC ceramics [7].

Using a combination of MA and HP method, the bulk ceramics with large dimensions can be easily prepared, and various mechanical properties can be conveniently evaluated. Hence, the results obtained here can be used for reference to some degree. The rapid improvement in density and mechanical property of the MA-Si_2BC_3N indicates sintering is most likely in its later intermediate stage. As indicated in Fig. 5.2, the pores in the as-prepared ceramics are significantly reduced, and the residual pores become small and isolated. The ceramic particles appear to be flattened, implying the elimination of the pores and the densification of the ceramic. Such high temperature and pressure may accelerate plastic deformation, viscous flow, and atomic diffusion in the ceramic matrix, and hence facilitate ceramic densification. The low Young's modulus of the hot-pressed MA-Si_2BC_3N ceramics generally means desirable shock resistance, thus facilitating the wide applications of MA-Si_2BC_3N as high-temperature structural components.

(2) *High-temperature mechanical properties*

Compared to the RT strength, the flexural strength of MA-Si_2BC_3N ceramics annealed at 1000 °C in air is improved by about 13%, while that at 1200 °C is close to room temperature strength [10]. This thus suggests the resulting Si_2BC_3N monoliths don't exhibit strength degradation at least at 1200 °C in air (Table 5.4).

Though the flexural strength of the MA-Si_2BC_3N ceramics at ambient and elevated temperatures (<1200 °C in air) has almost no change, but a little differ in their fracture morphologies (Fig. 5.3). After test in air at 1000–1200 °C, the ceramic particles have a near-spherical outline with the sizes similar to that of the primary particles in the as-sintered ceramics, but the particles seem more denser with rougher fracture surfaces. This may imply the further sintering of the interparticle neck even grain growth during high-temperature tests. Besides, surficial micropores and microcracks healed by oxides can also do a favor to the improving flexural strength at high-temperatures, which usually occurs in nitride ceramics of Si_3N_4 and h-BN.

Further annealing at 1800 °C in 1 bar N_2 for 3 h, the mass loss of the MA-Si_2BC_3N ceramics is ~ 3.55%, and coincidentally their bulk density, flexural strength and Young's modulus are reduced by about 3%, 13% and 30%, respectively, while the fracture toughness and Vickers' hardness are basically unchanged (Table 5.5). Besides, the as-annealed MA-Si_2BC_3N ceramics exhibit varying degrees of shrinkage, with a larger shrinkage rate parallel to the pressure direction of ~ 0.535%, and a relatively smaller shrinkage rate ~ 0.203% perpendicular to the pressure direction.

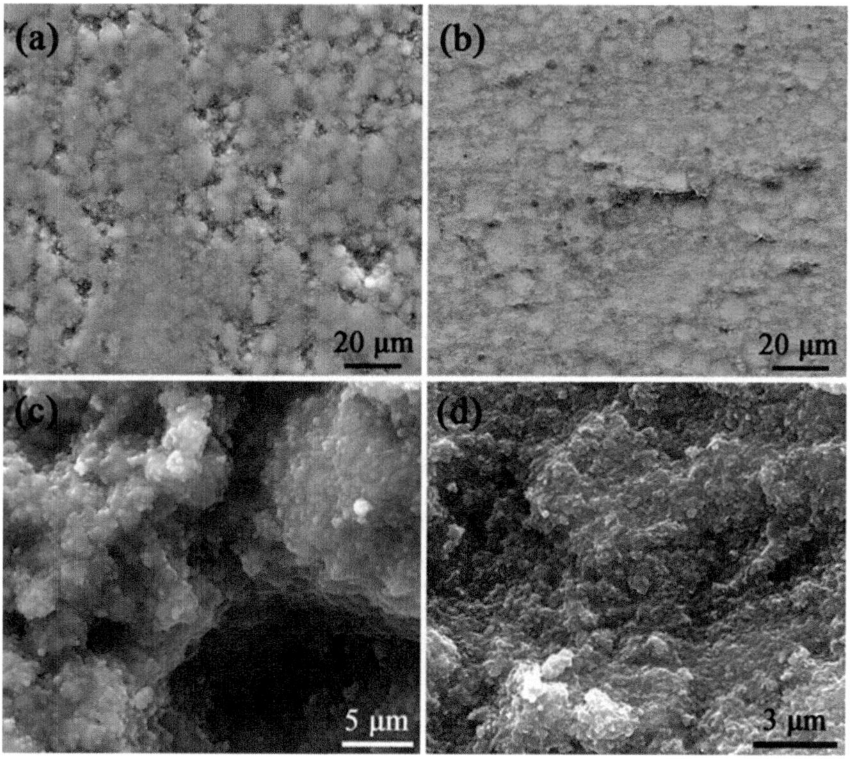

Fig. 5.2 The surface and fracture morphologies of the MA-Si_2BC_3N monoliths hot-pressed at different temperature: **a**, **c** 1800 °C; **b**, **d** 1900 °C [4]

Table 5.4 Three-pointed flexural strength of the hot-pressed MA-Si_2BC_3N monoliths after testing at 1000 or 1200 °C dwelling for 5 min in air atmosphere [10]

Test temperature	RT	1000 °C	1200 °C
Flexural strength σ (MPa)	241.5 ± 41.9	272.8 ± 63.5	237.2 ± 14.5

There is no significant change in the phase composition and relative content of each phase after annealing at 1800 °C in 1 bar N_2 (Fig. 5.4). The as-annealed MA-Si_2BC_3N ceramics are still composed of α/β-SiC and BN(C) though reduced content of α-SiC is observed. The average grain sizes of SiC are approximately 87.0 μm with equiaxed shape, while BN(C) are in tens of nanometers. The nanocapsule-like microstructures are still retained with unform distribution of each phase. Similar surface and fracture morphologies of the as-annealed ceramics are found comparable to the as-sintered ones (Fig. 5.5).

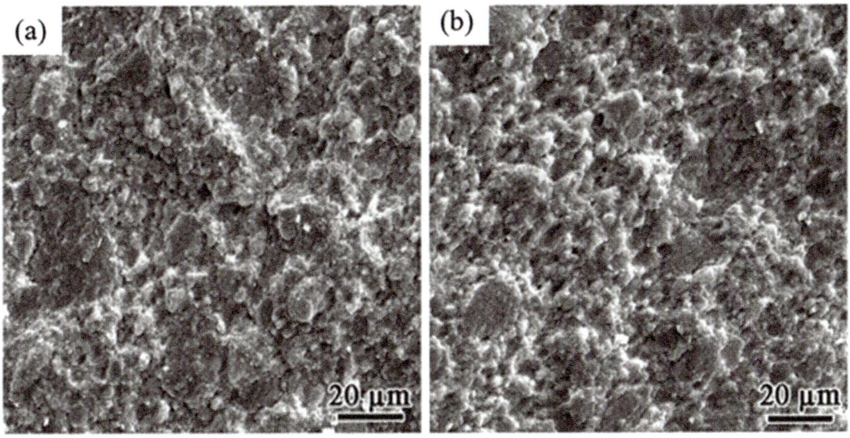

Fig. 5.3 Fracture morphologies of the hot-pressed MA-Si_2BC_3N ceramics after bending strength testing in air: **a** At 1000 °C; **b** At 1200 °C [10]

Table 5.5 The mass, bulk density and mechanical properties changes of the MA-Si_2BC_3N bulk ceramics after annealing at 1800 °C in 1 bar N_2 for 3 h [10]

Mass loss (%)	Bulk density (g/cm^3)	Relative density (%)	Flexural strength (MPa)	Young's modulus (GPa)	Fracture toughness (MPa $m^{1/2}$)	Vickers' hardness (GPa)
−3.55 ± 0.80	2.47	87.0	286.9	98.2 ± 5.1	2.74 ± 0.26	5.4 ± 0.3

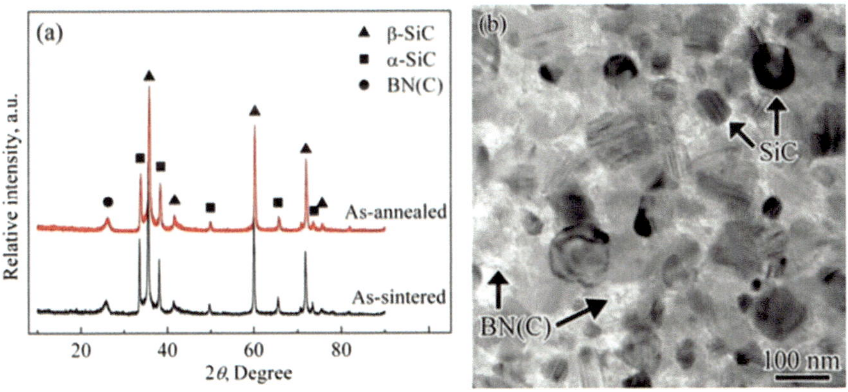

Fig. 5.4 Phase composition and microstructure of the hot-pressed MA-Si_2BC_3N ceramics after annealed at 1800 °C in 1 bar N_2: **a** XRD pattern; **b** Bright-field TEM image [10]

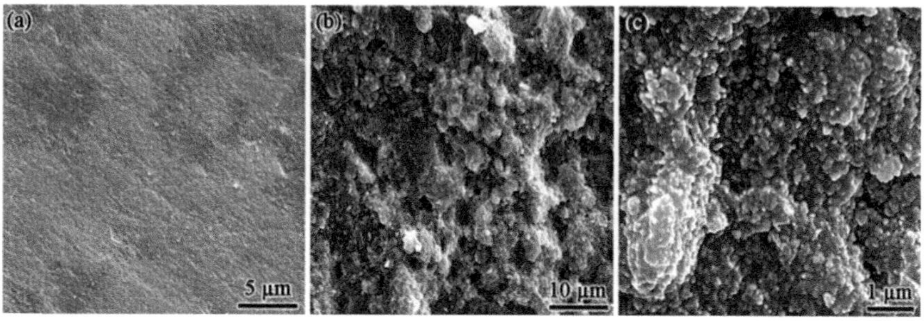

Fig. 5.5 Surface and fracture morphologies of the hot-pressed MA-Si_2BC_3N ceramics after annealing at 1800 °C in 1 bar N_2 for 3 h: **a** Surface morphology; **b** and **c** Fracture morphology [10]

Considering the above microstructural evolution, the mass loss is most likely caused by the decomposition of the materials since some amorphous structure or low crystallinity of BN(C) phases are retained in as-sintered bulk ceramics. During annealing at 1800 °C, further crystallization and/or resintering may happen accompanied with the change or breaking of the unstable bonds. For instance, carbothermal reduction reactions of [$Si_3N_4 + 3C \rightarrow 3SiC + 2N_2$], and/or self-decomposition of [$Si_3N_4 \rightarrow Si + 2N_2$] can occur when thermodynamic and kinetic conditions are both feasible, though Si_3N_4 are not found in hot-pressed MA-Si_2BC_3N ceramics.

The ball-milling method and sintering temperature both impact the high-temperature strength of the hot-pressed MA-Si_2BC_3N bulk ceramics [7]. After annealing at 1000 or 1400 °C in air, the CD2H1900 ceramics exhibit higher strength than those of CS2H1900 (Fig. 5.6 and Table 5.6). The high-temperature strength of the CD2H1900 ceramics is reduced ~ 11.3% at 1000 °C and ~ 38.5% at 1400 °C, respectively, but these values are obtained only ~ 8.2 and ~ 28.0% for CS2H1900 ceramics (Fig. 5.7).

Fig. 5.6 High-temperature strength for the hot-pressed MA-Si_2BC_3N annealed at 1000 or 1400 °C in air [4]

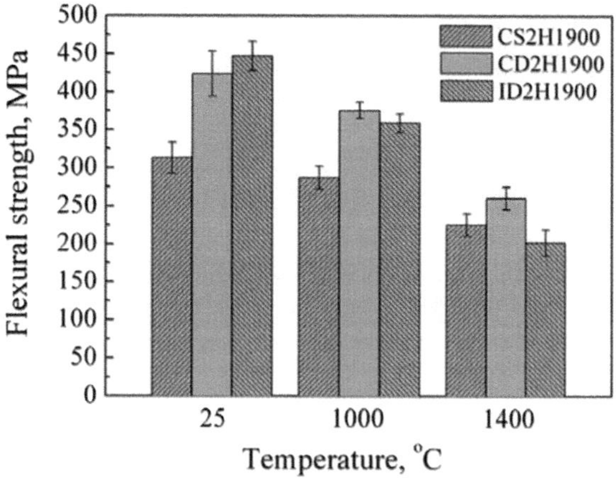

Table 5.6 High-temperature strength for the hot-pressed MA-Si_2BC_3N annealed at 1000 or 1400 °C in air [4]

Samples	High-temperature strength (MPa)/ strength retention ratio (%)	
	At 1000 °C	At 1400 °C
CS2H1900	287.2 (91.8%)	225.3 (72.0%)
CD2H1900	375.4 (88.7%)	358.7 (61.5%)
ID2H1900	260.4 (80.3%)	202 (45.2%)

Fig. 5.7 Surface and fracture morphologies of hot-pressed MA-Si_2BC_3N ceramics after annealing at 1000 °C (**a–c**) and 1400 °C (**d–f**) in air: **a**, **d** CS2H1900; **b**, **e** CD2H1900; **c**, **f** ID2H1900 [4]

(3) *High-temperature creep behavior and mechanism*

Hot-pressed nanocrystalline MA-Si_2BC_3N ceramics show excellent creep resistance in vacuum attributed to low self-diffusion coefficients [10]. At 1400–1600 °C and under a stress of 100–150 MPa, the nanocrystalline MA-Si_2BC_3N bulk ceramics exhibit primary creep of about 3 h followed by the steady-state creep. Temperature has a significant influence on steady-state creep (Fig. 5.8). At 1600 °C and 125 MPa in vacuum the steady-state creep rate is 8.5×10^{-8}/s, close to that of SiC, at the same order of magnitude as the PDCs-SiBCN. The measured activation energy for the creep at 1400–1600 °C in vacuum is 664.26 kJ/mol (Fig. 5.9), equal to SiC and at least 5 times higher than PDCs-SiBCN. Lattice change dominates creep in SiC. However, the stress exponent of the nanocrystalline MA-Si_2BC_3N materials is about 2 indicating grain boundary mechanism dominates creep, which is the same as that of PDCs-SiBCN. Consequently, diffusion and grain boundary sliding occur primarily during the creep of nanocrystalline MA-Si_2BC_3N coupled with plastic flow and viscous flow of grain boundary, dislocation slip and climb in the grain, and structural ordering and decomposition.

The amorphous PDCs-SiBCN and nanocrystalline PDCs-SiBCN show different creep behavior, though the deformation mechanisms are not yet well understood. Amorphous PDCs-SiBCN shrinks with a deformation of about 1.4% when held

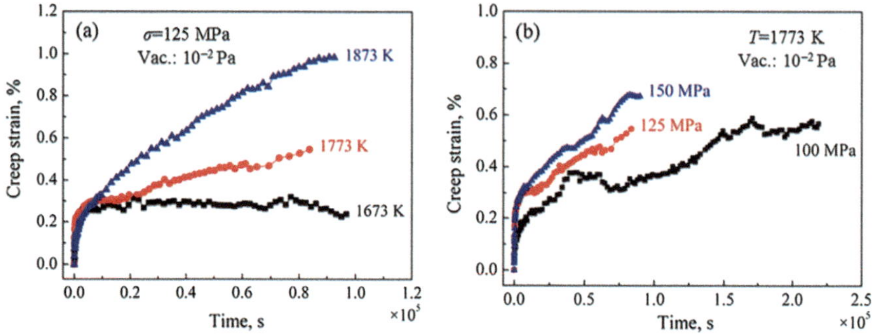

Fig. 5.8 Creep strain–time curves of hot-pressed (1900 °C/80 MPa) nanocrystalline MA-Si_2BC_3N bulk ceramics: **a** Under different temperature; **b** Under different load [10]

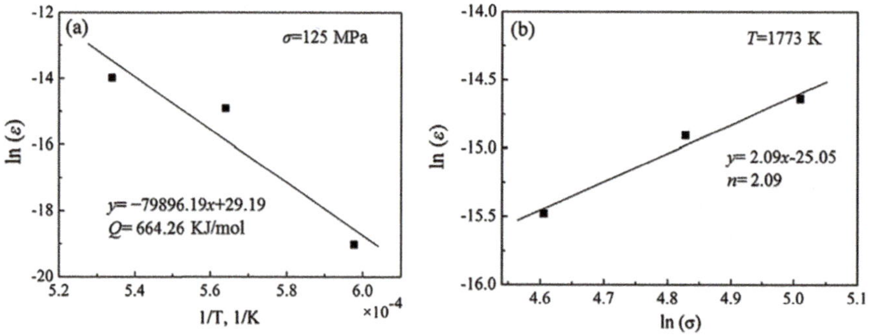

Fig. 5.9 The creep activation energy and stress exponent in the high-temperature steady-state creep stage of hot-pressed (1900 °C/80 MPa) MA-Si_2BC_3N bulk ceramics: **a** Creep activation energy; **b** Stress exponent [10]

at 1400 °C and 5 MPa for 300 h [11]. The deformation rate drops linearly and is always small with increasing time during creep deformation (Fig. 1.21). Shrinkage of amorphous materials during creep deformation most likely arises due to structural changes at high-temperatures [11]. In contrast, crystalline counterpart shows negligible shrinkage less than 0.1% even at 1500 °C for 300 h, much lower than 1.4% for amorphous PDCs-SiBCN.

Additionally, the deformation rate of nanocrystalline PDCs-SiBCN shows a strong temperature dependence during creep deformation with curves running parallel at temperatures as high as 1350–1500 °C [12]. This is quite contrary to amorphous PDCs-SiBCN ceramics where the temperature dependence decreases with time and diminishes beyond 50 h (Fig. 5.10). Coincidently, the calculated activation energy at 1350–1500 °C is determined to be ca. 100 kJ/mol, and the stress exponent is found to be 1.1, while amorphous PDCs-SiBCN ceramics exhibits more than twice, as much as 230 kJ/mol. The low activation energy hints that solid-state diffusion controls the creep deformation.

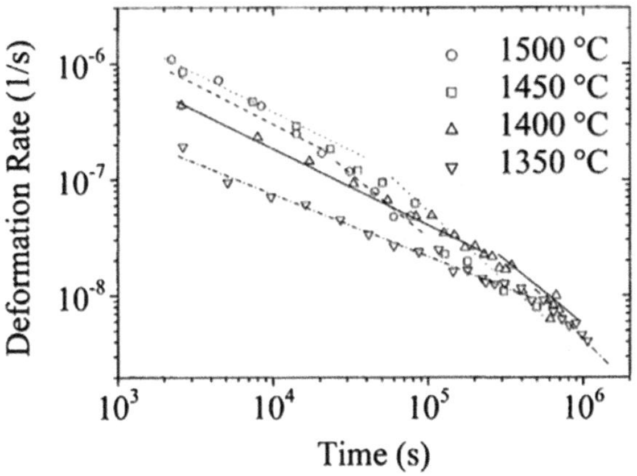

Fig. 5.10 Deformation rates as a function of time at 50 MPa and different temperature during isothermal creep deformation for the amorphous PDCs-SiBCN ceramics [12]

However, compression creep experiments on crystalline PDCs-SiBCN ceramics obtained from boron modified poly(vinyl)silazanes indicate only show primary creep during the entire creep time for all loads and temperature combinations [13, 14]. The maximum deformation at 1400 °C/100 MPa after 100 h is about 1.3% (Fig. 5.11). At all loads, primary creep is observed up to 105 s and above this time the strain rate becomes constant. A distinct dependence of strain rate on load is not observed at 75–150 MPa, though a slight load dependence of the minimum creep rate is observed. In another case, The PIP processing (C_f-SiC_p)/SiBCN CMCs show a low flexural creep deformation of ~ 0.25% at 1400 °C and 100 MPa for 60 h [15, 16].

The different creep behavior observed during creep deformation of amorphous PDCs-SiBCN and nanocrystalline PDCs-SiBCN ceramics depend most likely on their different structure. Continuous structural modifications occur during creep at high temperatures including atomic diffusion and grain boundary sliding leading to the crystallization and densification of materials thereby resulting eventually in

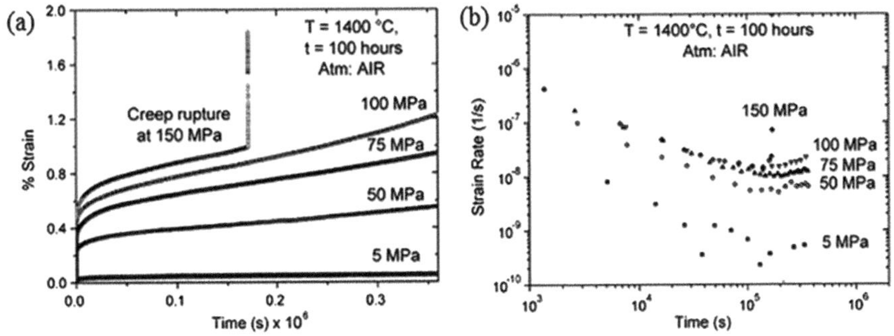

Fig. 5.11 Creep behavior of the nanocrystalline PDCs-SiBCN ceramics: **a** Deformation as a function of time for various loads at a temperature of 1400 °C; **b** Creep rates as a function of time during a creep time of 100 h deformation [13]

a macro-deformation. With increasing time during creep, the structure becomes gradually stable with the formation of capsule-like microstructure derived from SiC, Si_3N_4 and turbostratic BN(C). The special microstructure leads to the mutual restriction of atomic diffusion and grain growth thereby contributing to the excellent microstructural stability at high temperatures [17].

Consequently, the creep rate of PDCs-SiBCN is significantly lowered and the deformation becomes more difficult characterized by a small strain change at the final creep stage. In contrast, crystalline PDCs-SiBCN provides a stable and dense structure with a reduction of free volume compared to amorphous counterparts. Such processes as atomic diffusion and grain growth have taken place prior to conduction of creep experiments. The crystalline PDCs-SiBCN shows a dense homogenous microstructure and "clean" grain boundary coupled with the mutual encapsulation of nanosized grains contributing to small deformation indicative of an excellent creep resistance [12, 13].

Two deformation models that have been proposed to explain deformation in precursor-derived amorphous ceramics. These models are referred to as the 'two-step' rearrangement model and the free-volume model, which predict a linear increase in viscosity with time. According to the two-step rearrangement model, atomic rearrangement in a certain activation volume initiates viscous flow in the surrounding volume leading to creep deformation [18, 19]. The free-volume model explains viscosity of amorphous materials using formation of flow defects. Atomic transport for structural rearrangement is caused by flow defects that formed by fluctuations in the distribution of the free volume in the amorphous material. These defects are also available for atomic transport during creep deformation [20]. According to Aldinger et al. [12], the free-volume model fits the observed temperature dependence better than the 'two-step' rearrangement model, although both models accurately predict the time dependence of the deformation rate of amorphous PDCs-SiBCN ceramics.

(4) *Thermophysical properties*

The RT thermal conductivity of the hot-pressed MA-Si_2BC_3N ceramics is ~ 9.7 W/(m K), much lower than that of SiC ceramics (Fig. 5.12). The as-sintered MA-Si_2BC_3N ceramics are composed of α/β-SiC and BN(C) with grain sizes less than 100 nm, with numerous grain boundary interfaces. The presence of numerous grain boundaries significantly hinders phonon propagation thus decreasing the phonon heat conduction. Besides, the unique turbostratic structure of BN(C) phase, as well as the difference in lattice structures between SiC and BN(C), both result in great scattering of heat transfer caused by lattice vibration, further leading to the significant decrease in the thermal conductivity of the materials. At temperatures of 200 °C and above, the thermal conductivity of MA-Si_2BC_3N ceramics is stable and differs little with increasing annealing temperature.

The thermal expansion coefficient (CET) of the MA-Si_2BC_3N materials at 200, 800 and 1400 °C is 4.2×10^{-6} °C^{-1}, 4.9×10^{-6} °C^{-1} and 4.9×10^{-6} °C^{-1}, respectively. The average CET of the MA-Si_2BC_3N ceramics is 4.8×10^{-6} °C^{-1} at temperature range of 200–1400 °C, which is close to that of SiC ceramics. The relatively low CET, high strength and low Young's modulus must endow the MA-Si_2BC_3N ceramics

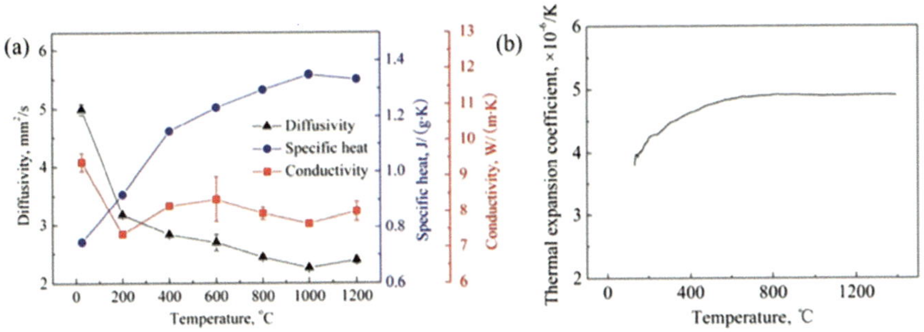

Fig. 5.12 Thermophysical properties of the hot-pressed MA-Si_2BC_3N bulk ceramics (1900 °C/80 MPa/30 min/8 bar N_2): **a** Thermal diffusivity, specific heat and thermal conductivity; **b** Thermal expansion coefficient [10]

Fig. 5.13 Thermal expansion coefficient of the MA-Si_2BC_3N ceramics using different milling method and raw materials (hot-pressed at 1900 °C/40 MPa/30 min/1 bar N_2) [4]

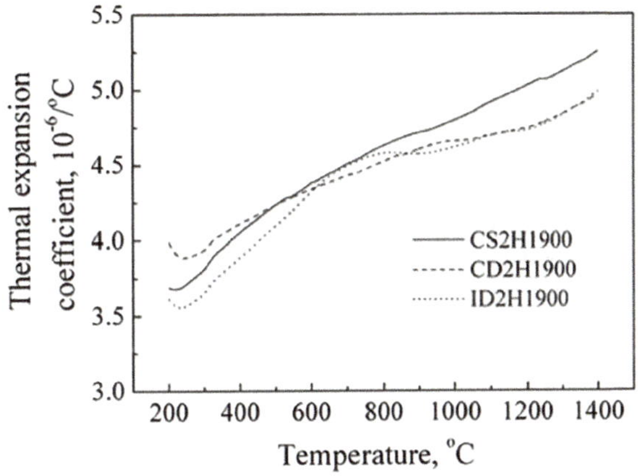

with good thermal shock resistance. The CET of the CS2H1900, CD2H1900, and ID2H1900 ceramics increases gradually from RT to 1400 °C. The average values of the three ceramics are 4.6×10^{-6} °C^{-1}, 4.6×10^{-6} °C^{-1} and 4.4×10^{-6} °C^{-1}, respectively, which are close to that of pure α-SiC or β-SiC ceramics (Fig. 5.13). At temperature higher than 800 °C, the one-step milling derived CS2H1900 ceramics possess higher CET than others.

5.1.2 *Spark Plasma Sintered Monoliths*

(1) *Room temperature mechanical properties*

As sintering pressure increases from 25 to 40 MPa, the mechanical properties of the spark plasma sintering (SPS) MA-Si_2BC_3N ceramics are greatly enhanced (Table 5.7). The flexural strength and Vickers' hardness of the SPS MA-Si_2BC_3N

ceramic sintered at 1800 °C/40 MPa are three times larger than those of ceramics sintered under 25 MPa, while their fracture toughness and Young's modulus are approximate 2 times than the counterparts, obtained at ~ 3.45 MPa $m^{1/2}$ and ~ 133.0 MPa [21].

The flexural strength and Young's modulus of the various MA-SiBCN bulk ceramics increase with the increasing Si/C ratio (Fig. 5.14). The maximum flexural strength and fracture toughness are attained to 502 MPa and 5.6 MPa $m^{1/2}$. The monolithic SiC ceramics (relative density being 98%) are also prepared by mechanical alloying combined SPS process, and their flexural strength, elastic modulus, and fracture toughness are of 520 MPa, 380 GPa and 3.6 MPa $m^{1/2}$, respectively [22]. It can be seen that the flexural strength of the SPS MA-Si_3BC_4N ceramics is slightly lower, but their fracture toughness is higher and elastic modulus is only ~ 40% of the SPS SiC ceramics (Tables 5.8 and 5.9).

(2) *Thermophysical properties*

For the SPS densified bulk ceramics with different Si/C ratio, their thermal diffusivities decrease while the specific heat increases with the increasing test temperature (Fig. 5.15). Besides, they both decline monotonously with increasing Si/C ratio at certain temperature. As a result, the thermal conductivity decreases with increasing temperature, and increasing Si/C ratio promotes the thermal conductivity most likely originating from the increasing SiC content in ceramics [23]. The thermal conductivity of the nanocrystalline bulk ceramics with different Si/C ratio is less than 55 W/(m K), while those of hot-pressed SiC, reaction-bonded SiC, and pressureless sintered SiC reach ~ 280 W/(m K), ~ 80 W/(m K) and ~ 100 W/(m K), respectively [24]. The average CET for the MA-$SiBC_2N$, MA-Si_2BC_3N and MA-Si_3BC_4N from

Table 5.7 Mechanical properties of the SPS densified MA-Si_2BC_3N bulk ceramics [21]

SPS parameters (1800 °C/5 min) (MPa)	Flexural strength (MPa)	Young's modulus (GPa)	Fracture toughness (MPa $m^{1/2}$)	Vickers' hardness (GPa)
25	91.8 ± 16.7	63.7 ± 11.9	1.38 ± 0.19	1.0 ± 0.2
40	311.5 ± 29.5	133.0 ± 3.0	3.45 ± 0.13	3.7 ± 0.2

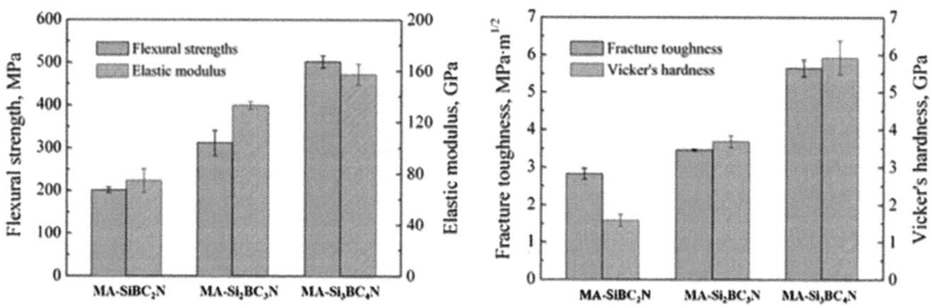

Fig. 5.14 Mechanical properties of the as-sintered bulk ceramics with different Si/C ratio [4]

Table 5.8 Mechanical properties of various bulk ceramics with different Si/C ratio by SPS at 1800 °C/40 MPa/3 min [4]

Samples	Flexural strength (MPa)	Young's modulus (GPa)	Fracture toughness (MPa $m^{1/2}$)	Vickers' hardness (GPa)
MA-$SiBC_2N$	202 ± 7	74 ± 9	2.8 ± 0.2	1.6 ± 0.2
MA-Si_2BC_3N	312 ± 30	133 ± 3	3.5 ± 0.1	3.7 ± 0.2
MA-Si_3BC_4N	502 ± 14	157 ± 9	5.6 ± 0.3	5.9 ± 0.5

Table 5.9 Mechanical properties comparison of the MA-Si_3BC_4N and SiC ceramics prepared by mechanical alloying combined SPS [22]

Mechanical properties	MA-Si_3BC_4N	SiC^{*1}	SiC^{*2}
Relative density (%)	95.4	98	~ 99
Flexural strength (MPa)	511.5	520	~ 530
Young's modulus (GPa)	157.3	380	~ 510
Fracture toughness (MPa $m^{1/2}$)	5.64	3.6	~ 4.5

SiC^{*1}: prepared by mechanical alloying of graphite and c-Si, and subsequent SPS densified. SiC^{*2}: SPS processing the commercial SiC powders with (B + C) additives. The SPS parameter is 1700 °C/40 MPa/10 min.

RT to 1800 °C are 3.61 × 10^{-6} °C^{-1}, 3.77 × 10^{-6} °C^{-1}, and 3.98 × 10^{-6} °C^{-1}, respectively. And the average CET for the hot-pressed MA-Si_2BC_3N and pure SiC ceramics are about 4.55 × 10^{-6} °C^{-1}.

5.1.3 *Hot-Isostatic Pressed Monoliths*

The relative densities and mechanical properties of the hot-isostatic pressing (HIP) sintered MA-Si_2BC_3N monoliths increase with the sintering temperature (Fig. 5.16). After HIP sintered at 1900 °C, the resulting MA-Si_2BC_3N bulk ceramics show the optimized bulk density ~ 2.80 g/cm^3, and also exhibit the highest flexural strength of 471.5 MPa, Young's modulus of 133.5 GPa, fracture toughness of 5.13 MPa $m^{1/2}$ and Vickers' hardness of 11.0 GPa, respectively (Table 5.10). Interestingly, the HIP MA-Si_2BC_3N sintered at 1700 °C show better mechanical performance than those ceramics HP sintered at 1900 °C.

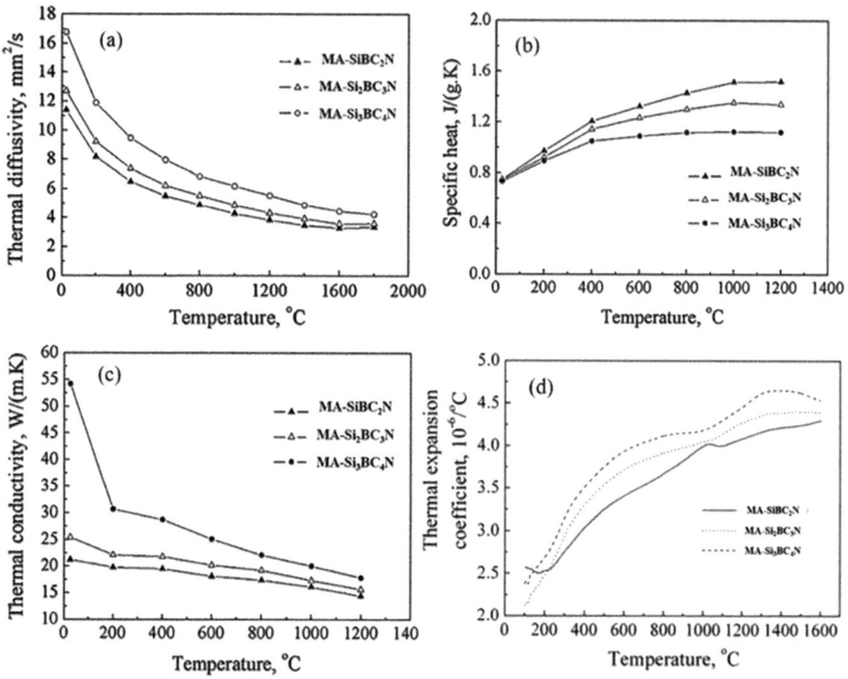

Fig. 5.15 Thermophysical properties of the various bulk ceramics with different Si/C ratio prepared by SPS at 1800 °C/40 MPa/3 min: **a** Thermal diffusivity; **b** Specific heat; **c** Thermal conductivity; **d** Thermal expansion coefficient [4, 22]

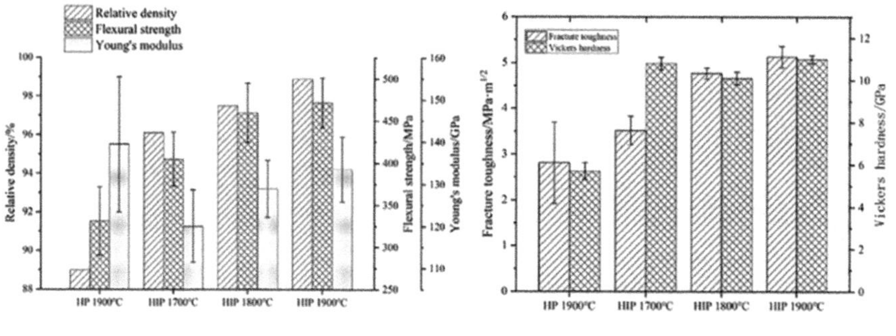

Fig. 5.16 Bulk density and mechanical properties of the hot-isostatic pressing (HIP) sintered MA-Si_2BC_3N monoliths

Table 5.10 Mechanical properties comparison of the MA-Si_2BC_3N monoliths prepared by hot-pressing (HP) and hot isostatic pressing (HIP), respectively

Sintering parameters	Relative density (%)	Flexural strength (MPa)	Young's modulus (GPa)	Fracture toughness (MPa $m^{1/2}$)	Vickers' hardness (GPa)
HP 1900 °C/40 MPa	89.0	331.0 ± 40.5	139.4 ± 16.0	2.81 ± 0.89	5.7 ± 0.4
HIP 1700 °C/ 190 MPa	96.1	404.4 ± 32.1	120.0 ± 8.6	3.52 ± 0.31	10.8 ± 0.3
HIP 1800 °C/ 190 MPa	97.5	459.5 ± 35.3	128.9 ± 6.7	4.77 ± 0.12	10.1 ± 0.3
HIP 1900 °C/ 190 MPa	98.9	471.5 ± 29.5	133.5 ± 7.7	5.13 ± 0.23	11.0 ± 0.2

5.2 Strengthening and "Deformation" Toughening in Amorphous MA-SiBCN Monoliths

(1) *Mechanical and thermophysical properties*

The nanohardness and Young's modulus of the high-pressure sintering (HPS) amorphous MA-Si_2BC_3N monoliths increase first and subsequently decrease with sintering temperature (Fig. 5.17). The dense amorphous MA-Si_2BC_3N ceramics sintered under 1100 °C/5 GPa exhibit the highest nano-hardness and Young's modulus of 29.4 GPa and 291.0 GPa, respectively [25]; however, nanocrystalline counterparts sintered under 1600 °C/5 GPa show nanohardness of 20.2 GPa and Young's modulus of 220.3 GPa, much lower than the amorphous MA-Si_2BC_3N (Table 5.11).

Consolidated at 1200 °C, the nanohardness and Young's modulus of the MA-Si_2BC_3N monoliths increase rapidly when sintering pressure applied from 1 to 3 GPa [25]. In sintering pressure range of 3–5 GPa at 1200 °C, a relatively slow growth rate of the mechanical properties is observed. The fracture toughness of as-sintered MA-Si_2BC_3N first decreases and then increases with the sintering temperature. But the

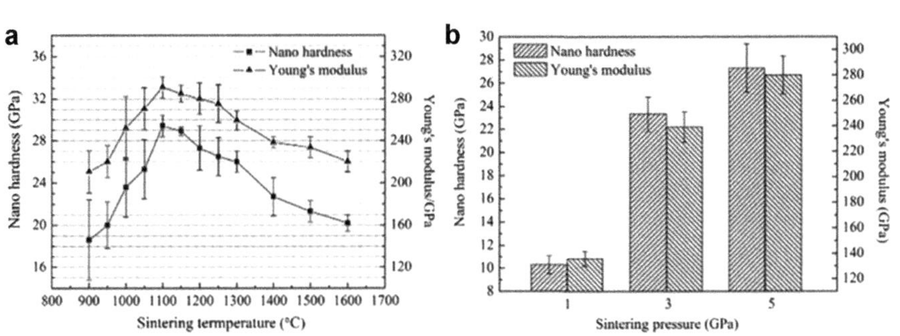

Fig. 5.17 Nanohardness and elastic modulus of the MA-Si_2BC_3N monoliths: HPS at **a** 900–1600 °C/5 GPa/30 min; **b** 1–5 GPa/1200 °C/30 min [25]

Table 5.11 Bulk density and mechanical properties of MA-Si_2BC_3N monoliths sintered at 5 GPa and different temperature [17]

Properties	RT	1000 °C	1100 °C	1600 °C
Bulk density (g/cm^3)	2.35 ± 0.02	2.64 ± 0.01	2.75 ± 0.01	2.83 ± 0.01
Nanohardness (GPa)	–	23.6 ± 2.8	29.4 ± 1.0	20.2 ± 0.8
Young's modulus (GPa)	–	252.0 ± 30	291.0 ± 10	220.3 ± 10

value change is not significant (Fig. 5.18). Dense amorphous MA-Si_2BC_3N monoliths sintered under 1100 °C/5 GPa exhibit the lowest fracture toughness of ~ 3.6 MPa m$^{1/2}$.

Under the sintering condition of 5 GPa/30 min, the MA-Si_2BC_3N monoliths sintered at 1100 or 1200 °C both show relatively low CET (<5 × 10^{-6} °C^{-1}). The difference is that the CET of the former (sintered at 1100 °C) sharply increases with testing temperature > 1200 °C (Fig. 5.19). Interestingly, the CET of the ceramics consolidated at 1600 °C increases first and then deceases, but finally increases with the test temperature. Generally, the nanocrystalline MA-Si_2BC_3N monoliths sintered at 1600 °C exhibit relatively high CET values than those of the amorphous ones. The change of CET is highly relevant to the material microstructural evolution, which can cause macroscopic changes in ceramic size, represented by the change of CET. For ceramics sintered under 1200 °C/5 GPa or 1600 °C/5 GPa, they remain basically unchanged microstructure and thus, little change in CET (200–1400 °C) is expected. The completely amorphous MA-Si_2BC_3N, may undergo adequate amorphous-to-crystalline transformation during test, leading to gradually increasing CET, especially beyond the crystallization temperature.

(2) *"Deformation" toughening and fracture*

The completely amorphous MA-Si_2BC_xN ($x = 1$–4) and MA-$Si_2B_yC_2N$ ($y = 1.5$–4) monoliths behave as ideally brittle materials, in which a propagating crack needs only

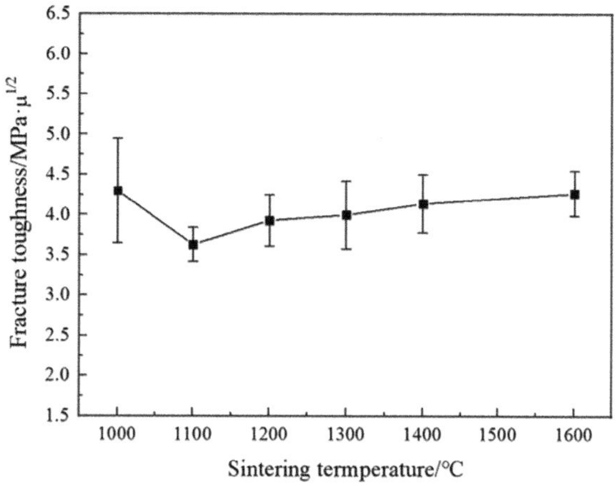

Fig. 5.18 Fracture toughness of MA-Si_2BC_3N monoliths HPS at 900–1600 °C/5 GPa/30 min [25]

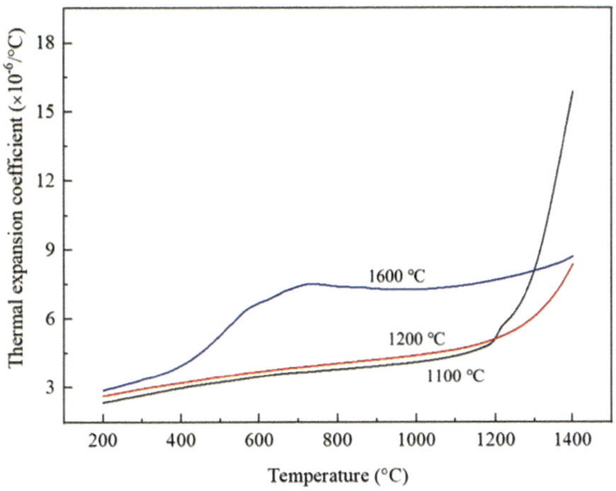

Fig. 5.19 Thermal expansion coefficient of the MA-Si_2BC_3N monoliths prepared by HPS under 1100–1600 °C/5 GPa/30 min [25]

overcome the surface energy of the monoliths. Actually, the fracture morphologies of the as-sintered dense amorphous monoliths are strongly impacted by the loading rate, leading to different morphological/microstructural evolution. It is well known that when the load changes abruptly, the stress waves will propagate through the materials and reflect off free surfaces. The reflecting stress waves thus impact the local cracks-tip stress and strain field, which in turn strongly affects the fracture behavior as well as fracture morphologies.

When trying to explain the "deformation" toughening behavior of the dense amorphous monoliths, two major classes of dynamic fracture problems should be considered: (i) fracture initiation as a result of slow or rapid loading rate, (ii) and propagation of a crack. In the latter case, the propagation of a crack may initiate either quasistatic or rapid expansion. In some cases, some cracks can be blocked after some amount of unstable propagation. Herein, with a very fast loading rate (e.g. instantaneous failure), dense amorphous MA-$Si_2B_{1.5}C_2N$ monoliths do not deform synergistically due to the limited relaxation time and strong covalent bonding. As a result, the initiation and propagation of a crack tend to take place simultaneously, leading to catastrophic failure and showing a smooth fracture morphology (Fig. 5.20). However, with a very slow loading rate (e.g. induced by nanoindentation test), limited deformation can occur at the crack-tip stress field zone since the neighboring stacking ceramic particles can deform, slip and rearrange to release partial stress so that some cracks may be arrested. In this case, considerable "shear bands" form, leading to the river-like structure (Fig. 5.21). When the driving force for a crack extension exceeds the materials resistance, the river-like structure becomes unstable and rapid crack propagation may occur. However, if the energy available for an incremental extension of a rapidly propagating crack falls down the material resistance, these racks will be arrested resulting in strengthening and toughening.

As mentioned above, high-pressure sintering can effectively crush the hard aggregates of ceramic particles and thus refine the powder particles. Obviously, high-pressure can eliminate porosity of the adjacent ceramic particles and facilitates

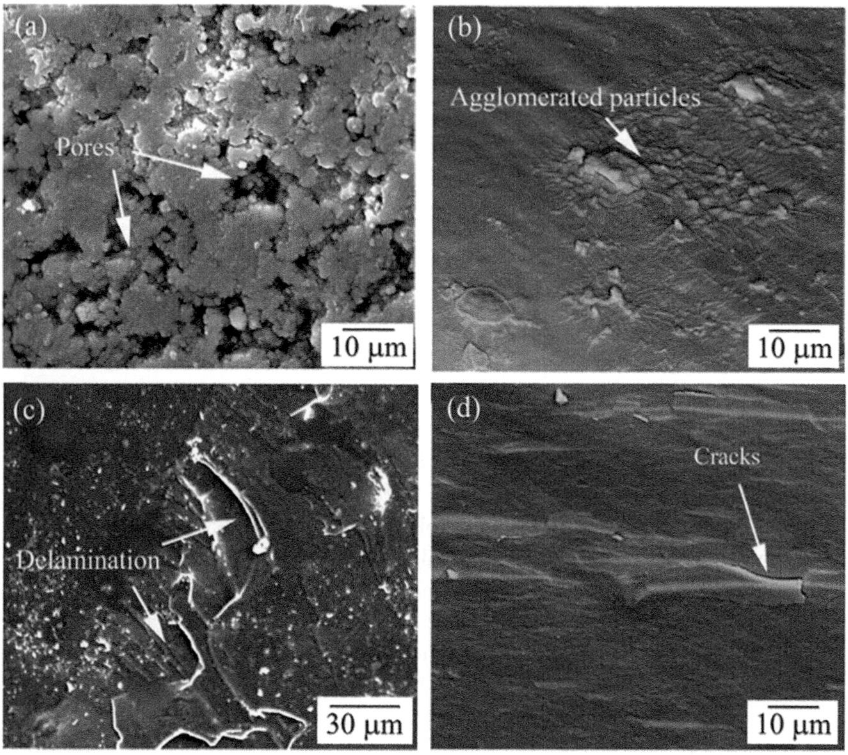

Fig. 5.20 Instantaneous failure surfaces of dense amorphous $Si_2B_{1.5}C_2N$ monoliths consolidated under different sintering pressure: **a** RT/5 GPa; **b** 1000 °C/3 GPa; **c** 1 000 °C/4 GPa; **d** 1000 °C/5 GPa (Fracturing with an estimated impact velocity of ~ 5.4 m/s)

short-range diffusion of atoms. From this point of view, increasing high-pressure can promote structure relaxation leading to the annihilation of free volume. It has been accepted that the activity energy for the formation of a "shear band" is strongly dependent on the pre-existing initial free volume, in which the "shear band" initiate and spread easily. Therefore, it can be expected that further increasing sintering pressure over 5 GPa will constrain the formation of the river-like shear bands. Generally, dense amorphous MA-Si_2BC_xN ($x = 1$–4) and MA-$Si_2B_yC_2N$ ($y = 1.5$–4) monoliths show brittle fracture at a macro view, but have limited plasticity on the microscopic scale. The fracture behavior for amorphous monoliths is closely related to the loading rate and deformable region, in which only a small number of local atoms (molecules) or atomic clusters can undergo severe deformation under a slow strain rate. However, the non-uniform deformation of the shear zone is unstable, which ultimately leads to the catastrophic failure.

Of special interest is that highly brittle amorphous SiBCN based materials are extremely sensitive to the in situ reactions (e.g., element enrichment, phase transformation, decomposition or cracks generation), especially when they come from

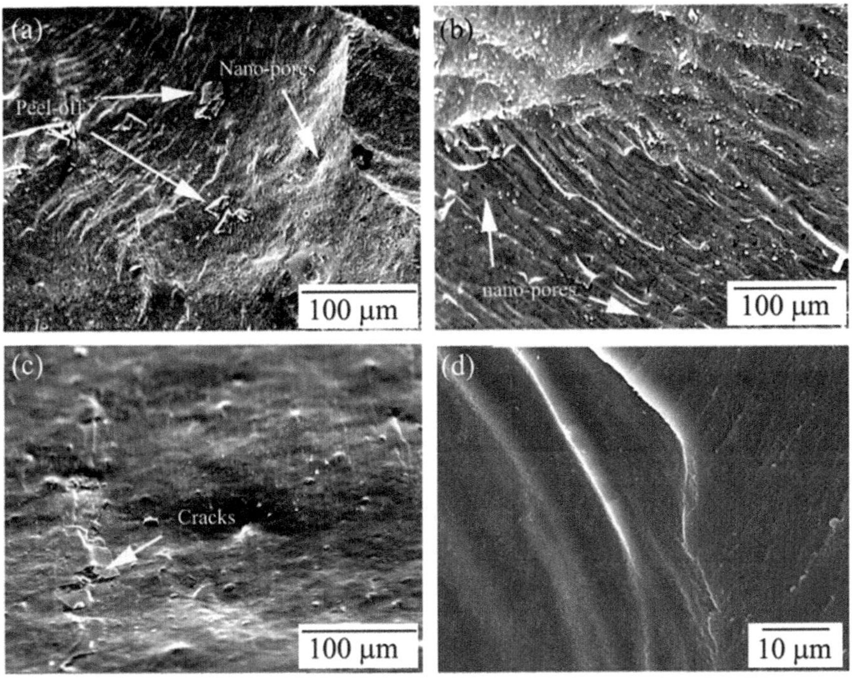

Fig. 5.21 Representative SEM-fracture morphologies for dense amorphous $Si_2B_{1.5}C_2N$ monoliths consolidated at 1 000 °C/30 min under different sintering pressure: **a** 3 GPa; **b** 4 GPa; **c** and **d** 5 GPa (Fracturing with an estimated press-in rate of ~ 10.0 nm/s)

elemental powders obtained by high-energy ball-milling, which already have energy levels above traditional powders, as preferential crystallization of certain chemical elements, generation of atomic groupings that can lead to anisotropy, severe thermal stresses due to the difference in coefficient of thermal expansion between phases (amorphous or nanocrystalline), and even spontaneous microcracking during cooling with controlled rates.

The strength and hardness of a nano material are closely related to its grain size based on Hall–Petch theory [26]. However, mechanical properties for amorphous ceramics mainly depend on bulk density, amorphous structure characteristics, chemical short-range order, types of chemical bonds and so on. Since there are no defects such as dislocations and grain boundaries, dense amorphous monoliths should possess high strength, high hardness and high modulus. For MA-$Si_2B_yC_2N$ ($y = 1.5$–4) monoliths consolidated at 3 GPa, the obtained nano hardness and Young's modulus are not satisfying although a well-developed amorphous structure is retained (Fig. 5.22). For monoliths consolidated at higher pressure (4–5 GPa), higher hardness and modulus are expected due to the increasing density as compared with the formers. In fact, their mechanical properties may be also dependent on the number and size of these preformed metal Si and BN(C) nanocrystals, or on the changes in

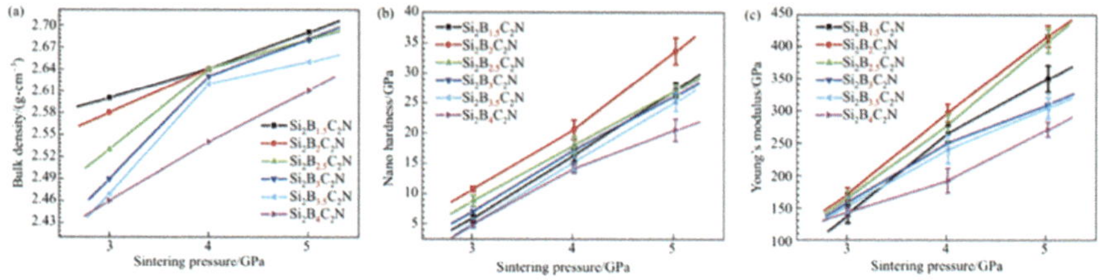

Fig. 5.22 Density and mechanical properties for dense amorphous MA-$Si_2B_yC_2N$ ($y = 1.5$–4) monoliths consolidated at 1000 °C/30 min under different sintering pressure: **a** Bulk density; **b** Nano hardness; **c** Young's modulus

the amorphous phase itself due to composition changes, relaxation of the state of short-range order and annihilation of the free volume.

Generally, the nano hardness for HPS densified MA-Si_2BC_xN ($x = 1$–4) and MA-$Si_2B_yC_2N$ ($y = 1.5$–4) monoliths can be regarded as a synergistic effect of amorphous matrix and β-SiC, BN(C), h-BN, graphite, metal Si crystalline phases. BN(C) nanocrystals are believed to has a low hardness, which are probably close to that of h-BN and/or graphite. Sintering at 1000 °C under 3 GPa, dense amorphous monoliths possess completely amorphous structure thereby showing a relatively high hardness. As sintering pressure further progresses, the initial highly packed amorphous network structure of Si–C, B–C, B–N, C–C, C–N and N–B–C bonds was broken due to the pressure induced crystallization. It is believed that dense amorphous 3-D network structure can against deformation doggedly, while nanocrystals within a glassy matrix can rotate and slid to offer limited deformation leading to the reduction of hardness and strength. It thus can be inferred that the growing rate of the nano hardness will decrease gradually as sintering pressure further increases (beyond 5 GPa) since crystallization of the amorphous phases is unavoidable at elevated sintering pressure. In this scenario, an increase in sintering pressure will deteriorate the mechanical performance of dense monoliths once crystallization takes place.

5.3 "Additive" Strengthened and Toughened MA-SiBCN Composite Ceramics

5.3.1 Effect of Metal Particles

(1) *Al particles addition*

The flexural strength of the $Si_2BC_3NAl_{0.6}$ (Al as the Al source) bulk ceramics hot-pressed at 1800 °C/40 MPa is obtained ~ 246.2 MPa [27]. However, it significantly increases to 500.1 MPa and subsequently decreases when sintered at 1950 °C. The noticeable improvement of flexural strength should be attributed to enhanced bulk density. The trend of fracture toughness of the $Si_2BC_3NAl_{0.6}$ composite ceramics is consistent with that of flexural strength, both obtaining maximum values after sintering at 1900 °C (Fig. 5.23). For $Si_2BC_3NAl_{0.6}$ bulk ceramics hot-pressed at 1800 or 1900 °C, Young's modulus of 124.1 and 232.2 GPa is obtained, but subsequently decreases at higher sintering temperature. The difference in Vickers' hardness of the MA-$Si_2BC_3NAl_{0.6}$ ceramics hot-pressed at 1800 or 1900 °C is negligible, while that of 1950 °C sintered ceramics is significantly larger than the formers.

Sintered at 1900 °C, the flexural strength, fracture toughness, Young's modulus and hardness of the MA-$Si_2BC_3NAl_{0.6}$ ceramics decrease as applied pressure raises from 40 to 50 MPa. Besides, MA-$Si_2BC_3NAl_{0.6}$ ceramics sintered under N_2 atmosphere (here referred to CN), present better mechanical properties than those of ceramics sintered under Ar (referred to CA), except for the similar Vickers' hardness. For example, the flexural strength and elastic modulus of CN are 1.2 times larger than those of CA. Fracture toughness of CN is 1.5 times larger than that of CA. Since the densities of the CA and CN are 2.77 and 2.90 g/cm^3, respectively, the higher flexural strength for CN materials can be attributed to the higher relative density (Table 5.12).

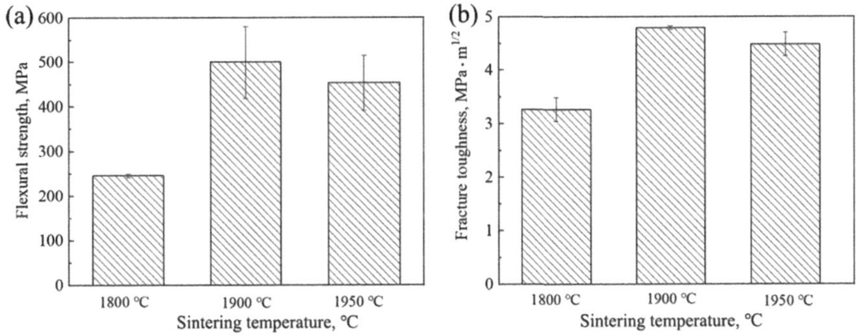

Fig. 5.23 The flexural strength and fracture toughness of the MA-$Si_2BC_3NAl_{0.6}$ (Al as the Al source) bulk ceramics hot-pressed at different temperature [27]

Table 5.12 Mechanical properties of the MA-$Si_2BC_3NAl_{0.6}$ (Al as the Al source) bulk ceramics hot-pressed at different pressure [27]

Hot-pressed sintering parameters	Flexural strength (MPa)	Fracture toughness (MPa m$^{1/2}$)	Young's modulus (GPa)	Vickers' hardness (GPa)
1900 °C/40 MPa	500.1 ± 81.2	4.80 ± 0.03	232.2 ± 7.4	8.2 ± 1.0
1900 °C/50 MPa	438.7 ± 69.8	4.19 ± 1.13	223.2 ± 3.0	12.9 ± 0.7
1800 °C/30 MPa/Ar	421.9 ± 27.3	3.40 ± 0.15	174.1 ± 10.2	12.7 ± 0.3
1800 °C/30 MPa/N_2	526.8 ± 10.4	5.25 ± 0.20	222.1 ± 27.7	11.6 ± 0.5

For MA-$Si_2BC_3NAl_{0.6}$ ceramics hot-pressed at different condition, they all show similar change trend of CET. Testing temperature below 800 °C, their thermal expansion coefficient rapidly increases, while the increasing trend slows down and finally reaches a stable state above 800 °C. At 1300 °C, the CET of the four ceramics is similar < 5.25×10^{-6} °C^{-1} (Fig. 5.24). Similarly, the CET of the MA-$Si_2BC_3NAl_{0.2}$, MA-$Si_2BC_3NAl_{0.6}$ (Al as Al source) and MA-$Si_2BC_3N_{1.6}Al_{0.6}$ (AlN as Al source) ceramics changes in the same way from 200 °C to 1400 °C. At 1300 °C, the CET of the three ceramics is in the range of $(4.8–5.1) \times 10^{-6}$ °C^{-1}. At temperatures above 400 °C, the CET of the three bulk ceramics is presented as MA-$Si_2BC_3NAl_{0.2}$ > MA-$Si_2BC_3NAl_{0.6}$ > MA-$Si_2BC_3N_{1.6}Al_{0.6}$ in descending order. The higher CET of the as-sintered $Si_2BC_3NAl_{0.6}$ bulk ceramics should be argued to the formation of highest proportion of AlON, which has the highest CET among crystallized phase (Table 5.13).

(2) *Mo particles addition*

With the increasing MA-Si_2BC_3N, the bulk density of the Mo/Si_2BC_3N composite ceramics shows a downward trend except for the composites with 50 vol.% MA-Si_2BC_3N [28, 29]. Similarly, the hardness of the composites increases from 7.3 GPa to 14.5 GPa as the content of MA-Si_2BC_3N increases from 10 vol.% to 30 vol%. The hardness decreases gradually when the MA-Si_2BC_3N content is higher than 30

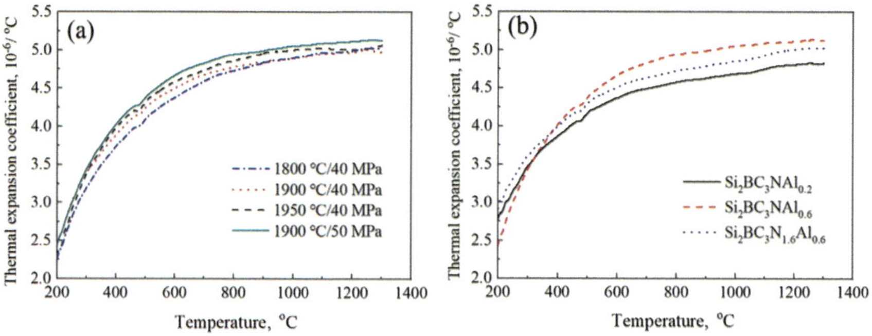

Fig. 5.24 Thermal expansion coefficient versus temperature curves of the as-sintered monoliths: **a** MA-$Si_2BC_3NAl_{0.6}$ hot-pressed at different condition; **b** Hot-pressed at 1900 °C/50 MPa [27]

Table 5.13 A comparison of thermal expansion coefficient for the different ceramics [27]

Ceramics	Test temperature (°C)	Thermal expansion coefficient ($\times 10^{-6}$ °C^{-1})
α-SiC	0–1000	4.7
β-SiC	0–1000	5.1
AlN	0–1000	4.5
BCN	1100–1300	5.5
AlON	25–1000	7.0

vol.%. Meanwhile, with the increasing MA-Si_2BC_3N, the Young's modulus of the composites first increases and then decreases (Fig. 5.25).

The bending strength of the composites is higher with 10–30 vol.% MA-Si_2BC_3N addition. For example, the bending strength of the Mo/Si_2BC_3N composites with 10 vol.% MA-Si_2BC_3N is 679.8 MPa, which is more than 13 times that of the pure MA-Si_2BC_3N. Obviously, the hardness, and bending strength of the composites increases first and then decreases with the increase of MA-Si_2BC_3N content, and the peak values are obtained with about 10–30 vol.% MA-Si_2BC_3N (Table 5.14).

Introduction of 10 wt.% $MgO–Al_2O_3–SiO_2$ (MAS) additive is beneficial to the densification, and the hardness, Young's modulus and bending strength of Mo/Si_2BC_3N-MAS composite ceramics are also obviously improved. The introduction of MAS additive increases the hardness of composite ceramics to 6.4 GPa, which is nearly 6 times higher than that of composites with 90 vol.% MA-Si_2BC_3N (~1.1 GPa). Additionally, the MAS additive in pure MA-Si_2BC_3N increases the hardness

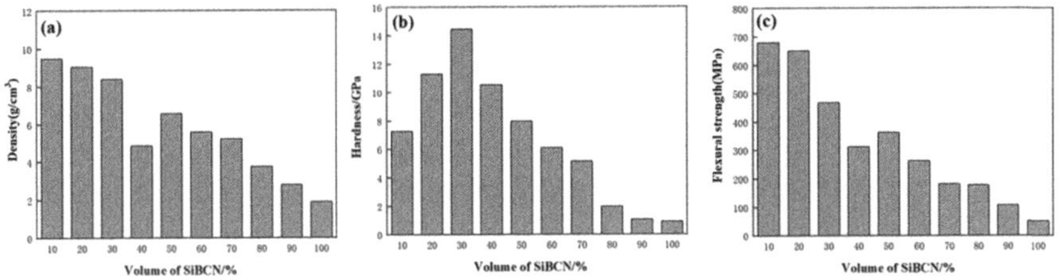

Fig. 5.25 Density and mechanical properties of the Mo/Si_2BC_3N composite ceramics with different volume fraction of MA-Si_2BC_3N: **a** Bulk density; **b** Vickers' hardness; **c** Flexural strength [28]

Table 5.14 Bulk density and mechanical properties of Mo/Si_2BC_3N composite ceramics [29]

Volume content of MA-Si_2BC_3N	Bulk density (g/cm^3)	Flexural strength (MPa)	Vickers' hardness (GPa)
10	9.48	679.8 ± 0.2	7.3 ± 2.3
20	9.05	651.4 ± 5.9	11.3 ± 1.0
30	8.41	468.4 ± 3.8	14.5 ± 0.9
40	4.85	312.2 ± 3.5	10.6 ± 0.4
50	6.57	363.3 ± 0.6	8.0 ± 2.0
60	5.6	262.8 ± 1.3	6.1 ± 1.2
70	5.25	181.3 ± 3.9	5.2 ± 0.6
80	3.76	178.1 ± 2.4	2.0 ± 0.3
90	2.80	108.1 ± 4.8	1.1 ± 0.1
90 + MAS	3.26	163.5 ± 5.2	6.4 ± 0.4
100	1.91	51.4 ± 0.4	0.9 ± 0.1
100 + MAS	2.44	216.6 ± 2.7	3.6 ± 0.4

to 3.6 GPa, which is nearly 4 times higher than additive-free composites (~0.9 GPa). The bending strength of additive-incorporating composite ceramics increases nearly 3 times (from 51.4 to 216.6 MPa).

The hot-pressed Mo/Si_2BC_3N composite ceramics all show brittle fracture behavior (Fig. 5.26). Generally, the glassy phase formed by the MAS additive has a certain fluidity, which can promote the movement of particles during sintering, thereby improving the bulk density and mechanical strength of the Mo/Si_2BC_3N composite ceramics. Nevertheless, the presence of glassy phase will reduce the comprehensive mechanical properties of the composites; however, no obvious aggregation of glassy phases is found in MA-Si_2BC_3N ceramic matrix.

The types and contents of Mo-compounds in MA-Si_2BC_3N matrix are strongly correlated with the thermal expansion coefficient (CET) of the composites. When the content of MA-Si_2BC_3N reaches 40 vol.%, the CET reaches the maximum value of 19.5. When MA-Si_2BC_3N content is in range of 50–100 vol.%, the CET of the composites decreases due to the relatively lower density and larger porosity. At a certain test temperature, the CET of the MAS additive containing composite ceramics decreases as compared to the counterparts (Fig. 5.27).

The CET of Li_2O–Al_2O_3–SiO_2 and CaO–Al_2O_3–SiO_2 are 3.3×10^{-6} °C^{-1} and 6.5×10^{-6} °C^{-1}, respectively [30]. Thus, the CET of MAS additives is supposed to be (3–6) $\times 10^{-6}$ °C^{-1}, which is close to the that of the additive-free 90 vol.% MA-Si_2BC_3N incorporating composite ceramics, but higher than that of pure MA-Si_2BC_3N ceramics (Table 5.15).

(3) *Zr particles addition*

The Zr and B powder mixtures are mechanical alloyed into the MA-Si_2BC_3N by one-step or two-step ball milling, subsequently hot-pressed at 1800–2000 °C/40 MPa/30 min [31]. The sintering temperature has a great impact on the mechanical

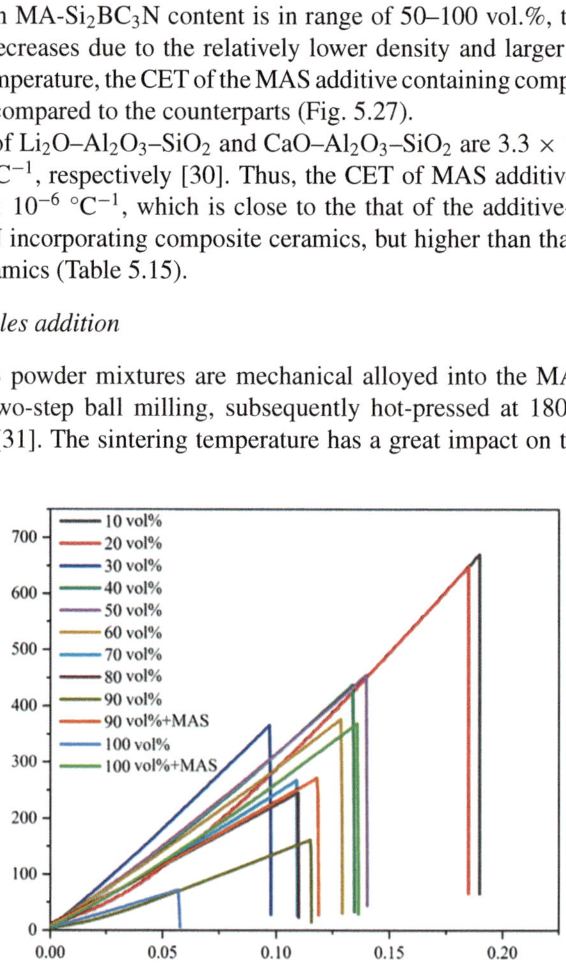

Fig. 5.26 Load–displacement curves of the various Mo/Si_2BC_3N composite ceramics [29]

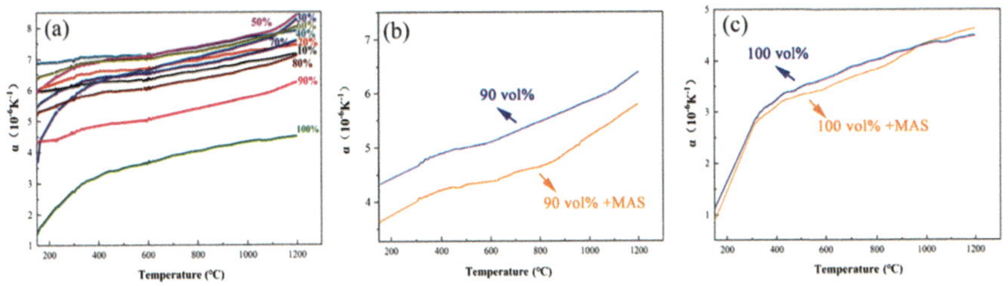

Fig. 5.27 Thermal expansion coefficient versus temperature curves of the Mo/Si_2BC_3N composite ceramics: **a** With different MA-Si_2BC_3N content; **b** and **c** With or without 10 wt.% MAS additive [29]

Table 5.15 Average thermal expansion coefficient of Mo/Si_2BC_3N composite ceramics [30]

MA-Si_2BC_3N content (vol.%)	Thermal expansion coefficient ($\times 10^{-6}$ °C^{-1})
10	8.64
20	9.35
30	16.00
40	19.50
50	15.50
60	11.90
70	10.90
80	10.80
90	9.93
90 + MAS	9.06
100	5.61
100 + MAS	4.04

properties of Zr/Si_2BC_3N composite ceramics due to higher density obtained by higher temperature. The fracture toughness, flexural strength and Vicker's hardness of the MA-$Si_2B_5C_3NZr_2$ bulk ceramics maintain a rising trend, while the Young's modulus declines first but increases rapidly at last (Fig. 5.28).

The flexural strength of MA-$Si_2B_5C_3NZr_2$ ceramics hot-pressed at 1800 °C is 222.1 MPa, while that sintered at 1900 °C and 2000 °C is 302.9 MPa and 400.0 MPa, respectively. The fracture toughness of MA-$Si_2B_5C_3NZr_2$ hot-pressed at 1800 °C and 1900 °C differs a little, that is 2.83 MPa $m^{1/2}$ and 2.91 MPa $m^{1/2}$. Ceramics prepared at 2000 °C exhibit the highest fracture toughness of 3.16 MPa $m^{1/2}$. The MA-$Si_2B_5C_3NZr_2$ hot-pressed at 1800–2000 °C show an undesirable fracture toughness may due to the complex phase composition and the mismatched CET of each phase (SiC, 4.9×10^{-6} °C^{-1}; ZrN, 7.2×10^{-6} °C^{-1}; ZrB_2, 6.9×10^{-6} °C^{-1}). The Young's modulus of ceramics hot-pressed at 1800–2000 °C exhibits an approximate linear relationship with sintering temperature, reaching to 142.1 GPa, 197.3 GPa and 251.6 GPa, respectively. The Vicker's hardness of the ceramics sintered at 2000 °C is 9.57 GPa, 60% higher than the formers.

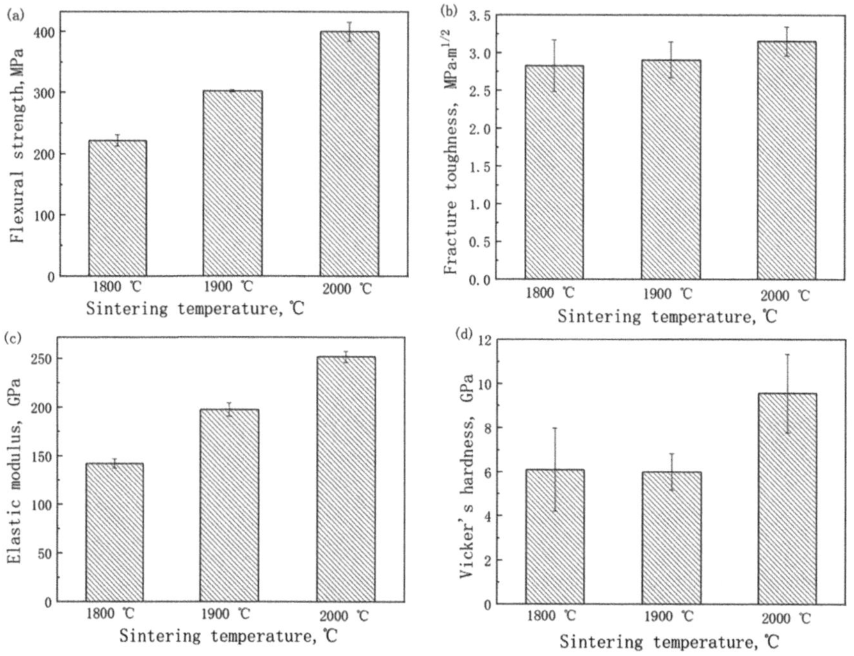

Fig. 5.28 Mechanical properties of MA-$Si_2B_5C_3NZr_2$ bulk ceramics hot-pressed at different temperature: **a** Flexural strength; **b** Fracture toughness; **c** Young's modulus; **d** Vickers' hardness [31]

With the increasing content of Zr and B powder mixtures, both fracture toughness and Vickers' hardness improve steadily while the flexural strength remains almost unchanged (Fig. 5.29). Meantime, the Young's modulus drops obviously from 242.1 ± 4.6 to 142.1 ± 4.6 GPa. However, it seems that the mechanical properties of the bulk ceramics prepared by two-step ball milling perform better than their counterparts though they have almost the same bulk density (Table 5.16). The MA-$Si_2B_5C_3NZr_2$ bulk ceramics hot-pressed at 2000 °C perform superior on mechanical properties and density, obtaining flexural strength of 400.0 ± 15.4 MPa, fracture toughness of 3.16 ± 0.19 MPa $m^{1/2}$, Young's modulus of 251.6 ± 5.8 GPa and Vickers' hardness of 9.6 ± 1.2 GPa, respectively.

(4) *Al_4SiC_4 particles addition*

Using amorphous MA-Si_2BC_3N and nano Al_4SiC_4 powders as raw materials, dense Al_4SiC_4/Si_2BC_3N composite ceramics are prepared by SPS technique [32]. The Al_4SiC_4 containing MA-Si_2BC_3N ceramics show lamellate structures of Al_4SiC_4 that are completely different from BN(C). The lamellate Al_4SiC_4 functions as reinforcement mainly through the "crack bridging" and/or "crack deflection" mechanisms, leading to a more tortuous crack propagation path (Fig. 5.30).

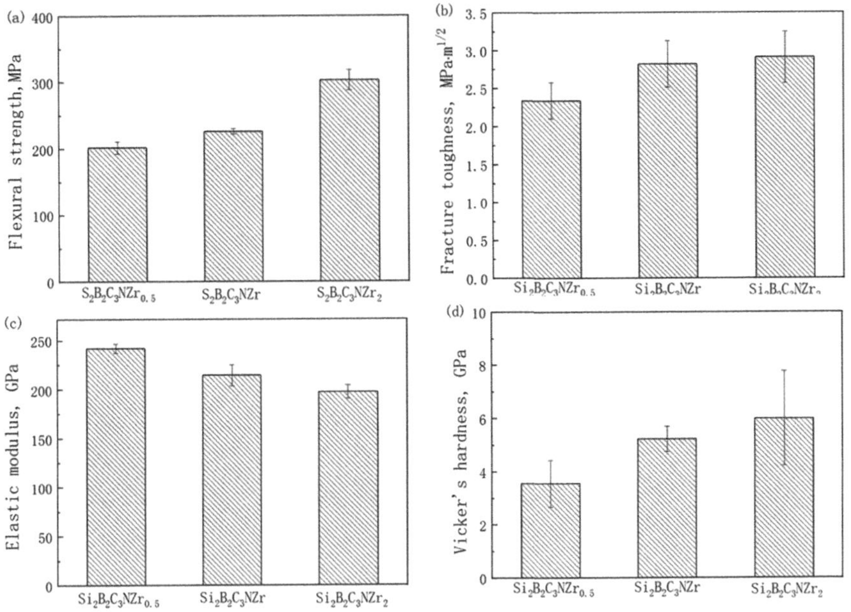

Fig. 5.29 Mechanical properties of hot-pressed bulk ceramics with different Zr-B powder mixtures: **a** Flexural strength; **b** Fracture toughness; **c** Young's modulus; **d** Vickers' hardness [31]

Table 5.16 Bulk density and mechanical properties of the hot-pressed bulk ceramics prepared by different milling condition and sintering parameter [31]

Composition	Bulk density (g/cm^3)	Flexural strength (MPa)	Fracture toughness ($MPa\ m^{1/2}$)	Young's modulus (GPa)	Vickers' hardness (GPa)
[a]$Si_2B_2C_3NZr_{0.5}$	3.24	202.0 ± 9.1	2.34 ± 0.24	242.1 ± 4.6	3.6 ± 0.9
[b]$Si_2B_2C_3NZr_{0.5}$	3.22	213.0 ± 7.8	2.57 ± 0.41	252.4 ± 5.8	3.3 ± 0.6
[a]$Si_2B_3C_3NZr$	3.65	226.0 ± 4.2	2.82 ± 0.31	214.5 ± 10.6	5.2 ± 1.2
[a]$Si_2B_5C_3NZ_2$ (sintered at 1800 °C)	3.96	222.1 ± 9.1	2.83 ± 0.34	142.1 ± 4.6	6.1 ± 0.5
[a]$Si_2B_5C_3NZr_2$ (sintered at 1900 °C)	3.97	302.9 ± 1.7	2.91 ± 0.24	197.3 ± 6.9	6.0 ± 1.9
[a]$Si_2B_5C_3NZr_2$ (sintered at 2000 °C)	4.08	400.0 ± 15.4	3.16 ± 0.19	251.6 ± 5.8	9.6 ± 1.2

[a]One-step milling: The c-Si, h-BN, graphite, boron and zirconium powder mixture are simultaneously milled for 20 h

[b]Two-step milling: The c-Si, graphite and h-BN powders in a mole ratio of 2:3:1 are milled for 20 h, and then the as-milled Zr and B powder mixtures (5 h) are added and milling continued for another 5 h

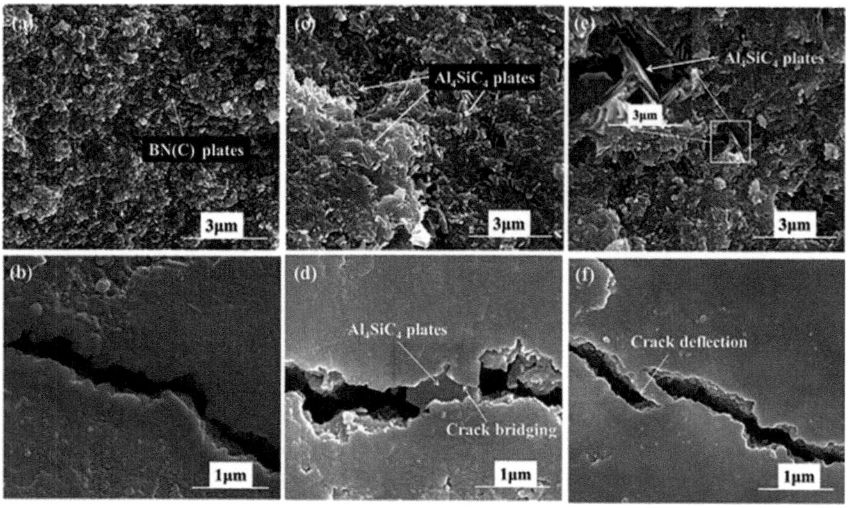

Fig. 5.30 SEM images of the MA-Si_2BC_3N ceramics with various Al_4SiC_4 additive: **a**, **b** Pure MA-Si_2BC_3N; **c**, **d** + 5 wt.% Al_4SiC_4; **e**, **f** + 10 wt.% Al_4SiC_4 [32]

Apparently, the bulk density and relative density of the Al_4SiC_4/Si_2BC_3N composite ceramics are improved with the incorporation of Al_4SiC_4. Consequently, their flexural strength, fracture toughness, Vickers' hardness and Young's modulus outperform the pristine MA-Si_2BC_3N ceramics (Table 5.17). The enhanced mechanical properties should be attributed to the strengthening and toughening effects of the well-developed BN(C) platelets and uniformly distributed Al_4SiC_4 platelets. Note that 10 wt.% Al_4SiC_4 addition has no further improvement on fracture toughness as compared with 5 wt.% Al_4SiC_4 addition.

(5) *(Cu-Ti) particles addition*

With 35 vol.% MA-Si_2BC_3N solid content, the bulk densities of all the MA-Si_2BC_3N porous ceramics annealed at 1900 °C are close to 1.01 g/cm^3. The open porosities of all the porous ceramics are ~ 40%, though different layer structures are observed.

Table 5.17 Densities and mechanical properties of the Al_4SiC_4/Si_2BC_3N composite ceramics by SPS at 1900 °C/40 MPa/5 min [32]

Composition	Bulk density (g/cm^3)	Relative density (%)	Flexural strength (MPa)	Fracture toughness (MPa m$^{1/2}$)	Young's modulus (GPa)	Vickers' hardness (GPa)
MA-Si_2BC_3N	2.58	91.1	329.1 ± 15.9	4.11 ± 0.22	104.0 ± 6.2	4.3 ± 0.2
+ 5 wt.% Al_4SiC_4	2.66	94.4	345.1 ± 7.9	5.44 ± 0.35	119.4 ± 3.1	6.9 ± 0.3
+ 10 wt.% Al_4SiC_4	2.65	94.3	451.5 ± 7.9	5.17 ± 0.28	128.4 ± 0.8	6.7 ± 0.2

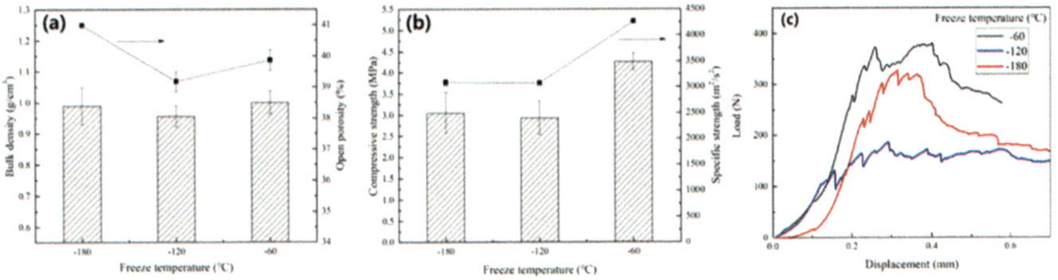

Fig. 5.31 Porous MA-Si_2BC_3N ceramics prepared at different freezing temperature: **a** Open porosity and bulk density; **b** Compressive strength and specific strength; **c** Load–displacement curve [33]

The porous MA-Si_2BC_3N ceramics frozen at −60 °C show the highest compressive strength and specific strength, reaching 4.3 ± 0.2 MPa and 4264.0 ± 36.3 m^2/S^2, respectively. Further deceasing the freezing temperature leads to the gradual decrease in mechanical properties (Fig. 5.31).

As the MA-Si_2BC_3N content increased from 30 to 40 vol.%, the density of the as-annealed porous ceramics increases from 1.43 ± 0.01 to 1.66 ± 0.01 g/cm^3, while the open porosity decreases from 55.5 to 48.7%. The compressive strength and specific strength of the porous ceramics increase linearly with the increasing solid content, obtained satisfying compressive strength of 23.1 ± 1.8 MPa and specific strength of 14,004.8 ± 6.4 m^2/s^2.

Mechanical alloyed nano ZrB_2 ~ 33 wt.% are introduced into the MA-Si_2BC_3N and, subsequently annealed at 2000 °C. The bulk density of all the as-annealed porous Si_2BC_3N-ZrB_2 ceramics increases ~ 12% with a slight decrease of porosity. In terms of mechanical properties, the compressive strength of the porous Si_2BC_3N-ZrB_2 ceramics increases ~ 37, ~ 12, and ~ 23% as compared to the porous MA-Si_2BC_3N. Besides, after introduction of ~ 33 wt.% nano ZrB_2, the specific strength of the 30 vol.% MA-Si_2BC_3N containing porous ceramics has been greatly improved ~ 51%, while that of the 40 vol.% MA-Si_2BC_3N containing ceramics is only increased ~ 7% (Fig. 5.32).

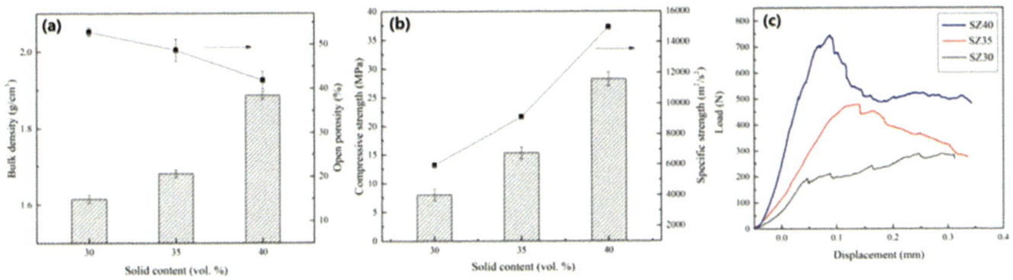

Fig. 5.32 Porous MA-Si_2BC_3N ceramics with ~ 33 wt.% nano ZrB_2 prepared at freezing temperature of −60 °C: **a** Open porosity and bulk density; **b** compressive strength and specific strength; **c** load–displacement curve [33]

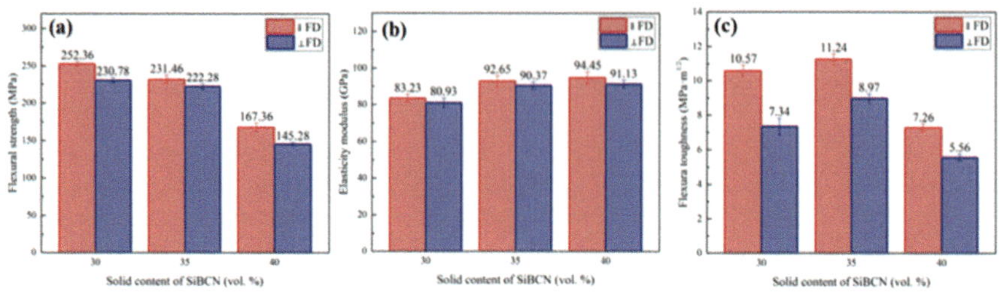

Fig. 5.33 Mechanical properties of the $Cu_{80}Ti_{20}/Si_2B_2C_3N$ composite ceramics with different MA-$Si_2B_2C_3N$ content: **a** Flexural strength; **b** Young's modulus; **c** Fracture toughness (FD: Freezing direction) [33]

The $Cu_{80}Ti_{20}$ compounds are used as the infiltrated alloys. With 30 vol.% MA-$Si_2B_2C_3N$ content, $Cu_{80}Ti_{20}/Si_2BC_3N$ composites exhibit maximum flexural strength of 252.4 ± 3.3 MPa parallel to the freezing direction (FD), while that perpendicular to FD is 230.8 ± 3.5 MPa [33]. The 30 vol.% MA-$Si_2B_2C_3N$ porous ceramics have a low closed porosity allowing sufficient metal infiltration, leads to moderate interfacial bonding conducive to the load transfer. With 35 vol.% MA-$Si_2B_2C_3N$, their flexural strength slightly decreases, which are 167.4 ± 5.9 MPa parallel to FD and 145.3 ± 2.3 MPa perpendicular to FD. Further increasing MA-$Si_2B_2C_3N$ to 40 vol.%, many closed pores are remained in the ceramic skeleton, leading to poor metal infiltration (Fig. 5.33).

The Young's modulus of the $Cu_{80}Ti_{20}/Si_2BC_3N$ composite ceramics increases with increasing ceramic content regardless of the loading direction. The composites with 40 vol.% MA-$Si_2B_2C_3N$ exhibit the highest Young's modulus of 94.5 ± 3.5 GPa parallel to FD and of 91.1 ± 2.2 GPa perpendicular to FD, respectively. The fracture toughness of the $Cu_{80}Ti_{20}/Si_2BC_3N$ composite ceramics increases first and then decreases with the increasing MA-Si_2BC_3N content. The composites with 35 vol.% MA-$Si_2B_2C_3N$ display optimized fracture toughness of 11.24 ± 0.31 MPa $m^{1/2}$ parallel to FD and of 8.97 ± 0.25 MPa $m^{1/2}$ perpendicular to FD. However, the composites with 30 vol.% solid content show relatively lower fracture toughness, obtaining 10.57 ± 0.32 MPa $m^{1/2}$ parallel to FD and 7.34 ± 0.47 MPa $m^{1/2}$ perpendicular to FD. The composites with 40 vol.% MA-$Si_2B_2C_3N$ exhibit the lowest fracture toughness. The debonding, pull-out and bridging of metal and/or ceramic platelets both improve the fracture toughness of the $Cu_{80}Ti_{20}/Si_2BC_3N$ composite ceramics.

(6) *Zr-Al particles addition*

The bulk density and relative density of (Zr-Al)/Si_2BC_3N composite ceramics are improved obviously comparing with the pristine MA-Si_2BC_3N ceramics [34]. However, the Vickers' hardness decreases gradually with the increasing Zr-Al content, but their Young's modulus remains almost unchanged. Expectedly, the flexural strength and fracture toughness of the (Zr-Al)/Si_2BC_3N are significantly

Table 5.18 Mechanical properties of the (Zr-Al)/Si_2BC_3N composite ceramics SPS at 1900 °C/40 MPa/5 min [34]

Composition	Flexural strength (MPa)	Fracture toughness (MPa m$^{1/2}$)	Young's modulus (GPa)	Vickers' hardness (GPa)
+ 1 mol.% (Zr-Al)	590.2 ± 28.1	4.93 ± 0.57	120.6 ± 3.6	5.8 ± 0.1
+ 3 mol.% (Zr-Al)	480.1 ± 29.1	5.42 ± 0.26	122.6 ± 3.3	5.3 ± 0.5
+ 5 mol.% (Zr-Al)	481.3 ± 28.7	5.60 ± 0.30	120.4 ± 3.8	5.2 ± 0.1

improved. In detail, the strength of 1 mol.% Zr-Al containing ceramics reaches 590 MPa, 79.3% higher than that of the pure ceramics. With 5 mol.% Zr-Al addition, the composite ceramics possesses the highest fracture toughness of 5.60 MPa m$^{1/2}$, 36.2% higher than that of the counterparts (Table 5.18).

The aforementioned results clearly suggest the reasonable Zr-Al addition leading to the effective strengthening and toughening of MA-Si_2BC_3N. The densification of the (Zr-Al)/Si_2BC_3N composite ceramics should be ascribed to the enhanced solid diffusion in oxides. The ZrO_2 and Al_2O_3 oxides both stimulate the solid diffusion of SiC and BN(C). Therefore, the (Zr-Al)/Si_2BC_3N composite ceramics with well-developed SiC and BN(C) structure showing desirable mechanical properties. Note that the nonchemical bonding between oxides and ceramic matrix may deteriorate the intimate bonding between SiC and BN(C), impairing the oxides-enhanced densification advantages. As a result, their flexural strength decreases with further increasing Zr-Al addition. In contrast, the pining effects of nano/submicron oxides increase the crack propagation paths thereby enhancing the fracture toughness.

5.3.2 *Effect of Oxide Sintering Additives*

The addition of 5 mol.% ZrO_2 or 5 mol.% AlN increases both the density and mechanical properties of the hot-pressed MA-Si_2BC_3N ceramics (Table 5.19). Hot-pressed at 1900 °C and under a pressure of 80 MPa for 30 min, the relative density, bending strength, elastic modulus, fracture toughness and Vickers' hardness of the as-prepared MA-Si_2BC_3N monoliths free from any additives is 88.7%, 331.1 MPa, 139.4 GPa, 2.81 MPa m$^{1/2}$, and 5.7 GPa, respectively [35]. However, with few AlN or ZrO_2 significantly increases the relative density to 96.9% or 97.7%, and improves the bending strength to 415.7 MPa or 575.4 MPa, respectively. Obviously, the improvement of mechanical performance is highly dependent on their increasing bulk densities.

With 10 wt.% MgO–ZrO_2–SiO_2 (MZS) or 10 wt.% MgO–$ZrSiO_4$ (MZ) sintering additive [36], the resulting composites ceramics show much better mechanical performance than the monoliths, obtained flexural strength of 470.4 MPa or 394.2 MPa, fracture toughness of 5.10 MPa m$^{1/2}$ or 5.86 MPa m$^{1/2}$, Vickers hardness of 7.3 GPa and 8.3 GPa, while the Young's modulus remains almost unchanged. (Table 5.20). In another cases 6 wt.% MgO–Al_2O_3–SiO_2 (MAS) addition, they effectively increase

Table 5.19 Bulk density and mechanical properties of hot-pressed MA-Si_2BC_3N ceramic with or without 5 mol.% ZrO_2 or 5 mol.% AlN (sintering condition: 1900 °C/80 MPa/30 min) [35]

Composition	Bulk density (g/cm^3)	Flexural strength (MPa)	Young's modulus (GPa)	Fracture toughness ($MPa\ m^{1/2}$)	Vickers' hardness (GPa)
MA-Si_2BC_3N	2.52	331.1 ± 40.5	139.4 ± 16.0	2.81 ± 0.89	5.7 ± 0.4
+ 5 mol.% ZrO_2	2.83	575.4 ± 73.7	159.2 ± 21.7	3.67 ± 0.01	6.7 ± 0.7
+ 5 mol.% AlN	2.74	415.7 ± 147.3	148.4 ± 8.3	4.08 ± 1.18	6.4 ± 1.2

Table 5.20 Mechanical properties of the hot-pressed MA-Si_2BC_3N monoliths with various oxide additives (sintering condition: 1900 °C/80 MPa/30 min) [36]

Composition	Flexural strength (MPa)	Young's modulus (GPa)	Fracture toughness ($MPa\ m^{1/2}$)	Vickers' hardness (GPa)
+ 10 wt.% (MgO–ZrO–SiO_2)	470.4 ± 71.1	134.9	5.10 ± 0.62	7.3 ± 0.5
+ 10 wt.% (MgO–$ZrSiO_4$)	394.2 ± 41.7	152.9	5.86 ± 0.86	8.3 ± 0.6
+ 6 wt.% (MgO–Al_2O_3–SiO_2)[a]	425.1 ± 11.4	–	–	6.2 ± 0.3

[a]HP sintered at 1800 ° C/60 MPa/60min

the bending strength of the hot-pressed MA-Si_2BC_3N monoliths (sintered at 1800 °C/60 MPa) to 425.1 MPa, obtained Vickers' hardness of 6.2 GPa.

5.3.3 *Effect of Ultrahigh Temperature Ceramic Components*

(1) *ZrC particles addition*

The flexural strength of the as-sintered ZrC/Si_2BC_3N composite ceramics is in a range of 150–230 MPa, while the Young's modulus is 113–142 GPa [37]. The Vickers' hardness increases slightly with the increases of ZrC content. The bulk ceramics with 10 wt.% ZrC have the highest fracture toughness ~ $2.25 \pm 0.37\ MPa\ m^{1/2}$ (Fig. 5.34). Generally, the as-sintered ZrC/Si_2BC_3N composite ceramics do not show desirable mechanical performance due to their low relative densities. It thus suggests that the introduction of sol–gel derived ZrC can strengthen and toughen the ceramic matrix only when higher hot-pressing temperature or pressure is adopted. Besides, homogenous distribution of ZrC and well-development of each phase in ceramic matrix are also required (Table 5.21).

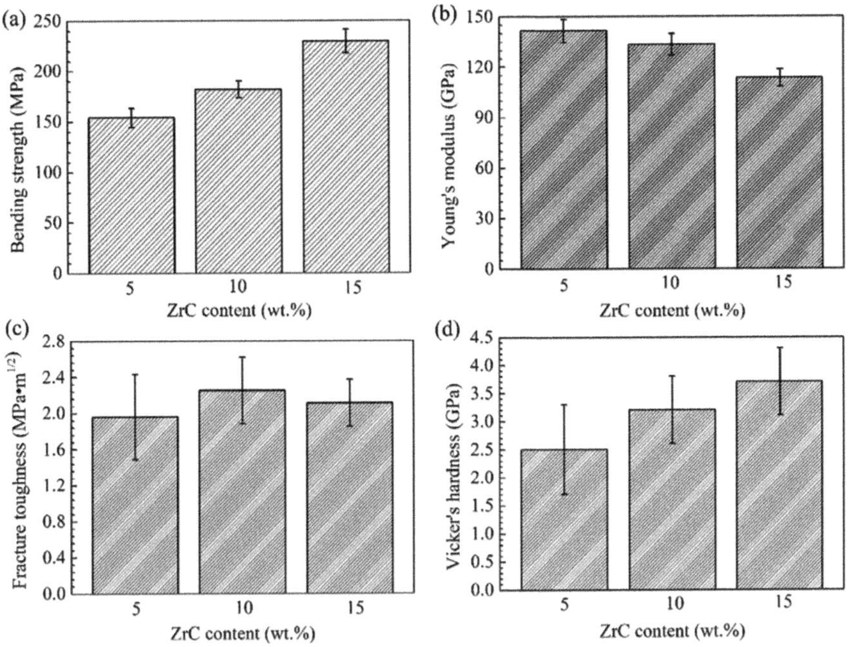

Fig. 5.34 Mechanical properties of the hot-pressed ZrC/Si_2BC_3N composite ceramics with different ZrC content (sintering condition: 1900 °C/60 MPa/60 min): **a** Flexural strength; **b** Young's modulus; **c** Fracture toughness; **d** Vickers' hardness [37]

Table 5.21 Mechanical properties of the various hot-pressed ZrC/Si_2BC_3N composite ceramics (sintering condition: 1900 °C/60 MPa/60 min) [37]

Composition	Flexural strength (MPa)	Fracture toughness (MPa $m^{1/2}$)	Young's modulus (GPa)	Vickers' hardness (GPa)
+ 5 wt.% ZrC	153.9 ± 9.6	1.96 ± 0.46	141.7 ± 6.8	2.5 ± 0.8
+ 10 wt.% ZrC	181.8 ± 8.3	2.25 ± 0.37	133.4 ± 6.4	3.2 ± 0.6
+ 15 wt.% ZrC	229.5 ± 11.6	2.11 ± 0.26	113.4 ± 5.2	3.4 ± 0.6

(2) *TaC particles addition*

The introduction of MA derived TaC nanocrystals strongly improves the relative density and of the TaC/Si_2BC_3N composite ceramics (Fig. 5.35). Besides, nano TaC also effectively promotes their flexural strength and fracture toughness [38]. The TaC/Si_2BC_3N composite ceramics with 10 wt.% TaC addition possess the best mechanical performance with flexural strength and fracture toughness reaching to 399.5 MPa and 4.26 MPa $m^{1/2}$, respectively, which are higher than 127.9 MPa and 1.82 MPa $m^{1/2}$ of the pristine MA-Si_2BC_3N (Table 5.22).

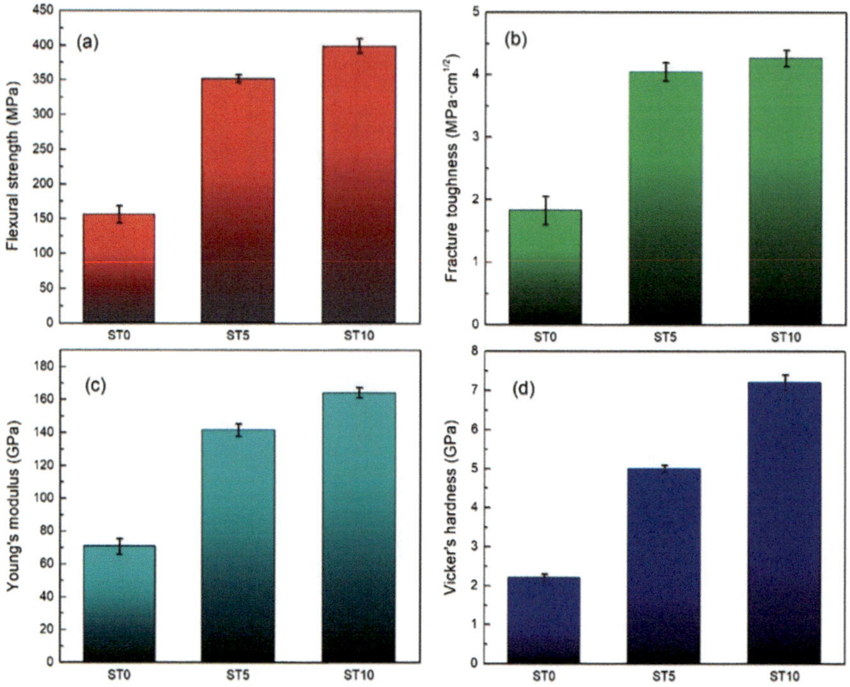

Fig. 5.35 Mechanical properties of various TaC/Si_2BC_3N composite ceramics hot-pressed at 1900 °C/60 MPa/60 min: **a** Flexural strength; **b** Fracture toughness; **c** Young's Modulus; **d** Vickers' hardness [38]

Table 5.22 Bulk densities and mechanical properties of the hot-pressed TaC/Si_2BC_3N composite ceramics (sintering condition: 1900 °C/60 MPa/60 min) [38]

Composition	Bulk density (g/cm^3)	Relative density (%)	Flexural strength (MPa)	Fracture toughness (MPa m$^{1/2}$)	Young's modulus (GPa)	Vickers' hardness (GPa)
MA-Si_2BC_3N	2.24	78.9	156.1 ± 12.6	1.82 ± 0.23	70.6 ± 4.7	2.2 ± 0.1
+ 5 wt.% TaC	2.61	88.2	351.3 ± 5.5	4.04 ± 0.15	141.4 ± 3.8	5.0 ± 0.1
+ 10 wt.% TaC	2.79	90.4	339.5 ± 10.3	4.26 ± 0.14	161.1 ± 3.1	7.2 ± 0.2

The MA derived TaC dispersed uniformly within MA-Si_2BC_3N matrix in the form of nanocrystals, acts as a critical role in grain refinement contributing to densification For example, the ceramic particles are gradually refined with increasing nano TaC addition. Moreover, the crack growth of the 10 wt.% TaC containing ceramics is more tortuous (Fig. 5.36). The underlying reasons for strengthening and toughening of nano TaC are the CET mismatch of each phase and the generation of residual stress field around the second reinforced phases with a fact that not possible reactions take

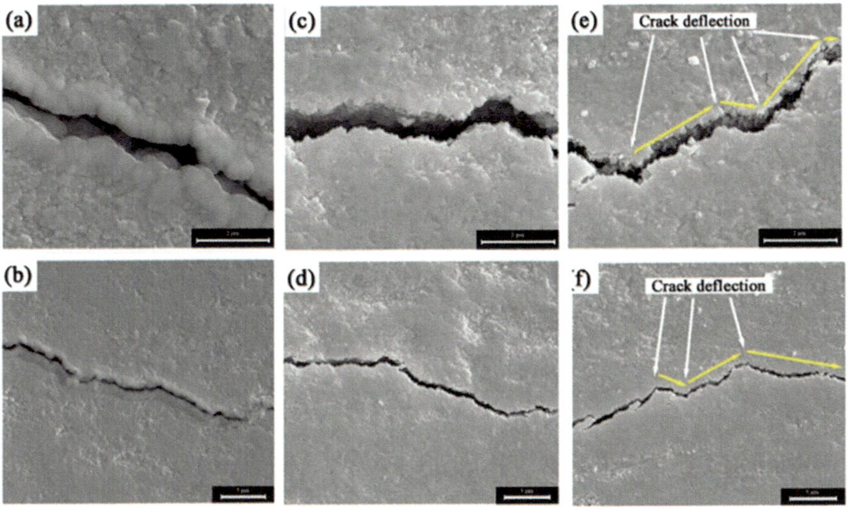

Fig. 5.36 The crack propagation morphologies of the hot-pressed TaC/Si_2BC_3N composite ceramics: **a**, **b** MA-Si_2BC_3N; **c**, **d** + 5 wt.% TaC; **e**, **f** + 10 wt.% TaC [38]

place among the components. Without doubt, TaC has a larger CET than that of MA-Si_2BC_3N matrix. Accordingly, the TaC particles endures tensile stress, while the MA-Si_2BC_3N matrix is subjected to tensile stress in radial direction and subjected to compressive stress in tangential direction. The cracks thus tend to bypass the TaC particles and continue to propagate, which can absorb extra energy to achieve the purpose of toughening.

(3) *Ta_4HfC_5 particles addition*

The introduction of MA derived Ta_4HfC_5 nanocrystals has effectively improved the flexural strength and fracture toughness of the MA-Si_2BC_3N monoliths (Fig. 5.37). For pure MA-Si_2BC_3N, they only show flexural strength of 156.1 MPa and fracture toughness of 1.82 MPa $m^{1/2}$ [39]. With 10 wt.% nano Ta_4HfC_5 addition, the composite ceramics possess optimized flexural strength 344.1 MPa, while composite ceramics with 5 wt.% Ta_4HfC_5 obtain highest fracture toughness of 4.52 MPa $m^{1/2}$. The satisfying mechanical properties of the as-sintered Ta_4HfC_5/Si_2BC_3N composite ceramics should be mainly argued to the uniform distribution of nanosized Ta_4HfC_5 and the reasonable phase composition (Table 5.23).

(4) *TiB_2-TiC particles addition*

The 10 vol.% TiB_2/Si_2BC_3N composite ceramics hot-pressed at 1900 °C/40 MPa/30 min exhibit the optimum comprehensive mechanical properties [40], with flexural strength of 311.2 ± 59.0 MPa, fracture toughness of 3.92 ± 0.39 MPa $m^{1/2}$, Young's modulus of 141.3 ± 8.7 GPa and Vickers hardness of 7.0 ± 0.2 GPa, respectively, while those of pure MA-Si_2BC_3N prepared under the same condition are 111.8 ±

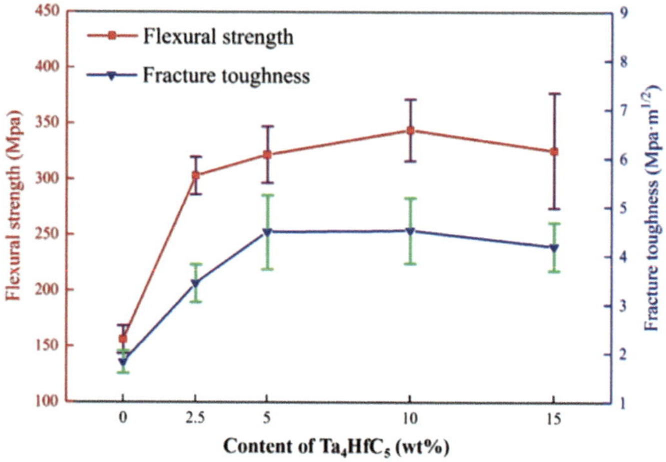

Fig. 5.37 The flexural strength and fracture toughness of the as-sintered Ta_4HfC_5/Si_2BC_3N composite ceramics hot-pressed at 1900 °C/60 MPa/60 min in vacuum [39]

Table 5.23 The calculated relative contents of each phase after complete crystallization of the as-sintered Ta_4HfC_5/Si_2BC_3N composite ceramics [39]

Composition	Ta_4HfC_5 (wt.%)	SiC (wt.%)	BN(C) (wt.%)
MA-Si_2BC_3N	0.0	52.1	47.9
+ 2.5 wt.% Ta_4HfC_5	2.5	50.8	46.7
+ 5.0 wt.% Ta_4HfC_5	5.0	49.5	45.5
+ 10.0 wt.% Ta_4HfC_5	10.0	46.9	43.1
+ 15.0 wt.% Ta_4HfC_5	15.0	44.3	40.7

7.9 MPa, 1.03 ± 0.23 MPa $m^{1/2}$, 58.0 ± 2.2 GPa and 2.1 ± 0.1 GPa, respectively (Fig. 5.38 and Table 5.24). Generally, the rod-like TiB in MA-Si_2BC_3N ceramic matrix show best toughening effect. Additionally, introduction of both TiB_2 and TiC also show better fracture toughness than that of pure TiC reinforcement. Except for the reinforcement types, the increases in sintering temperature can effectively improve their mechanical properties, showing optimal flexural strength of 319.2 ± 5.2 MPa, fracture toughness of 4.24 ± 0.32 MPa $m^{1/2}$, Young's modulus of 170.0 ± 2.8 GPa and Vickers hardness of 4.9 ± 0.1 GPa, respectively (Table 5.25).

The crack propagation path in pure MA-Si_2BC_3N ceramics is twisted with limited crack deflection. For 10 vol.% TiB_2 or 10 vol.% TiB containing ceramics, substantial crack branching is observed. In cases of the 10 vol.% TaC or 10 vol.% (TiB_2–TiC) addition, fracture behavior is mainly controlled by the crack deflection and crack propagation along the second phases. Besides, the extraction of the layered structure and the generation of a large number of fresh surfaces also cause significant resistance to crack propagation. The transgranular fracture and crack pinning are also the major factors of the improvements in mechanical properties (Fig. 5.39).

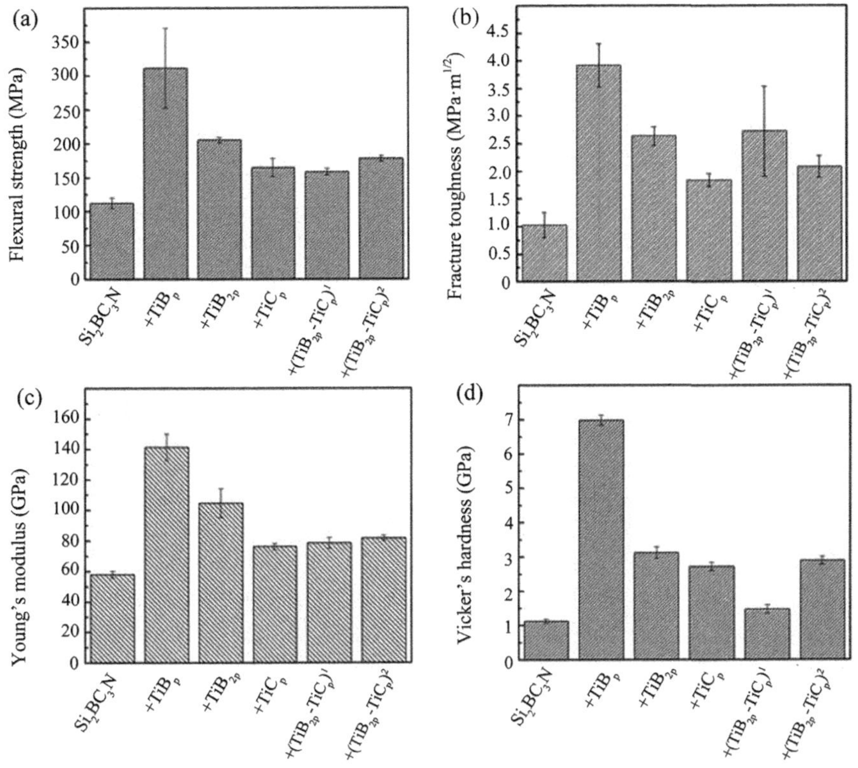

Fig. 5.38 Mechanical properties of (TiB_2–TiC)/Si_2BC_3N composite ceramics hot-pressed at 1900 °C/40 MPa/30 min: **a** Flexural strength; **b** Young's modulus; **c** Fracture toughness; **d** Vickers' hardness ([1] TiB_2/TiC mole ratio of 2: 1; [2] TiB_2/TiC mole ratio of 1: 2) [40]

Table 5.24 Mechanical properties of 10 vol.% (TiB_2–TiC)/Si_2BC_3N composite ceramics hot-pressed at 1900 °C/40 MPa/30 min [40]

Composition	Flexural strength (MPa)	Fracture toughness (MPa $m^{1/2}$)	Young's modulus (GPa)	Vickers' hardness (GPa)
MA-Si_2BC_3N	111.8 ± 7.9	1.03 ± 0.23	58.0 ± 2.2	2.1 ± 0.1
+ 10 vol.% TiB	311.2 ± 59.0	3.92 ± 0.39	141.3 ± 8.7	7.0 ± 0.2
+ 10 vol.% TiB_2	205.2 ± 4.2	2.64 ± 0.17	104.6 ± 9.5	3.1 ± 0.2
+ 10 vol.% TiC	164.9 ± 13.2	1.85 ± 0.12	76.1 ± 2.1	2.7 ± 0.1
+ 10 vol.% (TiB_2–TiC)[a]	158.2 ± 4.9	2.73 ± 0.81	78.5 ± 3.6	1.5 ± 0.1
+ 10 vol.% (TiB_2–TiC)[b]	177.9 ± 4.4	2.09 ± 0.20	81.6 ± 1.7	2.9 ± 0.1

[a] TiB_2/TiC mole ratio of 2: 1; [b] TiB_2/TiC mole ratio of 1: 2

Table 5.25 Mechanical properties of 10 vol.% (TiB_2–TiC)/Si_2BC_3N composite ceramics (TiB_2/TiC mole ratio of 2: 1) hot-pressed at 1900–2000 °C/80 MPa/30 min [40]

Hot-pressing parameters (80 MPa/30 min)	Flexural strength (MPa)	Fracture toughness (MPa $m^{1/2}$)	Young's modulus (GPa)	Vickers' hardness (GPa)
1900 °C	290.4 ± 2.8	3.57 ± 0.12	164.2 ± 3.9	4.8 ± 0.1
2000 °C	319.2 ± 5.2	4.24 ± 0.32	170.0 ± 2.8	4.9 ± 0.1

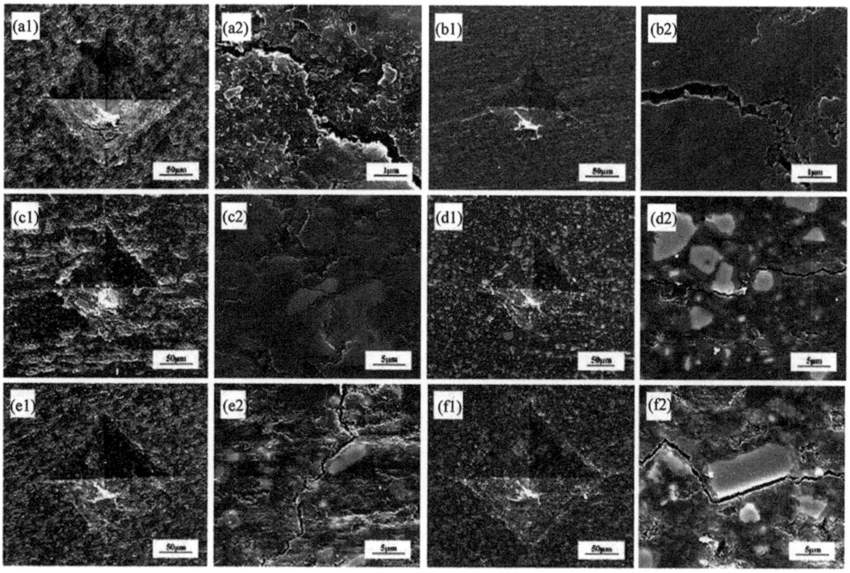

Fig. 5.39 Indentation crack of the various (TiB_2–TiC)/Si_2BC_3N composite ceramics: **a** MA-Si_2BC_3N; **b** + 10 vol.% TiB; **c** + 10 vol.% TiB_2; **d** + 10 vol.% TiC; **e** 10 vol.% (TiB_2–TiC) (TiB_2/TiC mole ratio of 2: 1); **f** + 10 vol.% (TiB_2–TiC) (TiB_2/TiC mole ratio of 1: 2) [40]

(5) *ZrB_2 particles addition*

The ZrO_2 precursor or ZrO_2, C and B_2O_3 precursor are introduced in MA-Si_2BC_3N ceramic matrix by sol–gel method, and subsequently SPS consolidated at 2000 °C/60 MPa/5 min to *in-situ* generate ZrB_2/Si_2BC_3N composite ceramics [41, 42]. Regardless of the precursor used, the flexural strength, fracture toughness and Vickers' hardness of the as-sintered ZrB_2/Si_2BC_3N are significantly improved as compared to pure MA-Si_2BC_3N. Actually, the ZrB_2/Si_2BC_3N composite ceramics derived from ZrO_2 precursor have the optimal comprehensive mechanical properties. The Young's modulus of two ZrB_2/Si_2BC_3N composite ceramics is similar, and their Vickers' hardness and flexural strength increase with the increasing *in-situ* ZrB_2 content.

For ZrO_2 precursor derived ZrB_2 reinforced MA-Si_2BC_3N ceramics, the fracture toughness improves owing to the well-developed laminated BN(C) [41]. Their fracture toughness, bending strength and Young's modulus are 4.9 ± 0.2 MPa $m^{1/2}$, 411 ±

16 MPa and 320 ± 16 GPa, respectively (Table 5.26). For ZrO_2, C and B_2O_3 precursor derived ZrB_2/Si_2BC_3N composite ceramics, their bulk density and relative density increase with the ZrB_2 content [42]. Besides, the flexural strength, fracture toughness, elastic modulus and Vickers' hardness also increase with the increasing ZrB_2. The measured flexural strength, fracture toughness, Young's modulus and Vicker's hardness of 20 wt.% ZrB_2 containing MA-Si_2BC_3N ceramics reach 351 ± 18 MPa, 4.5 ± 0.2 MPa $m^{1/2}$, 172 ± 8 GPa and 7.2 ± 0.2 GPa, respectively (Table 5.27).

The crack paths in the 15 wt.% ZrB_2 incorporating ceramics appear tortuous. Inspection of crack propagation further reveals crack deflection and crack bridging as a result from the ZrB_2 particles. These toughening mechanisms increase energy dissipation during crack propagation leading to a higher fracture toughness. The *in-situ* generated ZrB_2 clearly contributes to the ceramic matrix densification, thus pullout and bridging of lamellar BN(C) are found (Fig. 5.40). However, the *in-situ* generated ZrB_2 also favors the grain growth of each phase within the ceramic matrix.

The MA derived nanosized ZrB_2 reinforced MA-Si_2BC_3N ceramics also have high bulk density and relative density [43]. Compared with the formers, positive effects on the densification with direct mechanical alloyed nano ZrB_2 addition are found (Table 5.28). The SPS densified ZrB_2/Si_2BC_3N composite ceramics show

Table 5.26 Bulk density and mechanical properties of the ZrB_2/Si_2BC_3N composite ceramics prepared from ZrO_2 precursor (SPS condition: 2000 °C/60 MPa/5 min) [41]

Composition	Bulk density (g/cm^3)	Flexural strength (MPa)	Fracture toughness (MPa $m^{1/2}$)	Young's modulus (GPa)	Vickers' hardness (GPa)
MA-Si_2BC_3N	2.67	125 ± 16	3.5 ± 0.1	133 ± 6	3.1 ± 0.2
+ 5 wt.% ZrB_2	2.82	315 ± 18	4.6 ± 0.2	151 ± 8	5.2 ± 0.2
+ 10 wt.% ZrB_2	3.08	358 ± 15	4.7 ± 0.2	160 ± 6	6.0 ± 0.1
+ 15 wt.% ZrB_2	3.24	362 ± 16	5.0 ± 0.1	173 ± 5	7.5 ± 0.1
+ 20 wt.% ZrB_2	3.53	411 ± 16	5.1 ± 0.1	181 ± 6	7.8 ± 0.2

Table 5.27 Bulk density and mechanical properties of SPS ZrB_2/Si_2BC_3N composite ceramics prepared from ZrO_2, C and B_2O_3 precursor (SPS condition: 2000 °C/60 MPa/5 min) [42]

Composition	Bulk density (g/cm^3)	Relative density (%)	Flexural strength (MPa)	Fracture toughness (MPa $m^{1/2}$)	Young's modulus (GPa)	Vickers' hardness (GPa)
MA-Si_2BC_3N	2.57	94	211 ± 16	3.5 ± 0.1	133 ± 6	3.1 ± 0.2
+ 5 wt.% ZrB_2	2.77	96	236 ± 16	3.7 ± 0.1	144 ± 4	4.3 ± 0.2
+ 10 wt.% ZrB_2	2.94	97	291 ± 20	3.9 ± 0.2	155 ± 6	5.0 ± 0.1
+ 15 wt.% ZrB_2	3.12	98	338 ± 15	4.6 ± 0.3	160 ± 5	6.5 ± 0.1
+ 20 wt.% ZrB_2	3.22	96	351 ± 18	4.5 ± 0.2	172 ± 8	7.2 ± 0.2

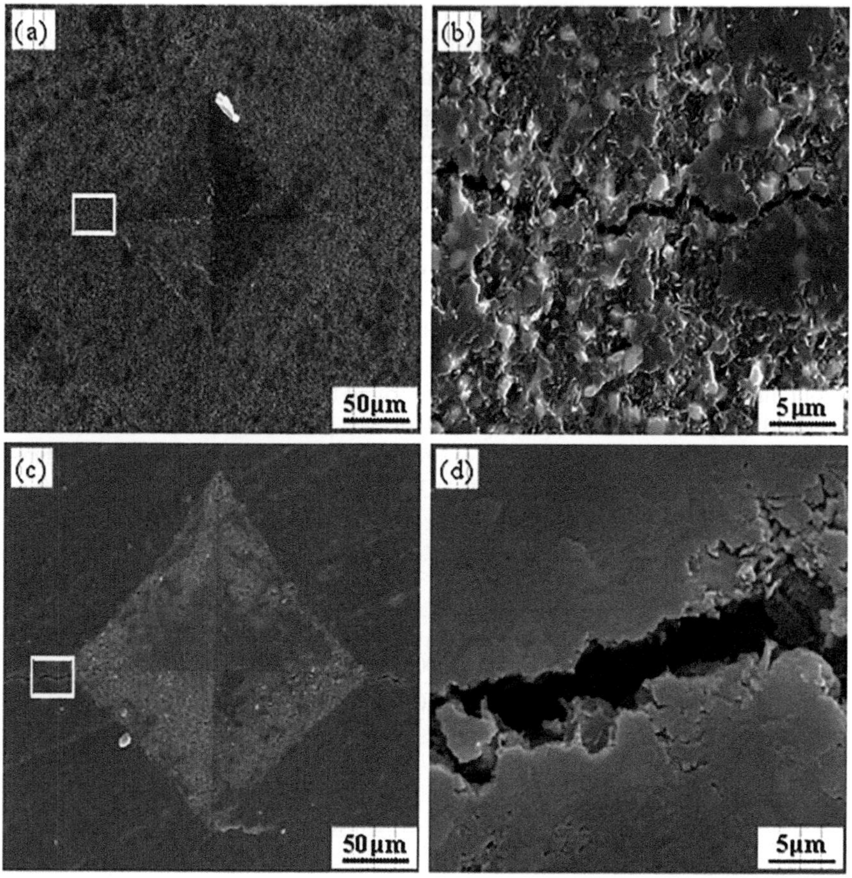

Fig. 5.40 Vickers' indentation morphologies of the as-sintered ZrB_2/Si_2BC_3N composite ceramics: **a**, **b** Using ZrO_2 precursor as the raw materials; **c**, **d** Using ZrO_2, C and B_2O_3 precursor as the raw materials [41, 42]

much higher strength and toughness than those of pristine MA-Si_2BC_3N. Also, their Young's modulus increases with the incorporation of nano ZrB_2. Especially, the addition of nano ZrB_2 greatly improves their fracture toughness compared with the sol–gel derived ZrB_2/Si_2BC_3N. In detail, the flexural strength of 10 wt.% nano ZrB_2 containing ceramics reaches 559.6 MPa, 61% higher than that (347.5 MPa) of 20 wt.% nano ZrB_2 containing ones. However, the latter possesses the highest fracture toughness at 6.77 MPa $m^{1/2}$, which is 32.7% higher than 5.1 MPa $m^{1/2}$ of the former.

After an indentation test, apparently, the 20 wt.% ZrB_2 containing ceramics present a relatively twisted crack propagation path. Namely, the cracks propagate alongside the interfaces between SiC and BN(C), with observed "pull-out" of BN(C) platelets, "crack bridging" and "crack deflection" (Fig. 5.41).

Table 5.28 Densities and mechanical properties of MA derived nanosized ZrB_2 reinforced MA-Si_2BC_3N composite ceramics (SPS condition: 2000 °C/60 MPa/5 min) [43]

Composition	Bulk density (g/cm^3)	Relative density (%)	Flexural strength (MPa)	Young's modulus (GPa)	Fracture toughness (MPa $m^{1/2}$)	Vickers' hardness (GPa)
+ 10 wt.% nano ZrB_2	2.78	93.2	559.6 ± 30.5	163.6 ± 2.2	6.71 ± 0.14	5.6 ± 0.09
+ 20 wt.% nano ZrB_2	2.99	93.4	512.3 ± 7.2	178.9 ± 4.1	6.77 ± 0.17	5.2 ± 0.10

Fig. 5.41 Micro-cracks propagations in the as-sintered ceramics after an indentation test: **a** Pure MA-Si_2BC_3N; **b** + 10 wt.% ZrB_2 [43]

(6) *HfB_2 particles addition*

SMA derived nano HfB_2 into MA-Si_2BC_3N matrix and subsequently hot-pressing sintered at 1900 °C/60 MPa/5 min to densify the HfB_2/Si_2BC_3N composite ceramics [44]. With increasing HfB_2 content (≤10 vol.%), the flexural strength, fracture toughness, Vickers' hardness and Young's modulus of the as-sintered HfB_2/Si2BC_3N composite ceramics is significantly improved (Fig. 5.42). Since the hardness of HfB_2 and HfC is higher than those of SiC and BN(C). Thus, the composite ceramics with higher HfB_2 and HfC display increasing Vickers' hardness. The composite ceramics with 30 vol.% HfB_2 exhibit the highest flexural strength of 176.1 MPa, fracture toughness of 4.17 MPa $m^{1/2}$, Vickers' hardness of 7.2 GPa and Young's modulus of 270.1 GPa, respectively (Table 5.29).

The crack in the pure MA-Si_2BC_3N propagates relatively straight without significant crack deflection. After introducing some nano HfB_2, the crack propagation path is more tortuous. The extraction of some layered BN(C) results in some fresh surfaces, causing significant resistance to crack propagation. The non-penetrating crack is essentially different from the penetrating ones when encountering reinforced phases, but both can help develop tortuous routes of crack propagation and the

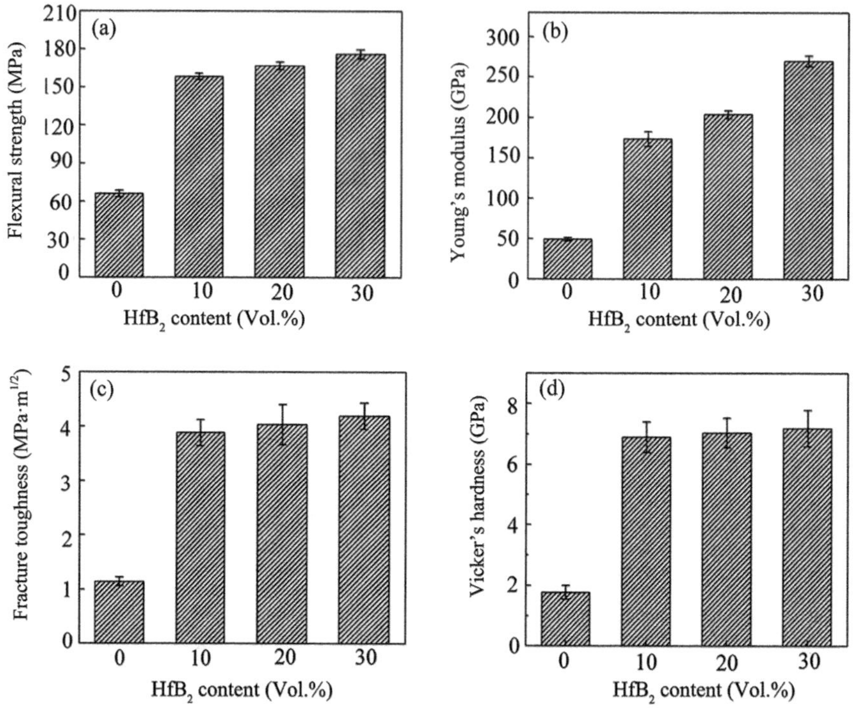

Fig. 5.42 Mechanical properties of the HfB_2/Si_2BC_3N composite ceramics hot-pressed under 1900 °C/60 MPa/30 min: **a** Flexural strength; **b** Young's modulus; **c** Fracture toughness; **d** Vickers' hardness [44]

Table 5.29 Mechanical properties of the HfB_2/Si_2BC_3N composite ceramics hot-pressed under 1900 °C/60 MPa/30 min [44]

Composition	Flexural strength (MPa)	Fracture toughness (MPa m$^{1/2}$)	Young's modulus (GPa)	Vickers' hardness (GPa)
MA-Si_2BC_3N	66.0 ± 2.6	1.14 ± 0.08	49.1 ± 2.2	1.8 ± 0.2
+ 10 vol.% nano HfB_2	158.2 ± 2.5	3.38 ± 0.18	173.5 ± 9.6	6.9 ± 0.5
+ 20 vol.% nano HfB_2	166.8 ± 3.1	4.03 ± 0.37	203.8 ± 5.4	7.0 ± 0.5
+ 30 vol.% nano HfB_2	176.1 ± 3.7	4.17 ± 0.24	270.1 ± 7.6	7.2 ± 0.6

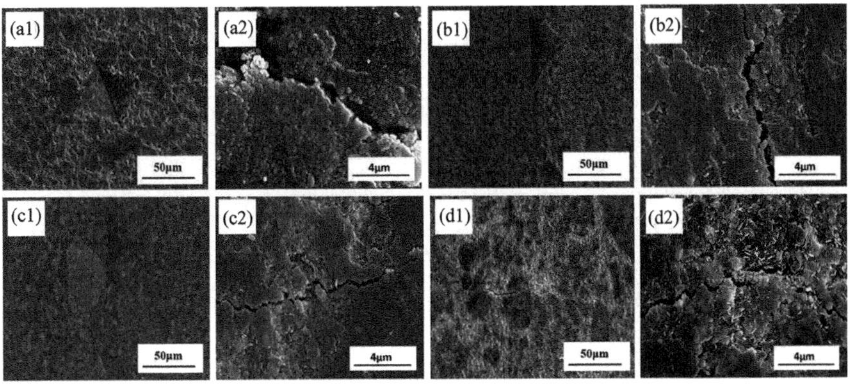

Fig. 5.43 Vickers' indentation morphologies of the various HfB_2/Si_2BC_3N composite ceramics: **a** Pure MA-Si_2BC_3N; **b** + 10 vol.% HfB_2; **c** + 20 vol.% HfB_2; **d** + 30 vol.% HfB_2 [44]

branching, deflecting and bridging. These toughening mechanisms jointly improve the fracture toughness of the HfB_{2p}/Si_2BC_3N composite ceramics (Fig. 5.43).

(7) *LaB_6 particles addition*

The LaB_6/Si_2BC_3N composite ceramics prepared by one-step ball-milling and subsequently SPS densified at 1800–1900 °C show high relative densities and desirable mechanical properties (Table 5.30). The relative densities show little change with the temperature increasing from 1800 to 1900 °C. However, higher sintering temperature drives significantly the grain growth. As a result, the subsequent coarsening of significant lamellar BN(C) and SiC leads to decrease in Vickers' hardness, flexural strength and elastic modulus [45, 46]. Additionally, the LaB_6/Si_2BC_3N composite ceramics sintered at 1800 °C show a relatively low toughness 2.53 MPa $m^{1/2}$ with a relatively straight crack propagation. However, the counterparts obtained at 1900 °C provide high toughness 7.87 MPa $m^{1/2}$. This can be attributed to the solid-solution of La in BN(C) grains promotes ordered arrangement of BN(C) along the (0002) plane causing *in-situ* growth of BN(C) thus leading to an increase in toughness by 40% (from 5.6 to 7.8 MPa $m^{1/2}$) (Fig. 5.44).

The mechanical properties of the LaB_6/Si_2BC_3N composite ceramics prepared by a combination of two-step ball-milling and SPS technique are listed in Table 5.31. As the sintering temperature increases, their bulk densities increase leading to the higher flexural strength, Young's modulus, and Vickers' hardness. After SPS densified at 1800 °C, the development of lamellar BN(C) structures is insufficient, obtained fracture toughness of only ~ 3.01 MPa $m^{1/2}$. At high-temperatures, the well-developed layered BN(C) phase leads to the more tortuous crack propagation path, thus effectively improving the fracture toughness. The same toughening mechanisms are frequently observed during facture, including grain pull-out, crack bridging and crack deflection comparable to the former (Fig. 5.45).

Table 5.30 Relative density and mechanical properties of the SPS densified LaB_6/Si_2BC_3N composite ceramics (sintering condition: 1800–1900 °C/50 MPa under 1 bar N_2) [45]

Properties	LaB_6/Si_2BC_3N		MA-Si_2BC_3N
	Sintered at 1800 °C	Sintered at 1900 °C	Sintered at 1900 °C
Relative density (%)	96.9 ± 0.5	97.2 ± 0.5	89.0 ± 0.5
Vickers' hardness (GPa)	5.69 ± 0.1	4.8 ± 0.2	4.8 ± 0.2
Flexural strength (MPa)	340.2 ± 30.8	273.0 ± 21.6	207.9 ± 15.0
Young's modulus (GPa)	182.3 ± 20.6	113.3 ± 10.8	102.3 ± 9.3
Fracture toughness (MPa $m^{1/2}$)	2.53 ± 0.03	7.87 ± 0.24	5.62 ± 0.18

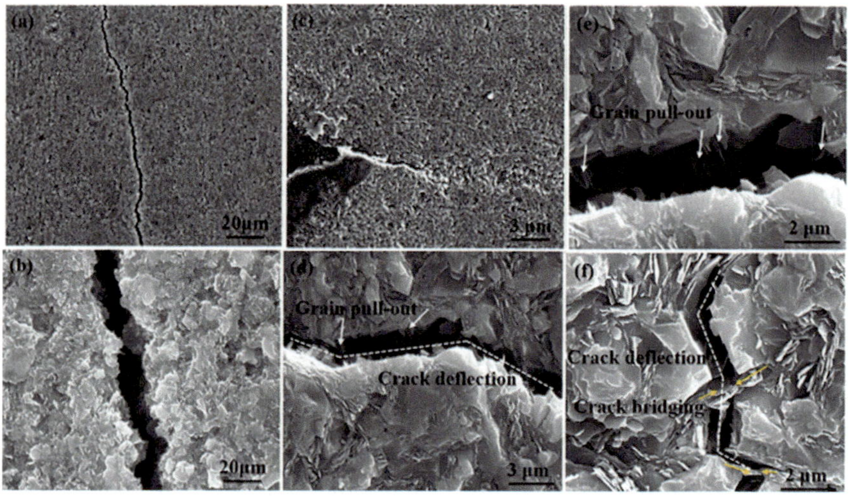

Fig. 5.44 SEM images for the indentation induced crack propagation in the as-sintered LaB_6/Si_2BC_3N composite ceramics: Sintered at **a** and **b** 1800 °C; **c–f** 1900 °C [45]

Table 5.31 Mechanical properties of LaB_6/Si_2BC_3N composite ceramics prepared by two-step ball milling and subsequently SPS consolidated under 1800–2000 °C/50 MPa [25]

Sintering temperature (°C)	Flexural strength (MPa)	Young's modulus (GPa)	Fracture toughness (MPa $m^{1/2}$)	Vickers' hardness (GPa)
1800	311.2 ± 19.5	155.4 ± 5.6	3.01 ± 0.26	6.1 ± 0.3
1900	372.4 ± 14.7	165.8 ± 17.7	4.23 ± 0.08	6.4 ± 0.2
2000	390.6 ± 27.1	184.7 ± 23.0	4.64 ± 0.19	6.6 ± 0.3

For LaB_6/Si_2BC_3N composite ceramics sintered at 2000 °C/50 MPa, the CET increases to 5.4×10^{-6} °C^{-1} with the increasing temperature to 1200 °C, and then remains almost unchanged to 1400 °C. The average value in range of 200–1400 °C is 4.2×10^{-6} °C^{-1}. The reported average value of the SPS densified MA-Si_2BC_3N is

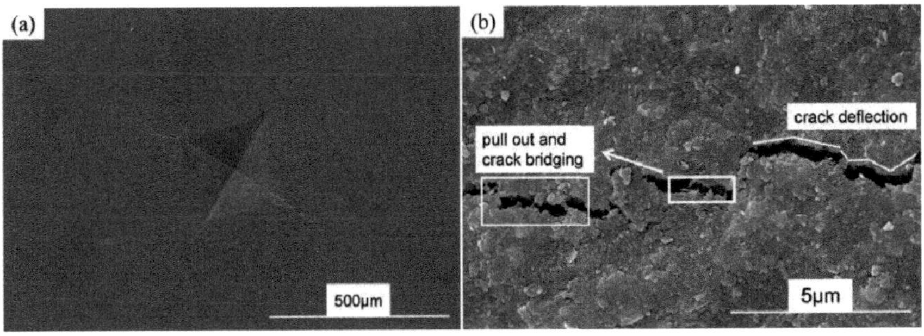

Fig. 5.45 SEM surface morphologies of the 15 wt.% LaB_6/Si_2BC_3N composite ceramics by two-step milling: **a** Indentation; **b** Microcracks deflection and bridging [37]

3.8×10^{-6} °C^{-1} [22], lower than 4.5×10^{-6} °C^{-1} [3] of the hot-pressed counterparts. However, LaB_6 results in the *in-situ* growth of lamellar BN(C) which probably has a higher CTE than that of La-free BN(C) with turbostratic structures thereby causing a high CET of LaB_6/Si_2BC_3N composite ceramics.

5.4 General Mechanical Behavior and Toughening of MA-SiBCN CMCs

5.4.1 C_f/Si_2BC_3N CMCs

(1) *C_f/Si_2BC_3N composites with PyC coating*

The 20 vol.% C_f/Si_2BC_3N CMCs hot-pressed at 1900–2000 °C/60 MPa presents pseudoplastic behavior: after elastic linear deformation, there are jagged stress steps similar to metal yield, due to the gradual pull-out and fracture of fiber bundles [47]. For C_f/Si_2BC_3N composites sintered at 1800 °C, fracture process has no obvious linear elastic deformation and yield point, showing typical pseudo-plasticity deformation behavior. The smooth curve without jagged platform indicates that too weak interface leads to little resistance of fiber debonding during fracture (Fig. 5.46).

With increasing sintering temperature (1800–2000 °C), the bending strength of the C_f/Si_2BC_3N composites is significantly improved (30.4–70.5 MPa), but the deformation amount is gradually reduced (0.22–0.075). The main reason is that higher sintering temperature activates more sintering mechanisms to significantly reduce defects and improve matrix strength. Besides, with increasing sintering temperature (1800–2000 °C), fiber coating bonded well with matrix causing higher interfacial strength. Due to higher elastic modulus of carbon fiber (~210 GPa) than that of MA-Si_2BC_3N matrix (~150 GPa), carbon fibers are loaded prior to the matrix, so stronger interfacial bonding is conducive to more loading of the fiber. Compared with MA-Si_2BC_3N monoliths hot-pressed under the same condition, the bending

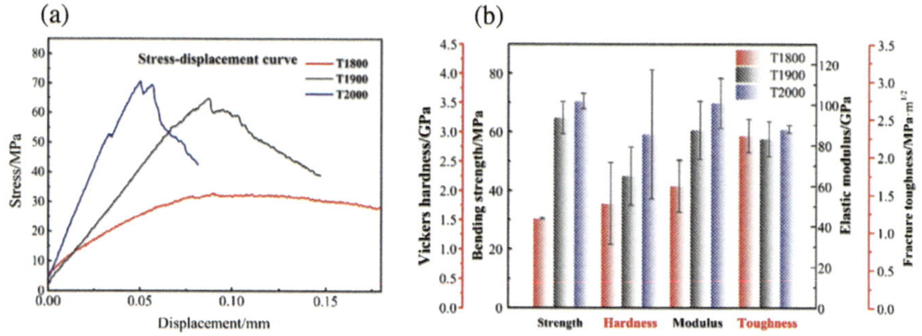

Fig. 5.46 Mechanical performance of the 20 vol.% C_f/Si_2BC_3N CMCs (fibers are coated with PyC) hot-pressed at 1800–2000 °C/60 MPa: **a** Bending stress-displacement curve; **b** Comprehensive mechanical properties [47]

strength of C_f/Si_2BC_3N CMCs is relatively lower, due to the fiber bridging, poor sintering property, and thermal stress release all leads to substantial defects such as micro-cracks and pores in the composites. However, the composites are not sensitive to the defects and therefore present pseudoplastic behavior during fracture.

In detail, the comprehensive mechanical properties of C_f/Si_2BC_3N composites are significantly improved with increasing sintering temperature (1800–2000 °C): bending strength from 30.4 to 70.5 MPa, Vickers' hardness from 0.91 to 2.25 GPa, and modulus from 20.3 to 41.6 GPa, while the change of fracture toughness is not obvious (2.24–2.38 MPa·$m^{1/2}$). The improvement of Vickers' hardness and elastic modulus is mainly due to the reduction of defects in the matrix, and the unchanged fracture toughness is caused by the offset effect of two factors. On one hand, increase of sintering temperature leads to higher matrix strength and stronger interface bonding, so greater external stress is needed for debonding and pull-out of fibers and fracture of matrix, resulting in the improvement of fracture toughness. On the other hand, with increasing sintering temperature, stronger interfacial bonding decreases the number and length of fiber pull-out, resulting in decrease of fracture toughness. Actually, fiber coating (PyC) has a more positive effect on fracture toughness of composites. Fracture toughness of the C_f/Si_2BC_3N composites without fiber coating is only about 1.47 MPa·$m^{1/2}$. This should be argued to the serious damage of fibers and short length pull-out of fibers leading to less fracture toughness.

The CET of the 20 vol.% C_f/Si_2BC_3N composites with PyC coating increases with the increasing hot-pressing temperature. The composites hot-pressed at 1800 °C has the lowest thermal expansion coefficient (CET) of 1.60×10^{-6} °C^{-1} at a test temperature of 1200 °C (Fig. 5.47). The CET of the C_f/Si_2BC_3N composites HP sintered at 1800 °C, 1900 °C and 2000 °C, all increases with testing temperature [(0.43–1.6) $\times 10^{-6}$ °C^{-1}, (1.60–3.18) $\times 10^{-6}$ °C^{-1}, (1.33–3.74) $\times 10^{-6}$ °C^{-1})] and shows a similar increasing trend.

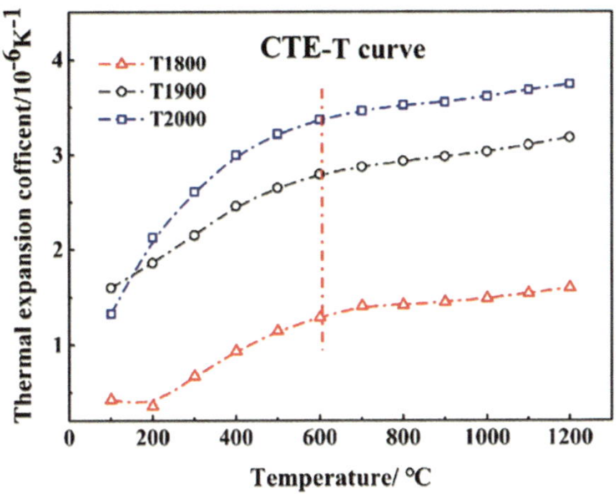

Fig. 5.47 Thermal expansion coefficient as a function of testing temperature for the 20 vol.% C_f/Si_2BC_3N composites (fibers are coated with PyC) hot-pressed at different temperature [47]

(2) *C_f/Si_2BC_3N composites with ZrO_2 additive*

For C_f/Si_2BC_3N CMCs free of any sintering additive, considerable micropores are retained in the ceramic matrix limiting their property improvement [48]. Besides, most of the aggregates and nano particles still show near spherical shape. That is, the amorphous MA-Si_2BC_3N powder particles are difficult to be deformed and densified. To reduce damage to carbon fibers, a relatively low pressure (40 MPa) and temperature (1800 °C) are required (Fig. 5.48).

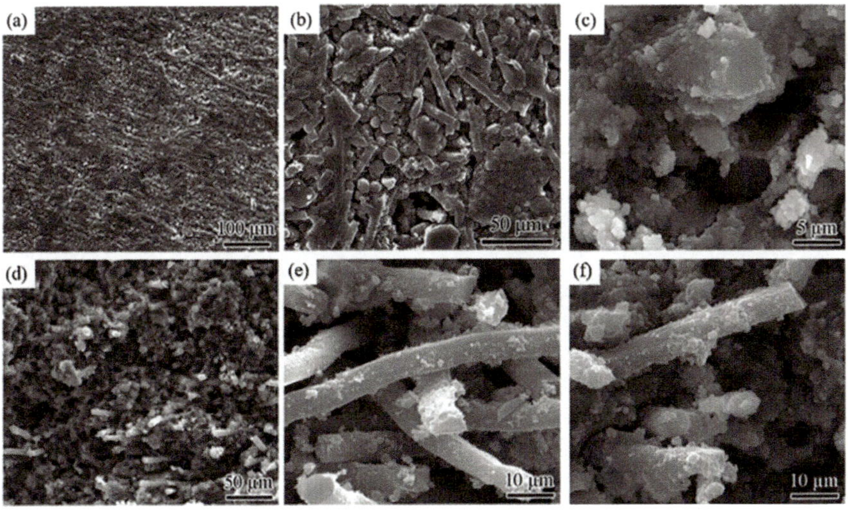

Fig. 5.48 Surface and fracture morphologies of the additive-free 20 vol.% C_f/Si_2BC_3N CMCs hot-pressed at 1800 °C/40 MPa: **a–c** Surface morphologies; **d–f** Fracture morphologies [48]

During HP sintering, many closed or partially closed micropores form in the fiber aggregation regions, where ceramic particles are difficult to enter. Therefore, both the inferior sinterability of the MA-Si_2BC_3N powder and the aggregations of carbon fibers result in the large porosity of the as-prepared composites. Additionally, a lot of ceramic particles adhere on the fiber pull-out surfaces, implying severe interface reactions occurring between carbon fibers and ceramic matrix.

After introduction of 7.5 mol.% ZrO_2 sintering aid, the flexural strength, elastic modulus and fracture toughness of the as-prepared 20 vol.% C_f/Si_2BC_3N composites are increased to 112.4 MPa, 111.1 GPa and 2.94 MPa $m^{1/2}$, respectively (Table 5.32). The addition of ZrO_2 will depress the interface reactions, reduce the damage to carbon fibers and promote the densification of the ceramic matrix. With this in mind, if more ZrO_2 (15–20 mol.%) added, the interface reactions and degradation of carbon fibers may be further reduced (Fig. 5.49). As expected, with a large amount of ZrO_2 ~ 15 mol.% addition, poll-out part of the carbon fibers with clean surface and clear grooves are observed on fracture surfaces, and no ceramic particles are adhered to the fiber surfaces. As a result, denser composites with higher relative density show good mechanical properties.

The as-prepared MA-Si_2BC_3N monoliths present a typical feature of brittle fracture (Fig. 5.50). The addition of carbon fibers slightly improves their strength and fracture work. The weak mechanical properties of MA-Si_2BC_3N and 20 vol.% C_f/

Table 5.32 Bulk density and mechanical properties of the 20 vol.% C_f/Si_2BC_3N CMCs hot-pressed under 1800 °C/40 MPa/30 min [48]

Composition	Bulk density (g/cm^3)	Flexural strength (MPa)	Young's modulus (GPa)	Fracture toughness (MPa $m^{1/2}$)
C_f/Si_2BC_3N	1.88	64.6 ± 3.2	14.7 ± 7.4	1.47 ± 0.09
C_f/Si_2BC_3N with 7.5 mol.% ZrO_2	2.39	112.5 ± 12.1	111.1 ± 23.3	2.94 ± 0.25

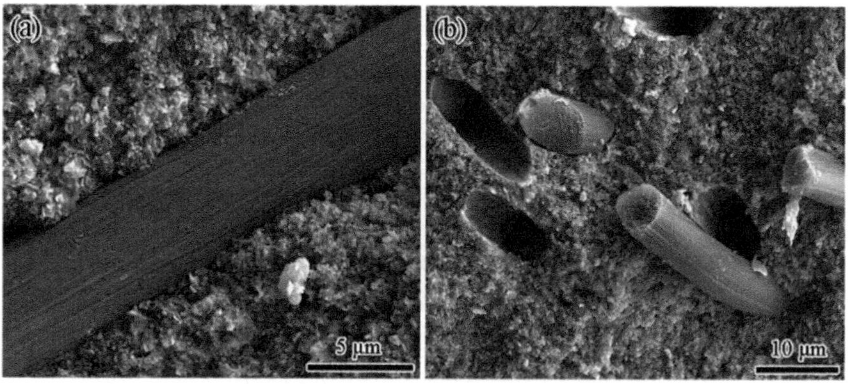

Fig. 5.49 The fracture morphologies of the hot-pressed 20 vol.% C_f/Si_2BC_3N CMCs containing a large amount of ZrO_2 (15–20 mol.%): **a** High-resolution; **b** Low-resolution [48]

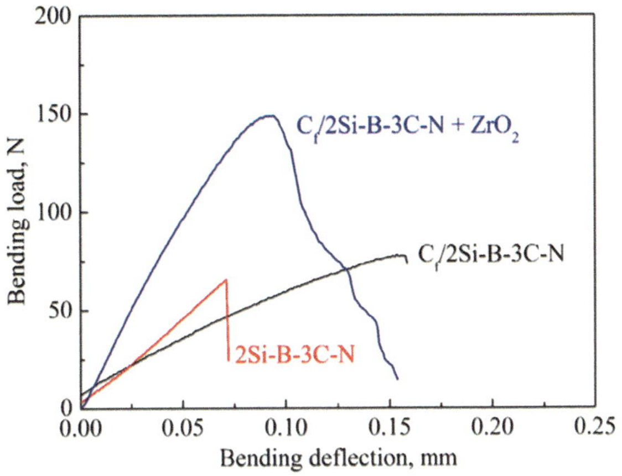

Fig. 5.50 Load–displacement curves of the hot-pressed 20 vol.% C_f/Si_2BC_3N CMCs and MA-Si_2BC_3N monoliths [48]

Si_2BC_3N composites are mainly caused by the poor sinterability of the ceramic matrix, the aggregation of carbon fibers, and the remaining large porosity. Surprisingly, with only 7.5 mol.% ZrO_2 addition, strongly improves their bending strength and fracture work showing a pseudoplastic fracture characteristic. Eventual fracture doesn't occur during testing, even though the bending load decreases to a value smaller ~ 25 N with a displacement ~ 0.15 mm. Most possibly, ZrO_2 promotes the densification and property improvement of the composites by facilitating the solid diffusion of most atoms.

In sum, the decreasing porosity and the moderate interface bonding both contribute to the property improvement of the 20 vol.% C_f/Si_2BC_3N-ZrO_2 materials. The possible toughening mechanisms may include the short fibers toughening, the particles toughening and the ZrO_2 phase transformation toughening. A higher strength means a larger resistance to fracture and hence is also beneficial to the increasing toughness.

(3) *C_f/Si_2BC_3N-MAS composites with PyC coating*

The bending strength, Vickers' hardness and fracture toughness of the C_f/Si_2BC_3N-MAS composites gradually decrease with the increasing fiber content (Fig. 5.51). For C_f/Si_2BC_3N-MAS composites with 5 vol.% fiber addition, their hardness, bending strength and fracture toughness show the best, reaching to 6.2 GPa, 400.2 MPa and 4.5 MPa $m^{1/2}$, respectively [49]; however, the composites with 20 vol.% carbon fiber exhibit hardness of 4.2 GPa, bending strength of 256.9 MPa and fracture toughness of 3.3 MPa $m^{1/2}$, respectively (Table 5.33). The decadent properties should be argued to the overlapping of the fibers unfavoring to the matrix densification, the decreasing fiber debonding resistance and the decreasing fiber pull-out length. Thus, the brittle fracture behavior for all the investigated C_f/Si_2BC_3N-MAS composites are found regardless of uniform PyC coating.

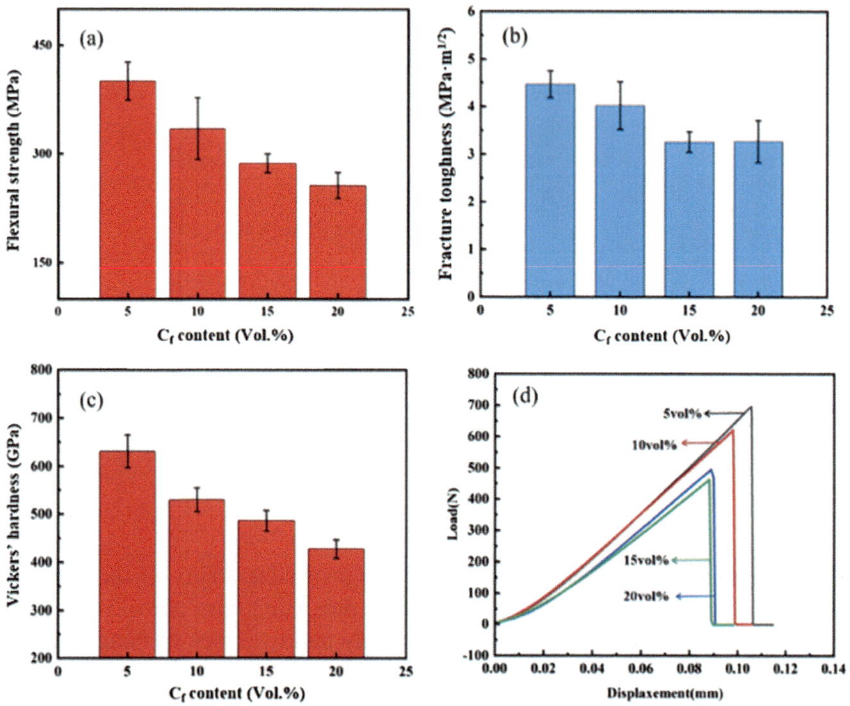

Fig. 5.51 Mechanical performance of the C_f/Si_2BC_3N-MAS composites (fibers coated with PyC) hot-pressed at 1800 °C/60 MPa/60 min: **a** Flexural strength; **b** Fracture toughness; **c** Vickers' hardness; **d** Load–displacement curve [49]

Table 5.33 Mechanical properties of the C_f/Si_2BC_3N-MAS composites (fibers coated with PyC) hot-pressed at 1800 °C/60 MPa/60 min [49]

The PyC coated C_f content (vol.%)	5	10	15	20
Vickers' hardness (GPa)	6.2	5.2	4.8	4.2
Flexural strength (MPa)	400.2	334.8	287.2	256.9
Fracture toughness (MPa $m^{1/2}$)	4.5	4.0	3.3	3.3

(4) *C_f/Si_2BC_3N-Zr composites with h-BN coating*

After incorporation of 5–20 vol.% carbon fibers, the mechanical properties and densities of the C_f/Si_2BC_3N-Zr CMCs various with the carbon fiber content (Fig. 5.52 and Table 5.34) [50, 51]. The flexural strength and fracture toughness of the C_f/Si_2BC_3N-Zr CMCs first increase and then decrease with increasing the carbon fiber content. And when the carbon fiber content is 10 vol%, the CMCs show the highest flexural strength (416.6 ± 34.7 MPa) and fracture toughness (5.25 ± 0.66 MPa $m^{1/2}$). However, with the increase of the carbon fiber content, the gradual decline

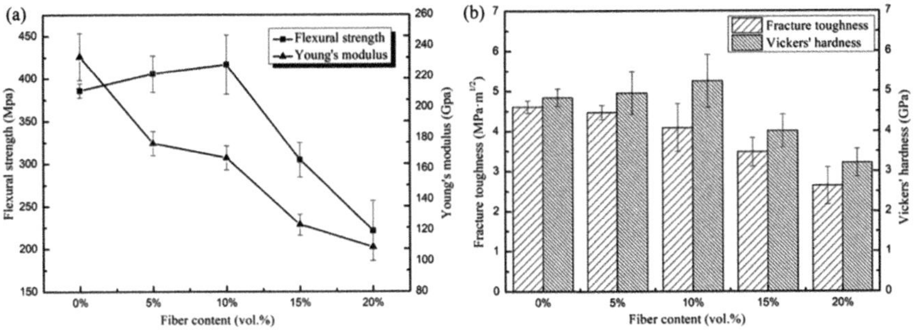

Fig. 5.52 Mechanical properties of various C_f/Si_2BC_3N composites (fibers coated with h-BN) hot-pressed under 1900 °C/60 MPa/30 min [50]

of the Young's modulus and Vickers' hardness of the CMCs can be observed. And when the carbon fiber content is 5 vol%, the CMCs exhibit the highest Young's modulus (176.4 ± 7.9 GPa) and Vickers' hardness (4.5 ± 0.2 GPa). The increasing fiber content results in the reduction of the bulk density. Besides, the agglomeration and degradation of carbon fibers both lead to insufficient local short-range diffusion, causing stress concentration and generation of initial cracks. After some Zr introduction, strong chemical bonds such as Zr-B, Zr-N and Zr-O have established during the MA process, and consequently forming new ultra-high temperature phases of ZrB_2, ZrN and ZrO_2 after hot-pressing sintering. The ultra-high temperature components together with the original matrix phases of BN(C) and SiC ensure lower modulus. The introduction of h-BN coated carbon fibers retards the fiber degradation and avoids strong interface bonding. The fracture model of the as-sintered C_f/Si_2BC_3N-Zr CMCs thereby changes from brittle failure to pseudoplastic fracture.

When crack interacts with the short carbon fibers, debonding and even pull-out of the fibers are both observed, with a crack deflection length of about 1.69 μm. However, once crack is arrested by longer fibers, a propagation length about 2.38 μm is measured due to the crack deflection (Fig. 5.53). Although the maximum load decreases with the increasing fiber content, the fracture strain gradually increases with increasing fiber content. Interestingly, the platform of pseudo-plastic deformation gradually appears as the fiber content increases (Fig. 5.54).

Table 5.34 Densities and mechanical properties of the C_f/Si_2BC_3N-Zr composites (fibers coated with h-BN) hot-pressed under 1900 °C/60 MPa/30 min [51]

The C_f content (vol.%)	Bulk density (g/cm^3)	Relative density (%)	Flexural strength (MPa)	Young's modulus (GPa)	Fracture toughness (MPa $m^{1/2}$)	Vickers' hardness (GPa)
+ 5	3.36	89.6	405.7 ± 21.4	176.4 ± 7.9	4.95 ± 0.53	4.5 ± 0.2
+ 10	3.16	86.7	416.6 ± 34.7	166.9 ± 7.8	5.25 ± 0.66	4.1 ± 0.6
+ 15	2.70	76.3	304.7 ± 20.5	123.4 ± 6.9	4.01 ± 0.41	3.5 ± 0.4
+ 20	2.22	65.4	221.2 ± 35.4	108.8 ± 8.9	3.21 ± 0.35	2.6 ± 0.5

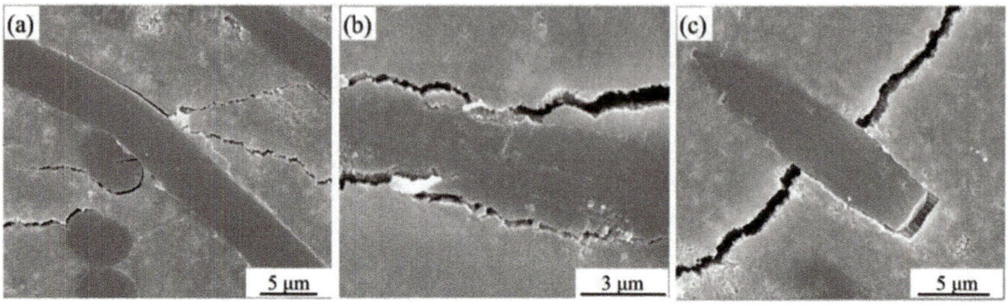

Fig. 5.53 Crack propagation of the hot-pressed C_f/Si_2BC_3N-Zr CMCs: **a** Crack deflection; **B** fiber debonding; **c** Fiber bridging [50]

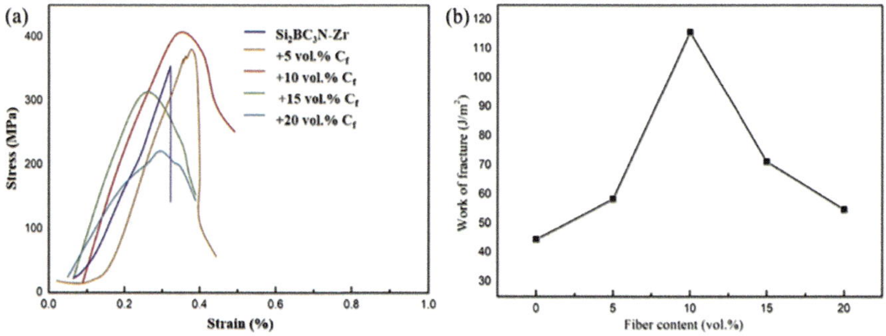

Fig. 5.54 The load–displacement curves and fracture work of the hot-pressed C_f/Si_2BC_3N-Zr ceramic composites: **a** Load–displacement curves; **b** Fracture work [50]

Noted that the energy consumption by pull-out of a fiber is much more than that of a particle, thus the fracture toughness of the C_f/Si_2BC_3N-Zr composites should be enhanced. Fibers pull-out toughened mechanism occurs only when the load-transfer behavior at the intimate interfaces of fibers and Si_2BC_3N-Zr matrix takes place. The load-transfer phenomenon is suggested to be controlled by an interfacial shear mechanism that originated from the mechanical interlocking of structural inhomogeneity and defects of the carbon fibers with the ceramic matrix and/or the formation of chemical bonding at the interfaces. The higher the interfacial shear resistance, the higher the extents of the load transfer from the matrix to the carbon fibers, which can even exceed the failure resistance of the fibers.

The crystallization of carbon fibers and subsequently forming interfacial structure during HP sintering should be responsible for the poorer performance for the C_f/Si_2BC_3N and C_f/Si_2BC_3N-Zr composites. Surface modification of carbon fibers at much lower temperature avoiding degradation or perhaps choose much higher thermal stability of carbon fibers or change the nature of reinforcements are the effective strategies for property improvement.

The high-temperature bending strength of the as-sintered C_f/Si_2BC_3N and C_f/Si_2BC_3N-Zr composites gradually decreases as the annealing temperature raises, while Si_2BC_3N and Si_2BC_3N-Zr bulk ceramics show higher bending strength at 1000–1200 °C. In all cases for the fibers containing composites, the strength retention rates at 1000 or 1200 °C are higher ~ 92% or 83%, respectively, showing desirable stress tolerance (Table 5.35). Besides, the nanohardness and elastic modulus of the as-sintered C_f/Si_2BC_3N and C_f/Si_2BC_3N-Zr CMCs gradually decline as the testing temperature increases form RT to 300 °C (Table 5.36).

Similar microstructures are observed for the Si_2BC_3N-Zr ceramics before and after high-temperature strength tests (Fig. 5.55). For the 10 vol.% C_f containing composites, the numbers of fibers pull-out reduce and their pull-out length is only 1–2 μm. However, degradation of carbon fibers seems not serious since fiber pull-out judged by the holes left on the fractured surface are still found. Besides, grooves and limited particles adhered on the fiber surfaces are rarely encountered. The introduced h-BN coating can prevent the strong bonding between fibers and ceramic matrix even at high testing temperature in air.

The thermal conductivities of the Si_2BC_3N-Zr and Si_2BC_3N bulk ceramics are both lower than those composites, which show increasing thermal conductivities with increasing carbon fiber content at each testing temperature (Fig. 5.56). The measured CET of the as-sintered Si_2BC_3N-Zr ceramics and their matrix composites is lower than those of SiC (4.7×10^{-6} °C^{-1}) and ZrB_2/Si_2BC_3N (5.2×10^{-6} °C^{-1}) ceramics. The deviation between the measured and the theoretical calculation values may be induced by internal defects and residual stress within the materials. Generally, the

Table 5.35 High-temperature mechanical properties of the hot-pressed C_f/Si_2BC_3N-Zr CMCs (fibers coated with h-BN) after testing at 1000–1200 °C dwelling for 5 min in air [50]

Composition	Bending strength (MPa)		
	RT	1000 °C	1200 °C
Si_2BC_3N	331.1 ± 19.1	377.5 ± 34.5	336.3 ± 26.7
Si_2BC_3N-Zr	386.1 ± 8.3	423.7 ± 14.8	403.9 ± 20.1
5 vol.% C_f/Si_2BC_3N	339.2 ± 20.5	336.4 ± 33.7	295.1 ± 15.8
10 vol.% C_f/Si_2BC_3N	360.6 ± 10.1	340.2 ± 18.2	301.8 ± 26.8
5 vol.% C_f/Si_2BC_3N-Zr	405.7 ± 21.4	373.2 ± 26.5	365.1 ± 14.9
10 vol.% C_f/Si_2BC_3N-Zr	416.6 ± 34.7	388.1 ± 40.2	348.7 ± 19.2

Table 5.36 Nanohardness and elastic modulus of the hot-pressed C_f/Si_2BC_3N-Zr ceramic composites (fibers coated with h-BN) after testing at RT and 300 °C in air [50]

Composition	Nanoindentation (RT)		Nanoindentation (300 °C)	
	Nanohardness (GPa)	Elastic modulus (GPa)	Nanohardness (GPa)	Elastic modulus (GPa)
10 vol.% C_f/Si_2BC_3N	5.9	126.6	5.2	118.5
10 vol.% C_f/Si_2BC_3N-Zr	35.3	178.4	26.8	165.1

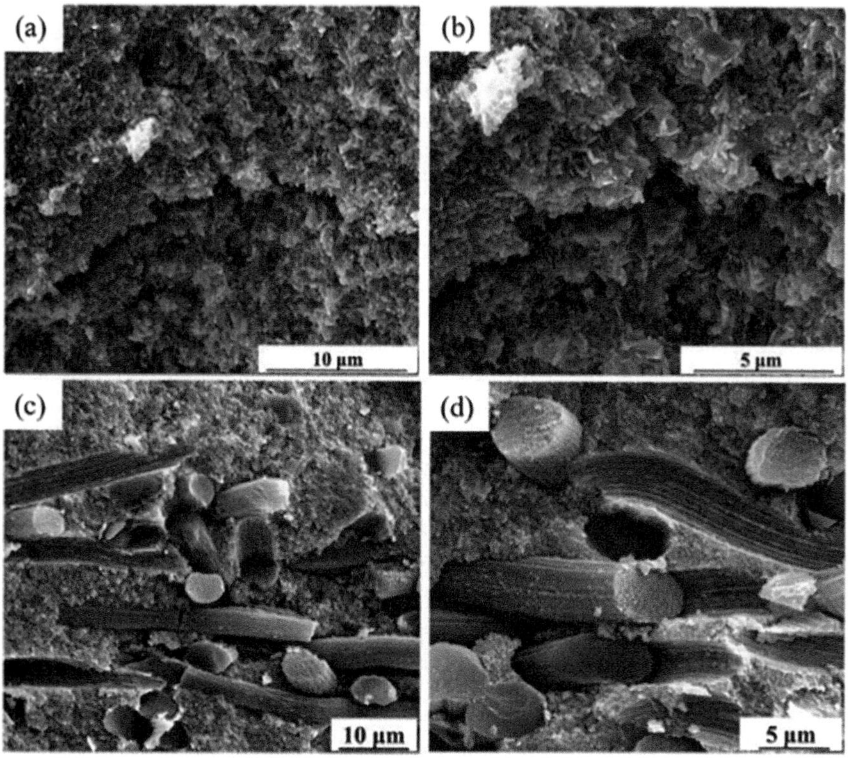

Fig. 5.55 Fracture morphologies of the Si_2BC_3N-Zr ceramics and 10 vol.% C_f/Si_2BC_3N-Zr composites after flexural strength test at 1200 °C dwelling for 5 min in air: **a** and **b** Si_2BC_3N-Zr ceramics; **c** and **d** 10 vol.% C_f/Si_2BC_3N-Zr composites [50]

average CET of the C_f/Si_2BC_3N-Zr composite materials decreases with increasing carbon fiber content (Table 5.37).

(5) *(C_f-ZrB_2)/Si_2BC_3N composites with h-BN coating*

With 15 wt.% ZrB_2, the bulk density of SPS densified 10 vol.% C_f containing (C_f-ZrB_2)/Si_2BC_3N composites is 2.83 g/cm^3, lower than that of ZrB_2/Si_2BC_3N bulk ceramics [52]. Besides, their flexural strength and elastic moduli decrease to 236 ± 16 MPa and 130 ± 4 GPa respectively, while the fracture toughness increases to 8.7 ± 0.1 MPa m$^{1/2}$. The thermal diffusivity of both ZrB_2/Si_2BC_3N ceramics and (C_f-ZrB_2)/Si_2BC_3N composites decreases with increasing testing temperature [53]. The (C_f-ZrB_2)/Si_2BC_3N composites exhibit higher room-temperature thermal diffusivity (equal to 24.3 mm^2/s) compared to the ZrB_2/Si_2BC_3N (equal to 16.5 mm^2/s). The CET of the two ceramic materials increases as testing temperature raises. However, at certain testing temperature, the CET of (C_f-ZrB_2)/Si_2BC_3N composites are lower than that of their counterparts (Fig. 5.57).

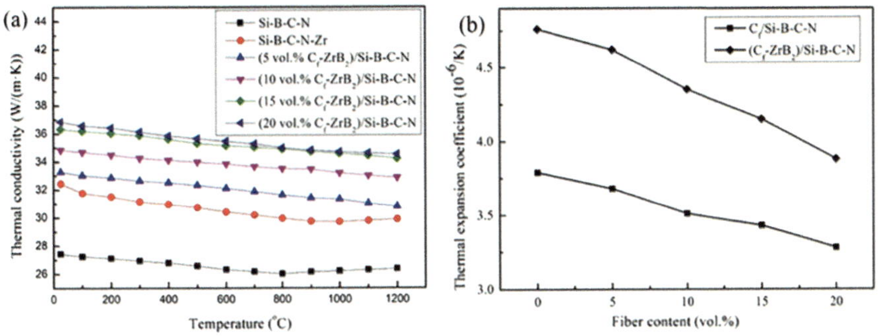

Fig. 5.56 Thermophysical properties of the Si_2BC_3N-Zr ceramics and C_f/Si_2BC_3N-Zr composites: **a** Thermal conductivity; **b** Average thermal expansion coefficient [50]

Table 5.37 The average thermal expansion coefficient of the hot-pressed ceramics and composites (fibers coated with h-BN) after testing in range of RT ~ 1200 °C [50]

Composition	Thermal expansion coefficient ($\times 10^{-6}$ °C^{-1})	
	Calculated values	Measured values
Si_2BC_3N	–	3.79
Si_2BC_3N-Zr	4.29	4.76
5 vol.% C_f/Si_2BC_3N	3.57	3.68
10 vol.% C_f/Si_2BC_3N	3.34	3.51
15 vol.% C_f/Si_2BC_3N	3.14	3.43
20 vol.% C_f/Si_2BC_3N	3.00	3.28
5 vol.% C_f/Si_2BC_3N-Zr	4.05	4.62
10 vol.% C_f/Si_2BC_3N-Zr	3.81	4.53
15 vol.% C_f/Si_2BC_3N-Zr	3.56	4.15
20 vol.% C_f/Si_2BC_3N-Zr	3.32	3.88

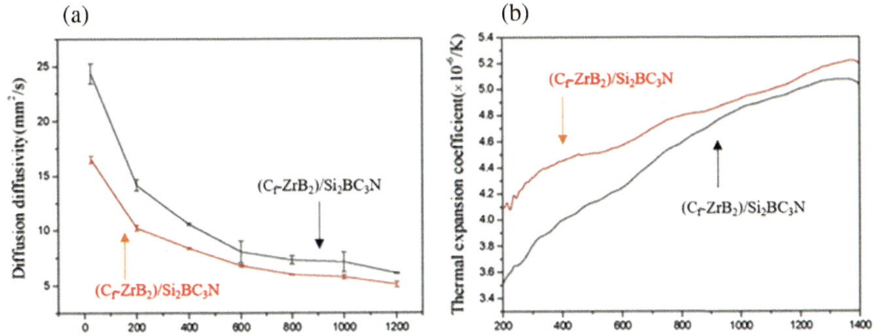

Fig. 5.57 Thermophysical properties of SPS densified ZrB_2/Si_2BC_3N ceramics and $(C_f$-$ZrB_2)/Si_2BC_3N$ composites (2000 °C/40 MPa/5 min): **a** Thermal diffusivity; **b** thermal expansion coefficient [53]

5.4.2 SiC_f/Si_2BC_3N CMCs

(1) *SiC_f/Si_2BC_3N composites with h-BN coating*

The SLFC1 SiC_f adopted are in density of 2.36 g/cm^3, diameter of 13 ± 0.5 μm, tensile strength of 1.5 ± 1.0 GPa, and Young's modulus of 140 ± 10 GPa [54]. The average thermal expansion coefficient of the SiC_f/Si_2BC_3N composites (fibers coated with h-BN) is slightly lower than that of the pure MA-Si_2BC_3N (~4.7 × 10^{-6} °C^{-1}). For composites hot-pressed at 1800 °C/60 MPa/30 min, their bulk density is low and micro-defects in the matrix constrain their mechanical properties improvement. With increasing sintering temperature or pressure, both the bulk density and mechanical properties gradually increase (Fig. 5.58). After hot-pressed at 1900 °C/80 MPa/30 min, the resulting SiC_f/Si_2BC_3N composites present bulk density, flexural strength, Young's modulus and fracture toughness of 2.70 g/cm^3, 208.0 ± 5.7 MPa, 114.3 ± 10.2 GPa and 3.32 ± 0.26 MPa·m$^{1/2}$, respectively, while those of 20 vol% C_f/Si_2BC_3N composites hot-pressed under same sintering condition are only 64.8 ± 5.5 MPa, 2.24 ± 0.23 MPa m$^{1/2}$ and 55.6 ± 3.2 GPa, respectively (Table 5.38).

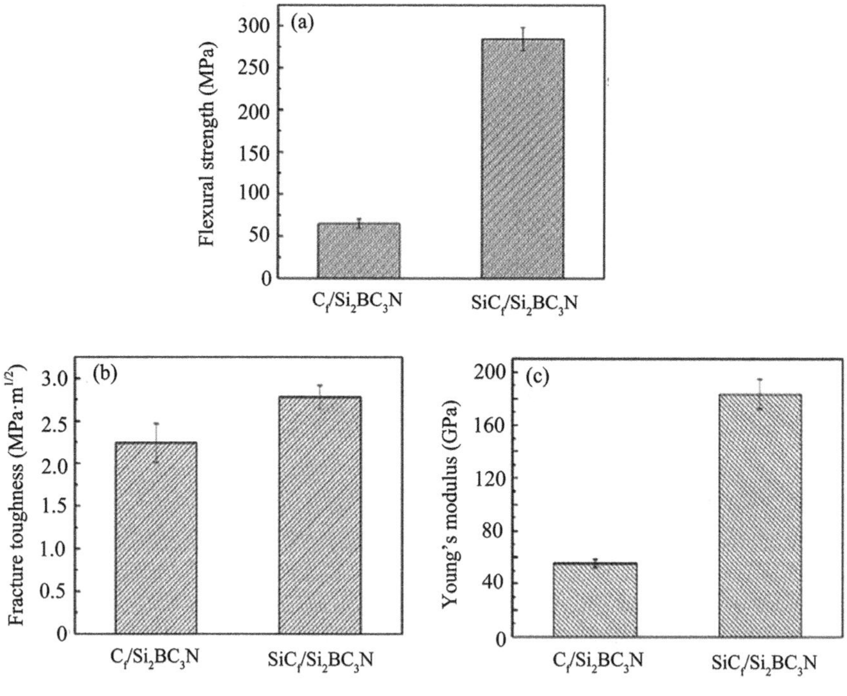

Fig. 5.58 Mechanical properties of the 20 vol.% C_f/Si_2BC_3N and 20 vol.% SiC_f/Si_2BC_3N composites (SLFC1 SiC_f coated with h-BN): **a** Flexural strength; **b** Fracture toughness; **c** Young's modulus [54]

Table 5.38 Bulk density and mechanical properties of the 20 vol.% SiC_f/Si_2BC_3N composites (SLFC1 SiC_f coated with h-BN) hot-pressed at different sintering parameter [54]

Hot-pressing condition	Bulk density (g/cm^3)	Flexural strength (MPa)	Fracture toughness (MPa $m^{1/2}$)	Young's modulus (GPa)
1800 °C/60 MPa/30 min	2.62	149.4 ± 13.3	2.50 ± 0.56	92.9 ± 3.6
1900 °C/60 MPa/30 min	2.67	193.4 ± 4.2	3.92 ± 0.45	154.1 ± 25.4
1900 °C/80 MPa/30 min	2.70	208.0 ± 5.7	3.32 ± 0.26	114.3 ± 10.2

In case the SLFC1 SiC_f are uncoated with h-BN, the flexural strength, fracture toughness and Young's modulus of the 20 vol.% SiC_f/Si_2BC_3N composites are also improved with the increasing sintering temperature. However, their bulk density and flexural strength show on obvious improvement as sintering pressure raises from 40 to 60 MPa, but their fracture toughness and Young's modulus are reduced (Table 5.39). Expectedly, examination of the fracture surface and load–displacement both indicate the stronger bonding between SiC_f and ceramic matrix leading to the catastrophic failure regardless of the h-BN coating (Fig. 5.59).

(2) *SiC_f/Si_2BC_3N composites with multilayer coating*

The adopted KD-II SiC_f with higher strength retention ratio (Ar/1 h) of ~ 90% at 1250 °C, ~ 90% at 1400 °C and ~ 70% at 1600 °C, are in density of 2.70 g/cm^3, diameter of 10–15 μm, RT tensile strength of 2.6 ± 0.3 GPa and Young's modulus of 180–200 GPa [55]. Prior to the sintering, the amorphous C, ZrB_2, and h-BN single-layer coatings as well as C/BN, C/ZrB_2, ZrB_2/BN, and C/ZrB_2/BN multilayer coatings are prepared on the KD-II SiC_f surfaces [56, 57].

For SiC_f/Si_2BC_3N composites with single-layer coating, their flexural strength increases to some extent [58]. For composites with multilayer coatings, their fracture

Table 5.39 Bulk density and mechanical properties of the 20 vol.% SiC_f/Si_2BC_3N composites (unmodified SLFC1 SiC_f) [54]

Hot-pressing condition	Bulk density (g/cm^3)	Flexural strength (MPa)	Fracture toughness (MPa $m^{1/2}$)	Young's modulus (GPa)
1800 °C/60 MPa/30 min	2.35	73.3 ± 2.4	1.04 ± 0.11	73.3 ± 3.0
1800 °C/40 MPa/30 min	2.46	70.2 ± 6.3	1.72 ± 0.11	64.1 ± 2.8
1900 °C/60 MPa/30 min	2.56	284.3 ± 17.9	2.78 ± 0.14	183.5 ± 11.1
1900 °C/60 MPa/60 min	2.57	171.8 ± 6.5	3.66 ± 0.08	107.3 ± 7.3

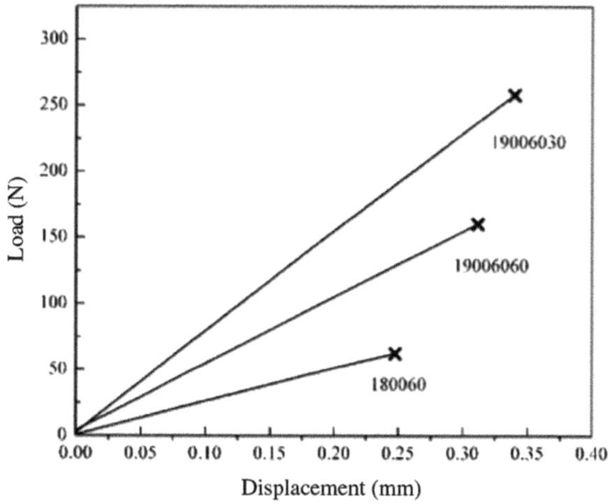

Fig. 5.59 Typical fracture behavior of the 20 vol.% SiC_f/Si_2BC_3N composites (SLFC1 SiC_f coated with h-BN) hot-pressed at different sintering parameter [55]

toughness is higher than those of composites with or without single-layer coatings (Fig. 5.60). The Young's modulus of all the investigated composites presents the same level of performance. Generally, the as-sintered $SiC_f(C/ZrB_2)/Si_2BC_3N$ composites have the best mechanical performance with bending strength of 171.3 ± 6.1 MPa, fracture toughness of 2.68 ± 0.09 MPa m^1, and Young's modulus of 87.0 ± 5.8 GPa (Table 5.40).

Obviously, all the investigated SiC_f/Si_2BC_3N composites exhibit catastrophic brittle fracture. Since the modified coating reinforced composites all show higher mechanical performance than that of their counterparts. Thus. the mainly possible reasons for their catastrophic fracture failure should be attributed to the serious crystallization of SiC_f leading to great strength loss, which is incapable of load transferring. Of failure mechanisms are different for SiC_f/Si_2BC_3N composites with different crystallization degree-bridged fibers and fiber pull-out for the lower but fiber fracture for the higher. The conventional toughening mechanisms are not applicable in the studied SiC_f/Si_2BC_3N composites no matter what coating used. (Fig. 5.61).

(3) *SiC_f/Si_2BC_3N-MAS composites with h-BN coating*

The new KD-SA SiC_f adopted are in density of 2.96 g/cm^3 and diameter of 10–15 μm. These KD-SA SiC_f can maintain their original tensile strength at 1800 °C for 5 h or at 1900 °C for 1 h under inert atmosphere, and they exhibit a high strength retention ~ 97% after exposure at 1300 °C for 1 h under air [59].

Without MAS sintering additive [60], the 20 vol.% h-BN coated SiC_f reinforced MA-Si_2BC_3N composites exhibit unsatisfying mechanical properties, obtaining flexural strength < 102.0 MPa, fracture toughness < 2.0 MPa $m^{1/2}$, and Vickers' hardness < 5.0 GPa (Fig. 5.62). With 10 wt.% MAS addition, the mechanical performance of the various SiC_f/Si_2BC_3N-MAS composites are greatly improved, showing flexural strength > 300.0 MPa, fracture toughness > 5.0 MPa $m^{1/2}$, and Vickers' hardness > 9.1

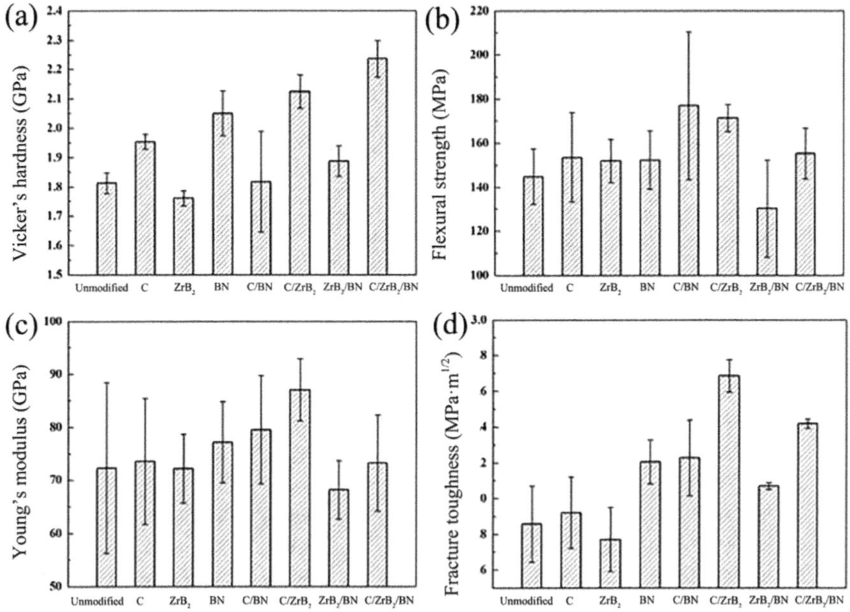

Fig. 5.60 Room-temperature mechanical properties of the 5 vol.% KD-II SiC_f reinforced MA-Si_2BC_3N composites with different coating: **a** Vickers' hardness; **b** Flexural strength; **c** Young's modulus; **d** Fracture toughness (HP condition: 1900 °C/60 MPa/30 min/1 bar Ar) [57]

Table 5.40 Mechanical properties of the 5 vol.% KD-II SiC_f reinforced MA-Si_2BC_3N composites with different coating (HP sintering condition: 1900 °C/60 MPa/30 min/1 bar Ar) [58]

Composites	Flexural strength (MPa)	Young's modulus (GPa)	Fracture toughness (MPa $m^{1/2}$)	Vickers' hardness (GPa)
SiC_f/Si_2BC_3N	144.7 ± 12.6	72.3 ± 16.1	1.86 ± 0.21	1.8 ± 0.1
$SiC_f(C)/Si_2BC_3N$	153.5 ± 20.3	73.5 ± 11.9	1.92 ± 0.20	2.0 ± 0.1
$SiC_f(ZrB_2)/Si_2BC_3N$	151.9 ± 9.8	72.2 ± 6.5	1.77 ± 0.18	1.8 ± 0.1
$SiC_f(BN)/Si_2BC_3N$	152.2 ± 13.2	77.1 ± 7.7	2.21 ± 0.12	2.1 ± 0.1
$SiC_f(C/BN)/Si_2BC_3N$	177.0 ± 33.6	79.5 ± 10.2	2.23 ± 0.21	1.8 ± 0.2
$SiC_f(C/ZrB_2)/Si_2BC_3N$	171.3 ± 6.1	87.0 ± 5.8	2.68 ± 0.09	2.1 ± 0.1
$SiC_f(ZrB_2/BN)/Si_2BC_3N$	130.2 ± 22.1	68.2 ± 5.5	2.07 ± 0.02	1.9 ± 0.1
$SiC_f(C/ZrB_2/BN)/Si_2BC_3N$	155.2 ± 11.5	73.2 ± 9.1	2.42 ± 0.03	2.2 ± 0.1

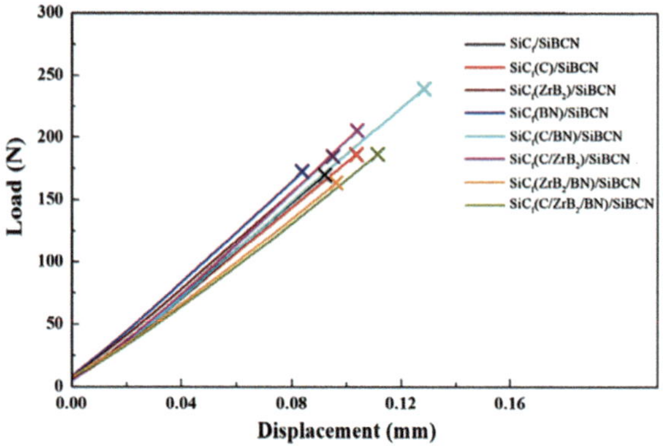

Fig. 5.61 The load–displacement curves of the various 5 vol.% SiC_f/Si_2BC_3N ceramic composites hot-pressed at 1900 °C/60 MPa/30 min [58]

GPa. The adding MAS greatly densifies the ceramic matrix, but has no improvement in the interfacial bonding of SiC fiber/matrix (Table 5.41).

Without h-BN coating, the SiC_f/Si_2BC_3N composites with 10 wt.% MAS show similar mechanical properties than those of their counterparts. The as-investigated SiC_f/Si_2BC_3N composites all display the brittle fracture characteristics except for the coating-free 20 vol.% SiC_f containing CMCs (Fig. 5.63). The h-BN coating modified KD-SA SiC_f seems unnecessary for MA-Si_2BC_3N composites.

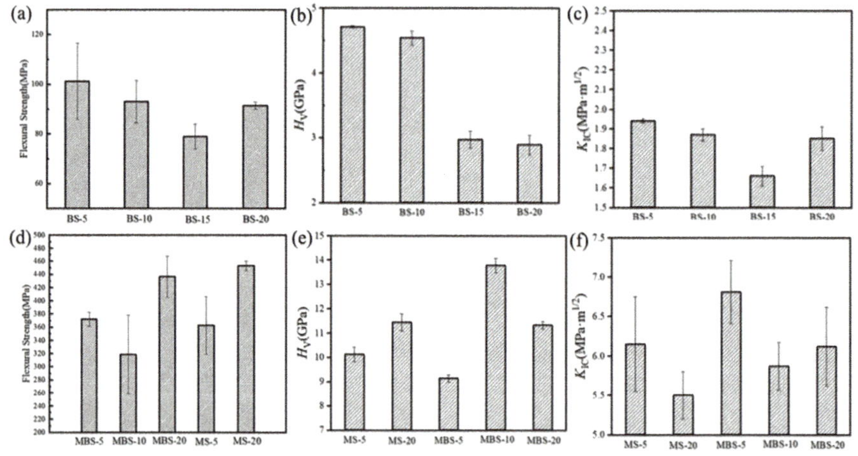

Fig. 5.62 Mechanical properties of the various KD-SA SiC_f reinforced MA-Si_2BC_3N composites hot-pressed at 1850–1900 °C/60 MPa/60 min: **a** Without MAS additive; **b** With MAS additive [60]

Table 5.41 Mechanical properties of various KD-SA SiC_f reinforced MA-Si_2BC_3N composites [60]

Composites	Flexural strength (MPa)	Fracture toughness (MPa $m^{1/2}$)	Vickers' hardness (GPa)
+ 5 vol.% SiC_f	101.2	1.94	4.7
+ 10 vol.% SiC_f	93.1	1.87	4.5
+ 15 vol.% SiC_f	79.0	1.63	3.0
+ 20 vol.% SiC_f	91.5	1.86	2.9
+ 5 vol.% SiC_f and 10 wt.% MAS	372.3	6.81	9.1
+ 5 vol.% SiC_f and 10 wt.% MAS*	363.1	6.15	11.2
+ 10 vol.% SiC_f and 10 wt.% MAS	318.7	5.87	13.8
+ 20 vol.% SiC_f and 10 wt.% MAS	436.6	6.12	11.4
+ 20 vol.% SiC_f and 10 wt.% MAS*	453.3	5.50	11.2

* SiC fibers are not coated with h-BN. The composites with MAS addition HP sintered at 1850 °C while their counterparts HP sintered at 1900 °C

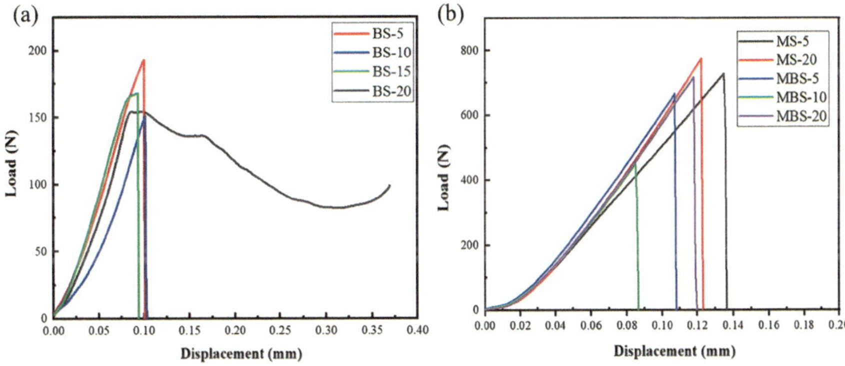

Fig. 5.63 The load–displacement curves for the various KD-SA SiC_f reinforced MA-Si_2BC_3N ceramic composites: **a** Without MAS additive; **b** With MAS additive [60]

5.4.3 (C_f-SiC_f)/Si_2BC_3N CMCs

Compared with the 20 vol.% C_f/Si_2BC_3N and 20 vol.% SiC_f/Si_2BC_3N (SLFC1 SiC_f) composites, the bulk density, flexural strength and Young's modulus of the 10 vol.% C_f and 10 vol.% SiC_f co-reinforced MA-Si_2BC_3N composites are similar with enhanced fracture toughness (Fig. 5.64) [61]. For hot-pressed (C_f-SiC_f)/Si_2BC_3N composites with Pyrolytic carbon (PyC) or h-BN coating, their flexural strengths are reduced from 97.2 MPa to 59.4 MPa or 69.2 MPa, Young's modulus decreasing from 83.8 GPa to 54.0 GPa or 72.9 GPa, and fracture toughness decreasing from 3.5 MPa $m^{1/2}$ to 2.0 MPa $m^{1/2}$ or 2.6 MPa $m^{1/2}$, respectively (Table 5.42).

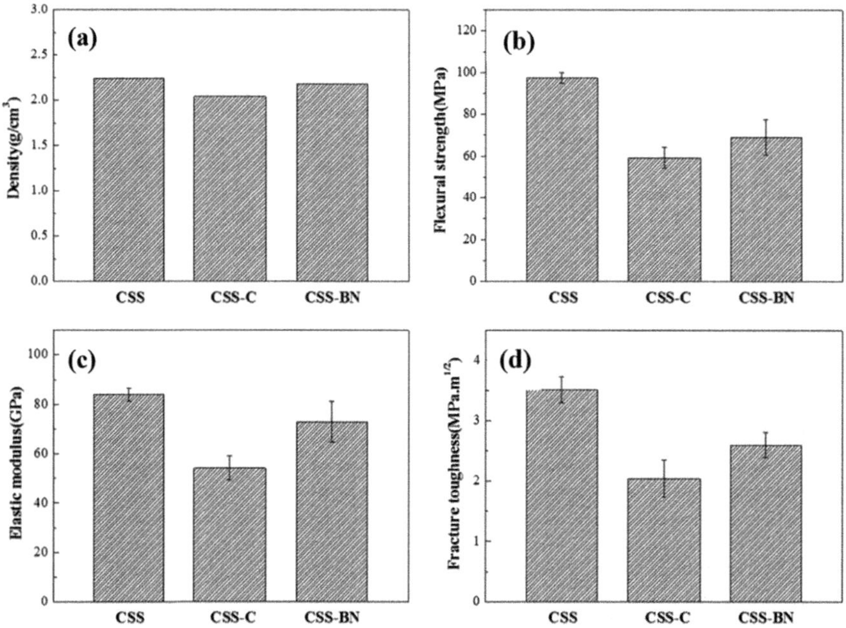

Fig. 5.64 Bulk density and mechanical properties of the various hot-pressed (C_f-SiC_f)/Si_2BC_3N ceramic composites (1900 °C/60 MPa/30 min): **a** Bulk density; **b** Flexural strength; **c** Young's modulus; **d** Fracture toughness [61]

Table 5.42 Density and mechanical properties of the various hot-pressed (C_f-SiC_f)/Si_2BC_3N ceramic composites (1900 °C/60 MPa/30 min) [61]

Composites*	Bulk density (g/cm^3)	Relative density (%)	Flexural strength (MPa)	Young's modulus (GPa)	Fracture toughness (MPa $m^{1/2}$)
^{a}(C_f-SiC_f)/ Si_2BC_3N	2.24	80.9	97.2 ± 2.6	83.8 ± 2.5	3.5 ± 0.2
^{b}(C_f-SiC_f)/ Si_2BC_3N	2.04	73.5	59.4 ± 5.1	54.0 ± 5.0	2.0 ± 0.3
^{c}(C_f-SiC_f)/ Si_2BC_3N	2.18	78.8	69.2 ± 8.5	72.9 ± 8.3	2.6 ± 0.2
^{d}(C_f-SiC_f)/ Si_2BC_3N-$ZrSiO_4$	2.58	93.1	235.3 ± 18.4	127.4 ± 1.7	5.3 ± 0.3

* The 10 vol.% SiC_f (SLFC1) and 10 vol.% C_f (TX-3) are adopted
^{a}SiC_f without any coating; ^{b}SiC_f with PyC coating; ^{c}SiC_f with h-BN coating; ^{d}Introduced 10 wt.% $ZrSiO_4$ with uncoated SiC_f

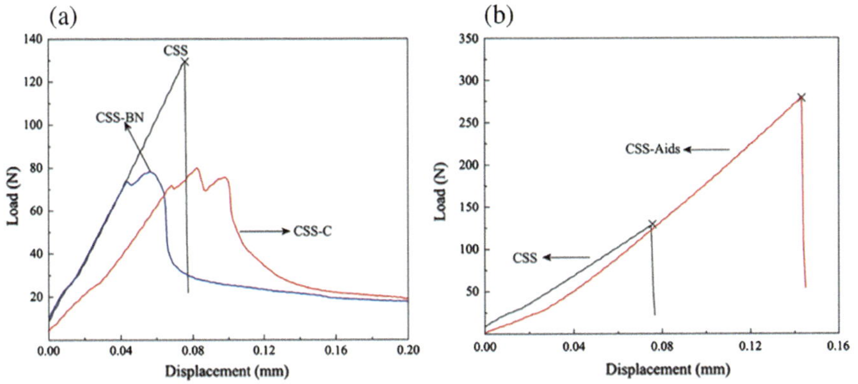

Fig. 5.65 Load–displacement curves of the various hot-pressed (C_f-SiC_f)/Si_2BC_3N composites: **a** SiC_f modified with various coating; **b** With or without 10 wt.% $ZrSiO_4$ additive [61]

The introduction of 10 wt.% $ZrSiO_4$ greatly promotes material densification, thereby reducing porosity and enhancing mechanical properties. The bulk density of the additive incorporating (C_f-SiC_f)/Si_2BC_3N composites is improved from 2.24 to 2.58 g/cm^3, flexural strength from 97.2 to 253.3 MPa, while their Young's modulus and fracture toughness is jumped by 43.6 GPa and 1.82 MPa m$^{1/2}$, respectively.

The introduction of weak interface layers, such as PyC and h-BN, both improves the interfacial strength and changes the fracture behavior of (C_f-SiC_f)/Si_2BC_3N composites from typical brittle failure to pseudoplastic fracture (Fig. 5.65). Nevertheless, introduction of 10 wt.% $ZrSiO_4$ does not change the fracture mode of the composites, but showing higher strength tolerance than that of the additive-free ones.

The thermal expansion rate of the various hot-pressed (C_f-SiC_f)/Si_2BC_3N ceramic composites shows near linear growth with testing temperature, and their thermal expansion coefficient tends to be stable at temperature beyond 300 °C (Fig. 5.66). For fibers modified with weak interface coating or introducing $ZrSiO_4$ additive, the CET of the composites is reduced. Among them, the composites with h-BN coating exhibit the lowest CET. The average CET of the four (C_f-SiC_f)/Si_2BC_3N ceramic composites are 4.15×10^{-6} °C^{-1}, 3.21×10^{-6} °C^{-1}, 3.05×10^{-6} °C^{-1} and 3.69×10^{-6} °C^{-1}, respectively (Table 5.43). The CET of the investigated (C_f-SiC_f)/Si_2BC_3N composites is all lower than that of 20 vol.% SiC_f/Si_2BC_3N (~4.29×10^{-6} °C^{-1}), but mostly higher than that of 20 vol.% C_f/Si_2BC_3N composites (~3.17×10^{-6} °C^{-1}).

5.4.4 *MWCNTs/Si_2BC_3N CMCs*

The flexural strength of 1 or 3 vol.% MWCNTs incorporating composites are 462.1 ± 51.5 MPa or 322.1 ± 6.4 MPa, respectively [62]. Apparently, the addition of only 1 vol.% MWCNTs significantly improves their bending strength and the Young's modulus of the MWCNT/Si_2BC_3N composites shows a similar variation trend. The Young's modulus increases with the introduction of 1 vol.% MWCNTs; however,

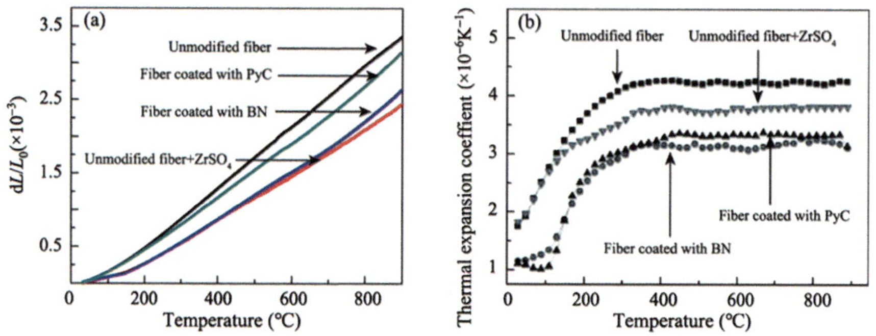

Fig. 5.66 Thermophysical properties of the hot-pressed various (C_f-SiC_f)/Si_2BC_3N ceramic composites: **a** Thermal expansion rate; **b** Thermal expansion coefficient [61]

Table 5.43 The average thermal expansion coefficient of the various hot-pressed (C_f-SiC_f)/Si_2BC_3N ceramic composites [61]

Composites*	Average thermal expansion coefficient ($\times 10^{-6}$ °C^{-1})
[a](C_f-SiC_f)/Si_2BC_3N	4.15
[b](C_f-SiC_f)/Si_2BC_3N	3.21
[c](C_f-SiC_f)/Si_2BC_3N	3.05
[d](C_f-SiC_f)/Si_2BC_3N-$ZrSiO_4$	3.69

* The 10 vol.% SiC_f (SLFC1) and 10 vol.% C_f (TX-3) are adopted
[a]SiC_f without any coating; [b]SiC_f with PyC coating; [c]SiC_f with h-BN coating; [d]Introduced 10 wt.% $ZrSiO_4$ with uncoated SiC_f

it decreases with higher MWCNTs. Additionally, the fracture toughness of 1 or 3 vol.% MWCNTs incorporating composites are 5.54 ± 0.45 or 4.31 ± 0.18 MPa $m^{1/2}$, respectively, both higher than the pristine MA-Si_2BC_3N (Fig. 5.67).

A relatively straight crack propagation path is found for pure MA-Si_2BC_3N ceramics. In contrast, much curved crack propagation paths are observed in MWCNTs/Si_2BC_3N composites with or without SiC coating. The main toughness mechanisms originate from the pullout of MWCNT and crack bridging (Fig. 5.68). Interestingly, with merely 1 vol.% MWCNTs addition, the composites show comparable fracture toughness and higher flexural strength than those fibers-containing MA-Si_2BC_3N. Besides, the Vickers' hardness of the composites decreases gradually with the increasing MWCNTs content, similar to the MWNCTs reinforced PDCs-SiBCN composites. The incorporation of 1 vol.% MWCNTs or 1 vol.% SiC coated MWCNTs has no apparent effect on the densification behavior [63, 64]. Similarly, the MWCNTs with or without SiC coating both improve the toughness significantly. For example, the introduction of 2 vol.% SiC coated MWCNTs improve the relative density and mechanical performance, obtained the highest bending strength of 532.1 MPa, Young's modulus of 144.0 GPa, and fracture toughness of 6.66 MPa $m^{1/2}$ (Table 5.44).

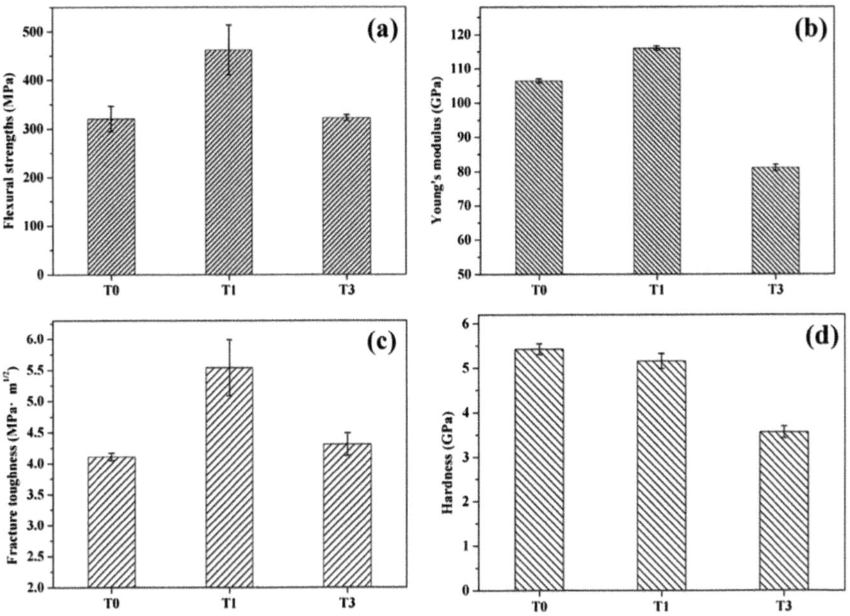

Fig. 5.67 Mechanical properties of the MWCNTs/Si_2BC_3N composites SPS sintered under 1900 °C/60 MPa/5 min: **a** Flexural strength; **b** Young's modulus; **c** Fracture toughness; **d** Vickers' hardness [62]

5.4.5 *GPLs/Si_2BC_3N CMCs*

The fracture toughness is greatly improved with the introduction of graphene platelets (GPLs) (Fig. 5.69) [65, 66]. The pure MA-Si_2BC_3N ceramics exhibit higher bulk density and relative density, but only obtained fracture toughness of 0.72 ± 0.15 MPa $m^{1/2}$. The GPL/Si_2BC_3N composites with 5 vol.% GPLs possess the highest fracture toughness of 5.40 ± 0.63 MPa $m^{1/2}$, but subsequently decreases to 3.71 ± 0.01 MPa $m^{1/2}$ with higher GPLs addition. The maximum fracture toughness of 5 vol.% GPL/Si_2BC_3N composites over the monoliths is as high as ~ 750%. Indeed, the obtained higher fracture toughness are believed to the improved dispersion of GPLs in MA-Si_2BC_3N matrix and the increasing contact areas between GPLs and the matrix. However, their Vickers' hardness performs a contrary tendency (Table 5.45).

The pull-out of GPLs, crack penetration of GPLs, crack bridging and crack deflection are observed in radial cracks originating from Vickers' hardness indents (Fig. 5.70). These toughening mechanisms are very similar for all the GPLs/Si_2BC_3N composites, except for the frequency of the occurrence when cracks meet GPLs. As is known to all, energy consumption of a tortuous path is much more than linear propagation to release stress, which also helps to enhance the fracture toughness. Crack bridging is generally observed after the interaction between larger GPLs and

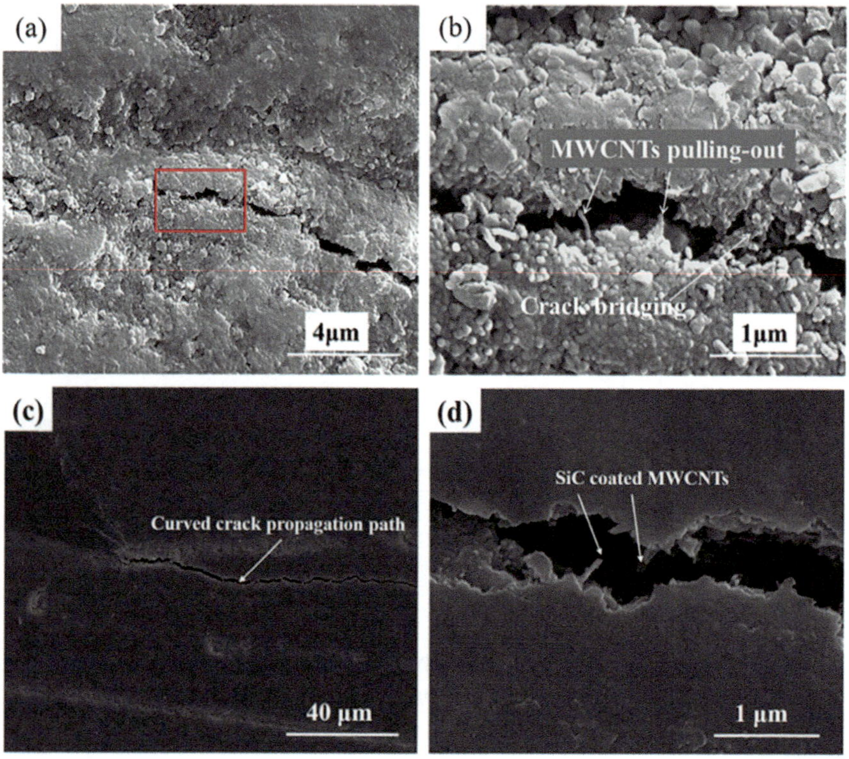

Fig. 5.68 Crack propagation paths after an indentation test: **a**, **b** + 1 vol.% unmodified MWCNTs; **c**, **d** + 1 vol.% SiC modified MWCNTs [64]

Table 5.44 Density and mechanical properties of the MWCNTs/Si_2BC_3N composites by SPS densified at 1900 °C/60 MPa/5 min [62, 64]

MWCNTs content	Bulk density (g/cm^3)	Relative density (%)	Flexural strength (MPa)	Young's modulus (GPa)	Fracture toughness ($MPa\ m^{1/2}$)
+ 1 vol.%	2.58	91.5	462.1 ± 35.2	111.6 ± 0.6	5.54 ± 0.45
+ 3 vol.%	2.41	85.8	322.1 ± 6.4	82.1 ± 0.8	4.31 ± 0.18
+ 1 vol.%*	2.59	91.7	390.9 ± 28.6	142.5 ± 6.5	5.80 ± 0.26
+ 2 vol.%*	2.61	93.1	532.1 ± 22.1	144.0 ± 1.8	6.66 ± 0.25

* MWCNTs are coated with SiC

cracks. The crack propagation direction and fracture surfaces reveal the different toughness mechanisms stemming from the present of GPLs.

Attention must focus on the phenomenon of crack penetrating GPLs (Fig. 5.71). Due to high strength and large specific surface area of GPLs, the crack may cross the interfaces between GPLs and ceramic matrix to form a three-dimensional route

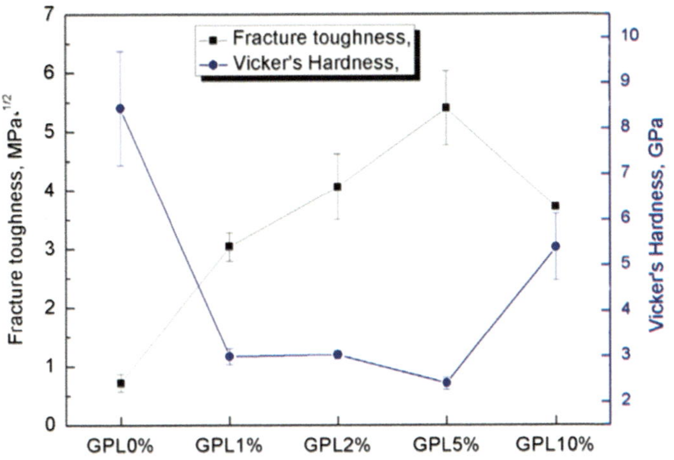

Fig. 5.69 Fracture toughness and Vicker's hardness of the SPS densified GPLs/Si_2BC_3N composites SPS consolidated under 1800 °C/40 MPa/5 min [66]

Table 5.45 Density and mechanical properties of the SPS densified GPLs/Si_2BC_3N composites SPS consolidated under 1800 °C/40 MPa/5 min [65, 66]

GPLs content	Bulk density (g/cm^3)	Relative density (%)	Flexural strength (MPa)	Young's modulus (GPa)	Fracture toughness (MPa m$^{1/2}$)	Vickers' hardness (GPa)
MA-Si_2BC_3N	2.7	97.3	64.8 ± 6.8	68.6 ± 6.8	0.72 ± 0.15	8.5 ± 1.3
+ 1 vol.%	2.4	85.9	153.4 ± 0.7	125.4 ± 5.0	3.04 ± 0.24	5.4 ± 0.7
+ 2 vol.%	2.5	86.6	168.9 ± 1.7	94.6 ± 3.9	4.06 ± 0.56	3.0 ± 0.1
+ 5 vol.%	2.4	86.2	135.3 ± 8.3	150.1 ± 2.9	5.40 ± 0.63	3.0 ± 0.2
+ 10 vol.%	2.2	77.2	196.6 ± 11.7	111.0 ± 9.3	3.71 ± 0.01	2.4 ± 0.1

instead of penetrating the GPLs when the interfacial bonding is strong enough. However, if the interfacial bonding is too strong, the matrix crack will penetrate the GPLs. In addition, composites with nano grain ceramic matrix can provide more contact chances with GPLs than micro grain matrix at the same density. So, cracks in composites with nano-sized grain matrix are more likely to take place to propagate in three dimensions rather than penetrate the GPLs.

For crack penetrating GPLs is less observed in MA-Si_2BC_3N matrix, therefore the three-dimensional cracks propagation may take place in composites. The mechanical response of structurally uniform GPLs/ceramic matrix strongly depends on the load carrying capacity of GPLs. The carrying ability of GPLs is determined first by the amount of load transfer at bonding interface and second by the intrinsic load bearing ability of GPLs after introduction into MA-Si_2BC_3N matrix. If the load transfer is poor, or the load bearing ability of GPLs is obviously reduced during sintering process, which usually involves high temperature, high pressure and surface

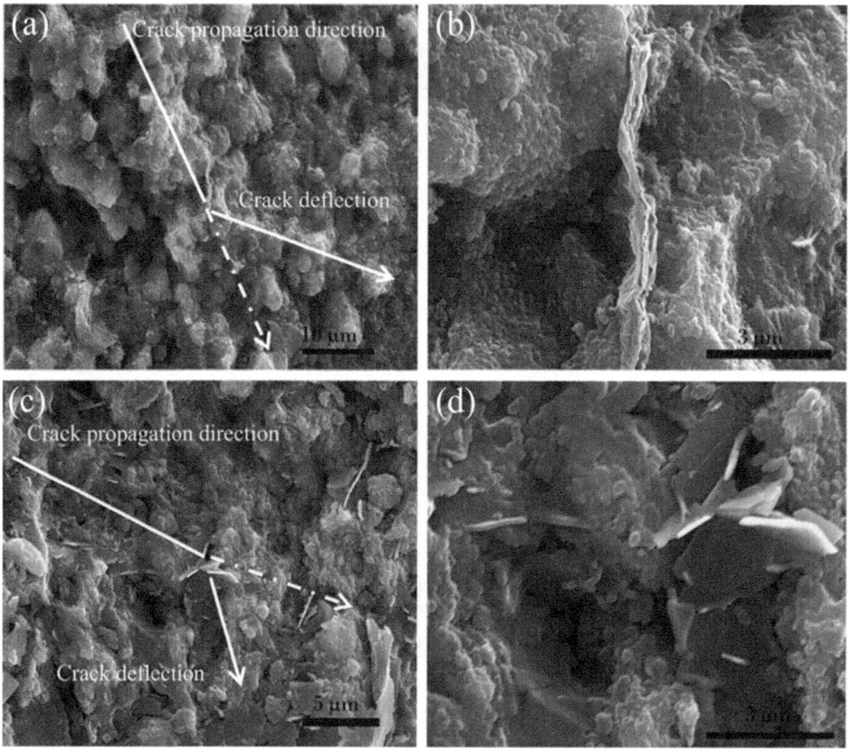

Fig. 5.70 SEM fracture morphologies of SPS densified GPLs/Si_2BC_3N CMCs: **a**, **b** + 5 vol.% GPLs; **c**, **d** + 10 vol.% GPLs [66]

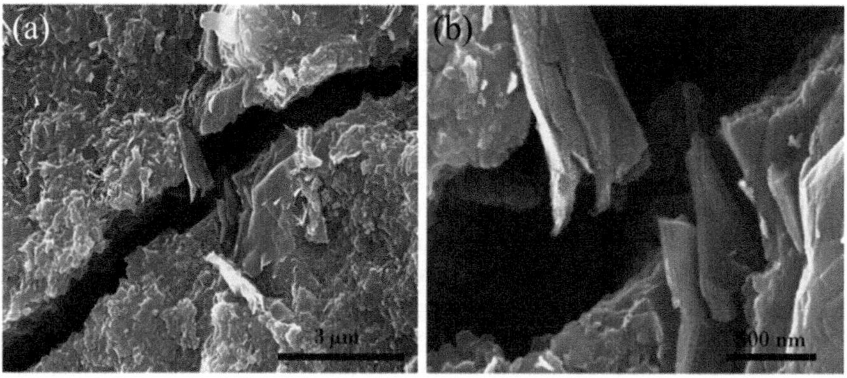

Fig. 5.71 Crack tip penetrates GPLs in the 2 vol.% GPLs containing CMCs: **a** Low-resolution; **b** High-resolution [66]

treatments, the GPLs may act a passive influence on mechanical properties even if they are uniformly distributed within the ceramic matrix with intimate interfaces. From the fracture surfaces, we can come to a conclusion that the load transfer at bonding interface is notable and the bearing ability of GPLs remains effectively after sintering. GPLs as nanofiller in MA-Si_2BC_3N ceramics can overcome the low fracture toughness effectively.

The GPLs with different dimensions and different pull-out lengths are evident on the fracture surfaces. This combination of deflection and pull-out mechanisms may take place at the interface between larger GPLs and crack tip. The reason lies in the fact that the crack deflection effect of small size GPLs is limited compared to the large size GPLs.

5.5 Strengthening and Toughening Mechanisms in SiBCN Materials

Both strength and toughness (damage tolerant) of the SiBCN based materials are a vital requirement for safety–critical applications under harsh condition; however, these properties are often-conflicting. For the brittle MA-SiBCN monoliths, the intrinsic toughening is the primary source of fracture resistance. Most of the MA-SiBCN monoliths with or without sintering additives consolidated under different sintering ways or sintering parameters are toughened in this manner, and are therefore subject to the initial conflict between strength and toughness, that is, the stronger the materials, the less deformation is available for intrinsic toughening. Coincidentally, if the sintering additive or sintering temperature/pressure both favors the ceramic matrix densification resistant to crack initiation and promotes the grains growth of matrix phases like α/β-SiC, especially the lamella BN(C) developments, the strong and tough MA-SiBCN monoliths are expected. There are instances, such as the metal Al, Zr, Al–Zr, Al_4SiC_4 or ceramic AlN, ZrO_2, MgO–ZrO_2–SiO_2 (MZS), MgO–Al_2O_3–SiO_2 (MAS), etc. particles reinforced MA-SiBCN composite ceramics, where the conflict of attaining both strength and toughness can be overcome (Fig. 5.72). The typical microstructural feature is the presence of brittle nanoscale glassy films along the grain boundaries or partly dissolved in matrix phases, thus promoting boundary cracking, grain bridging and hence the higher toughness.

In MA-SiBCN monolithic ceramics, as-mentioned above, intrinsic toughening requires changing the bonding strength; however, they can readily be toughened extrinsically by promoting crack deflection and grain bridging. In this case, the extrinsic toughening by ultra-high temperature components (UHTCs) is most effectively, and in many cases are the predominant toughening source. For instance, the addition of some ZrB_2 or LaB_6 both strengthen and toughen the ceramic matrix with higher Vickers' hardness and Young's modulus; in fact, they seem to be the most damage-tolerant MA-SiBCN bulk ceramics made (Fig. 5.73). The UHTCs/

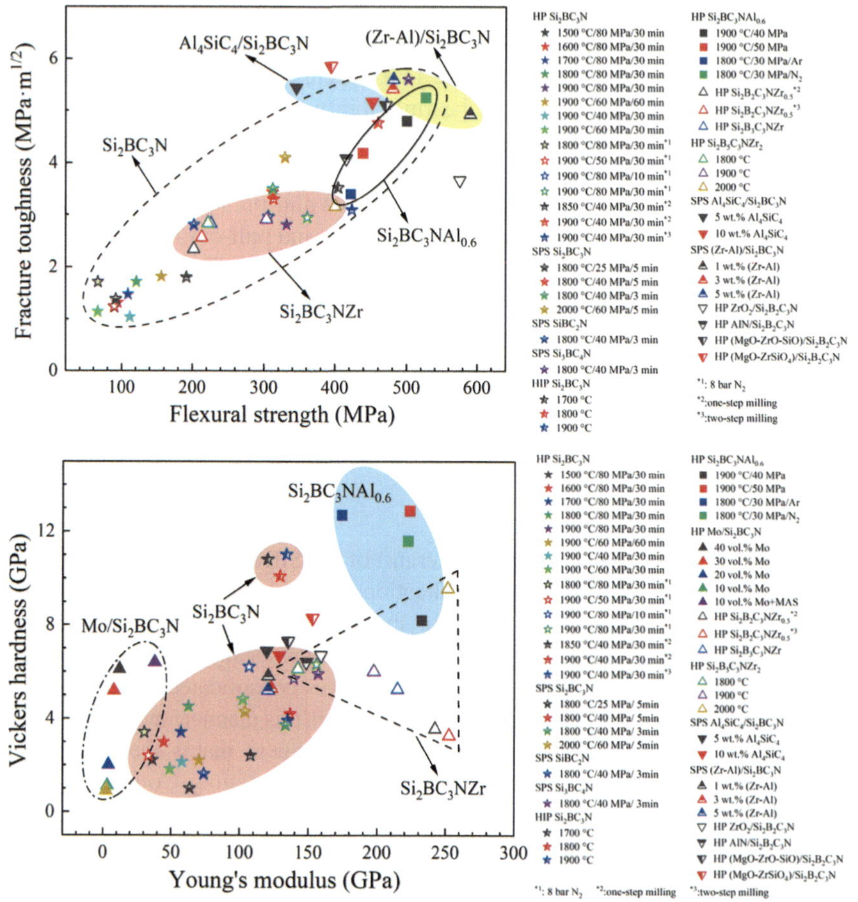

Fig. 5.72 Ashby plots showing toughness-strength and hardness-modulus relationships of the MA-SiBCN monoliths with or without sintering additive

Si_2BC_3N composite ceramics often encounter intergranular fracture, while transgranular fracture is occasionally observed. In most cases, the UHTCs both favors the grain growth of themselves and matrix phases; from a microstructural aspect, increasing the size of the reinforcing UHTCs grains will enhance contributions from both frictional bridging and pull-out. In addition, enhanced interfacial debonding, number of bridging reinforcements, and length of the debonded interfaces should all result in a more rapid rise in their energy dissipation of crack propagation, and therefore the toughness.

Though the attainment of both strength and toughness in MA-SiBCN composite ceramics is achieved, they all exhibit catastrophic fracture; however, the C_f/Si_2BC_3N composites offer pseudoplastic fracture behavior extremely beneficial to thermal

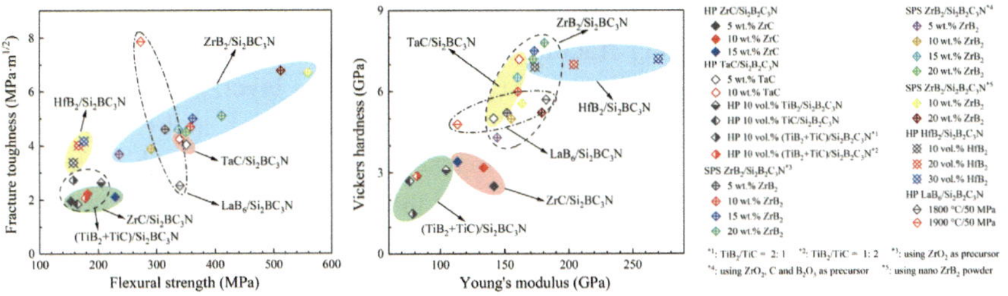

Fig. 5.73 The toughness-strength and hardness-modulus Ashby plots of the UHTCs/Si_2BC_3N composite ceramics prepared by a combination of mechanical alloying plus hot-pressing or spark plasma sintering techniques

shock resistance and therefore structural reliability (Fig. 5.74). Note that, simultaneously added additives like MAS or ZrO_2 to densify matrix, typical of the SiC_f/Si_2BC_3N-MAS CMCs, is often required since the ability of a dense matrix to undergo limited deformation is a critical aspect of conferring toughness, as this feature enables the local dissipation of high stresses that will otherwise cause the material to easy fracture. Interestingly, the addition of MAS or ZrO_2 into C_f/Si_2BC_3N CMCs not only promotes ceramic matrix densification but also prevents severe interfacial reactions thereby improving the mechanical properties of the final composites. Nevertheless, the SiC_f/Si_2BC_3N CMCs still show brittle fracture due to property deterioration of SiC fibers themselves during HP sintering irrespective of whether MAS additive is used or not. Actually, the strong interfacial strength between SiC_f and ceramic matrix also results in a brittle fracture. Coated h-BN layer on SiC_f surfaces effectively tailors the interfacial bonding between fibers and ceramic matrix enhancing the fracture toughness, but it still cannot substantially change the fracture mode of SiC_f/Si_2BC_3N and SiC_f/Si_2BC_3N-MAS CMCs. In contrast, distinct pseudoplastic fracture occurs in $(C_f$-$SiC_f)/Si_2BC_3N$ CMCs indicating improved fracture toughness.

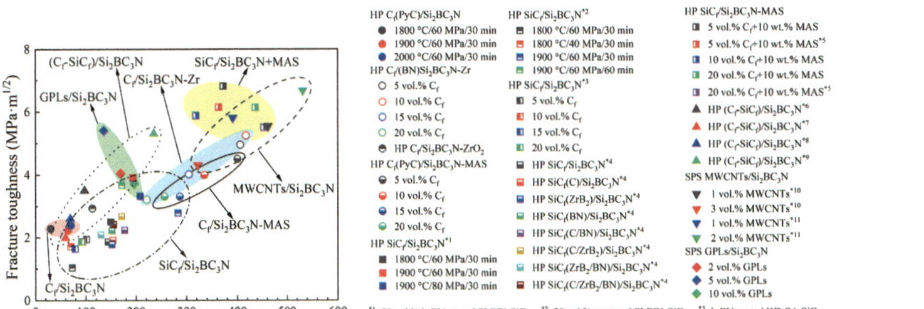

Fig. 5.74 The toughness-strength relationship for the 1-D or 2-D reinforcements reinforced MA-Si_2BC_3N ceramic matrix composites

Generally, both intrinsic and extrinsic toughening mechanisms are working in the MA-SiBCN CMCs during fracture. As mentioned earlier, two major factors influence their toughening mechanisms: (1) the interfacial layer, which affects the debonding and pullout of fibers or graphene platelets (GPLs), and (2) the well-dispersed grains in a fine, nano/submicron-grain-size matrix for crack deflection. However, these strengthening mechanisms are not the answer to all the problems for MA-SiBCN CMCs. The mechanical properties improvement by the addition of 1-D or 2-D reinforcements greatly influences the long-crack toughness but has little or no influence on the short-crack toughness. From a statistical viewpoint, the elastic modulus of the MA-SiBCN bulk ceramics and their CMCs generally increases with increasing hardness (Fig. 5.75), but it is not applicable to all structural ceramic materials. For example, ternary layered MAX phase ceramics have lower hardness but higher elastic modulus. In theory, there is no strong correlation between hardness and elastic modulus, but their relationship must vary with the energy dissipation capacity of a material.

When conventional depth-sensing indentation method of Oliver and Pharr is conducted on a solid material, the reduced modulus E_r and hardness H are correlated by a material parameter of recovery resistance R_s that represents the energy dissipation during indentation [67]. The local elastic–plastic deformation during loading determines its hardness and fracture work while the elastic recovery during unloading reflects its local energy dissipation and reduced elastic modulus. Namely, the greater the local energy dissipation of a material, the smaller hardness to elastic modulus ratio and smaller elastic recovery around the indentation can be obtained. Taking the C_f/Si_2BC_3N-Zr CMCs for example, they show lower ratio of hardness to elastic modulus than those of C_f/Si_2BC_3N counterparts [68], consequently shallower indentation depth and smaller indentation area are found (Fig. 5.76).

The mechanical properties of amorphous PDCs-SiBCN monoliths are closely correlated with their different chemical bonding [69]. Amorphous PDCs-SiBCN

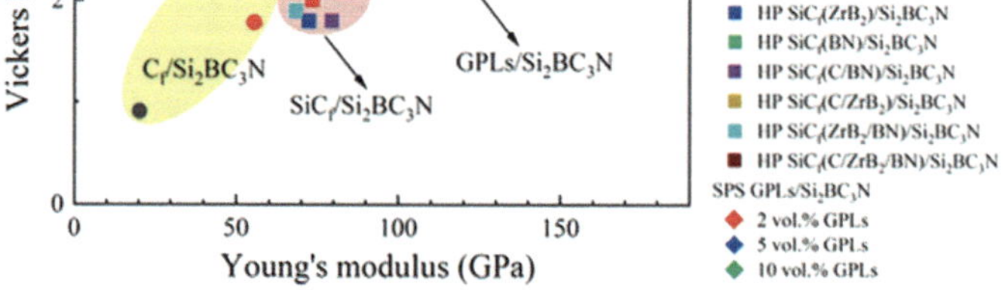

Fig. 5.75 The hardness-modulus Ashby plot for the 1-D or 2-D reinforcements reinforced MA-Si_2BC_3N ceramic matrix composites

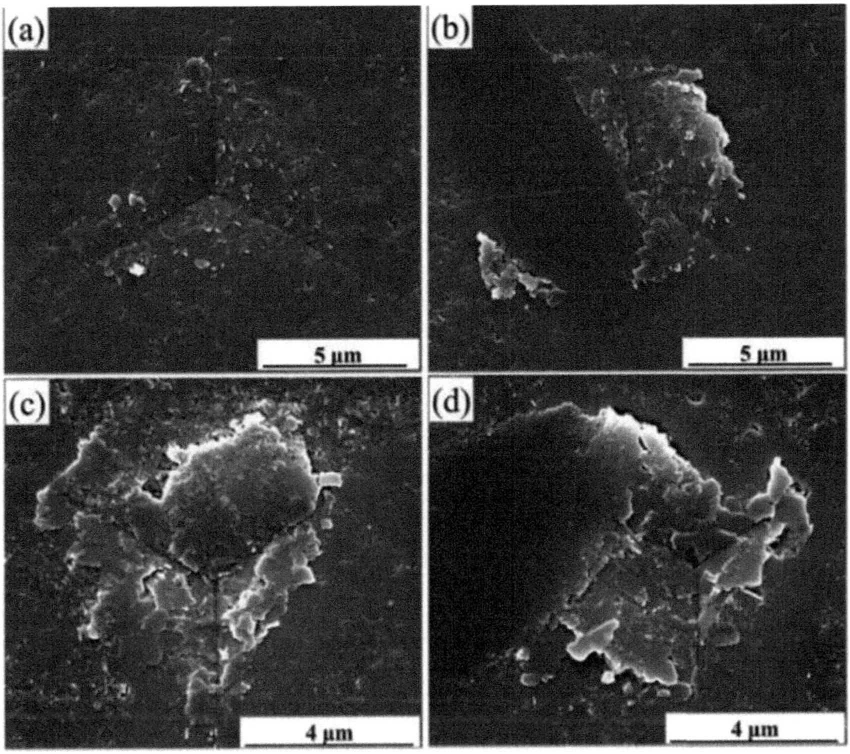

Fig. 5.76 Nanoindentation morphologies of the 10 vol.% C_f/Si_2BC_3N-Zr composites: **a** and **b** Measured at 300 °C; **c** and **d** Measured at RT [50]

derived from acyclic and cyclic single source precursors provides inorganic networks with different chemical composition. Carbon is incorporated into SiBCN via the methylenes bridging silicon and boron. Coincidently, borazine rings can be introduced using a borazine derivative as precursor molecule. Both the thermal stability and hardness are improved compared to the counterparts derived from precursors without cyclic units. The PDCs-$Si_3B_3C_7N_5$ and PDCs-$Si_3B_3C_5N_5$ monoliths are rich in carbon which bonds to four neighbors via strong covalent bonds instead of three for nitrogen thus exhibiting high hardness of about 15 GPa coupled with elastic modulus of ~130 GPa in comparison with PDCs-$Si_3B_3C_5N_7$ and PDCs-$Si_3B_3C_4N_7$ (Fig. 5.77). It also indicates more pronounced covalent character of Si–C bonds compared to Si–N bonds. On the other hand, the additional incorporation of borazine rings where strong p_π–p_π interactions exist between boron and nitrogen into the amorphous random network further increases the stiffness and strength thus leading to higher hardness of PDCs-$Si_3B_3C_4N_7$ (PDCs-$Si_3B_3C_5N_5$) derived from the borazine precursors than PDCs-$Si_3B_3C_5N_7$ (PDCs-$Si_3B_3C_7N_5$) derived from the acyclic precursors.

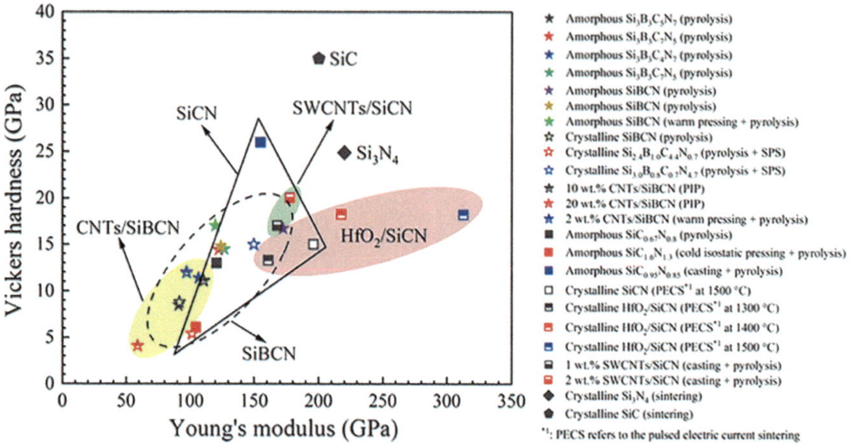

Fig. 5.77 The hardness-modulus Ashby plot of the PDCs-Si(B)CN based materials

Generally, the amorphous SiBCN monoliths exhibit higher hardness and elastic modulus compared to their nanocrystalline counterparts, most likely due to the dense, strong amorphous 3-D networks which ensures high mechanical properties. The similar results are reported by Kumar et al. [70] where the hardness and elastic modulus values of the as-thermolyzed amorphous PDCs-SiBCN monoliths are 16 GPa and 172 GPa, respectively. But the reduction in hardness to 9 GPa and elastic modulus to 92 GPa in the heat-treated nano ceramics is attributed to phase separation and precipitates of SiC and Si_3N_4. This is the same situation for the amorphous MA-Si_2BC_3N monoliths prepared by high pressure sintering (HPS) (1100 °C/5 GPa) show densities up to 2.80 g/cm^3, hardness up to 29.4 GPa and elastic modulus of 291 GPa, while the hardness and elastic modulus of nanocrystalline counterparts prepared by HPS at 1600 °C/5 GPa decrease to 20.2 GPa and 220 GPa, respectively. But thing changes in a PDCs-SiCN ceramic by Chen et al. [71] used nano-indentation studies to obtain a hardness around 30 GPa for crystalline monoliths and 22 GPa for its amorphous counterpart. The corresponding effective modulus is measured about 321.7 GPa for the crystalline ceramic and 164.4 GPa for amorphous one, respectively. The crystalline PDCs-SiCN bulk ceramics including SiC and Si_3N_4 show high hardness and elastic modulus compared to amorphous materials most likely due to the structural disorder of the random networks in amorphous materials, which leads to a compressible local free-volume, and thus to a certain softening.

The mechanical properties of crystallized nano PDCs-SiBCN monoliths are also impacted by their annealing atmospheres where amorphous PDCs-SiBCN powders are synthesized in nitrogen or ammonia atmosphere to modify their carbon content [72]. The nitrogen-treated amorphous powders subsequently sintered in the temperature range 1500–1900 °C lead to monoliths with a density increasing from 2.4 to 2.6 g/cm^3, Vickers's harness and Young's modulus ranging from 0.7 to 5.4 GPa and 48 to 102 GPa, respectively. In comparison, the ammonia-treated amorphous powders SPS consolidated at 1800 °C are mainly composed of Si_3N_4 and BN with a Vickers' hardness as high as 15 GPa.

The PIP processing CNTs/SiBCN martials [73, 74] show Vickers' hardness of pure SiBCN ~ 13.2 GPa, CNTs/SiBCN-10 ~ 11.5 GPa and CNTs/SiBCN-20 ~ 8.7 GPa under 3 N load. The nanoindentation measurement gives the hardness of ~ 11.1 GPa for CNTs/SiBCN-10 and ~ 4.1 GPa for CNTs/SiBCN-20, and the modulus of ~ 110.9 GPa for CNTs/SiBCN-10 and ~ 59.4 GPa for CNTs/SiBCN-20. Obviously, the hardness and modulus of the CNTs/SiBCN nanocomposites are lower than that of pure ceramics, possibly resulting from the limited chemical bonding between CNTs and SiBCN matrix.

Note that, the SiBCN films produced by ion beam sputtering have hardness values up to 22 GPa, comparable to the indentation hardness reported for amorphous SiBCN monoliths. High hardness is also reported for SiBCN films obtained by reactive magnetron sputtering where the target materials, substrate bias voltage, substrate temperature and the chemical composition of final film leads to the various hardness values. Discrepancies in reported hardness values are also found for the SiCN films and coatings due to differences in starting materials and processing conditions.

The high hardness of amorphous PDCs-SiBCN can be ascribed to incorporation of boron into the Si–C–N network which leads to an improved 3-D network connectivity of covalent bonds. Moreover, the formation of covalent B–C and B–N bonds in the system helps in increasing the structural rigidity of the network thereby resulting in high hardness of amorphous PDCs-SiBCN materials. Phase separation and crystallization upon heat-treatment at high temperatures result eventually in the depletion of amorphous network thereby affecting the rigidity between the bonds which leads to the low hardness of crystalline SiBCN materials. Coincidently, the presence of turbostratic phases such as BN(C) and graphite, can also result in the deterioration of the mechanical properties. As mentioned above, the debate between the mechanical performance of the two SiBCN monoliths (amorphous or crystalline) still exists. Only the two monolithic ceramics with the similar density and chemistry are to be compared.

The variety of starting precursors and processing routes also affects the elastic modulus and hardness values obtained for different PDCs-SiCN [6, 75–78] and PDCs-SiOC [79–83] based materials. For an amorphous PDCs-$Si_{0.67}CN_{0.8}$ [75], its mean hardness ~ 13 GPa, measured at a load of 250 mN—is approximately half that of polycrystalline Si_3N_4 (24.9 GPa) but higher than that of SiO_2 glass (8.9 GPa). The elastic modulus of the PDCs-$Si_{0.67}CN_{0.8}$, measured at a load of 250 mN, is ~ 121 GPa much lower than those of polycrystalline Si_3N_4 (220 GPa) and polycrystalline SiC (200 GPa). The spread of the bond angles and the lower atomic density are the possible reasons for its lower stiffness compared to the polycrystalline SiC or Si_3N_4.

The cross-linking and crushed powders are compacted by cold isostatic pressing at 630 MPa and pyrolyzed at 1050 °C in flowing argon to densify crack-free amorphous PDCs-$Si_{1.6}CN_{1.3}$ monoliths with open porosity ~ 9.6%, they show the elastic modulus of ~ 105 GPa measured by pulse-echo method, and Vicker's hardness of only ~ 6.1 GPa calculated from indentation at 98 MPa [6]. But a fully dense amorphous PDCs-$SiC_{0.95}N_{0.85}$ monolith by pressure casting process exhibits Young's modulus of 155

GPa, Vickers' hardness increasing from 15 GPa at an applied load of 1 N to 26 GPa at 3 N [76]. The fracture strength of up to 1.1 GPa is also measured for the PDCs-$SiC_{0.95}N_{0.85}$. The reported hardness values of the powder route specimens are in the range 6–16 GPa, significantly lower than the 25–27 GPa measured for the cast PDCs-$SiC_{0.95}N_{0.85}$. The microhardness of the cast specimens is unusually large, nearly equal to that of polycrystalline Si_3N_4 (24.9 GPa) and CVD SiCN coating. These values are particularly remarkable in view of the much lower modulus of the PDCs-$SiC_{0.95}N_{0.85}$, about one third that of the crystalline materials. Thus, the hardness to modulus ratio of the PDCs-$SiC_{0.95}N_{0.85}$ is two to three times higher than for polycrystalline SiC or Si_3N_4. The authors argue possible reasons as the molecular networks that support the majority of the load in a simple tensile elastic deformation are different from the molecular ensembles that support the load in an indentation experiment.

The introduction of homogeneously distributed octadecylamine-functionalized single-walled carbon nanotubes (SWCNTs) or hafnia incorporated PDCs-SiCN ceramics can both favor their mechanical performance [77]. For the former, the Young's modulus and the hardness of SWCNTs/SiCN composites show increase by ~ 30 and ~ 15%, and twofold enhancement of fracture toughness is measured. In the latter, the hardness and elastic modulus values of HfO_2/SiCN pyrolyzed at 1500 °C are ~15 GPa and ~196 GPa, respectively. The hardness value is in close agreement with that of powder route SiCN, whereas the elastic modulus is ~40% higher for the pure PDCs-SiCN [78], possibly due to the rigid structure obtained by the pulsed electric current sintering (PECS) technique. Further, for the HfO_2/SiCN ceramics [78], their elastic modulus is increased by ~85% as sintering temperature increases from 1300–1500 °C.

The PDCs-SiOC ceramics, regardless of their physical forms (monoliths or microspheres), residual phases (Si or free carbon) or microstructures (amorphous or nanocrystalline) [79–83], generally show relative lower hardness and Young's modulus as compared with the PDCs-SiBCN and PDCs-SiCN, most likely due to the composition of the amorphous oxycarbide network (formed by a random array of SiC_xO_{4-x} units ($0 \leq x \leq 4$) and carbidic carbon CSi_4 unit), the amount of free carbon, precipitated phase composition, etc. (Fig. 5.78). Interestingly, the obtained mechanical values are different once different testing method is adopted; however, the difference is not large. For example [79], the PDCs-SiOC microspheres are prepared by the ceramization of polysiloxane microspheres from processing the aqueous emulsion of polyhydromethylsiloxane, and show a hardness of 10.6–14.0 GPa and a high elastic modulus of 87.9–122.3 GPa, as measured for particles embedded in the resin matrix. However, the hardness and elastic modulus of particles embedded in glass (substantially harder than the resin matrix) are in the range 13.8–15.9 GPa and 122.2–144.6 GPa, respectively. The particles possessing an enhanced content of silicon-bonded carbon exhibit the largest hardness and elastic modulus regardless of their microstructure forms (amorphous or crystalline).

In most case, both the Young's modulus and the hardness of the PDCs-SiOC ceramics decrease with increasing free carbon content and follow a simple rule of mixture model [81]. In particular the elastic modulus decreases linearly from ~ 106

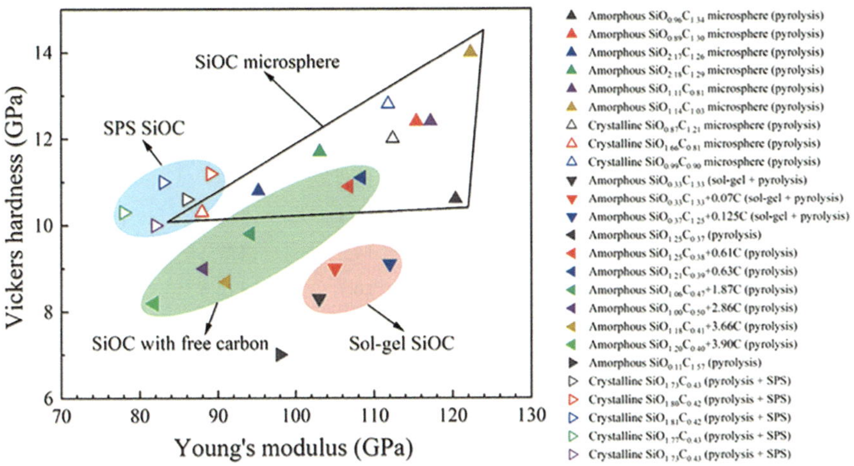

Fig. 5.78 The hardness-modulus Ashby plot of the PDCs-SiOC materials

GPa to ~ 80 GPa as the free carbon content increases from ~ 18 vol.% to ~ 60 vol%. In the same free carbon content range, the hardness decreases from ~ 11 GPa to ~ 8 GPa. However, for the sol–gel derived SiOC ceramics [80], their elastic modulus (E), Vickers' hardness (H_V) and flexural strength (MOR) increase markedly with the amount of carbon in the oxycarbide glasses reaching the maximum values (E = 115 GPa, H_V = 9 GPa and MOR = 550 MPa) for samples with the highest carbon content.

Besides, the elastic modulus, flexural strength and Vickers' hardness of the sol–gel derived SiOC ceramics increase with pyrolysis temperature and remarkably increase with an augmentation of the C/Si ratio, which correspond with the amount of SiC [82]. In comparison with silica glass, the presence of tetracoordinated C atoms in an amorphous SiO_2 network is the reason for the improvement of mechanical properties and thermal stability. The SiOC bulk ceramics SPS obtained at 1400 °C/40 MPa show the highest values of hardness (~12 GPa) and elastic modulus (~95 GPa) in both nano- and micro-indentation tests [83], while the lowest results displayed by the material sintered at 1700 °C due to phase separation and an incipient carbothermal reduction of SiO_2 (Table 5.46).

The continuous fiber reinforced Si(B)CN ceramic matrix composites (CMCs) can overcome the conflict of attaining strength and toughness [84–93], especially for the C_f/SiBCN and SiC_f/SiBCN CMCs (Fig. 5.79). The 3-D C_f/SiBCN CMCs [84] with an open porosity ~ 10% and uniform microstructure are obtained after only 7 cycles of PIP, and show flexural strength of 371 MPa and fracture toughness of 12.9 MPa $m^{1/2}$, respectively. However, for the C_f/ZrB_2-SiBCN composites prepared by dipping and winding combined with reactive hot-pressing [87], they show deceasing flexural strength and fracture toughness to 261 MPa and 11.96 MPa $m^{1/2}$, but obtaining good mass and linear ablation rates of 1.78 mg/s and 1.01 μm/s (plasma flame with a heat flux of 9.37 MW/m^2 for 300 s), respectively. In another contribution [88], the

Table 5.46 The hardness and Young's modulus of the PDCs-Si(B)CN and PDCs-SiOC materials

Composition	Preparation method	Hardness (GPa)	Young's modulus (GPa)	Reference
$Si_3B_3C_5N_7$ (amorphous)	Pyrolysis	8.5	91	[69]
$Si_3B_3C_7N_5$ (amorphous)		14.4	123	[69]
$Si_3B_3C_4N_7$ (amorphous)		11.4	107	[69]
$Si_3B_3C_5N_5$ (amorphous)		14.5	127	[69]
SiBCN (amorphous)		16.7	172.6	[70]
SiBCN (crystalline)		8.8	91.8	[70]
$Si_{2.4}B_{1.0}C_{4.4}N_{0.7}$ (crystalline)	Pyrolysis + SPS	5.4	102	[72]
$Si_{3.0}B_{0.8}C_{0.7}N_{4.7}$ (crystalline)		15	150	[72]
SiBCN (amorphous)	Pyrolysis	14.8	124.4	[73]
10 wt.% CNTs/SiBCN	PIP	11.1	110.9	[73]
20 wt.% CNTs/SiBCN		4.1	59.4	[73]
SiBCN (amorphous)	Warm pressing + pyrolysis	17	120	[74]
2 wt.% CNTs/SiBCN		12	97.5	[74]
$SiC_{0.67}N_{0.8}$ (amorphous)	Pyrolysis	13	121	[75]
$SiC_{1.6}N_{1.3}$ (amorphous)	Cold isostatic pressing + pyrolysis	6.1	105	[6]
$SiC_{0.95}N_{0.85}$ (amorphous)	Casting + pyrolysis	26	155	[76]
1 wt.% SWCNTs/SiCN		17	168	[77]
2 wt.% SWCNTs/SiCN		20	178	[77]
SiCN (crystalline)	[a]PECS at 1500 °C	15	196	[78]
HfO_2/SiCN (crystalline)	[a]PECS at 1300 °C	13.28	161	[78]
HfO_2/SiCN (crystalline)	[a]PECS at 1400 °C	18.20	218	[78]
HfO_2/SiCN (crystalline)	[a]PECS at 1500 °C	18.20	313	[78]
SiOC microspheres (amorphous)	Pyrolysis	10.6	120.4	[79]
		12.4	115.4	[79]
		10.8	95.2	[79]
		11.7	103.1	[79]
		12.4	117.2	[79]
		14.0	122.3	[79]
SiOC microspheres (crystalline)		12.0	112.4	[79]
		10.3	87.9	[79]
		12.8	111.8	[79]

(continued)

Table 5.46 (continued)

Composition	Preparation method	Hardness (GPa)	Young's modulus (GPa)	Reference
$SiO_{0.33}C_{1.33}$ without free C (amorphous)	Sol–gel + pyrolysis	8.3	103	[80]
$SiO_{0.33}C_{1.33}$ + 0.07C (amorphous)		9.0	105	[80]
$SiO_{0.37}C_{1.25}$ + 0.125C (amorphous)		9.1	112	[80]
$SiO_{1.25}C_{0.37}$ without free C (amorphous)	Pyrolysis	11.1	108.4	[81]
$SiO_{1.25}C_{0.38}$ + 0.61C (amorphous)		10.9	106.8	[81]
$SiO_{1.21}C_{0.39}$ + 0.63C (amorphous)		11.1	108.4	[81]
$SiO_{1.06}C_{0.47}$ + 1.87C (amorphous)		9.8	94.2	[81]
$SiO_{1.00}C_{0.50}$ + 2.86C (amorphous)		9.0	88.1	[81]
$SiO_{1.18}C_{0.41}$ + 3.66C (amorphous)		8.7	91.0	[81]
$SiO_{1.20}C_{0.40}$ + 3.90C (amorphous)		8.2	81.8	[81]
$SiO_{0.11}C_{1.57}$ (amorphous)		7.0	98	[82]
$SiO_{1.73}C_{0.43}$ (crystalline)	Pyrolysis + SPS	10.6	86	[83]
$SiO_{1.80}C_{0.42}$ (crystalline)		11.2	89	[83]
$SiO_{1.81}C_{0.42}$ (crystalline)		11.0	83	[83]
$SiO_{1.77}C_{0.43}$ (crystalline)		10.3	78	[83]
$SiO_{1.73}C_{0.43}$ (crystalline)		10.0	82.0	[83]
Si_3N_4 (crystalline)	Sintered	24.9	220	[6]
SiC (crystalline)		35.0	200	[6]

[a]PECS refers to the pulsed electric current sintering

well-established C_f/ZrB_2-SiBCN composites with distinct multilayer interphases are fabricated from a combination of slurry injection and PIP method. The composites with high-density fabric (~21 vol.%) and three coating layer ~ 360 nm offer the maximum fracture toughness of 6.34 MPa $m^{1/2}$. Whereas, their counterparts with low-density fabric (~8 vol%) and a monolayer coating ~ 200 nm present its maximum of only 4.03 MPa $m^{1/2}$ (Table 5.47).

The hydrothermal method is used to deposit a lanthanum monazite ($LaPO_4$) coating ~ 700 nm on SiC fiber surfaces, and subsequential the $SiC_f/LaPO_4$/SiBCN, SiC_f/BN/SiBCN, SiC_f/SiBCN CMCs are prepared using resin transfer molding

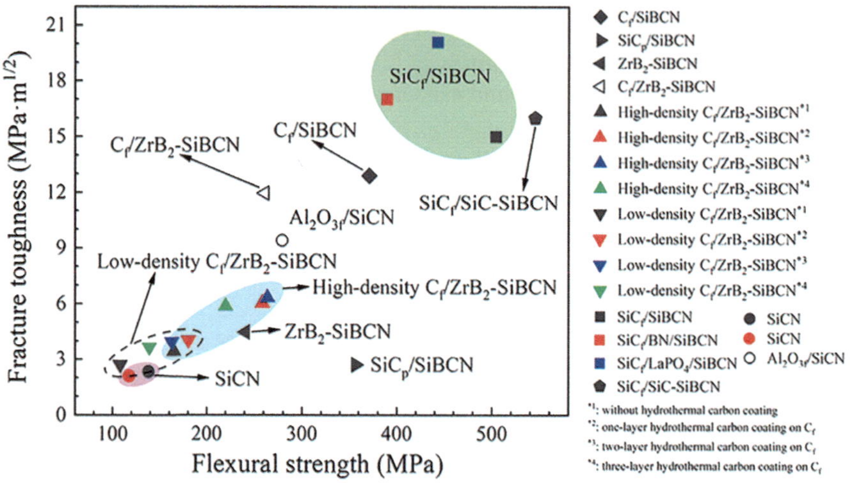

Fig. 5.79 The toughness-strength Ashby plots of the PDCs-Si(B)CN materials

(RTM) and polymer infiltration pyrolysis (PIP) techniques [89]. The resulting SiC_f/$LaPO_4$/SiBCN composites provide the almost identical flexure and tensile strength with the highest fracture toughness of 20.1 MPa $m^{1/2}$ at a cost of strength loss. The flexure strength retention ratio for the SiC_f/$LaPO_4$/SiBCN CMCs are measured to be higher than those of the SiC_f/SiBCN composites. The SiC_f/$LaPO_4$/SiBCN composites oxidized at 1350 °C for 50 h in air, exhibit an ultimate strength value of around 160 MPa with a retention ratio of 36.6%, followed by a fluctuating drop in stress values to ultimately reach zero. In another attempt [90], the chemical vapor infiltration (CVI) and/or PIP technologies are introduced to the SiC_f/SiC-SiBCN composites with a h-BN coating ~ 300 nm on fibers, obtaining flexural strength of 545.5 MPa thanks to the crack deflection between the CVI SiC and CVI SiBCN, as well as CVI SiBCN and PIP SiBCN matrix. The fracture toughness is measured 16.02 MPa $m^{1/2}$ owing to the extension of crack propagation path.

For the PIP + CVI Al_2O_{3f}/SiCN composites [93], their fracture toughness, bending modulus and flexural strength are tested to be 9.4 MPa $m^{1/2}$, 52 GPa and 279 MPa, respectively. The PIP + CVI Al_2O_{3f}/SiCN CMCs show better mechanical properties than other PIP densifying CMCs with a smaller open porosity of 4.5%. The fracture toughness 7.9 MPa $m^{1/2}$ and flexural strength 155 MPa for the PIP II-SiC_f/SiCN, while those of PIP I-SiC_f/SiCN are 10.2 MPa $m^{1/2}$ and 287 MPa, respectively. The SiCN monoliths [91, 92] show low strength of 118, 138.1 MPa, and fracture toughness of 2.10, 2.32 MPa $m^{1/2}$, also lower than those of ZrB_2-SiBCN and SiC_p-SiBCN monolithic ceramics [85, 86].

A high-performance PDCs-SiBCN fiber with average tensile strengths of 1.5 GPa and Young's modulus of 160 GPa was also reported [94]. A novel polyborosilazane with limited cross-link density was synthesized to obtain long polyborosilazane fibers by melt spinning. As-cured fibers were pyrolyzed in nitrogen at 1200 °C yielding

Table 5.47 The flexural strength and fracture toughness of the PDCs-Si(B)CN materials

Composition	Flexural strength (MPa)	Fracture toughness (MPa $m^{1/2}$)	References
C_f/SiBCN	371	12.9	[84]
SiC_p/SiBCN	357	2.7	[85]
ZrB_2-SiBCN	241	4.47	[86]
C_f/ZrB_2-SiBCN	261	11.96	[87]
High-density C_f/ZrB_2-SiBCN[a]	165	3.41	[88]
High-density C_f/ZrB_2-SiBCN[b]	259	6.02	[88]
High-density C_f/ZrB_2-SiBCN[c]	264	6.34	[88]
High-density C_f/ZrB_2-SiBCN[d]	220	5.88	[88]
Low-density C_f/ZrB_2-SiBCN[a]	108	2.71	[88]
Low-density C_f/ZrB_2-SiBCN[b]	181	4.03	[88]
Low-density C_f/ZrB_2-SiBCN[c]	163	3.96	[88]
Low-density C_f/ZrB_2-SiBCN[d]	139	3.67	[88]
SiC_f/SiBCN	504	15	[89]
SiC_f/BN/SiBCN	390	17	[89]
$SiC_f/LaPO_4$/SiBCN	443	20.1	[89]
SiC_f/SiC-SiBCN	545	16	[90]
SiCN	118	2.1	[91]
SiCN	138.1	2.32	[92]
Al_2O_{3f}/SiCN	279	9.4	[93]

[a] Without hydrothermal carbon coating
[b] One-layer hydrothermal carbon coating on C_f
[c] Two-layer hydrothermal carbon coating on C_f
[d] Three-layer hydrothermal carbon coating on C_f

PDCs-SiBCN fibers with average diameters of ca. 12 μm that remain amorphous up to 1700 °C. Baldus et al. [95] reported a PDCs-$SiBN_3C$ fiber prepared by pyrolysis of N-methylpolyborosilazane green fibers obtained by melt spinning. The resulting PDCs-$SiBN_3C$ fiber shows a low density of ca. 1.8 g/cm^3, good tensile strength of ca. 3 GPa, and elastic modulus of ca. 300 GPa at room temperature. Coincidently, it exhibits a low thermal expansion coefficient of 3×10^{-6} °C^{-1} and thermal conductivity of 3 W/(m K) at 1200 °C. It can easily withstand temperatures up to 1800 °C in inert atmospheres and up to 1500 °C under oxidizing conditions for long periods, while retaining almost all of their mechanical strength. For instance, it retains ~ 83% of its tensile strength and elastic modulus on air exposure to 1400 °C. A protective SiO_2/BN bilayer and the stability of the random network ensure that the fiber is corrosion resistant to liquid Si at high temperatures up to 1600 °C. The excellent combination of mechanical, thermophysical and high-temperature properties, especially oxidation resistance, superior to SiC fibers makes them to be promising reinforcements in high temperature CMCs [96, 97].

Nota that the pseudoplastic fracture are commonly observed in the C_f/SiBCN, C_f/ZrB_2-SiBCN and SiC_f/SiBCN CMCs, etc. prepared by the PDCs route [16, 68]. Taking the PIP processing C_f/SiBCN CMCs for example (Fig. 5.80), they show an average flexural strength of 255 MPa, and excellent creep resistance characterized by the low flexural creep strain of 0.55% at 1400 °C. The C_f/SiBCN composites retain most of its bending strength even at 1500 °C in Ar and no evidence for brittle fracture. Unfortunately, matrix crystallization and decomposition result in shrinkage and crack formation. In addition, the increasing residual stress in the CMCs with increasing heating temperature (due to differences in CTE between fibers and ceramic matrix) also leads to cracks to release residual stress during heating. Consequently, degradation of the PDCs-SiBCN matrix decreases strength together with changing fracture behavior after heating at above 1700 °C. The extra addition of SiC particulate fillers improves the thermal stability of the C_{fiber}/SiC_{filler}/$SiBCN_{matrix}$ CMCs up to 1700 °C leading to better mechanical properties (Fig. 5.81), retaining ~ 96% of its room-temperature strength and showing non-brittle fracture after heating at 1700 °C for 10 h in Ar [15, 16, 68]. In the SiC_p/SiCN CMCs [85], the composites even retain 88% of the original strength after heating at 2000 °C for 10 h although the fabrication temperature is only 1350 °C.

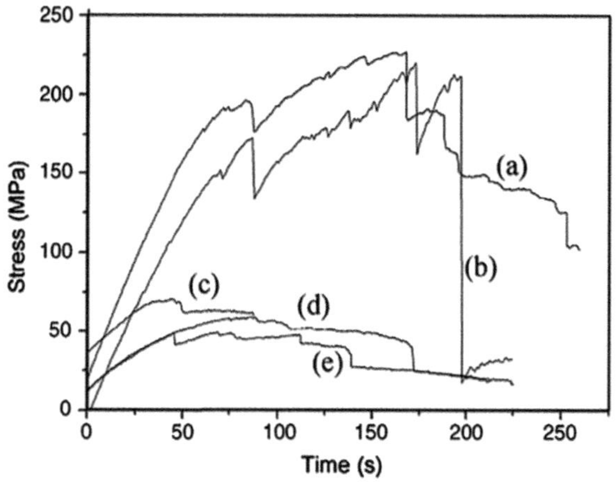

Fig. 5.80 Fracture behavior of the C_f/SiBCN CMCs consolidated by polymer infiltration pyrolysis technique: **a** As-fabricated; **b** *in-situ* bending at 1500 °C in Ar; heat treatment for 10 h in Ar at **c** 1700 °C; **d** 1900 °C; **e** 2000 °C [68]

Fig. 5.81 Fracture surfaces of the $C_{fiber}/SiC_{filler}/SiBCN_{matrix}$ CMCs: **a** As-fabricated; after heating at **b** 1700 °C for 10 h and **c** 2100 °C for 5 min in Ar [16]

The improved mechanical properties of the fiber/SiBCN CMCs are observed on a macro-scale through several mechanisms: fiber debonding, fiber pullout, crack bridging and/or crack deflection. Naturally, the common idea in these toughening mechanisms is to increase the energy needed to extend a crack. The above toughened behavior generally arises from the lower-strength boundaries of fibers/SiBCN matrix and is governed by the fiber length and volume fraction, though their Young's modulus and thermal expansion coefficient have a role.

A low thermal expansion coefficient (CTE) coupled with low elastic modulus and high flexural strength plays a positive role in improving thermal shock resistance and structural stability at high temperatures. The average CTE of MA-Si_2BC_3N monoliths is (4.0–4.8) × 10^{-6} °C^{-1}, which approaches to that of β-SiC. Thermal conductivity it is about 9.4 W/(m K) at room temperature but varies from 7.3 W/(m K) to 8.4 W/(m K) as temperatures increase from 200 to 1200 °C, far lower than pure SiC at room temperature. The thermal diffusion coefficient, thermal conductivity and CTE of MA-Si_2BC_3N monoliths prepared by SPS are lower than pure SiC due to the low degree of crystallinity or even the residual amorphous structures in the ceramic matrix. The increase in relative Si and C contents increases relative SiC content in sintered bulk ceramics eventually resulting in increases in thermal diffusion coefficient, thermal conductivity and CTE with decreases in specific heat. The values of these four thermophysical parameters for MA-Si_3BC_4N are lower than pure SiC.

Additionally, the source and content of Al have a great influence on CTE at 1000–1300 °C. The CTE values for MA-$Si_2BC_3NAl_{0.6}$ (5.1 × 10^{-6} °C^{-1}), MA-$Si_2BC_3N_{1.6}Al_{0.6}$ (4.9 × 10^{-6} °C^{-1}) and MA-$Si_2BC_3NAl_{0.2}$ (4.6 × 10^{-6} °C^{-1}) decrease in sequence most likely due to content increases in AlON (7.0 × 10^{-6} °C^{-1}) with higher CTE than AlN (4.5 × 10^{-6} °C^{-1}), BCN (5.5 × 10^{-6} °C^{-1}) and SiC (4.7 × 10^{-6} °C^{-1}). SiC fibers have CTEs almost equivalent to MA-Si_2BC_3N monolithic ceramics, while carbon fiber provides a lower CTE. Consequently, the CTE of SiC_f/Si_2BC_3N CMCs shows little change compared with MA-Si_2BC_3N, while the value for C_f/Si_2BC_3N CMCs decreases to 1.6 × 10^{-6} °C^{-1}.

The investigation of thermophysical properties of PDCs-Si(B)CN materials mainly centers on the ceramic films, coatings, fibers, monoliths and CMCs. The thermal expansion coefficient of crack-free amorphous SiCN monoliths heat treated up to 1400 °C increase from 3.08 × 10^{-6} °C^{-1} at 100 °C to 3.96 × 10^{-6} °C^{-1} at 1200 °C [6]. Coefficient of thermal expansion of amorphous SiCN(O) bulk ceramics [91] is as lower as (2.15–2.69) × 10^{-6} °C^{-1}. The thermal conductivity of CVI + PIP SiC_f/SiC-SiBCN CMCs in the thickness direction is 7.64 W/(m K) at 1200 °C with decreased electric resistivity to 1.53 × 103 Ω m [90]. The SiC coated needle-punched C_f fabrics reinforced PDCs-SiBCN composites enjoy a high compressive strength ~ 155.3 MPa, a light weight of 1.07–1.61 g/cm^3, and a relatively low thermal conductivity ~ 4.02 W/(m K) [98]. The PDCs-$SiBN_3C$ fibers show small CTE of 3 × 10^{-6} °C^{-1} [95], comparable to that of hot-pressed MA-Si_2BC_3N monoliths and PDCs-SiCN [99], and a thermal conductivity of ~ 3.0 W/(m K) at 1200 °C which is much lower than that of hot-pressed counterparts. The reported SiBCN coatings provide a low thermal expansion coefficient of (2.0–2.4) × 10^{-6} °C^{-1}, close to the carbon fibers, which makes them become possible candidates for high-temperature protection of novel materials.

References

1. Zhang PF, Jia DC, Yang ZH, Duan XM, Zhou Y. Crystallization and microstructural evolution process from the mechanically alloyed amorphous Si-B-C-N powder to the hot-pressed nano SiC/BN(C) ceramic. J Mater Sci. 2012;47:7291–304.
2. Zhang PF, Jia DC, Yang ZH, Duan XM, Zhou Y. Microstructural features and properties of the nano-crystalline SiC/BN(C) composite ceramic prepared from the mechanically alloy SiBCN powder. J Alloy Compd. 2012;537:346–56.
3. Yang ZH, Zhou Y, Jia DC, Meng QC. Microstructures and properties of $SiB_{0.5}C_{1.5}N_{0.5}$ ceramics consolidated by mechanical alloying and hot pressing. Mater Sci Eng A. 2008;489:187–92.
4. Yang ZH. Microstructure and high-temperature properties of the Si-B-C-N MA-powders and ceramics. Harbin: Harbin Institute of Technology; 2008.
5. Kumar NVR, Prinz S, Cai Y, Zimmermann A, Aldinger F, Berger F, Muller K. Crystallization and creep behavior of Si-B-C-N ceramics. Acta Mater. 2005;53(17):4567–78.
6. Kiryukhantsev-Korneev PV, Sheveyko AN, Levashov EA, Shtansky DV. Investigation of the Si-B-C-N coatings deposited by magnetron sputtering of SiBC targets. Russ J Non-Ferrous Met. 2015;56:540–7.
7. Kumar BVM, Roh MH, Kim YW, Kim W, Park SW, Seo W-S. Effect of additive composition on microstructure and mechanical properties of SiC ceramics sintered with small amount of RE_2O_3 (RE: Sc, Lu, Y) and AlN. J Mater Sci. 2009;44(21):5939–43.
8. Vishnyakov VM, Ehiasarian AP, Vishnyakov VV, Hovsepian P, Colligon JS. Amorphous boron containing silicon carbo-nitrides created by ion sputtering. Surf Coat Technol. 2011;206(1):149–54.
9. Houska J, Vlcek J, Potocky S, Perina V. Influence of substrate bias voltage on structure and properties of hard Si-B-C-N films prepared by reactive magnetron sputtering. Diam Relat Mater. 2007;16(1):29–36.
10. Zhang PF. Hot pressing behavior and high-temperature properties of mechanically alloyed 2Si-B-3C-N ceramic. Harbin: Harbin Institute of Technology; 2013.
11. Ravi-Kumar NV, Mager R, Cai Y, Zimmermann A, Aldinger F. High temperature deformation behaviour of crystallized Si-B-C-N ceramics obtained from a boron modified poly(vinyl)silazane polymeric precursor. Scripta Mater. 2004;51:65–9.
12. Christ M, Thurn G, Weinmann M, Bill J, Aldinger F. High-temperature mechanical properties of Si-B-C-N-precursor-derived amorphous ceramics and the applicability of deformation models developed for metallic glasses. J Am Ceram Soc. 2000;83:3025–32.

13. Kumar R, Phillipp F, Aldinger F. Oxidation induced effects on the creep properties of nano-crystalline porous Si-B-C-N ceramics. Mater Sci Eng, A. 2007;445–446:251–8.
14. Ravikumar N, Prinz S, Cai Y, Zimmermann A, Aldinger F, Berger F, Muller K. Crystallization and creep behavior of Si-B-C-N ceramics. Acta Mater. 2005;53(17):4567–78.
15. Lee SH, Weinmann M, Gerstel P, Aldinger F. Extraordinary thermal stability of SiC particulate-reinforced polymer-derived Si-B-C-N composites. Scripta Mater. 2008;59:607–10.
16. Lee SH, Weinmann M. C_{fiber}/SiC_{filler}/Si-B-C-N matrix composites with extremely high thermal stability. Acta Mater. 2009;57:4374–81.
17. Liang B, Yang ZH, Rao JC, Cai DL, Duan XM, Jia DC, Zhou Y, Yu DL, Tian YJ. Highly dense amorphous si_2bc_3n monoliths with excellent mechanical properties prepared by high pressure sintering. J Am Ceram Soc. 2015;98:3782–7.
18. Kosilov AT, Mikhailov VA, Sviridov VV, Khonik VA. Kinetics of isothermal creep in metallic glasses including the statistical distribution of activation parameters. Phys Solid State. 1997;39:1796–802.
19. Khonik VA, Kosilov AT, Mikhailov VA, Sviridov VV. Isothermal creep of metallic glasses: a new approach and its experimental verification. Acta Mater. 1998;46:3399–408.
20. Bagchi B. A molecular explanation of the transition from viscous to hopping mechanism of mass transport in the supercooled liquid near the glass transition. J Chem Phys. 1994;101:9946–55.
21. Yang ZH, Jia DC, Zhou Y, Zhang JX. Processing and characterization of $SiB_{0.5}C_{1.5}N_{0.5}$ produced by mechanical alloying and subsequent spark plasma sintering. Mate Sci Eng A. 2008;488:241–6.
22. Yang ZH, Jia DC, Duan XM, Sun KN, Zhou Y. Effect of Si/C ratio and their content on the microstructure and properties of Si-B-C-N ceramics prepared by Spark plasma sintering techniques. Mater Sci Eng, A. 2011;528:1944–8.
23. Daly BC, Antonelli GA, Maris HJ, Ford WK, Wong L, Andideh E. Measurements of the thermal conductivity of amorphous materials with low dielectric constants. Physica B. 2002;316:254–7.
24. Maruyama T, Harayama M. Relationship between dimensional changes and the thermal conductivity of neutron-irradiated SiC. J Nucl Mater. 2004;329:1022–8.
25. Liang B. Crystallization and high temperature oxidation mechanisms of amorphous Si_2BC_3N ceramics by high pressure sintering. Harbin: Harbin Institute of Technology; 2017.
26. Liang B, Yang ZH, Chen QQ, Wang SJ, Duan XM, Jia DC, Zhou Y, Luo K, Yu DL, Tian YJ. Crystallization behavior of amorphous Si_2BC_3N ceramic monolith subjected to high-pressure. J Am Ceram Soc. 2015;98:3788–96.
27. Ye D. Microstructure and oxidation resistance of the mechanically alloyed Si-B-C-N-Al powders and ceramics. Harbin: Harbin Institute of Technology; 2012.
28. Hou JN. Microstructural design and thermal shock resistance of Mo-SiBCN gradient composite materials. Harbin: Harbin Institute of Technology; 2016.
29. Zhang M, Hou JN, Li DX, Liang B, Niu ZB, Zhang Q, Yang ZH, Jia DC, Cai DL, Duan XM, He PG, Zhou Y. Mo-SiBCN metal-ceramic composites with enhanced and tunable thermophysical properties and thermal shock resistance. Ceram Int. 2022;48:5744–51.
30. Sun YH, Zhang YF, Guo JK. Effect of expansion coefficient of additive on mechanical properties of Y-TZP. J Inorg Mater. 2003;1:78–82.
31. Li DX, Yang ZH, Jia DC, Hu CC, Liang B, Zhou Y. Preparation, microstructures, mechanical properties and oxidation resistance of SiBCN/ZrB_2-ZrN ceramics by reactive hot pressing. J Eur Ceram Soc. 2015;35:4399–410.
32. Liao N, Jia DC, Yang ZH, Zhou Y, Li YW. Mechanical properties and thermal shock resistance of Si_2BC_3N ceramics with ternary Al_4SiC_4 additive. Ceram Int. 2018;44:9009–17.
33. Wang GY. Preparation and properties of SIBCN(Zr) gradient porous ceramics prepared by ice molding and their bicontinuous composites. Harbin: Harbin Institute of Technology; 2018.
34. Liao N, Jia DC, Yang ZH, Zhou Y, Li YW. Enhanced mechanical properties, thermal shock resistance and oxidation resistance of Si_2BC_3N ceramics with Zr-Al addition. Mater Sci Eng, A. 2018;725:364–74.
35. Zhang PF, Yang B, Lu Z, Jia DC. Effect of AlN and ZrO_2 on the microstructure and property of the 2Si-B-3C-N ceramic. Ceram Int. 2018;44:3406–11.

36. Li DX, Yang ZH, Mao ZB, Jia DC, Cai DL, Liang B, Duan XM, He PG, Rao JC. Microstructures, mechanical properties and oxidation resistance of SiBCN ceramics with the addition of MgO, ZrO_2 and SiO_2 (MZS) as sintering additives. RSC Adv. 2015;5:52194–205.
37. Yang ZH, Li DX, Zhao Y, Miao Y, Rao JC, Cai DL, Duan XM, He PG, Jia DC, Zhou Y. *In Situ* ZrC/Si-B-C-N monoliths prepared by sol-gel and reactive hot-pressing: processing, microstructure, mechanical properties and oxidation behavior. J Alloy Compd. 2019;811:151687.
38. Wang BZ, Yang ZH, Li DX, Cai DL, Niu B, Guan JY, Jia DC, Zhou Y. Effects of TaC addition on microstructure and mechanical properties of SiBCN composite ceramics. Ceram Int. 2019;45:22138–47.
39. Wang BZ, Li DX, Yang ZH, Jia DC, Guan JY, Peng H, Cai DL, He PG, Duan XM, Zhou Y, Zhang T, Gao CG. Microstructural evolution and mechanical properties of *in situ* nano Ta_4HfC_5 reinforced SiBCN composite ceramics. J Adv Ceram. 2020;9:739–48.
40. Liao XQ. Microstructure and properties of the (TiB_2+TiC)/SiBCN composites. Harbin: Harbin Institute of Technology; 2014.
41. Miao Y, Yang ZH, Liang B, Li Q, Chen QQ, Jia DC, Cheng YB, Zhou Y. A novel *in situ* synthesis of SiBCN-Zr composites prepared by a sol-gel process and spark plasma sintering. Dalton Trans. 2016;45:12739–44.
42. Miao Y, Yang ZH, Rao JC, Duan XM, He PG, Jia DC, Cheng YB, Zhou Y. Influence of sol-gel derived ZrB_2 additions on microstructure and mechanical properties of SiBCN composites. Ceram Int. 2017;43:4372–8.
43. Liao N, Jia DC, Yang ZH, Niu B, Zhou Y, Li YW. Enhanced mechanical properties, thermal shock resistance and ablation resistance of Si_2BC_3N ceramics with nano ZrB_2 addition. J Eur Ceram Soc. 2019;39:846–59.
44. Ao DF. The thermal shock and ablative properties of HfB_2/SiBCN multiphase ceramic. Harbin: Harbin Institute of Technology; 2018.
45. Liang B, Yang ZH, Miao Y, Li Q, Zhu QS, Liao XQ, Jia DC, Zhou Y. Microstructural evolution, mechanical and thermal properties of LaB_6 embedded in Si-B-C-N prepared by spark plasma sintering. Ceram Int. 2017;43:4814–20.
46. Liang B, Yang ZH, Miao Y, Zhu QS, Liao XQ, Li Q, Tian Z, Jia DC, Zhou Y. Si-B-C-N monoliths with LaB_6-induced well-developed BN(C) flakes. Mater Lett. 2017;187:36–9.
47. Niu ZB, Wang BZ, Pan LJ, Li DX, Jia DC, Yang ZH, Peng H, Ye SQ, Zhou Y. Mechanical and thermal shock properties of C_f/SiBCN composite: effect of sintering densification and fiber coating. J Am Ceram Soc. 2022;105:4321–35.
48. Zhang PF, Jia DC, Yang ZH, Yang B, Wang GX. Research on the interfacial structure and property improvement of the C_{fiber}/2Si-B-3C-N ceramic matrix composite. Mater Charact. 2018;142:59–67.
49. Zhoutang JE. Microstructure and high temperature thermal shock, ablation properties of the C_f/SiBCN-MAS composites. Harbin: Harbin Institute of Technology; 2023.
50. Zhu QS. Fabrication and high temperature damage mechanism of (C_f-ZrB_2)/Si-B-C-N ceramic composites. Harbin: Harbin Institute of Technology; 2018.
51. Zhu QS, Wang GY, Liao XQ, Li DX, Yang ZH, Jia DC, Duan XM, He PG, Zhou Y. Effects of Zr and chopped C fiber on microstructure and mechanical properties of SiBCN ceramics. Sci China Technol Sci. 2020;68:1520–30.
52. Miao Y. Preparation, oxidation resistance and ablation mechanisms of ZrB_2/SiBCN composites. Harbin: Harbin Institute of Technology; 2017.
53. Miao Y, Zhang FN, Yang ZH, Jia DC, Zhou Y. Incorporation of BN-coated carbon fibers into ZrB_2/SiBCN ceramic composites and their ablation behavior. J Eur Ceram Soc. 2020;40:78–1085.
54. Wang JY, Yang ZH, Duan XM, Jia DC, Zhou Y. Microstructure and mechanical properties of SiC_f/SiBCN ceramic matrix composites. J Adv Ceram. 2015;4:31–8.
55. Li DX, Wu DX, Yang ZH, Zhou YY, Wang SJ, Duan XM, Jia DC, Zhu QS, Zhou Y. Effects of in situ amorphous graphite coating on ablation resistance of SiC fiber reinforced SiBCN ceramics in an oxyacetylene flame. Corros Sci. 2016;113:31–45.

56. Wu DX. Effect of SiC_f surface amorphous carbon layer on the properties of SiC_f/SiBCN composite materials. Harbin: Harbin Institute of Technology; 2016.
57. Zhou YY. Preparation and characterization of composite coating on SiC fiber surface. Harbin: Harbin Institute of Technology; 2016.
58. Zhang M, Li DX, Zhou YY, Yang ZH, Jia DC, Cai DL, Niu ZB. Synthesis of coatings on SiC fibers and their effects on microstructure and mechanical properties of SiC_f/SiBCN composites. J Am Ceram Soc. 2021;104:6589–600.
59. Gou YZ, Jian K, Wang H, Wang J. Fabrication of nearly stoichiometric polycrystalline SiC fibers with excellent high-temperature stability up to 1900 °C. J Am Ceram Soc. 2018;101:2050–9.
60. Xiao JZ. Microstructure and thermal shock ablative behavior of the SiC_f/SiBCN composites. Harbin: Harbin Institute of Technology; 2023.
61. Li YT. Mechanical properties, thermal shock and ablation resistance of (C_f-SiC_f)/SiBCN composites. Harbin: Harbin Institute of Technology; 2014.
62. Liao N, Jia DC, Yang ZH, Zhou Y, Li YW. Strengthening and toughening effects of MWCNTs on Si_2BC_3N ceramics sintered by SPS technique. Mater Sci Eng, A. 2018;725:364–74.
63. Song N, Liu H, Fang JZ. Fabrication and mechanical properties of multi-walled carbon nanotube reinforced reaction bonded silicon carbide composites. Ceram Int. 2016;42:351–6.
64. Liao N, Jia DC, Yang ZH, Zhou Y. Enhanced mechanical properties and thermal shock resistance of Si_2BC_3N ceramics with SiC coated MWCNTs. J Adv Ceram. 2019;8:121–32.
65. Li DX, Yang ZH, Jia DC, Wu DX, Zhu QS, Liang B, Wang SJ, Zhou Y. Microstructure, oxidation and thermal shock resistance of graphene reinforced SiBCN ceramics. Ceram Int. 2016;42:4429–44.
66. Li DX. Spark plasma sintering graphene/SiBCN ceramic and high temperature properties. Harbin: Harbin Institute of Technology; 2014.
67. Bao YW, Wang W, Zhou YC. Investigation of the relationship between elastic modulus and hardness based on depth-sensing indentation measurements. Acta Mater. 2004;52:5397–404.
68. Lee SH, Weinmann M, Aldinger F. Processing and properties of C/Si-B-C-N fiber-reinforced ceramic matrix composites prepared by precursor impregnation and pyrolysis. Acta Mater. 2008;56:1529–38.
69. Jaschke T, Jansen M. Improved durability of Si/B/N/C random inorganic networks. J Euro Ceram Soc. 2005;25:211–20.
70. Sujith R, Kumar R. Experimental investigation on the indentation hardness of precursor derived Si-B-C-N ceramics. J Euro Ceram Soc. 2013;33:2399–405.
71. Chen LC, Chen KH, Wei SL, Kichambare PD, Wu JJ, Lu TR, Kuo CT. Crystalline SiCN: a hard material rivals to cubic BN. Thin Solid Films. 1999;355:112–6.
72. Bechelany MC, Salameh C, Viard A, Guichaoua L, Rossignol F, Chartier T, Bernard S, Miele P. Preparation of polymer-derived Si-B-C-N monoliths by spark plasma sintering technique. J Eur Ceram Soc. 2015;35:1361–74.
73. Han JJ, Su D, Zhao ZH, Li XL, Ji HM, Yan X, Jin ZG. Fabrication and toughening behavior of carbon nanotube (CNT) scaffold reinforced SiBCN ceramic composites with high CNT loading. Ceram Int. 2017;43:9024–31.
74. Katsuda Y. Reinforcement of precursor-derived Si-(B-)C-N ceramics with carbon nanotubes. Stuttgart: Max-Planck-Institut für Metallforschung; 2005.
75. Galusek D, Riley FL, Riedel R. Nanoindentation of a polymer-derived amorphous silicon carbonitride ceramic. J Am Ceram Soc. 2001;84(5):1164–6.
76. Shah RS, Raj R. Mechanical properties of a fully dense polymer derived ceramic made by a novel pressure casting process. Acta Mater. 2002;50:4093–103.
77. Shopova-Gospodinova D, Burghard Z, Dufaux T. Mechanical and electrical properties of polymer-derived Si-C-N ceramics reinforced by octadecylamine-modified single-wall carbon nanotubes. Compos Sci Technol. 2011;71(6):931–7.
78. Sujith R, Kumar R. Indentation response of pulsed electric current sintered polymer derived HfO_2/Si-C-N(O) nanocomposites. Ceram Int. 2013;39(8):9743–7.

79. Szymanski W, Lipa S, Fortuniak W. Silicon oxycarbide (SiOC) ceramic microspheres-structure and mechanical properties by nanoindentation studies. Ceram Int. 2019;45(9):11946–54.
80. Soraru GD, Dallapiccola E, D'Andrea G. Mechanical characterization of sol-gel-derived silicon oxycarbide glasses. J Am Ceram Soc. 1996;79:2074–80.
81. Soraru GD, Kunadanati L, Santhosh B. Influence of free carbon on the young's modulus and hardness of polymer derived silicon oxycarbide glasses. J Am Ceram Soc. 2019;102(3):907–13.
82. Walter S, Soraru GD, Brequel H, Enzo S. Microstructural and mechanical characterization of sol gel-derived Si-O-C glasses. J Eur Ceram Soc. 2002;22:2389–400.
83. Mazo MA, Soriano D, Rubio J. Mechanical response of silicon oxycarbide materials processed by spark plasma sintering. Ceram Int. 2023;49:12866–75.
84. Ding Q, Ni DW, Dong SM. 3D C_f/SiBCN composites prepared by an improved polymer infiltration and pyrolysis. J Adv Ceram. 2019;102:3630–40.
85. Lee SH. Thermal stability and mechanical properties of SiC particulate-reinforced Si-C-N composites after heating at 2000 °C. J Eur Ceram Soc. 2009;29:3387–93.
86. Li YJ, Zhang Y, Wang YS. Structure and oxidation behavior of high temperature ZrB_2-SiBCN ceramics with polyborosilazane as a sintering additive. J Ceram Soc Jpn. 2013;121:520–3.
87. Li ZL, Wu Y, Zhang LJ, Wei CC, Sui H, Sun ZY, Li DQ, Wen GW, Wang JQ. Preparation, mechanical properties and long-term ablation resistance of C_f/SiBCN-ZrB_2 composites. Ceram Int. 2023;49:19817–28.
88. Fang C, Bao Y, Hu, Dong S, Liu DZ, Wang HL, Zhang XH. Effects of multilayer hydrothermal carbon interphases on mechanical properties and thermal shock resistance of CF/ZrB_2-SiCBN. J Euro Ceram Soc, 2022;42:4759–69.
89. Tan X, Liu W, Cao LM, Dai SL. A lanthanum monazite coated silicon carbide fiber reinforced SiBCN composite with improved mechanical properties for high temperature applications. J Ceram Soc Jpn. 2018;126:632–40.
90. Song CK, Liu YS, Ye F, Cheng LF, Zhang PF, Chai N. Enhanced mechanical property and tunable dielectric property of SiC_f/SiC-SiBCN composites by CVI combined with PIP. J Adv Ceram. 2021;10:758–67.
91. Sun ZL, Zhou Y, Jia DC, Duan XM, Yang ZH, Ye D, Zhang PF, Zhang Q. Mechanical and thermal physical properties of amorphous SiCN(O) ceramic bulks prepared by hot-press sintering. Mater Lett. 2012;72:57–9.
92. Nishimura T, Haug R, Bill J, Thur G, Aldinger F. Mechanical and thermal properties of Si-C-N material from polyvinylsilazane. J Mater Sci. 1998;33:5237–41.
93. Liu XL, Xue JM, Cheng LF. Design and fabrication of Al_2O_{3f}/SiCN composite with microwave absorbing and mechanical properties. J Am Ceram Soc. 2020;103:6255–64.
94. Li WH, Wang J, Xie ZF, Wang H. A novel polyborosilazane for high-temperature amorphous Si-B-N-C ceramic fibres. Ceram Int. 2012;38:6321–6.
95. Baldus P. Ceramic fibers for matrix composites in high-temperature engine applications. Science. 1999;285:699–703.
96. Cooke TF. Inorganic fibers-a literature review. J Am Ceram Soc. 1991;74:2959–78.
97. Rishi R. Fundamental research in structural ceramics for service near 2000 °C. J Am Ceram Soc. 1993;76:2147–74.
98. Li YF, Yan LW, Song MJ, Chen SY, Guo AR, Hou F, Ye L, Han WJ, Liu JC. Integration of thermal protection and structural health monitoring for carbon fiber reinforced SiBCN composites with SiC coating enhanced interfacial performance. Ceram Int. 2023;49:21678–87.
99. Ziegler G, Richter I, Suttor D. Fiber-reinforced composites with polymer-derived matrix: processing, matrix formation and properties. Compos A. 1999;30:411–7.

Chapter 6
Thermal Shock Behavior and Ablation Corrosion Mechanisms of MA-SiBCN Materials

Long term stability of metastable ceramics and their ceramic matrix composites (CMCs) at high-temperatures is one of the great challenges for the modern aviation and aerospace technology developments. Up to present, increasing demands on higher engine power density, lower emissions and longer lifetime have imposed great challenges on the durability and reliability of the advanced aero- and space-engines due to the increased mechanical and thermal loading. Multicomponent ceramics such as SiBCN system gain a lot of interest for high temperature applications due to their higher stability of the 3-D amorphous covalent network structure.

Thermal shock is one of the main drawbacks in the utilization of SiBCN materials for high-temperature applications. Failure from thermal shock is generally caused by incompatible changes in dimension and failure occurs in regions where the strain locally exceeds the tensile failure strain of SiBCN. It generally occurs while the SiBCN materials are brittle and are rare at temperatures below 1000 °C where multiple strengthening and toughening mechanisms still take effect. As the C_f or C_f-SiC_f reinforced SiBCN CMCs have exhibited an improvement of fracture toughness and fracture work, one thus can envision these CMCs will also possess an increase in the resistance to fracture initiation and enhanced thermal shock resistance to catastrophic failure.

6.1 Ablation Behavior and Microstructure Evolution of MA-SiBN Monoliths

The oxyacetylene torch test is very aggressive involving high temperature, high velocity gas flow and a severely oxidizing environment. The ceramic surface is vertically exposed to the flame and heated at rates of up to 500 °C/s; moreover, the initial cooling rate is $\sim$ 1000 °C/s on removing the flame [1]. For the five as-sintered bulk ceramics, the matrix phase of Si_3N_4 content decreases gradually while

D. Li et al., *SiBCN Ceramics and Composites Prepared by Mechanical Alloying*,
https://doi.org/10.1007/978-981-97-3029-2_6

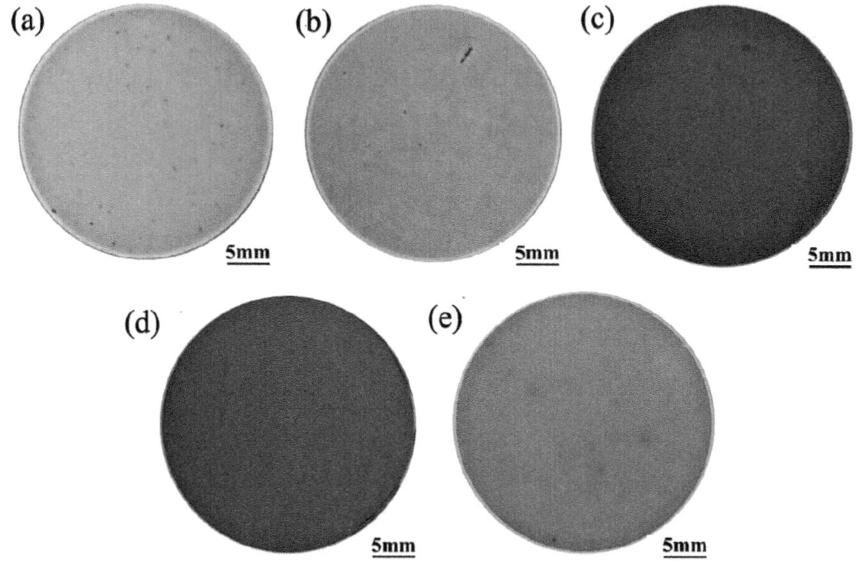

Fig. 6.1 Optical photographs of the as-sintered bulk ceramics before ablation: **a** MA-Si_3BN; **b** MA-Si_2BN; **c** MA-SiBN; **d** MA-SiB_2N; **e** MA-SiB_3N [1]

h-BN phase increases with the decreasing Si/B atomic ratio. Before oxyacetylene flame ablation, the surfaces of the hot-pressed MA-Si_3BN, MA-Si_2BN, MA-SiBN, MA-SiB_2N and MA-SiB_3N ceramics are smooth and dense free of any macrocracks and macropores (Fig. 6.1); however, the as-ablated monoliths show different ablation macroscopic morphologies after 10 s of ablation (GJB 323A-1996, heat flux ~ 4.18 WM/m^2) (Fig. 6.2).

Ablation pits with different size have appeared in the ablation centers for all the as-ablated bulk ceramics. Except for the hot-pressed MA-Si_2BN, obvious penetrating macrocracks are found on the surfaces for other ceramics. Coincidentally, hot-pressed MA-Si_3BN ceramics with the highest Si/B atomic ratio are split into four parts after 10 s of ablation. This thus suggest increasing h-BN content is beneficial to their thermal shock resistance.

During ablation, the maximum surface temperatures of the hot-pressed MA-Si_3BN, MA-Si_2BN, MA-SiBN, MA-SiB_2N and MA-SiB_3N monoliths are 1921 °C, 1759 °C, 1743 °C, 1979 °C and 2002 °C, respectively. The moderate Si/B atomic ratio leads to a better thermal insulation performance (Fig. 6.3). Expectedly, all the as-sintered bulk ceramics with different Si/B atomic ratio show low mass and linear ablation rates implying good ablation resistance (Table 6.1). The MA-Si_2BN monoliths have the lowest mass ablation rate ~ 0.12 mg/s, while the lowest linear ablation rate ~ 0.0031 mm/s is measured for MA-SiBN.

After ablation, their phase composition differs from ablation center to heat-affected region. For example, in ablation center of hot-pressed MA-Si_3BN ceramics, the weak diffraction peaks assigned to α-Si_3N_4, β-Si_3N_4, Si_2N_2O and h-BN are

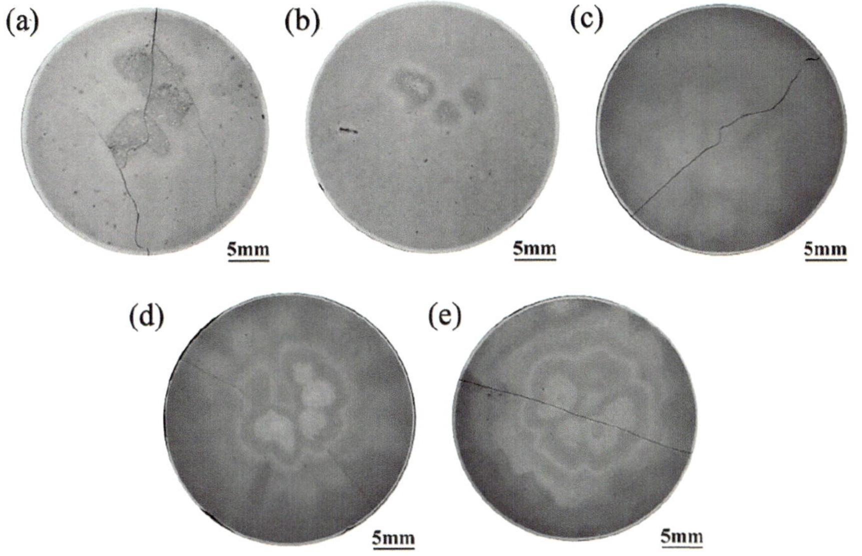

Fig. 6.2 Optical photographs of the as-sintered bulk ceramics after oxyacetylene flame ablation for 10 s: **a** MA-Si_3BN; **b** MA-Si_2BN; **c** MA-SiBN; **d** MA-SiB_2N; **e** MA-SiB_3N [1]

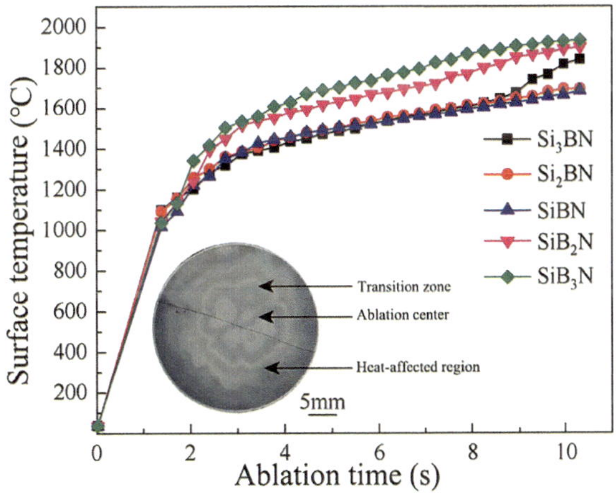

Fig. 6.3 Surface temperature curves of the various hot-pressed bulk ceramics (1800 °C /40 MPa/ 30 min) after oxyacetylene flame (heat flux ~ 4.18 WM/ m^2) ablation for 10 s [1]

observed, while those of the α-quartz, β-quartz and Si are stronger in intensity. This can be ascribed to the oxidation or decomposition of Si_3N_4, and decomposition of Si_2N_2O °C resulting in the formation of substantial α/β-quartz and Si, though part of the surface Si also rapidly oxidizes to form a protective SiO_2 (Fig. 6.4). In transition zone of MA-Si_3BN, the diffraction peaks of α/β-Si_3N_4, Si_2N_2O and h-BN are obviously stronger than those in the ablation center, and some amorphous phases are generated as indicated by the halo peaks in the range of $2\theta = 15°–25°$. Contrary to

Table 6.1 Ablation properties of the various hot-pressed bulk ceramics after oxyacetylene flame (heat flux ~ 4.18 WM/m^2) ablation for 10 s [1]

Composition	Mass ablation rate (mg/s)	Linear ablation rate (mm/s)
MA-Si_3BN	0.51	0.0100
MA-Si_2BN	0.12	0.0042
MA-SiBN	0.31	0.0031
MA-SiB_2N	0.52	0.0036
MA-SiB_3N	1.01	0.0049

expectations, there are still strong diffraction peaks of Si in this region. From this point of view, the bulk ceramics with higher Si/B atomic ratio are not conducive to their high-temperature wave-transparent properties. In the heat-affected zone, the peak intensities of the α/β-Si_3N_4, Si_2N_2O and h-BN are further enhanced accompanied with the disappearance of α-quartz and amorphous phases.

With the decreasing Si/B atomic ratio, all the as-ablated bulk ceramics show similar phase composition in ablation center, transition zone and heat-affected region. Note that the ablation kinetic process for the high Si/B atomic ratio containing ceramics, such as MA-Si_3BN and MA-Si_2BN is mainly governed by the oxidation and/or decomposition of α/β-Si_3N_4, while h-BN dominates the oxidation process for ceramics with low Si/B atomic ratio (MA-SiBN, MA-SiB_2N and MA-SiB_3N).

During ablation, the matrix phases of α/β-Si_3N_4 and h-BN oxidize to generate dominant SiO_2 and/or borosilicate glass, further soften and volatilize leading to larger micropores left on ablation center surfaces. Besides, the escape of N_2 and

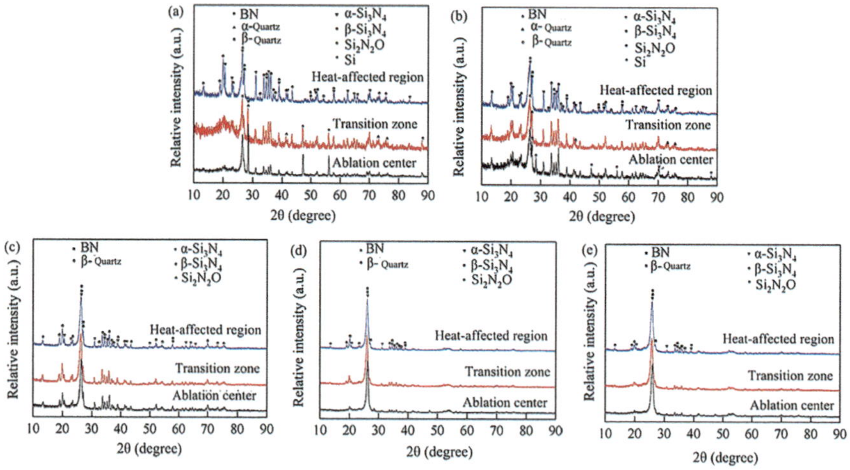

Fig. 6.4 XRD patterns of the a-sintered bulk ceramics after oxyacetylene flame ablation for 10 s: **a** MA-Si_3BN; **b** MA-Si_2BN; **c** MA-SiBN; **d** MA-SiB_2N; **e** MA-SiB_3N [1]

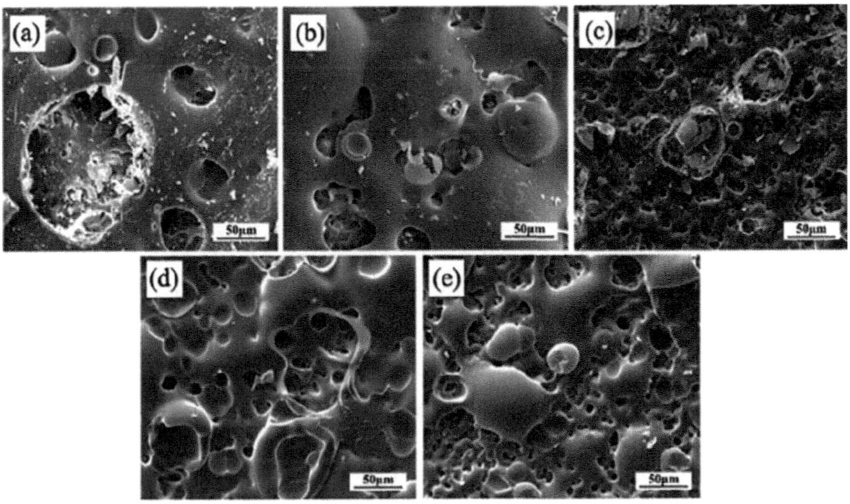

Fig. 6.5 SEM surface morphologies of the ablation center for the as-ablated bulk ceramics with different Si/B atomic ratio: **a** MA-Si_3BN; **b** MA-Si_2BN; **c** MA-SiBN; **d** MA-SiB_2N; **e** MA-SiB_3N [1]

volatilization of B_2O_3 and SiO also lead to this porous structure. Note that the high-speed gas flow is perpendicular to the ceramic surfaces, thus considerable preformed SiO_2 are scoured away. The softening and volatilization of oxides, and escape of gaseous byproducts will absorb part of the heat, thus reducing the ablation rates of the materials. Since the oxidation reactions of α/β-Si_3N_4 take place prior to h-BN, more protective oxidation products are easily generated in the ablation center with the increasing Si/B atomic ratio (Fig. 6.5).

In transition zone, finer holes constructed by the escape of sub-oxides and gases can be seen. The MA-SiB_2N and MA-SiB_3N monoliths with the higher h-BN content show more porous transition zone with larger micropores. In this area, MA-Si_2BN and MA-SiBN ceramics also show a small number of spherical SiO_2 particles. Strong airflow scouring will bring some SiO_2 to the transition zone, but liquid SiO_2 has poor fluidity and wettability on ceramic matrix at relative low ablation temperature. To minimize the Gibbs free energy, such liquid SiO_2 solidifies into spherical particles after cooling (Fig. 6.6).

The surface temperatures of the heat-affected region are lower than those of ablation center and transition zone. In this case, the viscosity of the preformed SiO_2 increases, thus the bubbles formed inside are hard to diffuse and escape through the outermost SiO_2 oxide layer. For MA-SiB_2N and MA-SiB_3N, abundant unbroken bubbles are observed, and the oxide layer is mainly composed of amorphous SiO_2. MA-Si_3BN ceramics with the highest Si_3N_4 content display a large number of damaged bubbles throughout the oxide layer (Fig. 6.7). The cross-sections of all the as-ablated ceramics can be divided into three regions: SiO_2 outermost layer, partially oxidized heat-affected zone and unoxidized ceramic matrix. The outermost

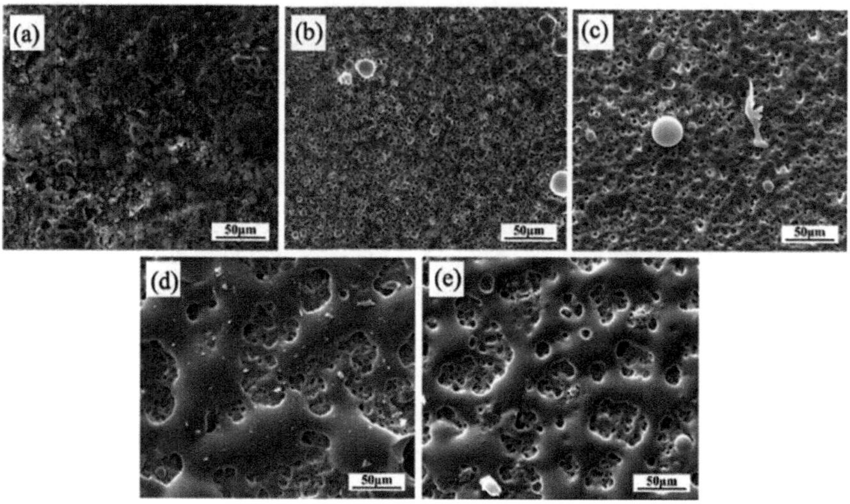

Fig. 6.6 SEM surface morphologies of the transition region for the as-ablated bulk ceramics with different Si/B atomic ratio: **a** MA-Si_3BN; **b** MA-Si_2BN; **c** MA-SiBN; **d** MA-SiB_2N; **e** MA-SiB_3N [1]

layer of fused SiO_2 on the surfaces is relatively dense, but beneath the SiO_2 oxide layer are the porous and loose structures of the heat-affected zone (Fig. 6.8).

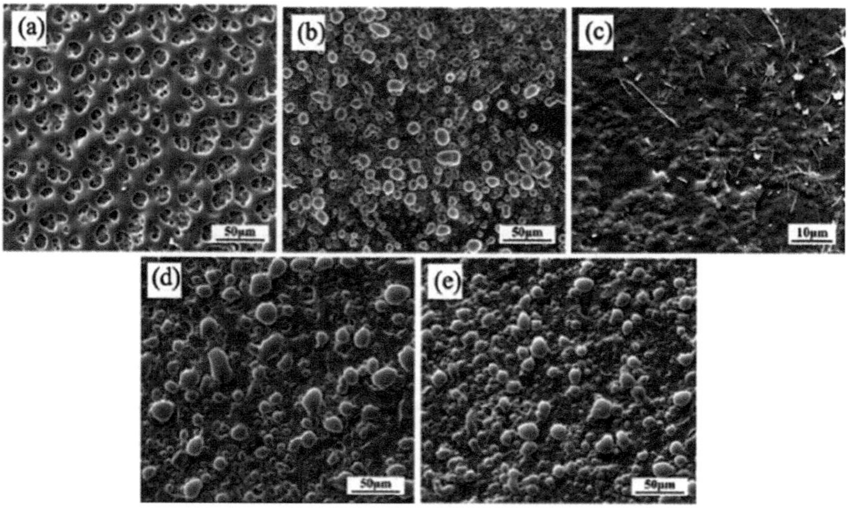

Fig. 6.7 SEM surface morphologies of the heat-affected zone for the as-ablated bulk ceramics with different Si/B atomic ratio: **a** MA-Si_3BN; **b** MA-Si_2BN; **c** MA-SiBN; **d** MA-SiB_2N; **e** MA-SiB_3N [1]

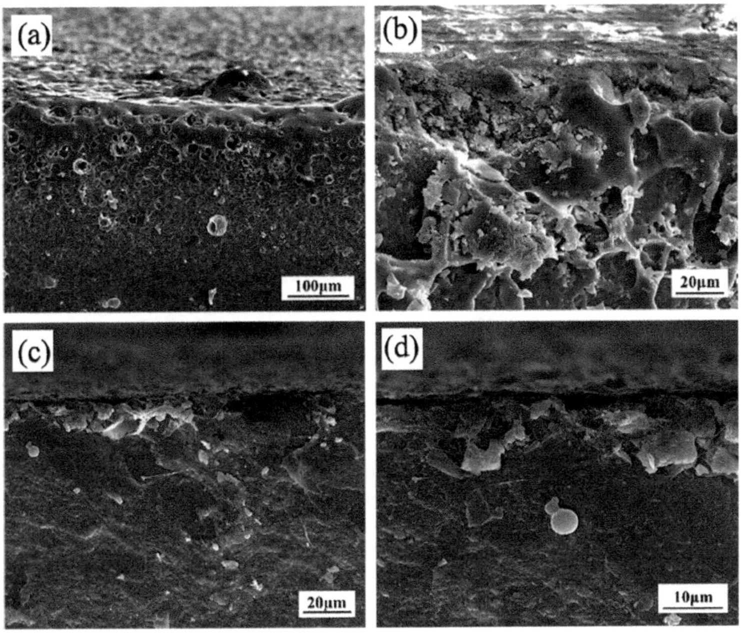

Fig. 6.8 SEM cross-section morphologies of the as-ablated bulk ceramics: **a** MA-Si_3BN; **b** MA-SiBN; **c** MA-SiB_2N; **d** MA-SiB_3N [1]

6.2 Thermal Shock and Ablation Behavior of MA-SiBCN Composite Ceramics

(1) *Mo/Si_2BC_3N composite ceramics*

Generally, the residual bending strength of the Mo/Si_2BC_3N composite ceramics decreases with the addition of varying MA-Si_2BC_3N. But for the composite ceramics with only 10–20 vol% MA-Si_2BC_3N, they both show increasing residual bending strength at a temperature difference (ΔT) of 900 °C. However, the residual bending strength of the composite ceramics with 30 vol% MA-Si_2BC_3N decreases to one-third of that with 20 vol% MA-Si_2BC_3N. For composites with 90 vol% MA-Si_2BC_3N and 10 wt% MgO-Al_2O_3-SiO_2 (MAS) additive, the residual bending strength increases to 179.6 MPa with strength retention ratio of 109.8%, much higher than the additive-free counterparts (Fig. 6.9 and Table 6.2).

At $\Delta T = 1600$ °C, the residual strength and bulk density of the composite ceramics with higher MA-Si_2BC_3N content also drop dramatically. The residual strength of composite ceramics with 20 vol% MA-Si_2BC_3N is 955.0 MPa, almost three times that of composites with 30 vol% MA-Si_2BC_3N. As mentioned above, when 30–40 vol% MA-Si_2BC_3N is added, Mo disappears and $Mo_{4.8}Si_3C_{0.6}$ replaces Mo_3Si. Therefore, the changes in the thermal shock resistance of the Mo/Si_2BC_3N materials is closely related to the formation of $Mo_{4.8}Si_3C_{0.6}$. The $Mo_{4.8}Si_3C_{0.6}$ is the only stable ternary

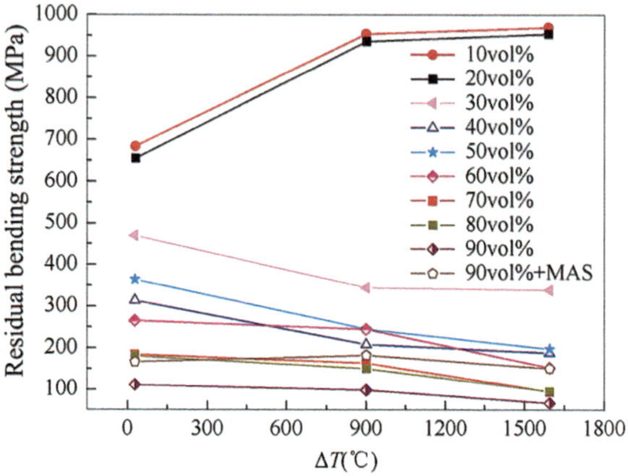

Fig. 6.9 Residual strength of the various Mo/Si_2BC_3N composite ceramics after thermal shock at different temperature difference [2]

Table 6.2 Residual strength and strength retention rate of the various Mo/Si_2BC_3N composite ceramics after thermal shock at different temperature difference [2]

MA-Si_2BC_3N content	$\Delta T = 900$ °C		$\Delta T = 1600$ °C	
	Residual strength (MPa)	Strength retention rate (%)	Residual strength (MPa)	Strength retention rate (%)
+ 10 vol% MA-Si_2BC_3N	954.4 ± 5.6	140.4	970.1 ± 3.4	142.7
+ 20 vol% MA-Si_2BC_3N	936.5 ± 4.9	143.8	955.0 ± 3.1	146.7
+ 30 vol% MA-Si_2BC_3N	342.0 ± 2.8	73.0	338.8 ± 1.3	72.3
+ 40 vol% MA-Si_2BC_3N	206.0 ± 0.5	66.0	187.4 ± 0.9	60.0
+ 50 vol% MA-Si_2BC_3N	243.1 ± 0.2	66.9	197.7 ± 0.2	54.4
+ 60 vol% MA-Si_2BC_3N	242.0 ± 0.7	92.1	151.3 ± 2.4	57.6
+ 70 vol% MA-Si_2BC_3N	160.8 ± 2.6	88.7	92.9 ± 0.6	83.5
+ 80 vol% MA-Si_2BC_3N	147.3 ± 3.6	82.7	94.4 ± 0.7	53.0
+ 90 vol% MA-Si_2BC_3N	95.1 ± 0.2	87.9	63.6 ± 1.3	58.8
+ 90 vol% MA-Si_2BC_3N and 10 wt% MAS	179.6 ± 2.1	109.8	149.5 ± 2.3	91.4

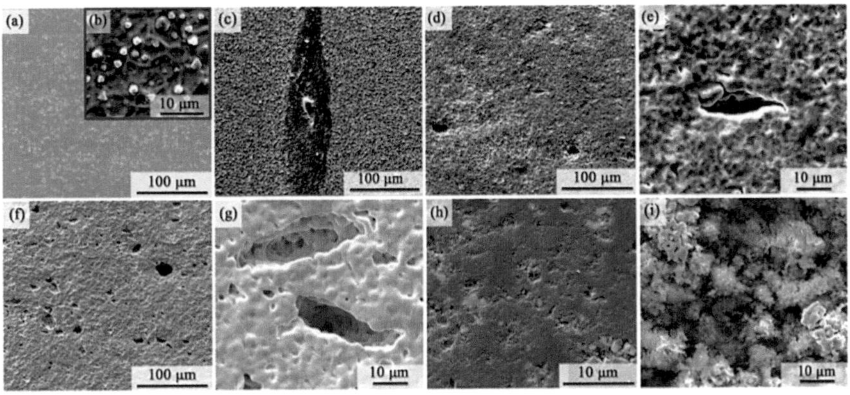

Fig. 6.10 SEM surface morphologies of the various Mo/Si_2BC_3N composites after thermal shock at $\Delta T = 1600$ °C: **a**, **b** + 10 vol% MA-Si_2BC_3N; **c** + 20 vol% MA-Si_2BC_3N; **d**, **e** + 30 vol% MA-Si_2BC_3N; **f**, **g** + 40 vol% MA-Si_2BC_3N; **h**, **i** + 90 vol% MA-Si_2BC_3N [2]

phase in Mo-Si–C system, and it has better creep resistance, fracture toughness and high-temperature mechanical properties than those of $MoSi_2$ and Mo_5Si_3 [3, 4].

At $\Delta T = 1600$ °C, the sample surfaces of the 10 vol% MA-Si_2BC_3N incorporating composite ceramics basically maintain a high density showing dense oxide scale with some oxide particles. With 20 vol% MA-Si_2BC_3N, there cracks appear on the surfaces. With 30–40 vol% MA-Si_2BC_3N, due to the increase in Mo-compounds, the formed big pores at high-temperatures gradually interconnect with each other. These pores may develop into cracks under thermal shock stress leading to failure. With 90 vol% MA-Si_2BC_3N addition, the surfaces of the Mo/Si_2BC_3N composite ceramics show dense region and porous zone with many agglomerating particles (Fig. 6.10).

(2) *Al_4SiC_4/Si_2BC_3N composite ceramics*

Before thermal shock, the SPS sintered Al_4SiC_4/Si_2BC_3N composite ceramics with 10 wt% Al_4SiC_4 additive show the highest flexural strength due to the synergistic effects of well-developed BN(C) and Al_4SiC_4 plates. After thermal chock at $\Delta T =$ 1000 °C, the composites with 5 wt% Al_4SiC_4 additive present the highest residual strength and the residual strength ratio. This should be attributed to a proper amount of Al_4SiC_4 additive can enhance the toughness and stress resistance; however, the higher thermal expansion coefficient of Al_4SiC_4, and formed AlN/Al_4C_3 may induce higher thermal stress during thermal shock (Fig. 6.11).

Though the Al_4SiC_4 additive may cause some negative effects on the oxidation resistance of MA-Si_2BC_3N ceramics, it enhances the ablation resistance at a flame temperature around 3000 °C with an extremely high heating rate. The red lines indicate the edges of the ablation pits. Apparently, that the ablation area and ablation depth of pure MA-Si_2BC_3N are obviously larger than those of 5 wt% Al_4SiC_4 containing ones. Besides, a white oxide layer mainly composed of SiO_2-B_2O_3-Al_2O_3 covers and adheres to the ablation region of the composite ceramics. Generally, the thermal

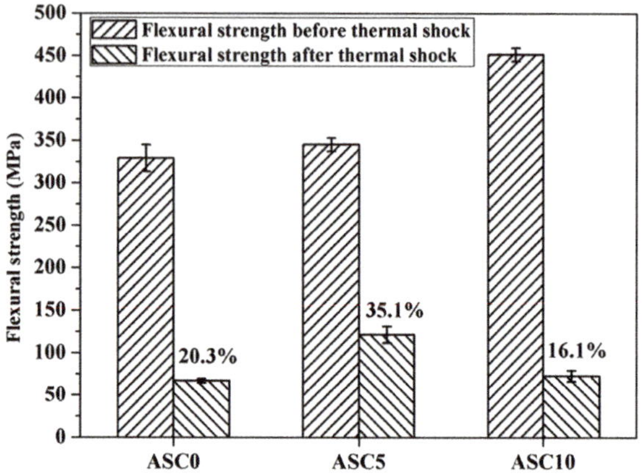

Fig. 6.11 Flexural strength of the Al_4SiC_4/Si_2BC_3N composite ceramics before and after thermal shock test at $\Delta T = 1000$ °C [5]

shock and ablation resistance are improved with the Al_4SiC_4 additive mainly due to their higher relative density and well-developed structures (Fig. 6.12).

(3) *(Zr-Al)/Si₂BC₃N composite ceramics*

The calculated R parameter is taken to evaluate its thermal shock fracture resistance, while the R'''' or R_{st} parameters are applied to indicate the thermal shock damage resistance [6–8]. The R parameter reaches 660 °C for 1 mol% Zr-Al incorporating composites, much higher than 470 °C of the pristine MA-Si_2BC_3N. The increased R indicates Zr–Al additive improving the ability to resist the fracture initiation. Nevertheless, the higher amount of additive may weaken its advantage due to their lower strength. Similarly, the thermal stress damage resistance is enhanced

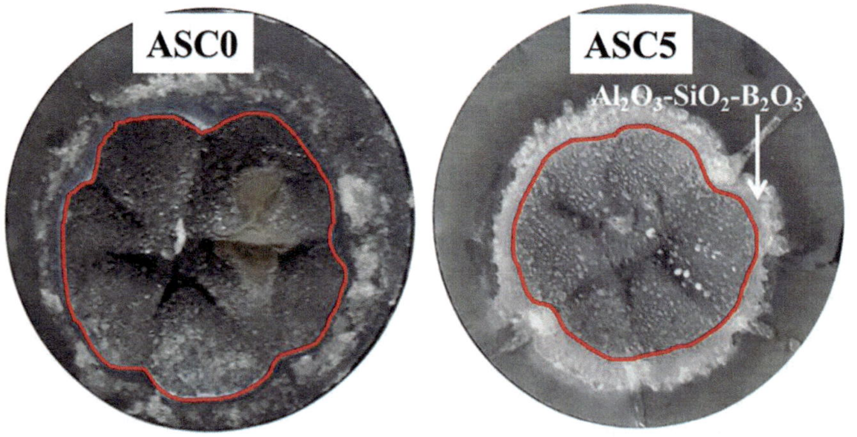

Fig. 6.12 Surface morphologies of the pure MA-Si_2BC_3N ceramics and Al_4SiC_4/Si_2BC_3N composites after oxyacetylene torch ablation test (heat flux ~ 4.2 MW/m^2) at 3000 °C for 30 s [5]

with the Zr-Al introduction. The R'''' varies in accordance with the fracture toughness. Another thermal shock parameter, R_{st} also suggests Zr-Al additive modified composites should have a better thermal shock resistance (Table 6.3).

When thermal shock test at $\Delta T = 600$ °C, the Zr-Al additive containing composite ceramics have much higher residual strength than the pure ceramics, and even present higher strength than their original values. This phenomenon is matched well with the calculated R parameter, indicative of 1 mol% Zr-Al containing composites should have no catastrophic crack initiation, since their R parameters are larger than 600 °C. The improvement of residual strength should be attributed to the formation of a thin oxidation layer on ceramic surfaces. At $\Delta T = 800$ °C, the residual strength decreases obviously while 1 mol% Zr–Al containing composites still remain the highest strength ~ 257.9 MPa, which is twice of the strength of pristine MA-Si_2BC_3N. At $\Delta T = 1000$ °C, The 1 mol% Zr-Al containing composites also possess the highest residual strength of ~ 158.6 MPa, ~ 69.8% higher than 93.4 MPa of the counterparts. However, the higher Zr-Al additive shows no further positive effect.

The thermal shock fracture and damage resistance are always associated with their strength, Young's modulus and thermal expansion coefficient. The Zr-Al additive significantly improves the strength while the Young's modulus slightly increases, favoring to higher resistance to the crack initiation. Therefore, the R parameter increases from 470 °C to 660 °C. Simultaneously, the additive containing composites still have much higher fracture toughness, which guarantees better thermal shock damage resistance. Therefore, the 1 mol% Zr-Al additive containing composites offer the optimized thermal shock resistance, outperforming the 5 wt% Al_4SiC_4 containing composites [5].

(4) *(Cu-Ti)/*Si_2BC_3N *composite ceramics*

The bulk density of the $Cu_{80}Ti_{20}/Si_2BC_3N$ bicontinuous composite ceramics with 35 vol% MA-Si_2BC_3N is close to the theoretical value. The mass ablation rate and

Table 6.3 The calculated thermal shock parameters and measured residual strength of the (Zr-Al)/Si_2BC_3N composite ceramics SPS at 1900 °C/40 MPa/5 min [8]

Composition	α ($\times 10^{-6}$ °C^{-1})	R (°C)	R'''' (mm)	R_{st} (K·m$^{1/2}$)	Residual strength (MPa)/strength retention rate (%)			
					RT	$\Delta T =$ 600 °C	$\Delta T =$ 800 °C	$\Delta T =$ 1000 °C
MA-Si_2BC_3N	4.51	470	1.9	23.5	329.1	245.1/ 74.5	125.6/ 38.2	93.4/ 28.4
+ 1 wt% Zr-Al	4.65	660	2.1	34.6	590.2	626.8/ 106.2	257.9/ 44.0	158.6/ 26.9
+ 3 wt% Zr-Al	4.80	517	2.7	31.2	480.1	533.6/ 111.1	136.9/ 28.5	88.2/ 18.4
+ 5 wt% Zr-Al	5.12	493	2.8	30.0	481.3	406.5/ 84.5	93.9/ 19.5	72.9/ 15.1

linear ablation rate of the $Cu_{80}Ti_{20}/Si_2BC_3N$ composites are ~ 60 mg/s and ~ 0.25 mm/s, respectively, after oxyacetylene flame ablation for 10 s [9]. Due to the short ablation time, part of the metal liquid and/or oxides with relatively high viscosity are retained on the sample surfaces before being completely scoured away by the high-speed airflow.

The macro surfaces of the as-ablated $Cu_{80}Ti_{20}/Si_2BC_3N$ bicontinuous composites can still be divided into three areas, but differing from the traditional non-(micro-) ablative materials. Here, the ablation center A is the main area of mass loss, that is, the height of ablation region is smaller than that of the as-sintered sample. Transition zone B is the high-height area formed by the liquid metal scoured from zone A. The heat-affected zone C refers to the region between the outer edge of zone B and the edge of the sample. The heat-affected surfaces are relatively smooth without serious ablation, where its height is basically the same as that of the as-sintered sample (Fig. 6.13).

During ablation, the surface temperature of the $Cu_{80}Ti_{20}/Si_2BC_3N$ bicontinuous composite ceramics increases to ~ 1700 °C within 3 s, and subsequently fluctuated in the range of 2000–2025 °C, and decreased slightly at the end of ablation. The main oxides are rutile TiO_2 (R-TiO_2) and a small amount of anatase TiO_2 (A-TiO_2). The diffraction peak intensity of R-TiO_2 gradually increases from the ablation center to the outside. Few SiO_2, CuO and a small amount of unoxidized Cu and Cu_xSi_y compounds are also found.

Due to the short-time high-enthalpy airflow, most of the $Cu_{80}Ti_{20}$ metal in the ablation center are oxidized which follows scouring of oxidation products, thus directly exposed the ceramic lamellas to the oxidizing air flow. The ceramic lamellar surfaces

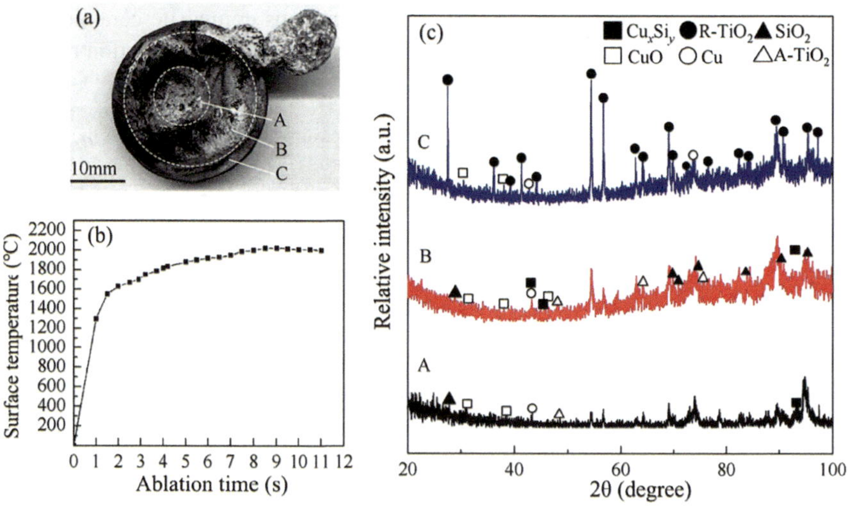

Fig. 6.13 Ablation behavior of the $Cu_{80}Ti_{20}/Si_2BC_3N$ composites after oxyacetylene flame (heat flux ~ 4.2 WM/m^2) ablation for 10 s: **a** Macro morphology; **b** surface temperature curve; **c** XRD patterns (A: ablation center, B: transition zone, C: heat-affected zone [9]

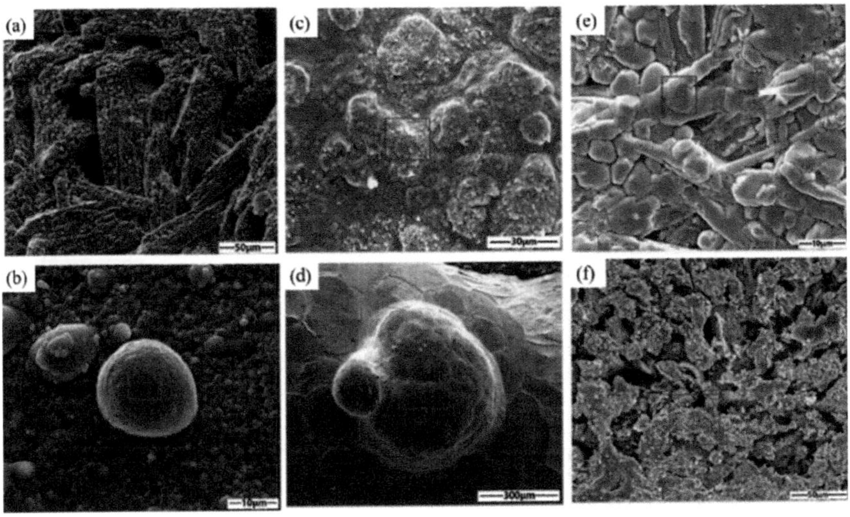

Fig. 6.14 SEM surface morphologies of the $Cu_{80}Ti_{20}/Si_2BC_3N$ bicontinuous composite ceramics after oxyacetylene flame (heat flux ~ 4.2 WM/m^2) ablation for 10 s: **a**, **b** Ablation center; **c**, **d** transition zone; **e**, **f** heated-affected region [9]

are mainly composed of TiO_2, SiO_2 and unoxidized C. Some spherical metal droplets with diameters of 10–30 μm are found in the ablation center. The main components of these solidified droplets are TiO_2, C, CuO and a small amount of SiO_2 (Fig. 6.14).

The transition zone of the $Cu_{80}Ti_{20}/Si_2BC_3N$ bicontinuous composites are covered by the stacking molten oxides and/or metal scouring from the ablation center. In this rough and loose region, substantial particles and micropores containing C, O, Ti and a small amount of Cu elements are detected. In other area, many bulges with diameters of ~ 500 μm, mainly contains C, O, Ti and a small amount of Si elements. For the heat-affected zone, porous structure with a large amount of rod-like and spherical TiO_2 on the surfaces are observed.

(5) *Ta_4HfC_5/Si_2BC_3N composite ceramics*

The surface temperature of the 15 wt% Ta_4HfC_5/Si_2BC_3N composite ceramics rapidly increases with exposing time. Pure MA-Si_2BC_3N ceramics exhibit deeper corrosion pits in the ablation center covered with some oxides. With some Ta_4HfC_5 addition, their ablation pits are relatively shallow and the surfaces of the ablation center are covered by a grey oxide layer (Fig. 6.15). The linear ablation rates of the 5 wt% or 15 wt% Ta_4HfC_5 containing MA-Si_2BC_3N are 0.176 mm/s or 0.123 mm/s, which are lower than those of MA-Si_2BC_3N [10].

The ablation layer of Ta_4HfC_5/Si_2BC_3N composite ceramics are mainly composed of Ta_2O_5, $Hf_6Ta_2O_{17}$ and tantalum hafnium silicate glass. After ablation, considerable amorphous phases are detected in transition and heat-affected zones for all the composite ceramics, especially for the 15 wt% Ta_4HfC_5 containing composites (Fig. 6.16).

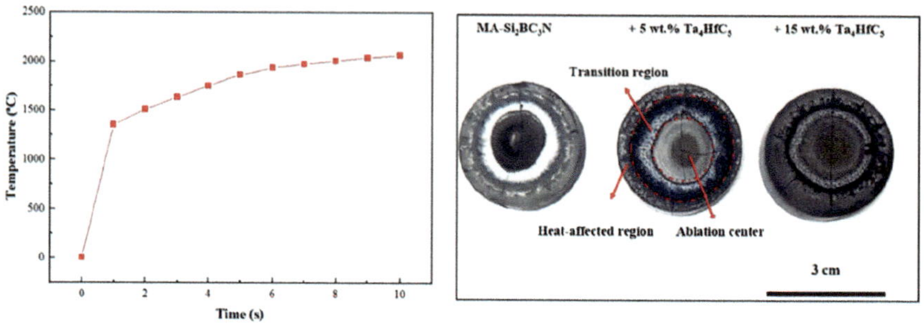

Fig. 6.15 Ablation behavior of the Ta_4HfC_5/Si_2BC_3N composite ceramics after oxyacetylene flame (heat flux ~ 4.2 WM/m^2) ablation for 10 s: **a** Sample surface temperature curve; **b** macro morphology [10]

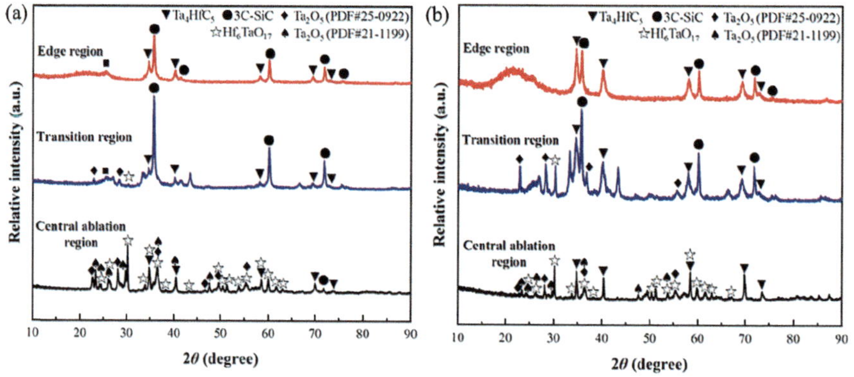

Fig. 6.16 XRD patterns of the Ta_4HfC_5/Si_2BC_3N composite ceramics after oxyacetylene flame ablation for 10 s: **a** + 5 wt% Ta_4HfC_5; **b** + 15 wt% Ta_4HfC_5 [10]

For 5 wt% Ta_4HfC_5/Si_2BC_3N composite ceramics, a porous structure is observed in ablation center area, which contains Si, Ta, Hf, and O elements. The transition zone is relatively dense, while the morphologies of the heat-affected zone is similar to that of pure MA-Si_2BC_3N, typical of hollow spheres with some micropores. With increasing Ta_4HfC_5, the composite ceramics show similar structure comparable to the formers, but differs in the oxide content and sizes of the micropores (Fig. 6.17).

From the cross-section morphologies, in both cases, the porous oxide layer covered on the ceramic surfaces are found in the ablation center. In transition region, double-layer oxide structure forms. The outermost layer is dense and thinner, while the porous inner layer is distributed with substantial micropores. A relatively dense and crack-free scale adheres on the heat-affected region (Fig. 6.18).

Ta_2O_5 is in form of mainly interlaced rod-shape, while that of $Hf_6Ta_2O_{17}$ is mainly interconnected spherical grains (Fig. 6.19). Ta_2O_5 and $Hf_6Ta_2O_{17}$ form a three-dimensional skeleton structure, and tantalum hafnium silicate glass fills the

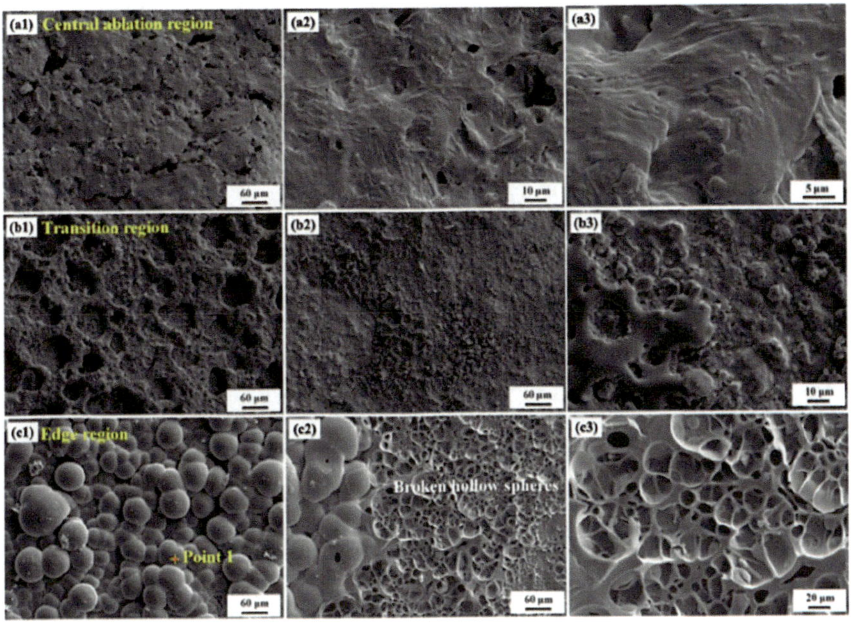

Fig. 6.17 Surface morphologies of the 15 wt% Ta_4HfC_5/Si_2BC_3N composite ceramics after oxyacetylene flame ablation for 10 s: **a** Ablation center; **b** transition region; **c** heat-affected area [10]

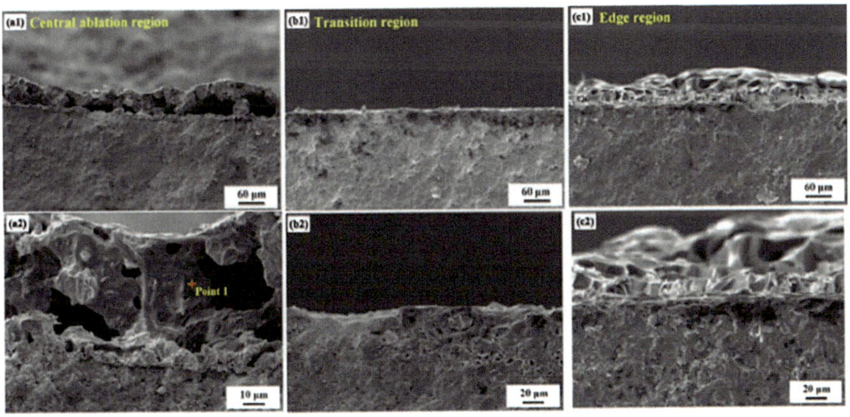

Fig. 6.18 Cross-section morphologies of the 15 wt% Ta_4HfC_5/Si_2BC_3N composite ceramics after oxyacetylene flame ablation for 10 s: **a** Ablation center; **b** transition region; **c** heat-affected region [10]

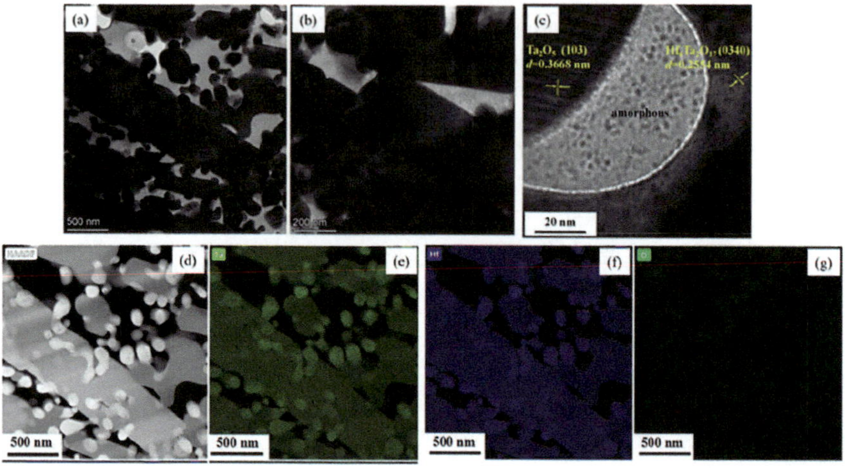

Fig. 6.19 TEM analysis of the 15 wt% Ta_4HfC_5/Si_2BC_3N composite ceramics after oxyacetylene flame ablation for 10 s: **a–c** Bright-field images in different region; **d** HAADF image; **e** Ta mapping; **f** Hf mapping; **g** O mapping [10]

inside of the skeleton. This oxide structure slows down the volatilization of tantalum hafnium silicate glass and strengthens the structural stability of ablation layer.

(6) *ZrB_2/Si_2BC_3N composite ceramics*

Thermal shock resistance

Reinforcements of *in-situ* ZrB_2 are generated by ZrO_2 precursor or by ZrO_2, C and B_2O_3 precursor via the sol–gel method [11]. In another case, the mechanical alloyed nano ZrB_2 are also incorporated in MA-Si_2BC_3N [12]. For sol–gel derived ZrB_2/Si_2BC_3N composite ceramics, their diffusion coefficient decreases with testing temperature, while the thermal expansion coefficient behaves opposite result obtained less than 5.2×10^{-6} °C^{-1} (Fig. 6.20).

After thermal shock test, the residual strength of the two sol–gel derived ZrB_2/Si_2BC_3N decreases gradually (Fig. 6.21). For ZrO_2 precursor derived ceramics, their residual strengths are higher than that of the counterparts at each thermal shock temperature difference. For example, at $\Delta T = 1100$ °C, the residual strength of the former decreases from 358.1 MPa to 230.5 MPa, reduced by 35.6%; however, the latter decreases from 338.6 MPa to 183.4 MPa, 45% lower than the RT value.

After thermal shock at $\Delta T = 800$–1100 °C, the composite ceramics show similar phase composition but differs in the relative content of the individua phase. At $\Delta T =$ 800 °C, no trace of oxides is found or below the limits of detection. With increasing ΔT, ZrO_2 oxides appear with higher intensity (Fig. 6.22). The sol–gel derived ZrB_2/Si_2BC_3N composite ceramics both show similar surface and fracture morphologies at certain thermal shock temperature difference. At $\Delta T = 800$ °C, a smooth and dense structure without obvious oxidation layer is found. At $\Delta T = 1000$ °C, the composite

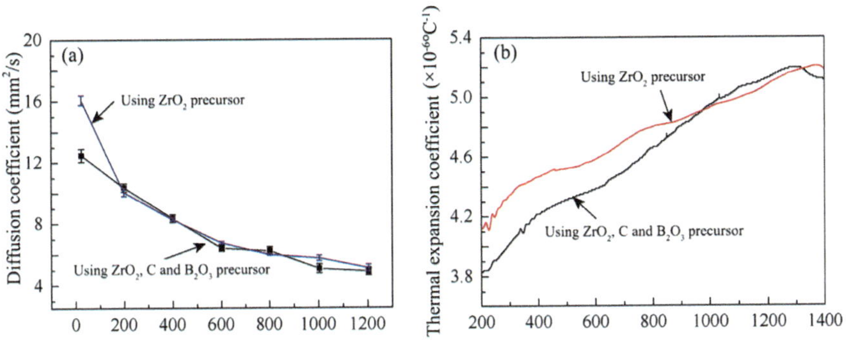

Fig. 6.20 Thermophysical properties of 15 wt% ZrB_2/Si_2BC_3N composite ceramics SPS densified at 2000 °C/40 MPa/5 min: **a** Diffusion coefficient; **b** CET [11]

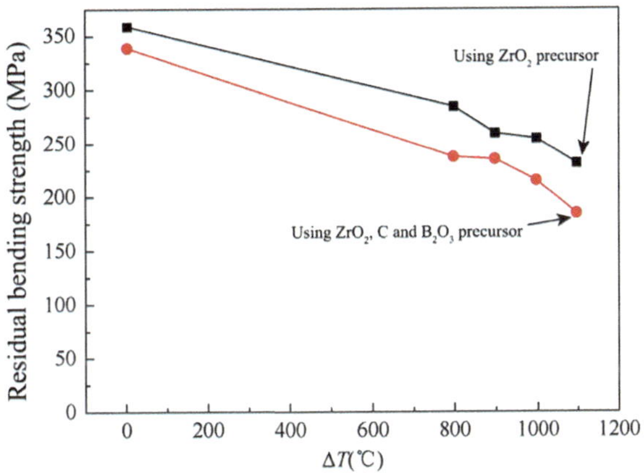

Fig. 6.21 Residual strength of the SPS densified 15 wt% ZrB_2/Si_2BC_3N composite ceramics after thermal shock at different temperature difference [11]

ceramics become loose and porous, and the oxidation products are SiO_2 and ZrO_2. When ΔT increases to 1100 °C, obvious oxide layer with substantial micropores appear on the ceramic surfaces (Fig. 6.23 and Fig. 6.24).

Mechanical alloyed nano ZrB_2 can also improve their thermal shock resistance. Theoretical calculation of the R parameter reaches 599 °C for 10 wt% nano ZrB_2 containing ceramics, much higher than 431 °C of the pristine MA-Si_2BC_3N. The higher nano ZrB_2 addition deteriorate this advantage due to their lower strengths, higher Young's modulus and thermal expansion coefficient. Similarly, the calculated R'''' varies in accordance with the fracture toughness and calculated R_{st} parameter also suggests improved thermal shock resistance with nano ZrB_2 addition (Table 6.4). After thermal shock test at $\Delta T = 800$ °C or $\Delta T = 1000$ °C, the nano ZrB_2 containing ceramics show much higher residual strength than that of pristine MA-Si_2BC_3N. At $\Delta T = 1000$ °C, the 10 wt% nano ZrB_2 containing ceramics remain 64.9% of its strength, much higher than 32.2%, 28.1%, and 26.8% of the 20 wt% nano ZrB_2 containing ones, pure MA-Si_2BC_3N, and 1 mol% Zr-Al incorporating ceramics.

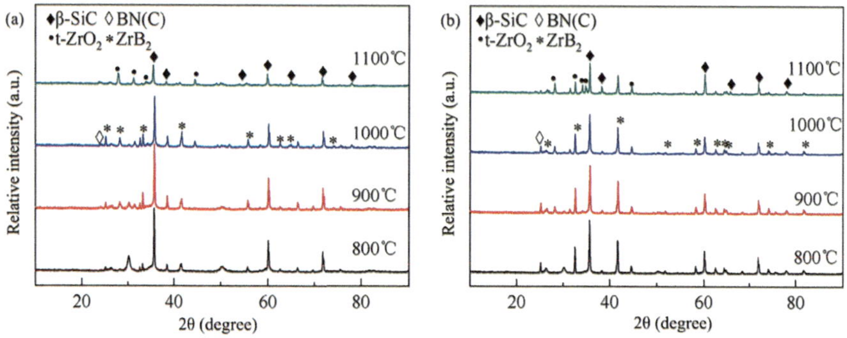

Fig. 6.22 XRD patterns of the SPS densified ZrB_2/Si_2BC_3N composite ceramics after thermal shock at different temperature difference: *In-situ* generated ZrB_2 by **a** ZrO_2 precursor; **b** ZrO_2, C and B_2O_3 precursor [11]

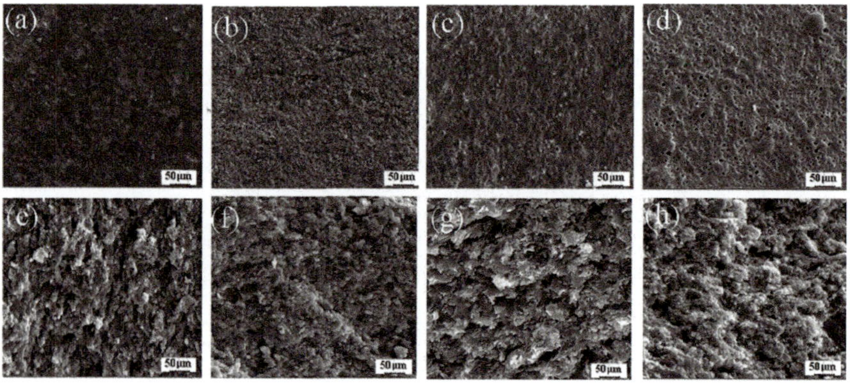

Fig. 6.23 SEM surface and fracture morphologies of the ZrB_2/Si_2BC_3N composite ceramics (using ZrO_2 the precursor) after thermal shock at different temperature difference: **a, e** $\Delta T = 800$ °C; **b, f** $\Delta T = 900$ °C; **c, g** $\Delta T = 1000$ °C; **d, h** $\Delta T = 1100$ °C [11]

Ablation resistance

The as-ablated pure MA-Si_2BC_3N has been damaged by the high-temperature high-speed airflow. However, the 15 wt% ZrB_2 containing composite ceramics (using ZrO_2 precursor) remain intact with some surface oxides [13]. The size of the ablation zone increases with the exposing time. Compared with the pure MA-Si_2BC_3N, the addition of *in-situ* ZrB_2 significantly improve their ablation resistance, obtaining linear rate of 0.0041 mm/s and mass ablation rates of 0.48 mg/s, respectively (Fig. 6.25).

During oxyacetylene flame (heat flux ~ 4.2 MW/m^2) ablation, the sample surface temperature increases sharply to 1600 °C and then leveled off to around 2300 °C. The oxide productions of $ZrSiO_4$, SiO_2 and ZrO_2 are evident; however, no trace of boron oxides is found. The peak intensities of the $ZrSiO_4$ and ZrO_2 in central region are much higher than other phases (Fig. 6.26). A discontinuous glassy layer covers

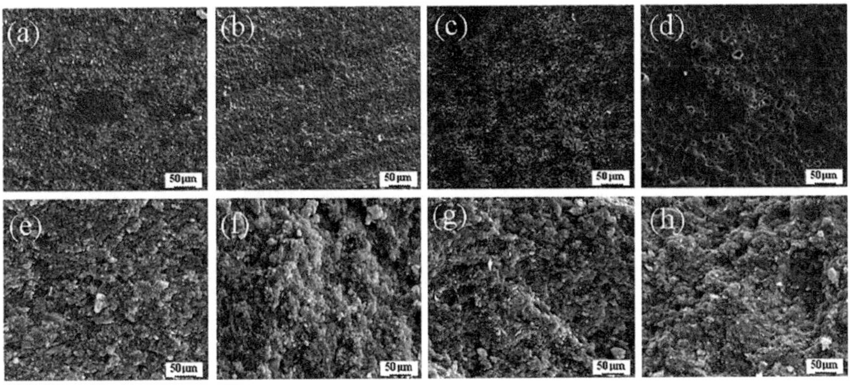

Fig. 6.24 SEM surface and fracture morphologies of the ZrB_2/Si_2BC_3N composite ceramics (using ZrO_2, C and B_2O_3 precursor) after thermal shock at different temperature difference: **a**, **e** $\Delta T = 800$ °C; **b**, **f** $\Delta T = 900$ °C; **c**, **g** $\Delta T = 1000$ °C; **d**, **h** $\Delta T = 1100$ °C [11]

Table 6.4 The calculated thermal shock parameters and measured residual strength of the mechanical alloyed nano ZrB_2 reinforced MA-Si_2BC_3N composite ceramics [12]

Nano ZrB_2 content (wt%)	α (10^{-6} °C^{-1})	R (°C)	R'''' (mm)	R_{st} (K $m^{1/2}$)	Residual strength (MPa)/ Strength retention rate (%)		
					RT	$\Delta T =$ 800 °C	$\Delta T =$ 1000 °C
+ 0	4.21	431	0.26	8.0	347.5	196.2/ 56.5	97.7/28.1
+ 10	4.28	599	0.27	11.34	559.6	446.6/ 79.8	363.6/ 65.0
+ 20	4.80	447	0.28	8.71	512.3	212.6/ 41.5	165.2/ 32.2

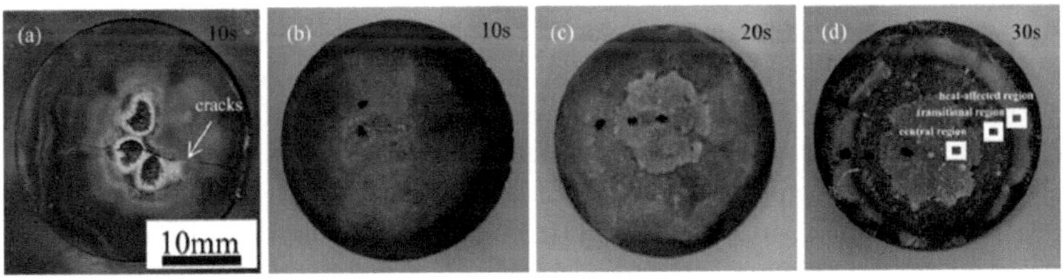

Fig. 6.25 Surface macro morphologies of the ZrB_2/Si_2BC_3N composite ceramics (using ZrO_2 precursor) after oxyacetylene flame (heat flux ~ 4.2 MW/m^2) ablation for different time: **a** Pure MA-Si_2BC_3N, 10 s; **b** ZrB_2/Si_2BC_3N, 10 s; **c** ZrB_2/Si_2BC_3N, 20 s; **d** ZrB_2/Si_2BC_3N, 30 s [13]

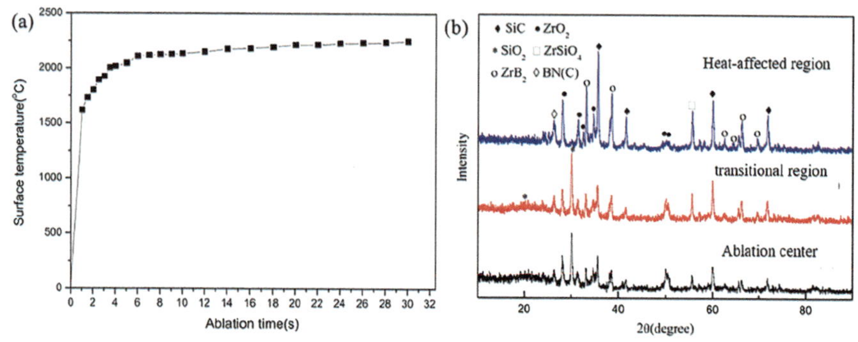

Fig. 6.26 Surface temperature curve and micro XRD patterns of the ZrB_2/Si_2BC_3N composite ceramics (using ZrO_2 precursor) after oxyacetylene flame (heat flux ~ 4.2 MW/m^2) ablation for 30 s: **a** Surface temperature curve; **b** micro XRD patterns [13]

on the ablation center surfaces, but no penetrating cracks are found. The glassy layer contains Zr, Si and O, mainly composed of ZrO_2, SiO_2 and/or $ZrSiO_4$. The total thickness of the porous oxide layer is about 380.2 μm (Fig. 6.27).

In transitional region, substantial fused SiO_2 are covered on the surfaces. No noticeable elemental content changes from the center to transition region. A few of broken bubbles are presented and some striped protuberances appear under the broken bubbles. The thickness of oxide layer is about 149.7 μm, less than the oxide thickness of the central region. Some bubbles ~ 30 μm consisted mainly of Si and O elements are distributed on the surfaces of the heat-affected region doe to the high viscosity of SiO_2 at relatively low ablation temperatures. These bubbles also position at the interfaces of oxide scale and ceramic matrix, suggesting undesirable binding strength (Fig. 6.28).

The surface macro morphologies of the ZrB_2/Si_2BC_3N composite ceramics (using ZrO_2, C and B_2O_3 precursor) are similar after 10 s or 20 s oxyacetylene flame (heat flux ~ 4.2 MW/m^2) ablation. Beyond 30 s of ablation, shallow ablation pits appear on the ceramic surfaces. After 60 s ablation, their linear and mass ablation rates are 0.0005 mm/s and 0.0071 mg/s, much lower than the formers. However, macroscopic visible cracks appear on the ceramic surfaces (Fig. 6.29).

After ablation, the phase composition of the composite ceramics consists of SiC, BN(C), ZrB_2, SiO_2, ZrO_2, and $ZrSiO_4$, (Fig. 6.30). Again, the diffraction peaks of $ZrSiO_4$ in the ablation center are stronger than that in the transition and heat-affected areas. Under the scouring of high-temperature airflow, SiO_2 and ZrO_2 can react quickly to form relatively stable $ZrSiO_4$. The loose and porous structure of ZrO_2, SiO_2 and/or $ZrSiO_4$ forms on the ablation center. There are a large number of micropores and holes left by bubble rupture in the transition zone. The heat-affected surface is covered with an oxide layer of hollow spherical SiO_2 (Fig. 6.31).

The larger ablation pits in the pristine MA-Si_2BC_3N ceramics suggests a low ablation resistance under oxyacetylene flame (heat flux ~ 4.2 MW/m^2). Contrast to

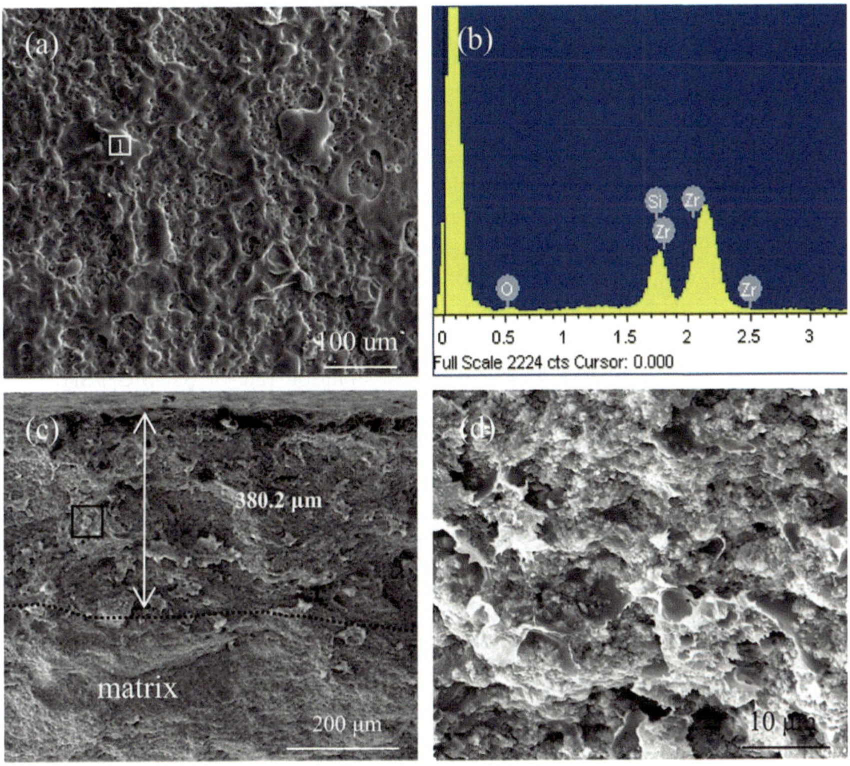

Fig. 6.27 Ablation center morphologies of the ZrB_2/Si_2BC_3N composite ceramics (using ZrO_2 precursor) after oxyacetylene flame ablation for 30 s: **a** Surface morphology; **b** EDS results; **c**, **d** fracture morphologies [13]

expectation, the mechanical alloyed nano ZrB_2 containing composites present much shallower and smaller ablation pits (Fig. 6.32).

The linear and mass ablation rates of pristine MA-Si_2BC_3N reach 0.047 mm/s and 20.80 mg/s, respectively. In comparison, the linear and mass ablation rates of 10 wt% nano ZrB_2 containing ceramics decrease significantly to 0.0091 mm/s and 1.91 mg/s, respectively. With 20 wt% nano ZrB_2, the linear and mass ablation rates are 0.0071 mm/s and 2.10 mg/s (Table 6.5). Their ablation properties of the SPS densified ZrB_2/Si_2BC_3N composites are lower than those sol–gel derived counterparts.

As for the pristine MA-Si_2BC_3N, main phases of α/β-SiC are determined in the ablation and transition zones. In comparison, relatively lower intensities of α/β-SiC are detected in the edge zone, and broad peaks assigned to cristobalite and amorphous ZrO_2 are also found, implying the scouring of ablation products to the other zones. In contrast, relatively higher diffraction peaks of α/β-SiC are presented in the nano ZrB_2 containing ceramics, together with detectable ZrB_2, ZrO_2 and $ZrSiO_4$ phases. Interestingly, the diffraction peak intensities of the ZrO_2 and $ZrSiO_4$ are the main differences in the three zones. In the ablation zone, relatively strong peaks of the

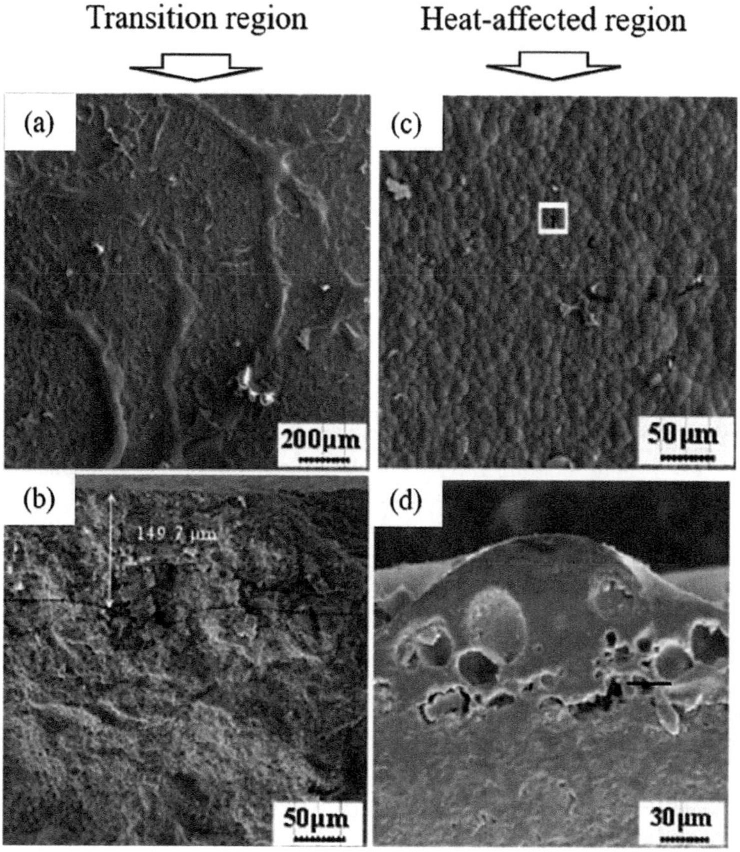

Fig. 6.28 SEM morphologies of the ZrB_2/Si_2BC_3N composite ceramics (using ZrO_2 precursor) after oxyacetylene flame ablation for 30 s: **a**, **c** Surface morphologies; **b**, **d** fracture morphology [13]

ZrO_2 and $ZrSiO_4$ are detected, but increasing ZrO_2 and decreasing $ZrSiO_4$ remain in the transition zone. ZrO_2 in heat-affected zone are reduced but much higher peaks of $ZrSiO_4$ are observed (Fig. 6.33).

In general, the ablation surfaces of the tested composite ceramics are similar regardless of the nano ZrB_2 content. Both porous ablation center surfaces with substantial micropores are distinguished for pure MA-Si_2BC_3N and 20 wt% ZrB_2/Si_2BC_3N composite ceramics, while their transition zones show relatively denser structure. However, the nano ZrB_2 containing ceramics exhibit many coarser particles on surfaces. Interestingly, these spherical particles of mainly SiO_2 are also observed at the heat-affected zone, and the spherical particles in nano ZrB_2 containing ceramics show smooth surfaces with smaller sizes (Fig. 6.34).

For pure MA-Si_2BC_3N, ablation center is covered with an oxide layer thinner than that in transition zone. In comparison, a thick layer of SiO_2 covers on the surfaces of

Fig. 6.29 Macroscopic morphologies of the ZrB_2/Si_2BC_3N composite ceramics (using ZrO_2, C and B_2O_3 precursor) after oxyacetylene flame (heat flux ~ 4.2 MW/m^2) ablation for different time: **a** 10 s; **b** 20 s; **c** 30 s; **d** 60 s [13]

heat-affected zone. As for the ablation zone, a thin layer ~ 60 μm of ZrO_2, SiO_2 and $ZrSiO_4$ oxides is observed for the 20 wt% ZrB_2/Si_2BC_3N, while the heat-affected zone are mainly composed of SiO_2 with limited Zr content (Fig. 6.35).

(7) *HfB_2/Si_2BC_3N composite ceramics*

Thermal shock resistance

The residual strength of pure MA-Si_2BC_3N increases first and then decreases with increasing thermal shock temperature difference (ΔT) [14]. After the introduction of mechanical alloyed nano HfB_2, the residual strength of composite ceramics decreases gradually with ΔT. The phase composition of pure MA-Si_2BC_3N are β-SiC, BN(C) and a small amount of SiO_2 at different ΔT. The composition in HfB_2 containing composite ceramics are basically the same, but the oxidation products

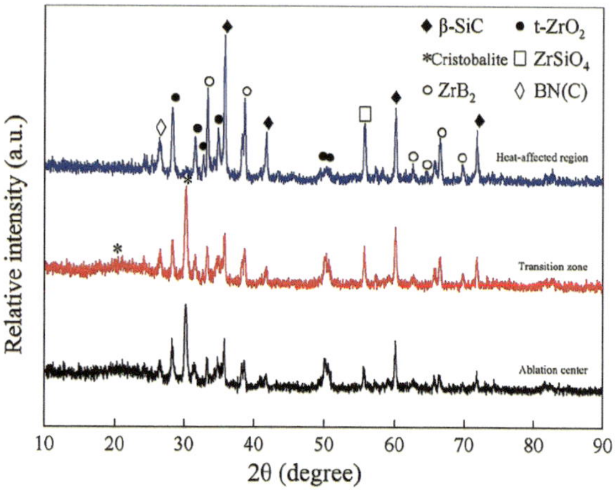

Fig. 6.30 XRD patterns of the ZrB_2/Si_2BC_3N composite ceramics (using ZrO_2, C and B_2O_3 precursor) after oxyacetylene flame ablation [13]

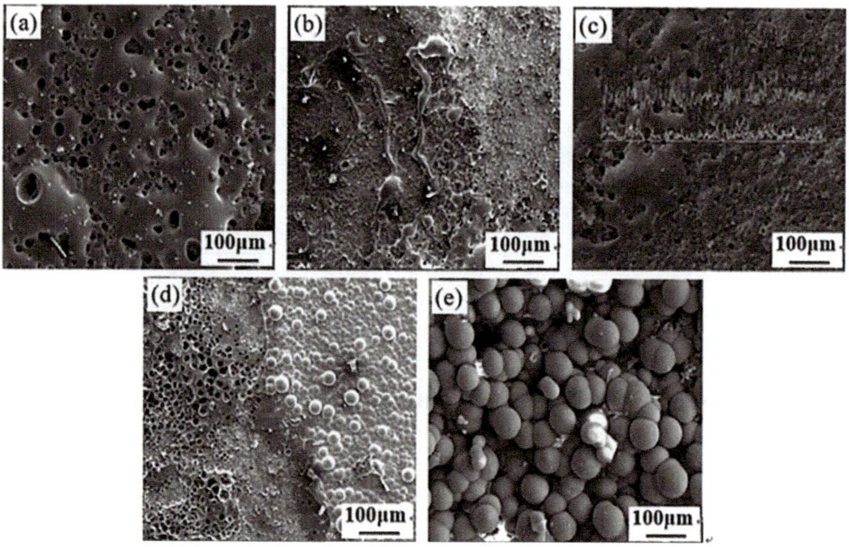

Fig. 6.31 Surface morphologies of the ZrB_2/Si_2BC_3N composite ceramics (using ZrO_2, C and B_2O_3 precursor) after oxyacetylene flame ablation for 60 s: **a** Ablation center; **b** interfaces between the ablation center and the transition zone; **c** transition zone; **d** interfaces between the transition zone and the heat affected region; **e** heat-affected zone [13]

are mainly HfO_2 and $HfSiO_4$ (Fig. 6.36). The residual strength of the 30 vol% nano HfB_2 containing composites shows the highest value at each ΔT. However, the 20 vol% HfB_2/Si_2BC_3N composite ceramics should have the optimized thermal shock resistance according to the strength retention rate (Table 6.6).

At $\Delta T = 800$ °C, slight oxidation occurs and there are some micropores on the surfaces of pure MA-Si_2BC_3N. At $\Delta T = 1000$–1200 °C, substantial micropores

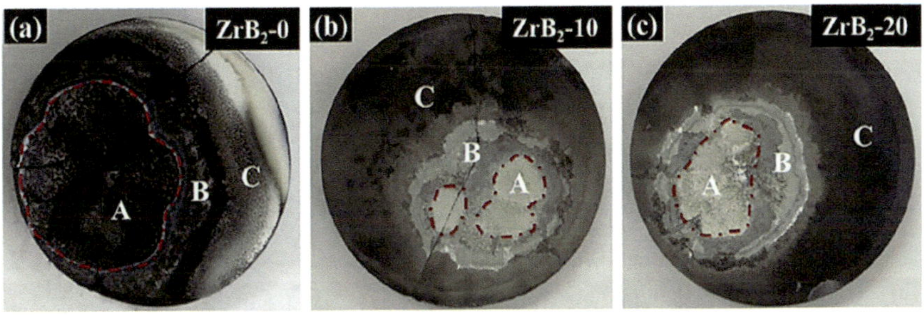

Fig. 6.32 Macroscopic morphologies of the SPS densified ZrB_2/Si_2BC_3N composite ceramics (using mechanical alloyed nano ZrB_2) after oxyacetylene flame (heat flux ~ 4.2 MW/m^2) ablation for 30 s: **a** Pure MA-Si_2BC_3N; **b** + 10 wt% nano ZrB_2; **c** + 20 wt% nano ZrB_2 (A: ablation center; B: transition region; C: heat-affected zone) [12]

Table 6.5 Ablation properties of the SPS sintered ZrB_2/Si_2BC_3N composite ceramics after oxyacetylene flame (heat flux ~ 4.2 MW/m^2) ablation for different time [12]

Composition	Ablation time (s)	Mass ablation rate (mg/s)	Linear ablation rate (mm/s)
MA-Si_2BC_3N	30	20.80	0.0470
+ 15 wt% ZrB_2[a]	30	0.48	0.0041
+ 15 wt% ZrB_2[b]	60	0.0071	0.0005
+ 10 wt% nano ZrB_2[c]	30	1.91	0.0091
+ 20 wt% nano ZrB_2[c]	30	2.10	0.0071

[a] Using ZrO_2 precursor; [b] using ZrO_2, C and B_2O_3 precursor; [c] using mechanical alloyed nano ZrB_2

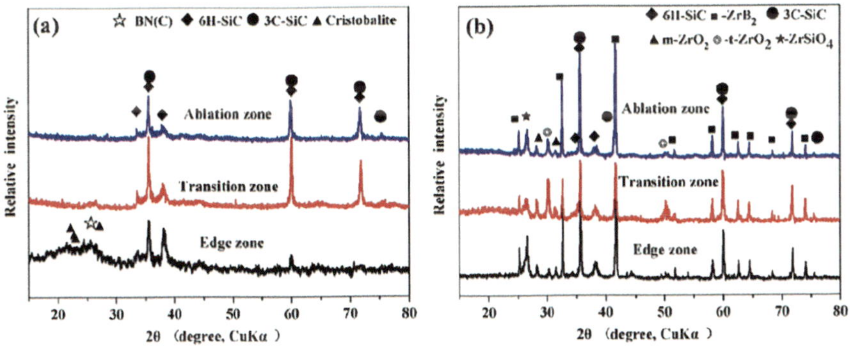

Fig. 6.33 Micro XRD patterns of the ZrB_2/Si_2BC_3N composite ceramics (using mechanical alloyed nano ZrB_2) after oxyacetylene flame ablation for 30 s: **a** Pure MA-Si_2BC_3N; **b** + 20 wt% nano ZrB_2 [12]

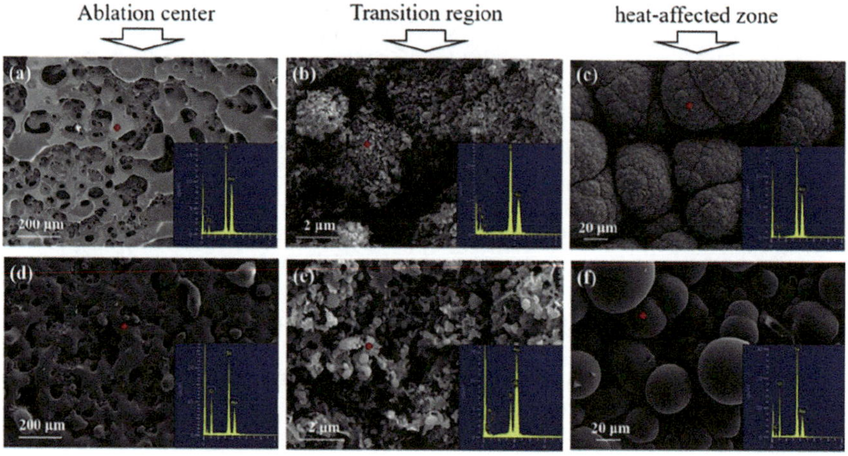

Fig. 6.34 SEM surface morphologies of the composite ceramics after oxyacetylene flame ablation for 30 s: **a–c** Pure MA-Si_2BC_3N and **d–f** + 20 wt% nano ZrB_2 [12]

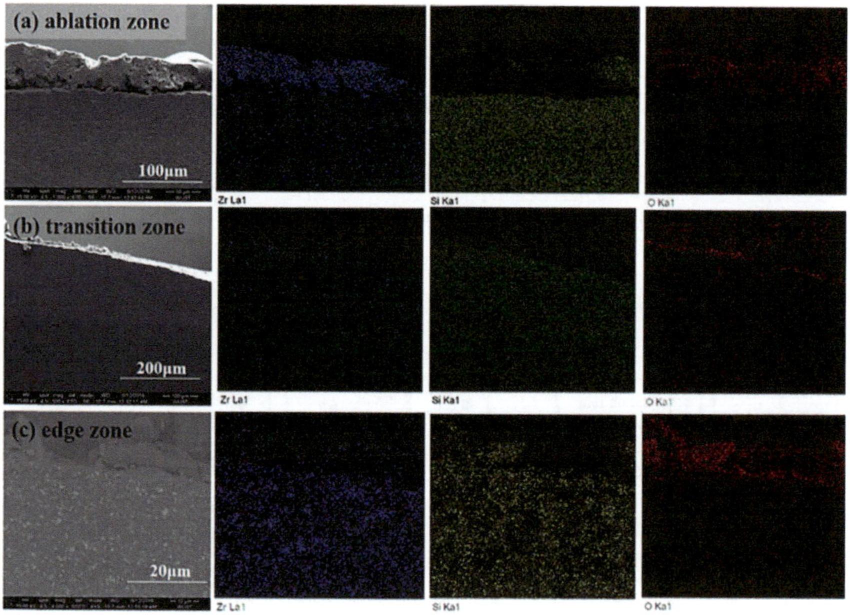

Fig. 6.35 SEM cross-section morphologies and corresponding element mappings of the 20 wt% nano ZrB_2 containing composite ceramics after oxyacetylene flame ablation for 30 s: **a** Ablation center; **b** transition zone; **c** heat-affected area [12]

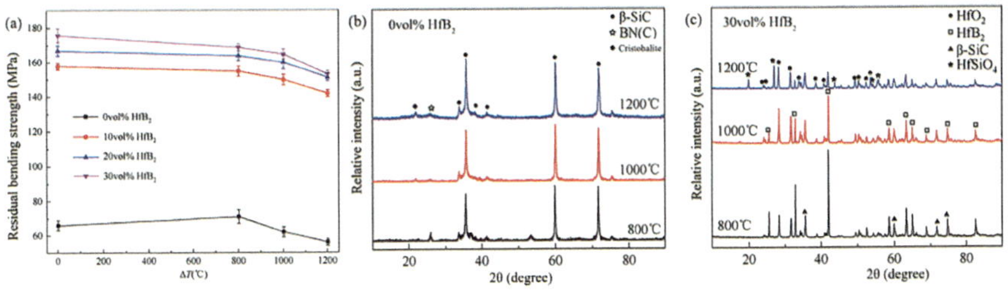

Fig. 6.36 Residual strength and XRD patterns of the HfB_2/Si_2BC_3N composite ceramics after thermal shock at different temperature difference: **a** Residual strength; **b** XRD patterns [14]

Table 6.6 Residual strength and strength retention rate of the HfB_2/Si_2BC_3N composite ceramics after thermal shock at different temperature difference [14]

Nano HfB_2 content (vol%)	$\Delta T = 800$ °C		$\Delta T = 1000$ °C		$\Delta T = 1200$ °C	
	Residual strength (MPa)	Strength retention rate (%)	Residual strength (MPa)	Strength retention rate (%)	Residual strength (MPa)	Strength retention rate (%)
MA-Si_2BC_3N	71	107.6	62	93.9	56	84.8
+ 10	155	98.1	150	94.9	142	89.9
+ 20	164	98.2	160	95.8	151	90.4
+ 30	169	96	165	93.8	153	86.9

appear on the surfaces, and part of the micropores are converted into microcracks. Some diffusion channels left by gas escape on the surfaces are also found (Fig. 6.37). The 30 vol% HfB_2/Si_2BC_3N composite ceramics have adverse variation trend in surface morphologies, showing gradual denser and smooth structure with increasing ΔT. Note that some oxide precipitates have distributed beneath the dense outermost oxide scale. After thermal shock at different ΔT, pull-out of BN(C) plates is found, and not obvious morphological change is observed (Fig. 6.38).

Ablation resistance

Hot-pressed MA-Si_2BC_3N ceramics show highest surface temperature ~ 1816 °C after 10 s of oxyacetylene flame (heat flux ~ 4.2 MW/m^2) ablation [14], which are ~ 1942 °C, ~ 2070 °C and ~ 2364 °C for the various HfB_2/Si_2BC_3N composite ceramics after 60 s of ablation. Interestingly, the 30 vol% HfB_2/Si_2BC_3N composite ceramics show lower surface temperature ~ 1579 °C than that of pure MA-Si_2BC_3N after 10 s of ablation, possibly resulting from the higher thermal conductivity of HfB_2 and HfC (Fig. 6.39).

After 10 s ablation, the mass and linear ablation rates of pure MA-Si_2BC_3N are the highest, obtaining 0.99 mg/s and 0.0293 mm/s, respectively. The addition of some nano HfB_2 significantly improves their ablation resistance. With the increasing nano HfB_2, the linear ablation rate decreases gradually reaching to 0.00055 mm/s, while the mass ablation rate is contrary to the change of linear ablation rate (Table 6.7).

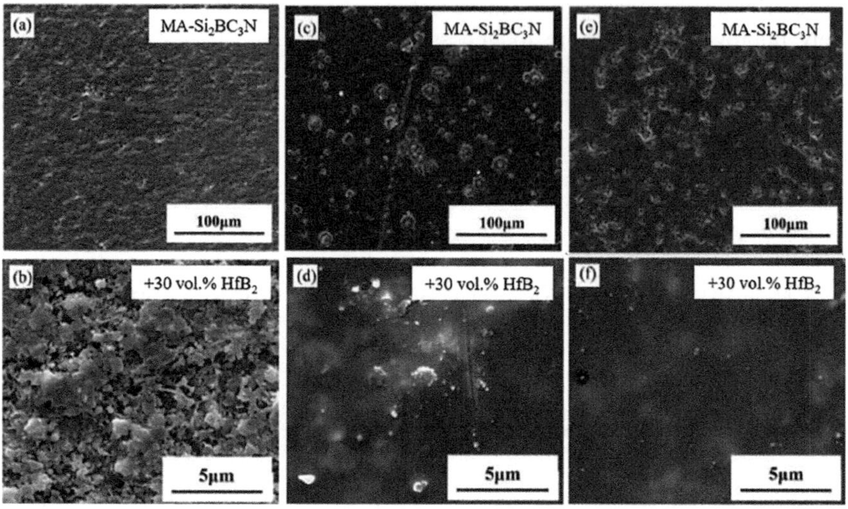

Fig. 6.37 SEM surface morphologies of pure MA-Si_2BC_3N and 30 vol% HfB_2/Si_2BC_3N ceramics after thermal shock at different temperature difference: **a**, **b** $\Delta T = 800$ °C; **c**, **d** $\Delta T = 1000$ °C; **e**, **f** $\Delta T = 1200$ °C [14]

Fig. 6.38 SEM fracture morphologies of 30 vol% HfB_2/Si_2BC_3N composite ceramics after thermal shock at different temperature difference: **a** $\Delta T = 800$ °C; **b** $\Delta T = 1000$ °C; **c** $\Delta T = 1200$ °C [14]

After ablation, penetrating cracks with different size are observed for some as-ablated bulk ceramics. The pure MA-Si_2BC_3N ceramics have been completely crushed with deeper ablation pits after 10 s of ablation. The ablation pit area further decreases with the increasing HfB_2. With increasing exposing time from 10 to 60 s, oxides gradually cover the ablation region, and no rupturing are found for the 30 vol% nano HfB_2 incorporating ceramics (Fig. 6.40).

After 10 s of oxyacetylene flame ablation, XRD patterns suggest some quartz and few cristobalites form on the oxide layer for pure MA-Si_2BC_3N ceramics. The peak intensities of the quartz decrease gradually from the center to heat-affected region. No trace of boron oxides can be detected. For HfB_2 containing ceramics, oxides of HfO_2 and $HfSiO_4$ are generated, and the peak intensities of the HfO_2 and $HfSiO_4$ in heat-affected zone are lower than those in center and transition regions (Figure 6.41).

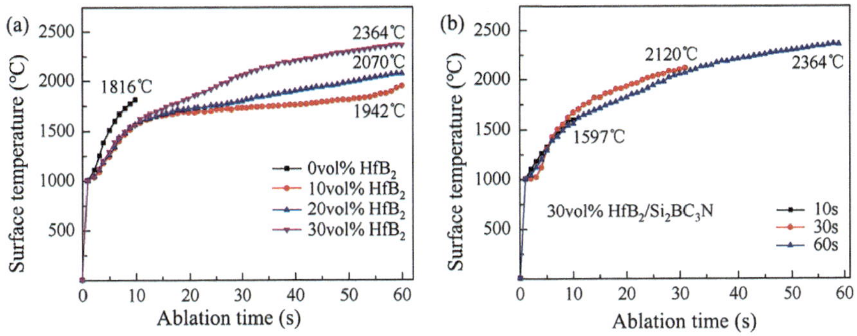

Fig. 6.39 Surface temperature curves of the hot-pressed HfB_2/Si_2BC_3N composite ceramics after oxyacetylene flame (heat flux ~ 4.2 MW/m^2) ablation: **a** Various composites ablated for 60 s; **b** composites with 30 vol% nano HfB_2 ablated for different time [14]

Table 6.7 Mass and linear ablation rates of the hot-pressed HfB_2/Si_2BC_3N composite ceramics after oxyacetylene flame (heat flux ~ 4.2 MW/m^2) ablation [14]

Nano HfB_2 content (vol%)	Ablation time (s)	Mass change (mg)	Mass ablation rate (mg/s)	Thickness change (mm)	Linear ablation rate (mm/s)
+ 0	10	9.89	0.99	0.293	0.02930
+ 10	60	5.37	0.091	0.643	0.01057
+ 20	60	9.76	0.16	0.554	0.00923
+ 30	10	0.25	0.025	0.004	0.00040
+ 30	30	1.20	0.041	0.014	0.00047
+ 30	60	11.11	0.19	0.033	0.00055

The ablation center becomes denser and continuous with increasing nano HfB_2 content. Besides, oxides of mainly SiO_2, $HfSiO_4$ and HfO_2 precipitate within the oxide scale. In all cases, the uneven and rough ablation surfaces are found (Fig. 6.42). Since the surface temperature of the transition zone is lower than that of the ablation center, a relatively dense and low-porosity surface structure is built in transition zone. Two kinds of morphologies in transition zone are compared, one is the wave-pattern structure constructed by the airflow scouring, and another is the porous structure with considerable micro-holes and bubble bursting. For pure MA-Si_2BC_3N, the transition zone is larger with an average width of ~ 893.7 μm, lower than their counterparts (Fig. 6.43).

The clear and apparent boundaries separate the transition zone and the heat affected area. Generally, the heat-affected surfaces are porous and smooth, distributed with average size of bubbles ~ 10 μm. During ablation, liquid oxides of SiO_2, $HfSiO_4$ and HfO_2 with relatively high viscosity and poor wettability, can heal cracks and pores automatically but powerless (Fig. 6.44).

(8) *(ZrB_2-ZrN)/Si_2BC_3N composite ceramics*

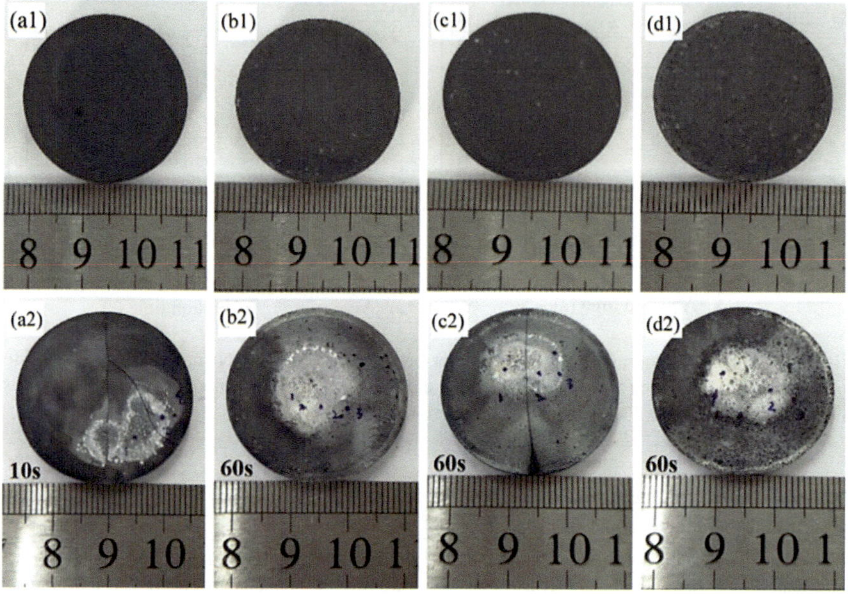

Fig. 6.40 Surface macro morphologies of the hot-pressed HfB_2/Si_2BC_3N composite ceramics before and after oxyacetylene flame (heat flux ~ 4.2 MW/m^2) ablation for different time: **a** MA-Si_2BC_3N; **b** + 10 vol% nano HfB_2; **c** + 20 vol% nano HfB_2; **d** + 30 vol% nano HfB_2 [14]

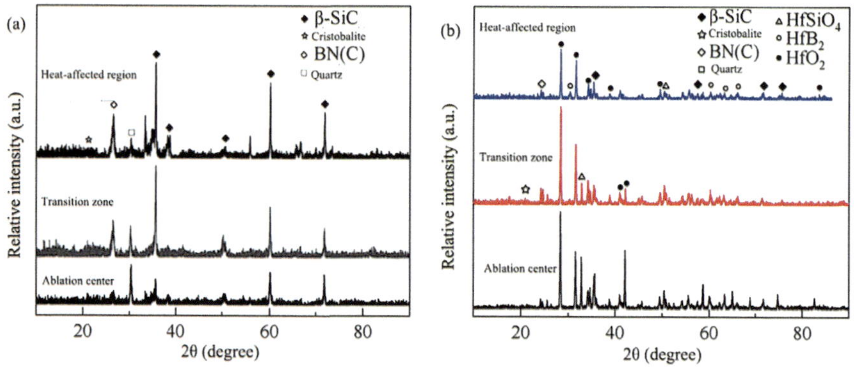

Fig. 6.41 Micro XRD patterns of the hot-pressed HfB_2/Si_2BC_3N composite ceramics after 10 s of oxyacetylene flame ablation: **a** MA-Si_2BC_3N; **b** + 30 vol% nano HfB_2 (A: ablation center; B: transition zone; C: heat-affected zone) [14]

For all the hot-pressed $(ZrB_2\text{-}ZrN)/Si_2BC_3N$ composite ceramics, the obvious macro-cracks appear on the sample surfaces with ablation pits. The deep ablation pits in composites sintered at 1700–1800 °C are not connected, possibly resulting from the instability of the oxyacetylene flame center during ablation (Fig. 6.45). With the increasing sintering temperature, the ablation resistance of the $(ZrB_2\text{-}ZrN)/Si_2BC_3N$

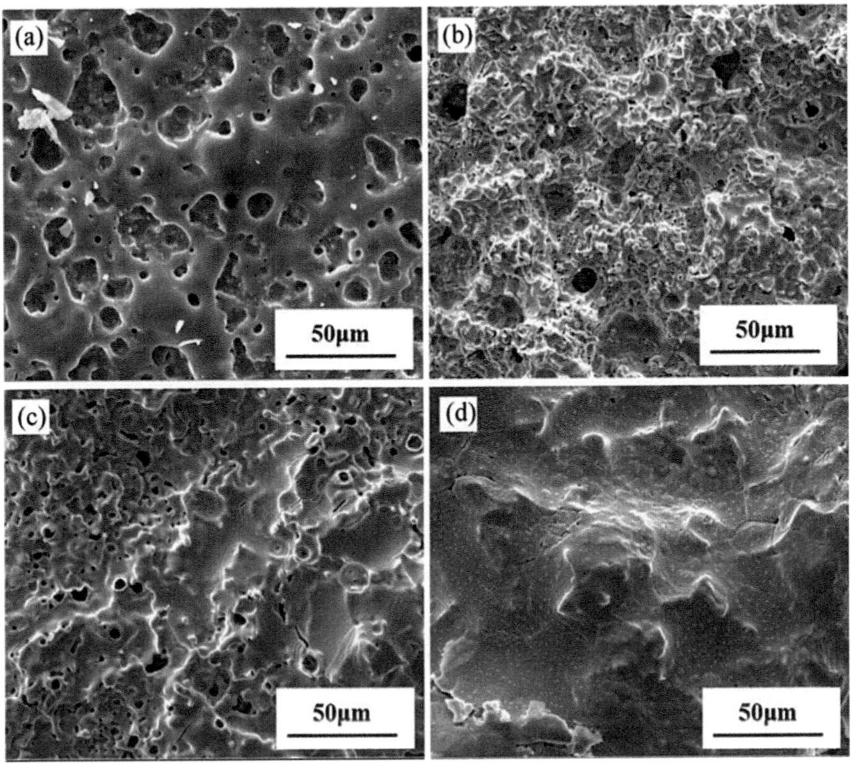

Fig. 6.42 SEM surface morphologies of the ablation center for the hot-pressed HfB_2/Si_2BC_3N composite ceramics after oxyacetylene flame ablation for 10 s: **a** MA-Si_2BC_3N; **b** + 10 vol% nano HfB_2; **c** + 20 vol% nano HfB_2; **d** + 30 vol% nano HfB_2 [14]

composite ceramics increases gradually. With the prolongation of ablation time, their mass ablation rate increases with deceasing linear ablation rate (Table 6.8). After 10 s of oxyacetylene flame (heat flux ~ 4.2 MW/m^2) ablation, the mass and linear ablation rates of the hot-pressed (ZrB_2-ZrN)/Si_2BC_3N composite ceramics are 0.066 mg/s and 0.0081 mm/s, respectively.

The main phases on the ablated surfaces are those of SiC, ZrO_2, cristobalite and some amorphous phases for all the as-ablated (ZrB_2-ZrN)/Si_2BC_3N. Diffraction peaks corresponding to the cristobalite and t-ZrO_2 oxides also appear in the unablated zone. In all regions, oxides and matrix phases are detected, but differs in their relative content (Fig. 6.46).

The ablated center surfaces of the (ZrB_2-ZrN)/Si_2BC_3N composite ceramics are loose and porous, distributed with a large number of larger pores and holes. There are a lot of spherical SiO_2 bubbles in the transition zone. This spherical structure is caused by the airflow scouring and SiC oxidation in the transition zone. Bubbles also appear in the heat-affected zone, and their average size varies from 10 μm to 30 μm. During ablation, most of the gaseous byproducts by internal oxidation are arrested

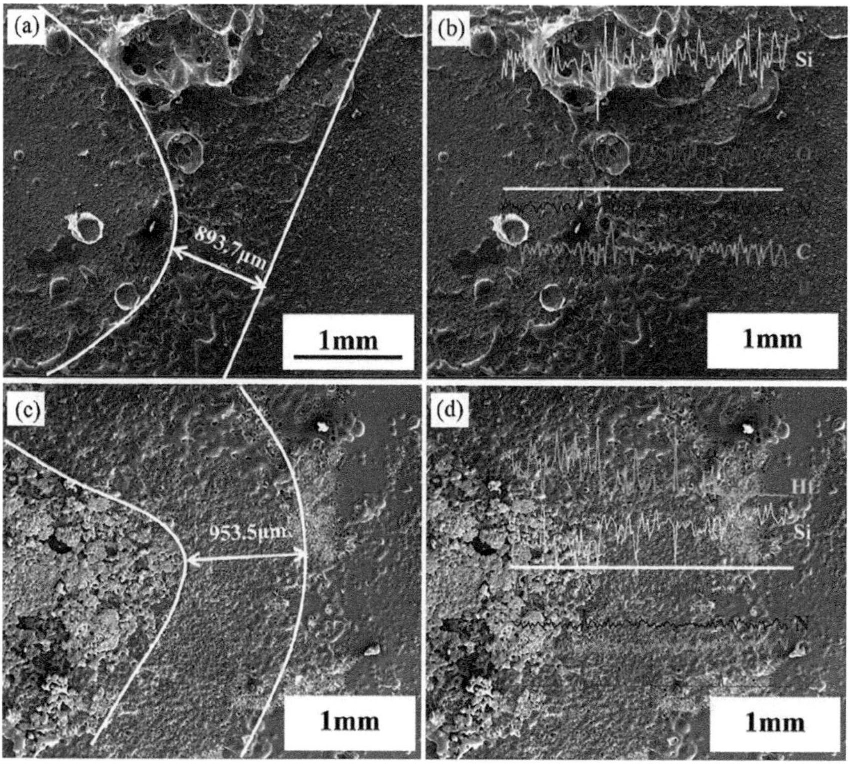

Fig. 6.43 SEM surface morphologies of transition zone for the hot-pressed HfB_2/Si_2BC_3N composite ceramics after oxyacetylene flame ablation for 10 s: **a**, **b** MA-Si_2BC_3N; **c**, **d** + 30 vol% nano HfB_2 [14]

thereby forming such bubble structure. Far away from the ablation center, bubbles tend to rupture (Fig. 6.47).

6.3 Thermal Shock and Ablation Behavior of MA-SiBCN CMCs

(1) *C_f/Si_2BC_3N CMCs*

The non-isothermal oxidation curves of all C_f/Si_2BC_3N composites have a similar variation trend, where the oxidation process can be divided into three stages [15]. In low-temperature section (<620 °C), water volatilization and gas desorption both cause obvious weight loss (5 wt%) at 180–260 °C, and afterward weight remains almost unchanged. In medium-temperature region (620–1100 °C), weight reduces rapidly, reaching the minimum at about 800–900 °C, and then rapidly increases with

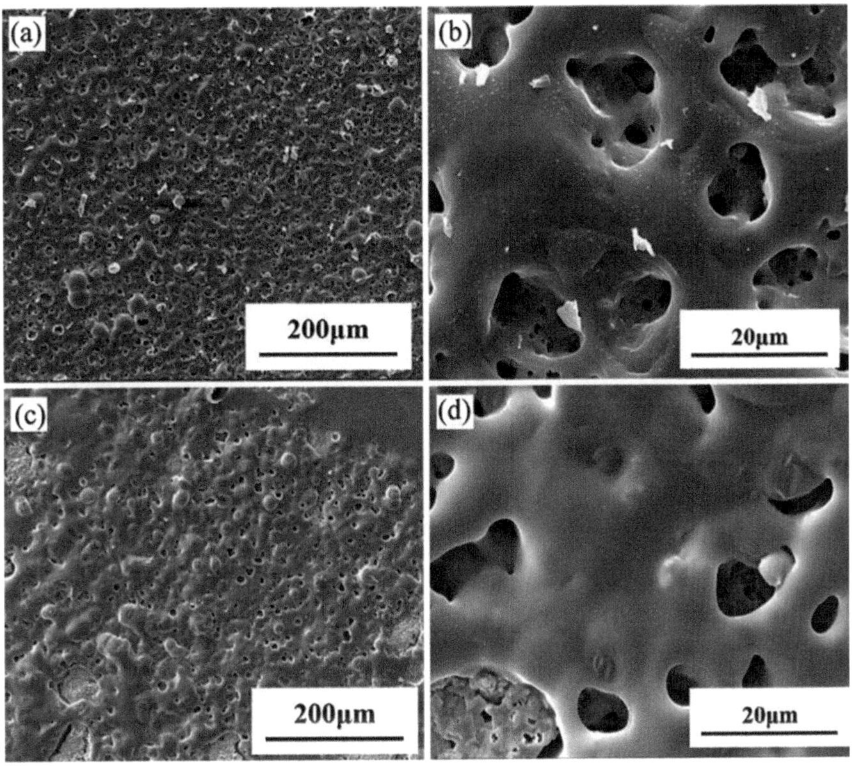

Fig. 6.44 SEM surface morphologies of heat-affected zone for the hot-pressed HfB_2/Si_2BC_3N composite ceramics after oxyacetylene flame ablation for 10 s: **a**, **b** MA-Si_2BC_3N; **c**, **d** + 30 vol% nano HfB_2 [14]

the increasing temperature from 900 to 1100 °C. Compared with T1900 (hot-pressed at 1900 °C) and T2000 (hot-pressed at 2000 °C) composites, the T1800 (hot-pressed at 1800 °C) composites start to gain weight at a lower temperature (860 °C) with less weight loss ~ 17%. In high-temperature region (1100–1400 °C), the weight of all the composites increases linearly, but the T1800 composites still maintain a relatively higher weight gain rate (Fig. 6.48).

After thermal shock at $\Delta T = 700$ °C, the bending strength of the T1900 and T2000 composites is reduced by 40.9% and 43.1%, respectively (Table 6.9). For T1900 and T2000 composites, due to stronger interface bonding, the thermal stress at the interface is hard to be alleviated by fiber debonding, and the tensile stress in the matrix is more likely to accumulate to the yield strength of the matrix causing more cracks in the matrix and decreasing mechanical properties.

At $\Delta T = 700$ °C, although carbon fibers are less damaged, the resin-derived carbon coating as interfacial phase has been damaged seriously due to oxidation and thermal stress, so the interfaces with weakened bonding strength cannot effectively transfer the load to fibers, resulting in a significant decline in mechanical properties.

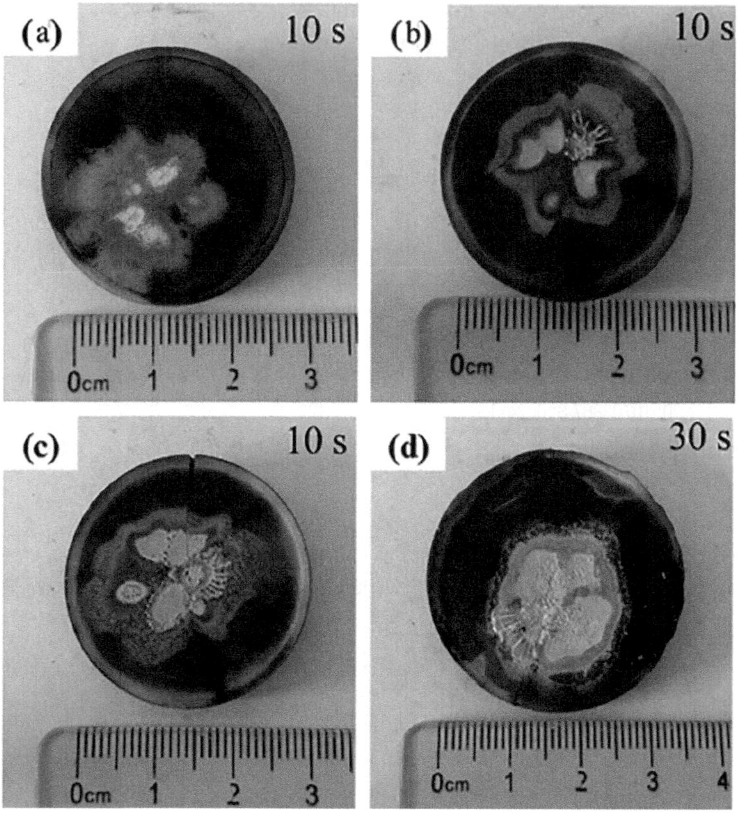

Fig. 6.45 Surface macro morphologies of as-ablated (ZrB_2-ZrN)/Si_2BC_3N composite ceramics hot-pressed sintered at **a** 1700 °C; **b** 1800°C; **c**, **d** 1900 °C

Table 6.8 Mass ablation rate and line ablation rate of the hot-pressed (ZrB_2-ZrN)/Si_2BC_3N composite ceramics after oxyacetylene flame ablation for 10 s

Hot-pressing condition (60 MPa/30 min)	Ablation time (s)	Mass ablation rate (mg/s)	Linear ablation rate (mm/s)
1700 °C	10	0.092	0.0032
1800 °C	10	0.086	0.0061
1900 °C	10	0.066	0.0081
	30	0.11	0.0077

However, after thermal shock at $\Delta T = 700$ °C, the strength of T1800 composites reduces by only 20.4%. Further increasing $\Delta T = 900$ °C or $\Delta T = 1100$ °C, the strength of the composites decreases, but the decrease is limited.

After thermal shock at $\Delta T = 700$ °C, no obvious oxidation of the ceramic matrix can be seen. Carbon fibers are only slightly oxidized: the active points of the carbon

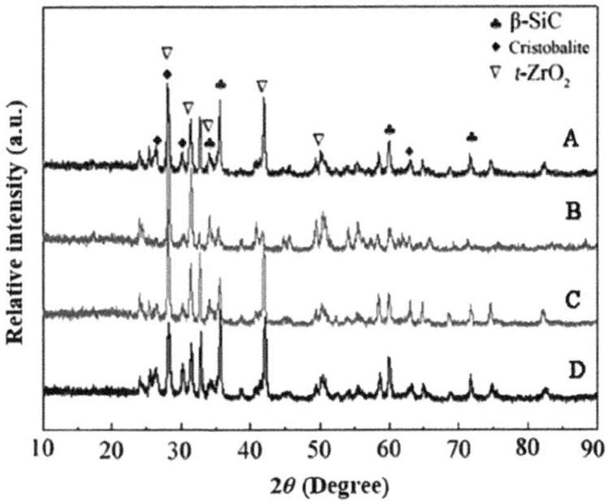

Fig. 6.46 Micro XRD patterns of the (ZrB_2-ZrN)/Si_2BC_3N composite ceramics (hot-pressed at 1900 °C) after oxyacetylene flame ablation for 10 s (A: ablation center; B: transition zone; C: heat-affected zone; D: ceramic matrix)

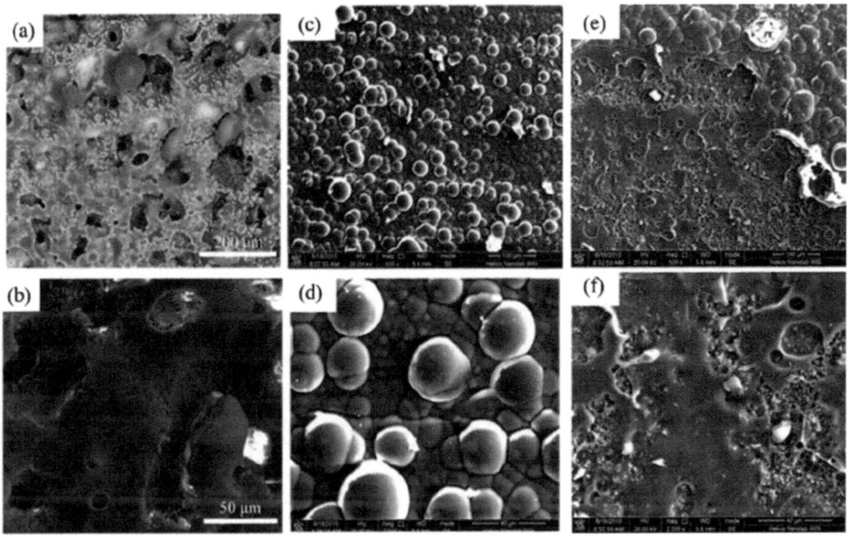

Fig. 6.47 SEM surface morphologies of the (ZrB_2-ZrN)/Si_2BC_3N composite ceramics (hot-pressed at 1900 °C) after oxyacetylene flame ablation for 10 s: **a**, **b** Ablation center; **c**, **d** transition zone; **e**, **f** heat-affected region

fibers are oxidized to form pits. Besides, obvious oxidation of the resin-derived carbon interface forms gaps between fibers and ceramic matrix; however, the fiber pull-out length increases considerably, due to the failure of the interface (Fig. 6.49).

At $\Delta T = 900$ °C, the carbon fibers on the surfaces are oxidized seriously to a shape of bamboo shoots, and even some are completely consumed to leave many holes as channel of oxygen internal diffusion. At fracture surfaces, the carbon fibers near to the sample surfaces are completely consumed, but the carbon fibers away from the surface are not oxidized obviously. Further thermal shock at $\Delta T = 1100$ °C, the carbon fibers on the surface are completely consumed to leave holes. The

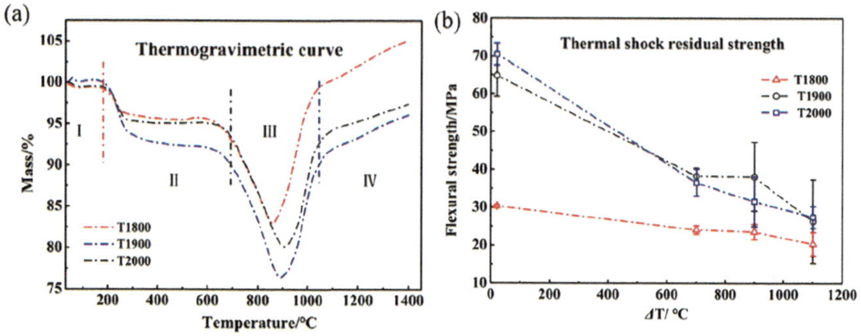

Fig. 6.48 Thermogravimetric curves and residual strength of the hot-pressed C_f/Si_2BC_3N composites: **a** Non-isothermal oxidation; **b** residual strength after thermal shock [15]

Table 6.9 Mechanical performance of the hot-pressed 20 vol% C_f/Si_2BC_3N composites after thermal shock test at different temperature difference [15]

Thermal shock temperature difference	Residual strength (MPa)/strength retention rate (%)		
	Sintered at 1800 °C	Sintered at 1900 °C	Sintered at 2000 °C
$\Delta T = 700$ °C	24.2/79.6	18.0/59.1	17.3/56.9
$\Delta T = 900$ °C	50.5/78.0	38.2/58.9	36.4/56.2
$\Delta T = 1100$ °C	47.4/67.3	28.3/40.1	27.4/38.9

matrix is obviously oxidized to form molten SiO_2 that tends to cover on the surfaces and hence healing pores and cracks. The deeper carbon fibers are also obviously oxidized, indicating an increasing oxygen diffusion depth.

Obviously, the oxidation degree gradually decreases with depth. The fibers of the T2000 composites near the sample surfaces are almost completely oxidized, leaving large holes in the matrix. Besides, the matrix oxidizes to form 3–4 μm oxidized layer. At depth of ~ 100 μm, fibers are oxidized seriously, with a bamboo-shoot shape, and the fiber surfaces are covered by molten oxides of mainly SiO_2. Similarly, the spherical molten oxides are also formed on the fiber surfaces. At depth of ~ 200 μm, fiber front becomes significantly thinner, the interface phases are completely oxidized with clear gaps between fibers and ceramic matrix. Due to the weak bonding or even complete debonding of fibers and matrix, the lengths of fiber pull-out are significantly greater than the formers, even larger than the facture length at RT. However, at a depth of ~ 300 μm, the fibers still maintain their initial morphologies and have a good bonding with the matrix (Fig. 6.50).

Generally, damage to the interfaces makes fibers failure to carry load effectively causing the decrease in the strength of T2000 composites. Due to the high activity of the resin-derived carbon, the strength of all the composites is significantly degraded after thermal shock at a lower temperature ($\Delta T = 700$ °C). The T1800 composites have the best thermal shock resistance, due to their lower CTE, lower fiber loading ratio, less stress concentration at the interface, and better defect healing effect.

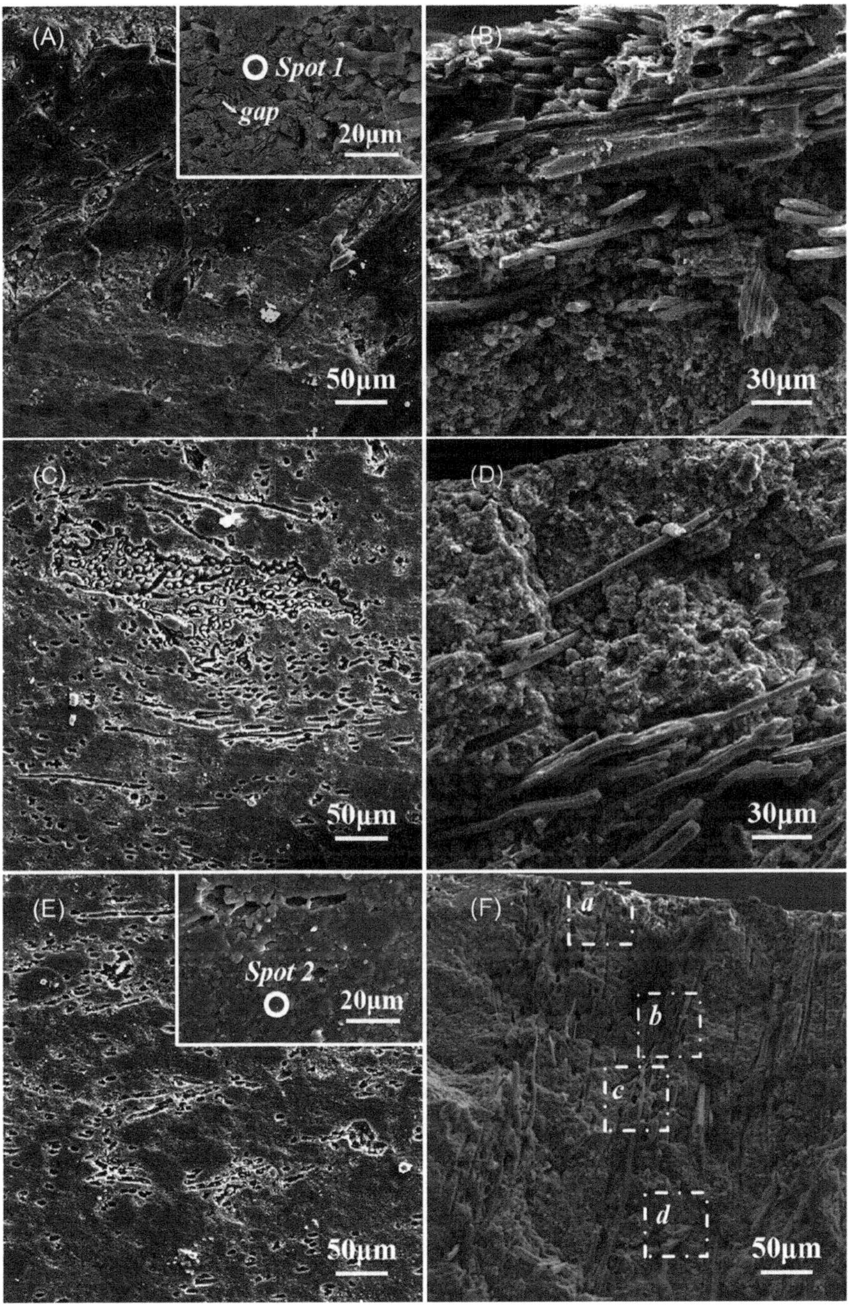

Fig. 6.49 SEM surface and fracture morphologies of the T2000 composites after thermal shock test: **a**, **b** $\Delta T = 700$ °C; **c**, **d** $\Delta T = 900$ °C; **e**, **f** $\Delta T = 1100$ °C [15]

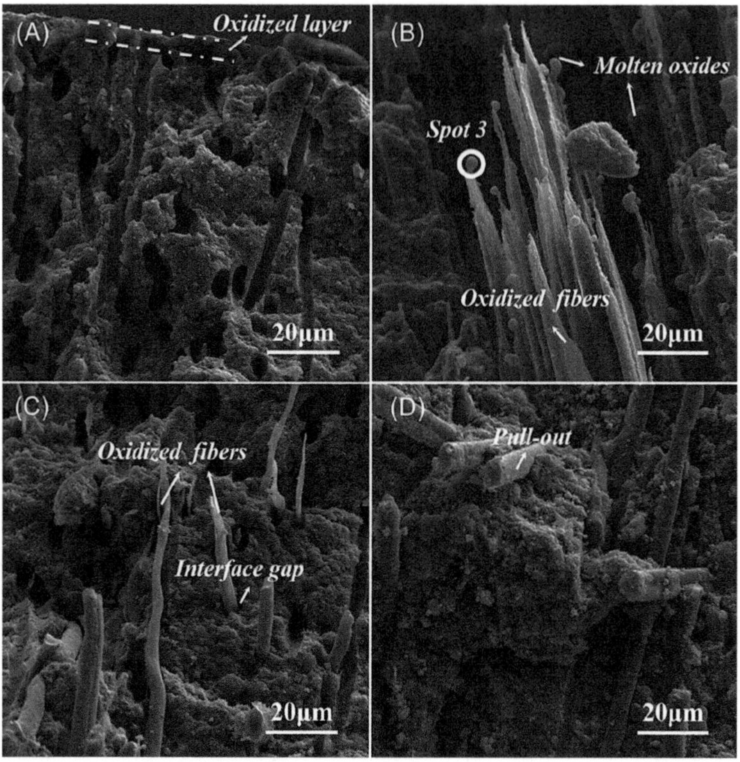

Fig. 6.50 An enlarged view of the fracture morphologies from regions in Fig. 6.49 f: **a** Region "a"; **b** region "b"; **c** region "c"; **d** region "d" [15]

The surface temperature of the three C_f/Si_2BC_3N composites reaches ~ 1900 °C after only 2 s of oxyacetylene flame (heat flux ~ 4.2 MW/m^2) ablation [16]. The highest surface temperature of the hot-pressed T1800, T1900 and T2000 composites are ~ 2150 °C, ~ 2220 °C and ~ 2180 °C, respectively. After ablation, a deeper ablation pit is found for all the composites without penetrating macro-cracks and structure fragmentation (Fig. 6.51).

With increasing hot-pressing temperature, the bulk density of the composites increases, but the linear and mass ablation rates decrease. The linear and mass ablation rates of the T1800 composites are 0.0731 mm/s and 25.5 mg/s, respectively, while the those of T2000 reach to 0.0422 mm/s and 13.7 mg/s (Table 6.10). After ablation, obvious microcracks appear on the ablation center surfaces, which are covered by a porous oxide layer. Besides, substantial discontinuous SiO_2 nanowires emerge and distribute on the oxides after cooling (Fig. 6.52). Unexpectedly, many carbon fibers are exposed directly on the transition zone surfaces. Obviously, these fibers are severely damaged with needle-like shape. Due to the protective SiO_2 on fiber surfaces, the ablation rate of the ceramic matrix is higher than that of C_f in this area. In heat-affected zone, the SiO_2 oxides will not be completely scoured away by

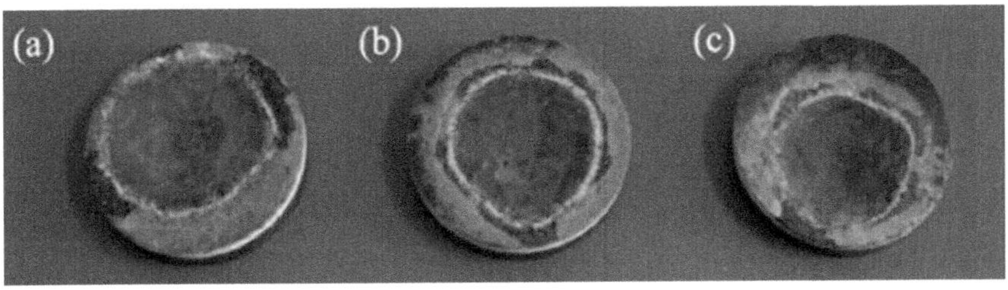

Fig. 6.51 The macro morphologies of the 20 vol% C_f/Si_2BC_3N composites after oxyacetylene flame (heat flux ~ 4.2 MW/m^2) ablation for 10 s: **a** T1800; **b** T1900; **c** T2000 [16]

the high-speed airflow, but pinned to the sample surfaces in the form of dense film, though some micropores retained (Fig. 6.53).

(2) *C_f/Si_2BC_3N-MAS CMCs*

After thermal shock test at $\Delta T = 1000$–1400 °C, the bending strength of the C_f/Si_2BC_3N-MAS composites decreases to some extent [17]. For composites with 10 vol% fiber addition, their bending strength obtains at 63.7 MPa ($\Delta T = 1000$ °C), 52.3 MPa ($\Delta T = 1200$ °C) and 57.8 MPa ($\Delta T = 1400$ °C), respectively. For 20 vol% fiber incorporation, the composite materials display residual strength of 59.2 MPa ($\Delta T = 1000$ °C), 77.5 MPa ($\Delta T = 1200$ °C) and 63.0 MPa ($\Delta T = 1400$ °C). However, the residual Vickers' hardness of C_f/Si_2BC_3N-MAS composites gradually

Table 6.10 Ablation properties of the hot-pressed 20 vol% C_f/Si_2BC_3N composites after oxyacetylene flame (heat flux ~ 4.2 MW/m^2) ablation for 10 s [16]

Composites	Bulk density (g/cm^3)	Linear ablation rate (mm/s)	Mass ablation rate (mg/s)
T1800	1.77	0.0731	25.5
T1900	1.97	0.0455	15.2
T2000	2.07	0.0422	13.7

Fig. 6.52 SEM surface morphologies of ablation center for the 20 vol% C_f/Si_2BC_3N composites after oxyacetylene flame ablation for 10 s: **a** Low resolution; **b** enlarged region of box "1"; **c** enlarged region of box "2" [16]

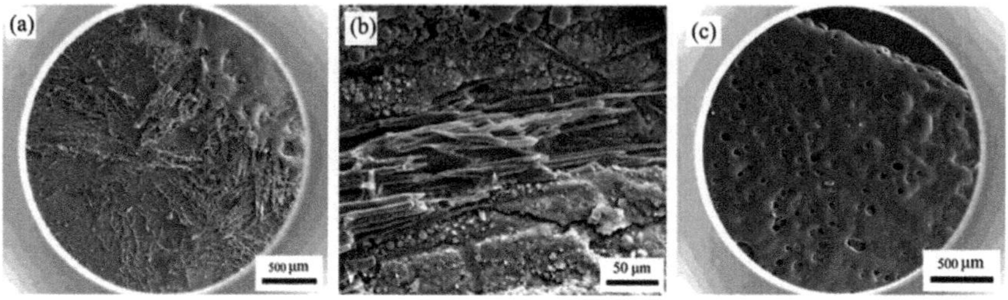

Fig. 6.53 SEM surface morphologies of the 20 vol% C_f/Si_2BC_3N composites after oxyethylene flame ablation for 10 s: **a**, **b** Transition zone; **b** heat-affected region [16]

increases below $\Delta T \leq 1200$ °C, finally decreases to 3.2 GPa and 3.4 GPa for the 10 vol% C_f and 20 vol% C_f containing composites, respectively (Fig. 6.54 and Table 6.11).

At $\Delta T = 1000$ °C, some micropores and grooves are retained, but pull-out of fibers is still working. Further increasing ΔT to 1200 °C, many micropores and grooves on the surfaces are effectively healed showing a dense oxide structure. Expectedly, the pull-out length of fibers is larger than the former. At $\Delta T = 1400$ °C, a smooth

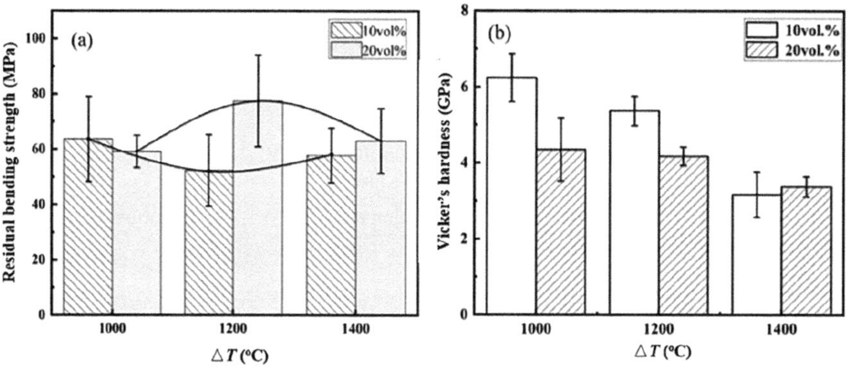

Fig. 6.54 The mechanical properties of the hot-pressed C_f/Si_2BC_3N-MAS ceramic composites after thermal shock at different temperature difference [17]

Table 6.11 The mechanical properties of the hot-pressed C_f/Si_2BC_3N-MAS ceramic composites after thermal shock at different temperature difference [17]

Composition	Residual strength (MPa)/residual strength ratio (%)			Vickers' hardness (GPa)/residual hardness ratio (%)		
	$\Delta T =$ 1000 °C	$\Delta T =$ 1200 °C	$\Delta T =$ 1400 °C	$\Delta T =$ 1000 °C	$\Delta T =$ 1200 °C	$\Delta T =$ 1400 °C
+ 10 vol% C_f	63.7/19.0	52.3/15.6	57.8/17.3	6.2/119.2	5.4/103.8	3.2/61.5
+ 20 vol% C_f	59.2/23.0	77.5/30.1	63.0/24.5	4.4/104.7	4.2/100.0	3.4/81.0

and pores-free dense oxide layer is built on material surfaces. In this case, few fibers are pulled out with shorter length (Fig. 6.55). After thermal shock at various ΔT, all the C_f/Si_2BC_3N-MAS composites show brittle fracture characteristics (Fig. 6.56).

After 10 s of oxyacetylene flame (heat flux ~ 4.2 MW/m^2) ablation, the C_f/Si_2BC_3N-MAS composite materials show varying degrees of ablation pits on the macroscopic surfaces. However, all the as-ablated composites have split into fractions (Fig. 6.57). Generally, the surface temperature climbs rapidly to about 1000 °C within 1 s, but gradually increases to ~ 2080 °C, ~ 2158 °C, ~ 2152 °C and ~ 2170 °C for the 5 vol%, 10 vol%, 15 vol% and 20 vol% fiber containing composites, respectively. The mass and linear ablation rates of the C_f/Si_2BC_3N-MAS composites

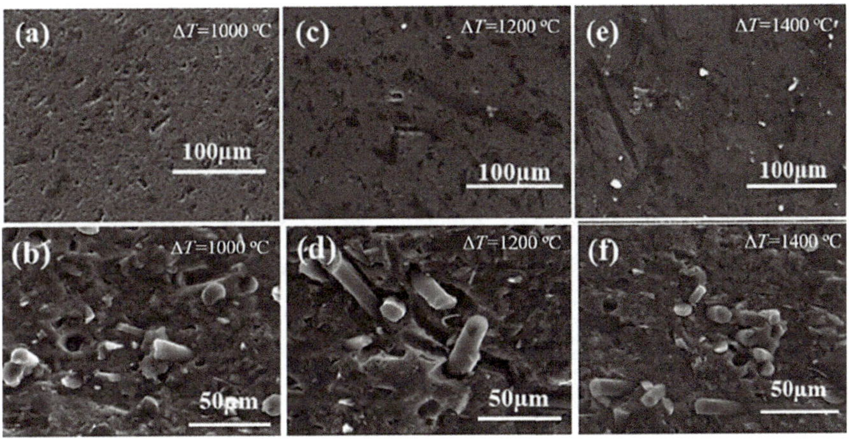

Fig. 6.55 The surface and fracture morphologies of the hot-pressed C_f/Si_2BC_3N-MAS ceramic composites after thermal shock at different temperature difference: **a**, **c**, **e** Surface morphologies; **b**, **d**, **f** fracture morphologies [17]

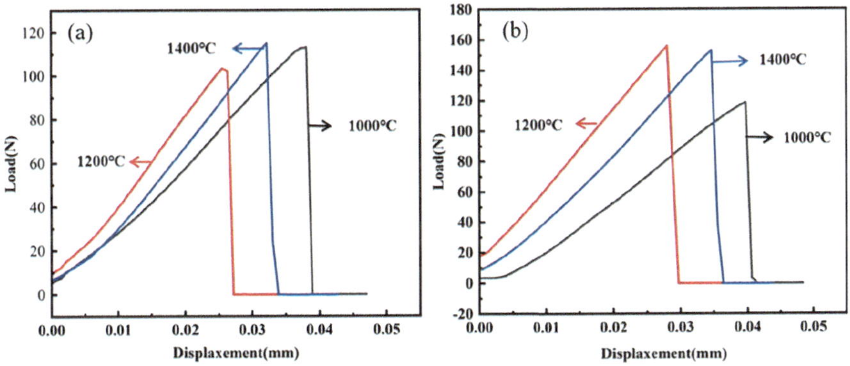

Fig. 6.56 Load–displacement curves of the hot-pressed C_f/Si_2BC_3N-MAS composites after thermal shock at different temperature difference: **a** + 10 vol% C_f; **b** + 20 vol% C_f [17]

change significantly with the fiber content. For composites with 5–20 vol% C_f, they show mass ablation rate of 0.013 g/s, 0.017 g/s, 0.016 g/s, and 0.026 g/s, respectively, while the linear ablation rates are 0.048 mm/s, 0.037 mm/s, 0.052 mm/s, and 0.038 mm/s (Fig. 6.58 and Table 6.12).

After ablation, the various C_f/Si_2BC_3N-MAS composites exhibit similar phase composition, both containing SiC, BN(C) and amorphous SiO_2. In all case, the amorphous SiO_2 are found in heat-affected region, while no oxides on sample surfaces are detected for other regions (Fig. 6.59).

There are obvious micropores in sizes of 10–100 μm distributed in ablation center, and the sizes of the pores increase with increasing fiber content. The surface fibers are

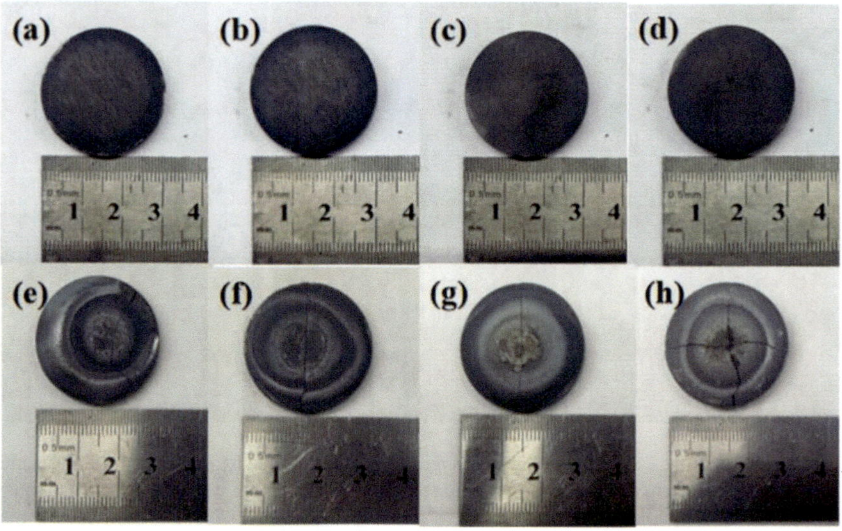

Fig. 6.57 Optical photos of the hot-pressed C_f/Si_2BC_3N-MAS composites after oxyacetylene flame (heat flux ~ 4.2 MW/m^2) ablation for 10 s [17]

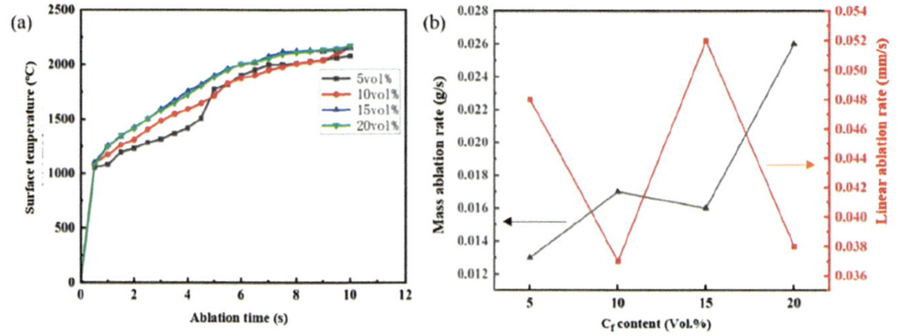

Fig. 6.58 Ablation behavior of the hot-pressed C_f/Si_2BC_3N-MAS composites: **a** Surface temperature curves; **b** ablation properties [17]

Table 6.12 Mass and linear ablation rates of the hot-pressed C_f/Si_2BC_3N-MAS ceramic composites after oxyacetylene flame (heat flux ~ 4.2 MW/m^2) ablation for 10 s [17]

Composition	Mass ablation rate (mg/s)	Linear ablation rate (mm/s)
+ 5 vol% C_f	~ 13.0	0.048
+ 10 vol% C_f	~ 17.0	0.037
+ 15 vol% C_f	~ 16.0	0.052
+ 20 vol% C_f	~ 26.0	0.038

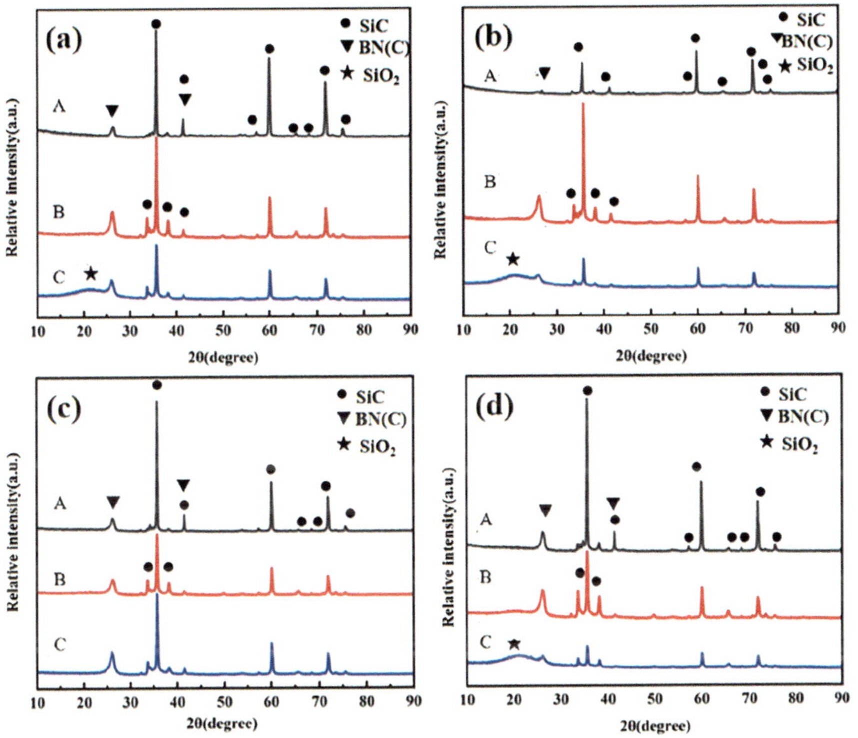

Fig. 6.59 Micro XRD patterns of the as-ablated C_f/Si_2BC_3N-MAS composites: **a** + 5 vol% C_f; **b** + 10 vol% C_f; **c** + 15 vol% C_f; **d** + 20 vol% C_f (A: ablation center; B: transition zone; C: heat-affected zone) [17]

completely oxidized left behind some diffusion channels. In transition zone, the pore sizes are reduced due to the healing of SiO_2, but still presents porous surface structure with some rod-like oxides. The heat-affected region both show relatively dense surface morphologies though some bubbles and grooves are still retained (Fig. 6.60).

(3) *C_f/Si_2BC_3N-Zr CMCs*

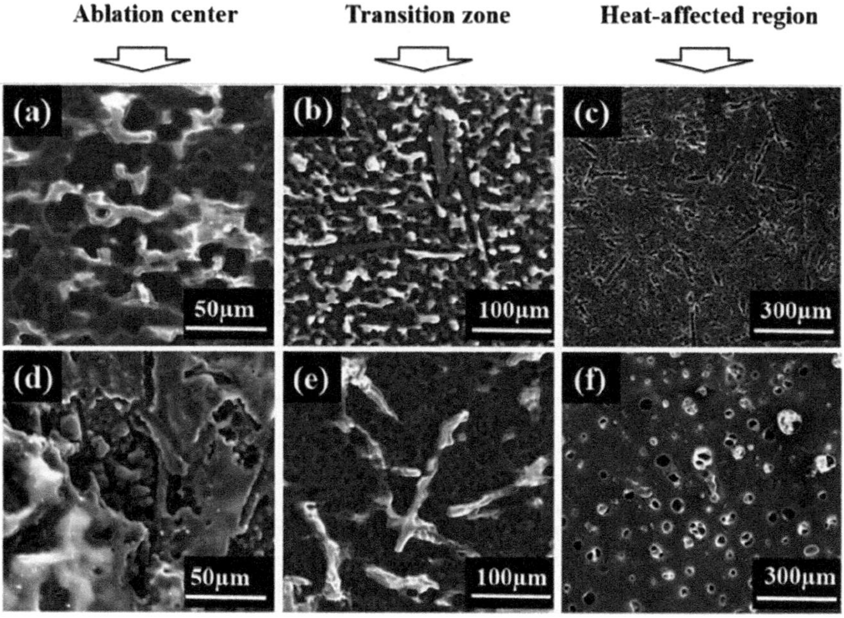

Fig. 6.60 Surface morphologies of the as-ablated C_f/Si_2BC_3N-MAS composites: **a–c** + 15 vol% C_f; **d–f** + 20 vol% C_f [17]

The thermal shock residual strength of the Si_2BC_3N-Zr ceramics is significantly higher than that of pure MA-Si_2BC_3N. For example, MA-Si_2BC_3N ceramics tested at $\Delta T = 1000$ °C show residual strength retention rate of only 52.4%. The addition of carbon fiber results in a great increase in the residual strength of the C_f/Si_2BC_3N CMCs and C_f/Si_2BC_3N-Zr CMCs. At $\Delta T = 1200$ °C, these composites can retain strength of ~ 69.7% and ~ 74.3%. At varying thermal shock temperature difference, phases of ZrB_2, SiC, BN(C) and ZrO_2 are detected for all the C_f/Si_2BC_3N-Zr CMCs, but differs in their relative content (Fig. 6.61 and Table 6.13).

After thermal shock test, limited fiber extraction is still observed at fracture morphologies, suggesting degradation of the carbon fibers. Besides, the pull-out length of carbon fiber is only a few micrometers. A thin oxide layer mainly composed of ZrO_2 is observed from the cross-section. The formed dense oxide scale effectively heals defects on the surfaces, thereby reducing thermal stress concentration (Fig. 6.62).

During oxyacetylene torch (heat flux ~ 4.2 MW/m^2) ablation, the surface temperature of the Si_2BC_3N-Zr ceramics sharply increases to approximately 1600 °C within 2 s and eventually reaches approximately 2150 °C. Coincidentally, the final surface temperature of the 20 vol% C_f/Si_2BC_3N-Zr composites is lower than those of the monoliths and 20 vol% C_f incorporated CMCs.

After 10 s ablation, the Si_2BC_3N-Zr bulk ceramics present good ablation properties, with linear and mass ablation rates of 0.021 mm/s and 0.0011 mg/s, respectively.

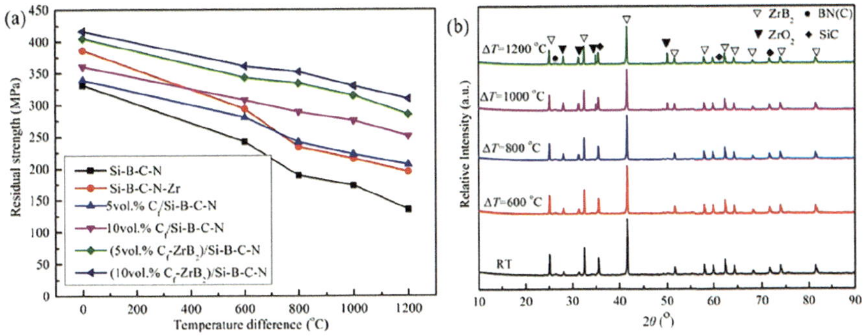

Fig. 6.61 Residual strength and phase composition of the hot-pressed C_f/Si_2BC_3N-Zr CMCs after thermal shock at different temperature difference: **a** Residual strength; **b** XRD patterns [18]

Table 6.13 The calculated thermal shock parameters and measured residual strength of the hot-pressed C_f/Si_2BC_3N-Zr CMCs after thermal shock at different temperature difference [18]

Composition	Calculated R (°C)	Calculated R'''' (μm)	Residual strength (MPa)/residual strength ratio (%)			
			$\Delta T = 600$ °C	$\Delta T = 800$ °C	$\Delta T =$ 1000 °C	$\Delta T =$ 1200 °C
Si_2BC_3N	472.5	96.0	242.4/73.2	189.7/57.3	173.5/52.4	135.1/40.8
Si_2BC_3N-Zr	360.0	132.2	294.2/76.2	234.7/60.8	215.8/55.9	195.4/50.6
5 vol% C_f/Si_2BC_3N	716.0	153.5	281.2/82.9	242.2/71.4	222.5/65.6	206.2/60.8
10 vol% C_f/Si_2BC_3N	776.2	168.0	308.0/85.4	289.6/80.3	275.5/76.4	251.3/69.7
5 vol% C_f/Si_2BC_3N-Zr	594.0	164.5	343.2/84.6	334.3/82.4	314.8/77.6	285.2/70.3
10 vol% C_f/Si_2BC_3N-Zr	661.0	174.3	361.2/86.7	352.1/84.5	329.9/79.2	309.5/74.3

After 30 s ablation, the Si_2BC_3N-Zr ceramics show linear and mass ablation rates of 0.030 mm/s and 0.0034 mg/s, much lower than those of the MA-Si_2BC_3N monoliths. Unexpectedly, with the addition of BN-coated carbon fiber, they all show decreasing ablation resistance (Table 6.14). After 10 s of ablation, visible radial surface cracks are observed in Si_2BC_3N-Zr monoliths and the 10 vol% C_f/Si_2BC_3N-Zr CMCs. However, no fatal cracks appear on the surfaces of the 20 vol% C_f containing ones (Fig. 6.63). No trace of penetrable cracks on the ablation surface suggests that such cracks initiate in the superficial ablation surface but are arrested during propagation.

Before ablation, the as-sintered Si_2BC_3N-Zr monoliths and C_f/Si_2BC_3N-Zr CMCs are mainly composed of ZrB_2, ZrN, α/β-SiC, BN(C), and a small amount of ZrO_2. The ZrO_2 forms, partly as a result from the surface oxides of the raw materials. After

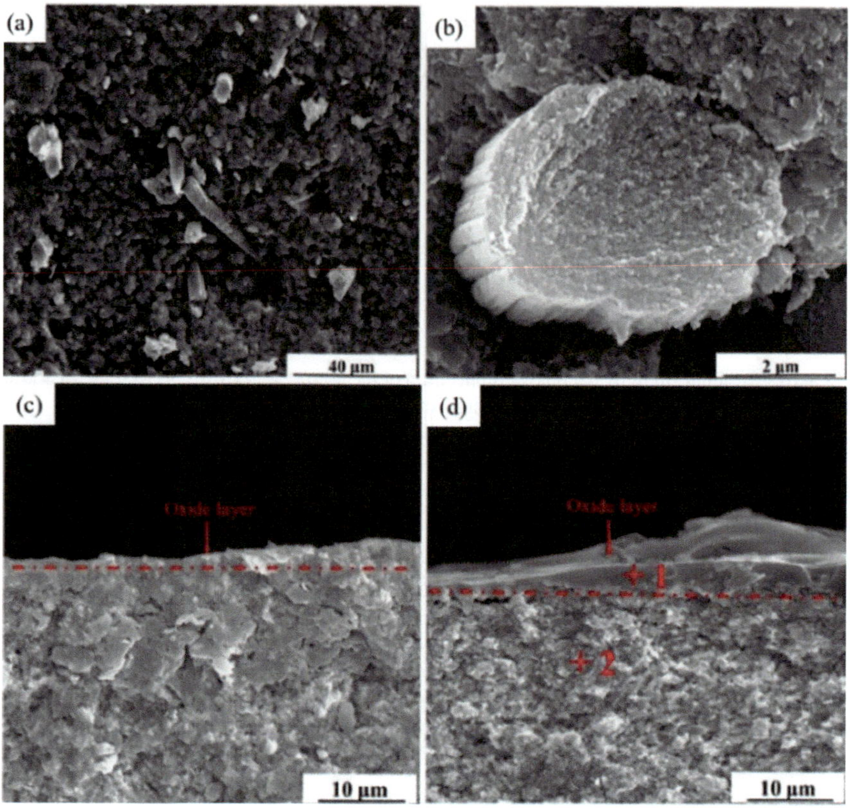

Fig. 6.62 The fracture and cross-section morphologies of the hot-pressed **a**, **b** C_f/Si_2BC_3N-Zr CMCs and **c**, **d** Si_2BC_3N-Zr ceramics after thermal shock at different temperature difference: **a**, **c** $\Delta T = 1000$ °C; **b**, **d** $\Delta T = 1200$ °C [18]

ablation, the ablation center surfaces for all the investigated materials are composed of t/m-ZrO_2, α/β-SiC, SiO_2 and cristobalite. As compared, the diffraction intensities of β-SiC in the non-ablated area or the heat-affected region are much higher. In the transition zone, diffraction peaks assigned to amorphous SiO_2 are observed in the 2θ range of 20° ~ 25°. In addition, amorphous SiO_2 peaks significantly decrease in the center region, and cristobalite appears at $2\theta = 24°$ in the transition and center regions. Cristobalite may have been the result of the SiC oxidation or transformation from the amorphous SiO_2. From the heat-affected region to the center region, the diffraction peaks of ZrN and ZrB_2 gradually disappear, whereas the diffraction peaks of ZrO_2 appear (Fig. 6.64).

The ablation center of the Si_2BC_3N-Zr monoliths presents a loose and porous network structure (Fig. 6.65). Additionally, a thermal shock crack spreads from the ablation center to the transition and heat-affected regions. During the ablation and scouring process, oxides do not form a protective glassy and dense layer; this

Table 6.14 The mass and linear ablation rates of the hot-pressed C_f/Si_2BC_3N-Zr ceramic composites after oxyacetylene flame (heat flux ~ 4.2 MW/m^2) ablation for 10 s [19]

Composition	Bulk density (g/cm^3)	Mass ablation rate (mg/s)	Linear ablation rate (mm/s)
Si_2BC_3N	2.74	0.0326	0.013
Si_2BC_3N-Zr	3.67	0.0011	0.021
Si_2BC_3N-Zr[a]	3.67	0.0034	0.030
5 vol% C_f/ Si_2BC_3N-Zr	3.36	0.0042	0.032
10 vol% C_f/ Si_2BC_3N-Zr	3.16	0.0053	0.037
15 vol% C_f/ Si_2BC_3N-Zr	2.70	0.0058	0.043
20 vol% C_f/ Si_2BC_3N-Zr	2.22	0.0061	0.048

[a] The ablation time is 30 s

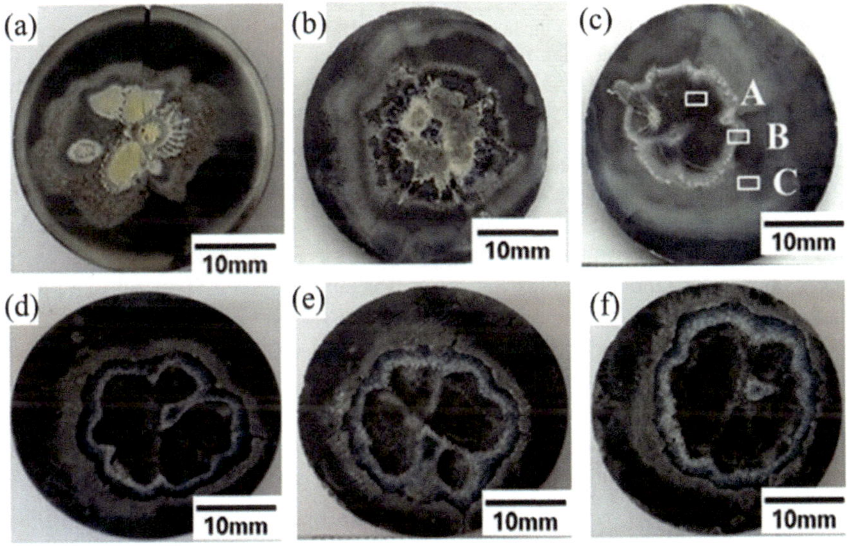

Fig. 6.63 Macro morphologies of the hot-pressed C_f/Si_2BC_3N-Zr CMCs after oxyacetylene flame (heat flux ~ 4.2 MW/m^2) ablation for 10 s: **a** 5 vol% C_f/Si_2BC_3N-Zr; **b** 10 vol% C_f/Si_2BC_3N-Zr; **c** 15 vol% C_f/Si_2BC_3N-Zr; **d** 5 vol% C_f/Si_2BC_3N; **e** 10 vol% C_f/Si_2BC_3N; **f** 15 vol% C_f/Si_2BC_3N [19]

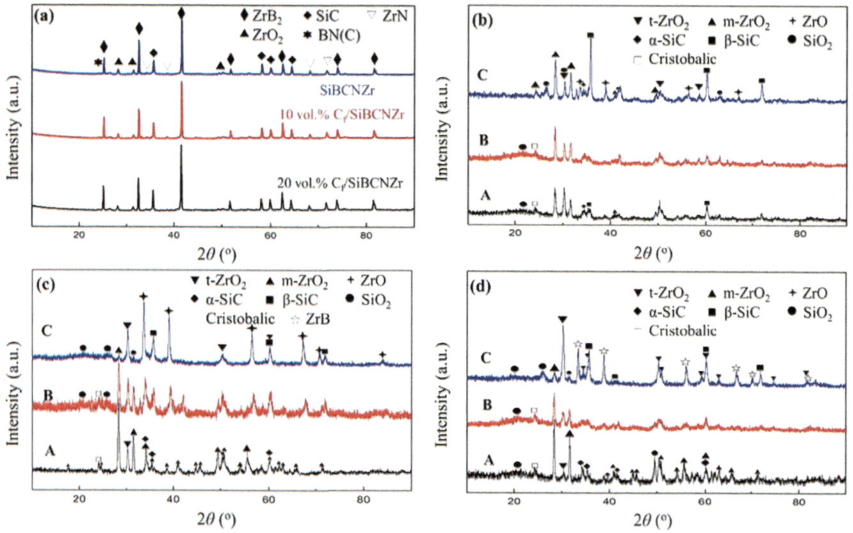

Fig. 6.64 Micro XRD patterns of the hot-pressed C_f/Si_2BC_3N-Zr CMCs after oxyacetylene flame (heat flux ~ 4.2 MW/m^2) ablation for 10 s: **a** Before ablation; **b** Si_2BC_3N-Zr; **c** 10 vol% C_f/Si_2BC_3N-Zr; **d** 20 vol% C_f/Si_2BC_3N-Zr (A: ablation center; B: transition region; C: heat-affected region) [19]

together with the emission of various gases, both lead to a porous oxide structure and formation of ablation pits on the ablation surfaces. A dense oxide layer assigned to SiO_2 and ZrO_2 covers on the transition region of the Si_2BC_3N-Zr monoliths, where a few micropores and some globular projections are observed. The rod-like phases mainly contain Si, Zr, and O elements, in which the Si/O atomic ratio was close to 1. In the heat-affected region, a dense oxide layer with the orbicular structure forms. The orbicular bubbles of SiO_2 approximately 10 μm in size likely result from the high-viscosity molten surface layer that captured gases emitted from the interior during ablation. Note that nearly all the fused SiO_2 in the heat-affected region are scoured from the transition region and/or ablation center.

After 10 vol% C_f addition, some C_f are oxidized and/or scoured off by the high-speed oxidized airflow, and some belt-like stripes form in the center region (Fig. 6.66). In some zones, some C_f are denuded to form a sponge-like structure containing mainly of Si, O, and C. Thus, the formed SiO_2 covers and enwraps the C_f. A thermal shock crack also appears and propagates from the ablation center to the edge of the composites. During crack propagation, pull-out of some C_f from the ceramic matrix is detected. The surface of the pulled-out C_f is relatively smooth, but some ceramic particles are adhered to the surfaces. To some extent the weak h-BN interfacial layer can reduce the strong interfacial bonding and degradation of C_f. The fiber pull-out consumes more energy than fiber fracture, favoring to the thermal shock and ablation resistance. Therefore, Si_2BC_3N-Zr CMCs with 20 vol% C_f show no cracks on the ablation surfaces. In transition region of the C_f/Si_2BC_3N-Zr CMCs, a dense oxide

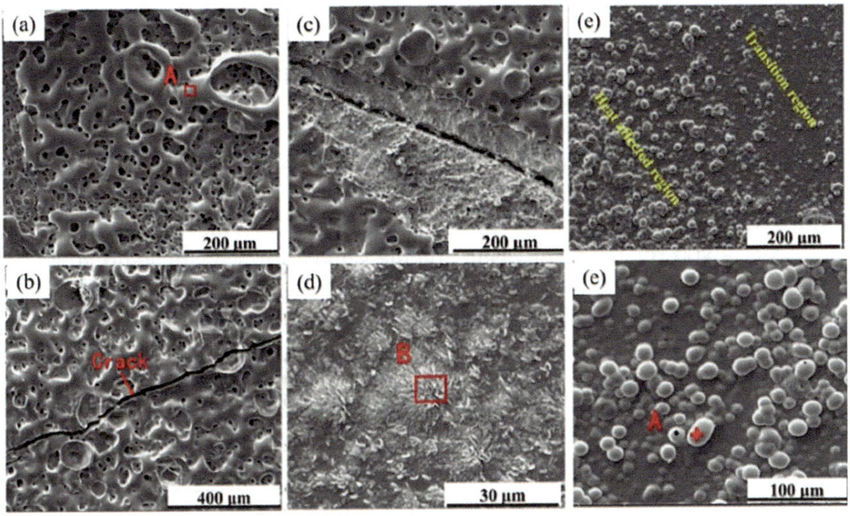

Fig. 6.65 Surface morphologies of the hot-pressed Si_2BC_3N-Zr monoliths after oxyacetylene flame (heat flux ~ 4.2 MW/m^2) ablation for 10 s: **a**, **b** Ablation center; **c**, **d** transition zone; **e**, **f** heat-affected region [19]

layer with a few larger holes is found. Apparently, the highly viscous SiO_2 cannot fill the pores effectively, and gas byproduct emission leads to a porous morphology in the heat-affected region.

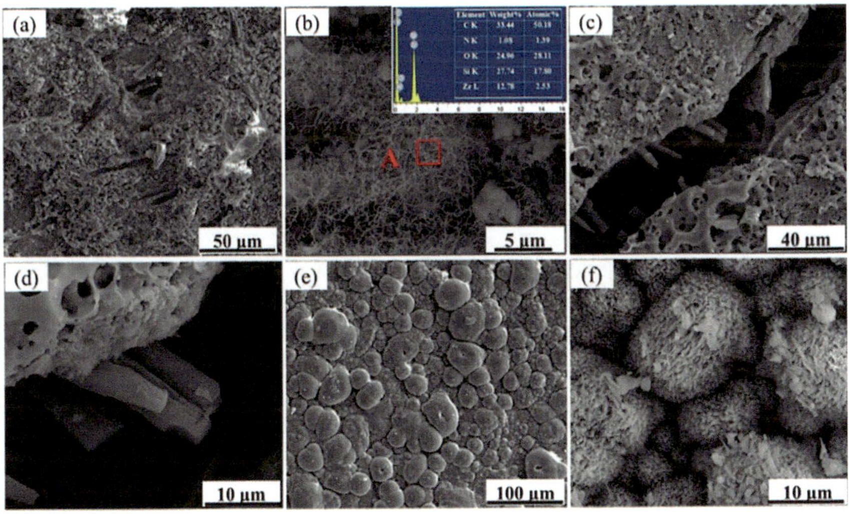

Fig. 6.66 Ablation morphologies of hot-pressed 10 vol% C_f/Si_2BC_3N-Zr CMCs after oxyacetylene flame (heat flux ~ 4.2 MW/m^2) ablation for 10 s: **a** Ablation center; **b** sponge-like C_f; **c** crack bridging; **d** pulling out of C_f; **e** transition region; **f** heat-affected region [19]

(4) *(C_f-ZrB_2)/Si_2BC_3N-Zr CMCs*

As ΔT increases, the residual strength of the (C_f-ZrB_2)/Si_2BC_3N CMCs decreases gradually except for the composites tested at $\Delta T = 900$ °C. At $\Delta T = 800$ °C, the (C_f-ZrB_2)/Si_2BC_3N CMCs show residual strength of ~ 209 MPa with a high strength retention ratio ~ 87.1%. Further increasing ΔT to 900 °C, their residual strength slightly climbs to ~ 227 MPa, but subsequently decreases to ~ 188 MPa at $\Delta T =$ 1000 °C or ~ 174 MPa at $\Delta T = 1100$ °C, respectively (Fig. 6.67).

At $\Delta T = 800$ °C, oxidation occurs on the sample surfaces. In addition, a large number of micropores are distributed in the oxide layer. As the ΔT further increases, some oxides form to partially fills the micropores and cracks, but still leads to a porous oxide structure. As a result, diameter of the pores increases as well. After thermal shock, the fracture morphologies of all the (C_f-ZrB_2)/Si_2BC_3N CMCs do not show significant change (Fig. 6.68).

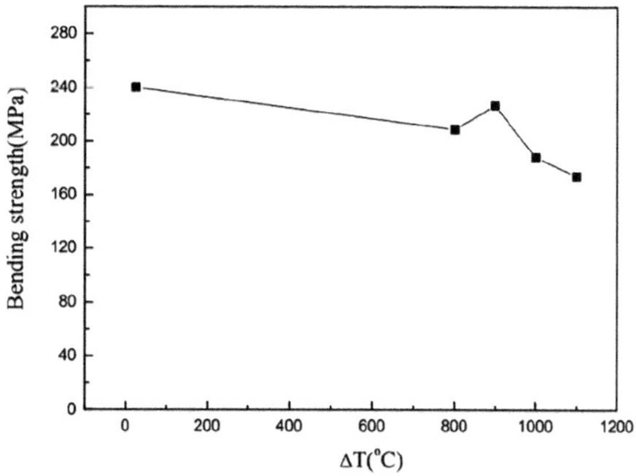

Fig. 6.67 Residual strength as a function of thermal shock temperature difference of the SPS densified (C_f-ZrB_2)/Si_2BC_3N CMCs after thermal shock at different temperature difference [19]

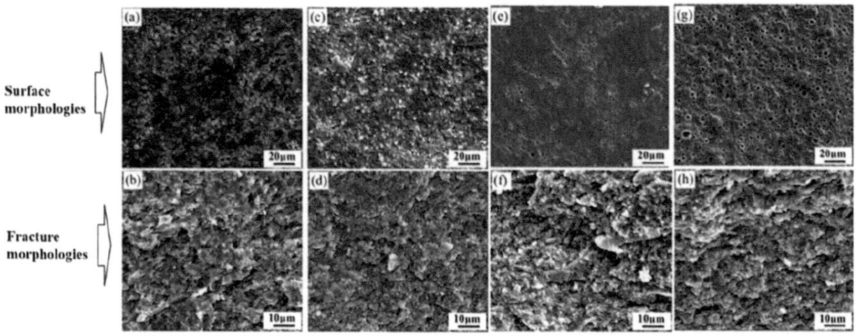

Fig. 6.68 SEM surface and fracture morphologies of the SPS sintered (C_f-ZrB_2)/Si_2BC_3N CMCs after thermal shock at different temperature difference: **a** $\Delta T = 800$ °C; **b** $\Delta T = 900$ °C; **c** $\Delta T =$ 1000 °C; **d** $\Delta T = 1100$ °C [19]

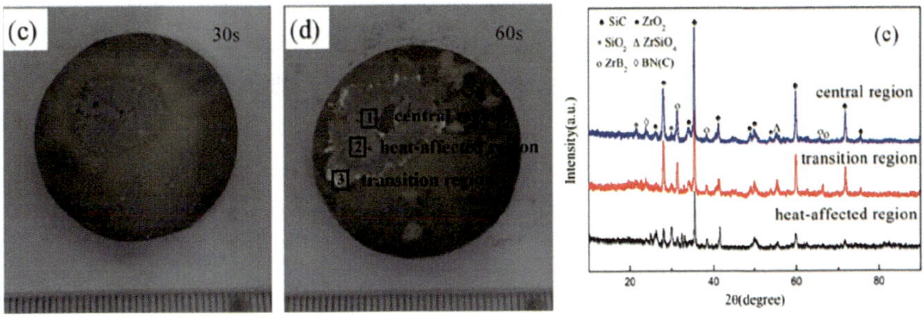

Fig. 6.69 Macro morphologies and phase composition of as-ablated (C_f-ZrB_2)/Si_2BC_3N CMCs after oxyacetylene flame (heat flux ~ 4.2 MW/m^2) ablation: **a** Ablation for 30 s; **b** Ablation for 60 s; **c** micro XRD patterns [19]

No catastrophic cracks are observed for the (C_f-ZrB_2)/Si_2BC_3N composites after oxyacetylene flame (heat flux ~ 4.2 MW/m^2) ablated for 30 s or 60 s. Prior to ablation, the main phases are ZrB_2, BN(C) and SiC. After exposure to the high-speed airflow, oxides of ZrO_2, SiO_2, and $ZrSiO_4$ form on surface regions. In these zones, some amorphous SiO_2 have transformed to cristobalite while B_2O_3 is not detected due to its high evaporation rates (Fig. 6.69).

During ablation, some ZrB_2 and SiC have been oxidized to form ZrO_2 and SiO_2. During which ZrO_2 can react with SiO_2 forming Zr–Si–O glass and/or $ZrSiO_4$. Some fibers and bubbles are observed from the ablation center surfaces, and the resulting porous oxide layer thickness is ~500 μm. The ablation airflow induces clearly visible corrosion in the transition zone, leading to the appearance of micropores and holes caused by the ruptured bubbles and formation of fiber-like features. An oxide layer ~200 μm, is thinner than the former. Some intact bubbles ~ 30 μm form on the surfaces of the heat-affected zone, but the oxide layer thickness in the heat-affected zone is only 100–150 μm (Fig. 6.70).

(5) *C_f/(ZrB_2-ZrN)/Si_2BC_3N CMCs*

The surface temperature of the (ZrB_2-ZrN)/Si_2BC_3N ceramics and C_f/(ZrB_2-ZrN)/Si_2BC_3N CMCs is basically the same. After 2 s of oxyacetylene flame (heat flux ~ 4.2 MW/m^2) ablation, the surface temperature of the three materials raises sharply to 1300 °C and fluctuated in range of 1600–1700 °C. The introduction of C_f does not reduce the surface temperature of (ZrB_2-ZrN)/Si_2BC_3N ceramics (Fig. 6.71). After ablation, part of surface SiC oxidize to amorphous SiO_2 positioned on ablation center and transition zone. Part of amorphous SiO_2 may crystallize to some cristobalite. The diffraction peaks of ZrN and ZrB_2 gradually disappear, while the diffraction peaks of ZrO_2 increase from sample edge to ablation center. The surface BN(C) have oxidized to liquid or gaseous B_2O_3, but diffraction peaks of BN(C) are also detected in all ablation areas. The as-ablated (ZrB_2-ZrN)/Si_2BC_3N ceramics and C_f/(ZrB_2-ZrN)/Si_2BC_3N CMCs all show satisfying ablation resistance, without any trace of macro structure fragmentations (Fig. 6.72).

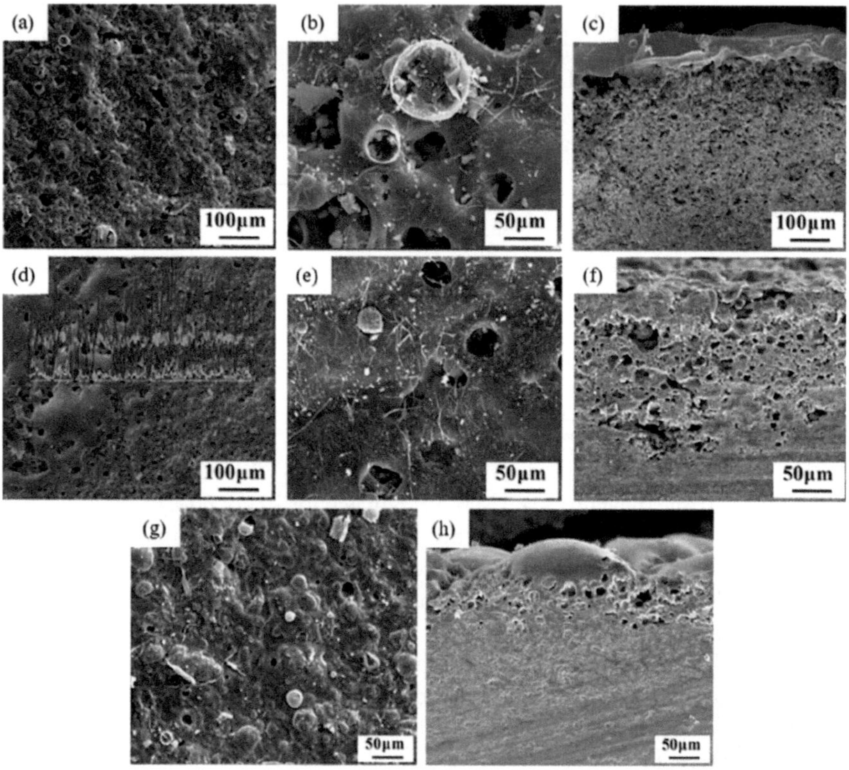

Fig. 6.70 SEM surface and cross-section morphologies of the SPS densified (C_f-ZrB_2)/Si_2BC_3N CMCs after oxyacetylene flame (heat flux ~ 4.2 MW/m^2) ablation for 60 s: **a–c** Ablation center; **d–f** transition area; **g–h** heat-affected region [19]

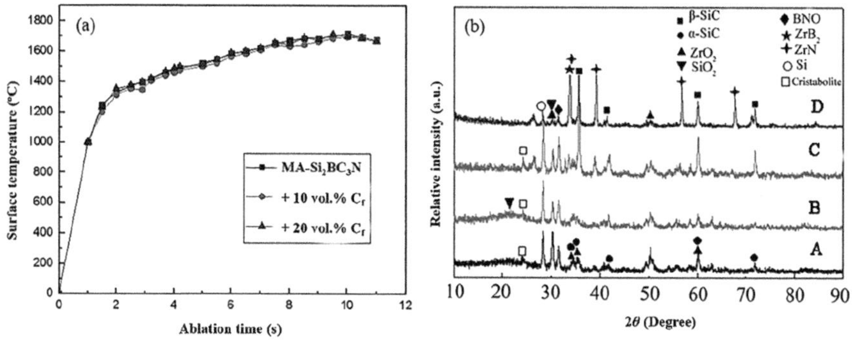

Fig. 6.71 Ablation behavior of the hot-pressed C_f/(ZrB_{2p}-Zr_{Np})/Si_2BC_3N composites: **a** Surface temperature curve; **b** micro XRD patterns (A: ablation center; B: transition zone; C: heat-affected zone; D ceramic matrix)

Fig. 6.72 Surface macro morphologies of the C_f/(ZrB_2-ZrN)/Si_2BC_3N CMCs before and after oxyacetylene flame ablation for 10 s: **a** (ZrB_2-ZrN)/Si_2BC_3N; **b** + 10 vol% C_f; **c** + 20 vol% C_f

Contrary to expectation, after 10 s of oxyacetylene flame ablation, the resulting 10 vol% C_f/(ZrB_2-ZrN)/Si_2BC_3N CMCs show decreasing mass ablation rate of 0.27 mg/s and linear ablation ratio of 0.002 mm/s, as compared to the (ZrB_2-ZrN)/Si_2BC_3N ceramics. Further increasing carbon fiber content is unfavorable to the ablation improvement. Interestingly, a negative linear ablation rate is measured due to the formation of considerable oxides in the ablation region (Table 6.15).

(6) *SiC_f/Si_2BC_3N ceramic CMCs.*

Thermal shock resistance

The thermal shock resistance of 20 vol% SiC_f/Si_2BC_3N CMCs can be slightly improved by h-BN coated SiC_f (SLFC1) (Fig. 6.73 and Table 6.16). At $\Delta T = 600$ °C, the residual strength of the coating-free composites decreases sharply with strength retention rate of only 39.0%, while those of the modified composites are 134.7 MPa and 77.9%, much higher than the pure MA-Si_2BC_3N. When tested at $\Delta T = 1200$ °C, the residual strength of the two C_f/Si_2BC_3N CMCs decreases by ~ 71.2% and ~ 69.9% as compared to the RT strength, respectively [20].

After thermal shock at $\Delta T = 1000$ °C, oxides on the two SiC_f/Si_2BC_3N composites are found, observed visible micropores and even extended cracks (Fig. 6.74). But at $\Delta T = 800$ °C, there are basically no obvious microcracks on the surfaces. Further increasing ΔT to 1100 °C, the surfaces of the two composites both show a porous oxide scale with some bubbles and microcracks. At $\Delta T = 1200$ °C, the obvious microcracks can be also observed on the surfaces of the two composites

Table 6.15 Ablation properties of the C_f/(ZrB_2-ZrN)/Si_2BC_3N composites after oxyacetylene flame (heat flux ~ 4.2 MW/m^2) ablation for 10 s

Properties	Adding C_f content (vol%)		
	0	10	20
Mass ablation rate (mg/s)	0.066	0.27	0.38
Linear ablation rate (mm/s)	– 0.003	0.002	0.004

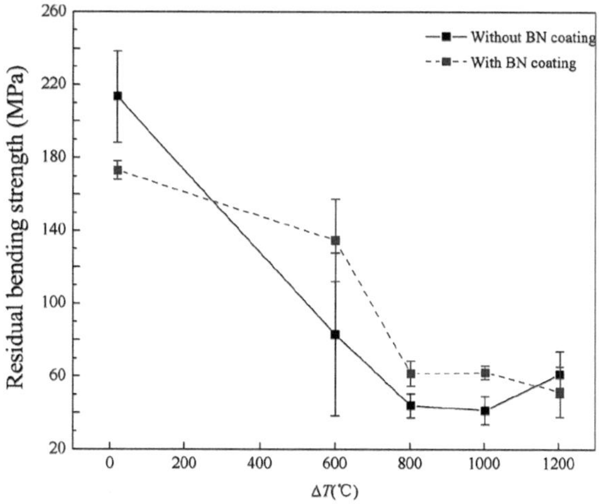

Fig. 6.73 Residual strength of the hot-pressed 20 vol% SiC_f (SLFC1) containing SiC_f/Si_2BC_3N composites after thermal shock at different temperature difference [20]

Table 6.16 Residual strength and strength retention rate of the hot-pressed 20 vol% SiC_f/Si_2BC_3N composites after thermal shock at different temperature difference [20]

Composites	Residual strength (MPa)/strength retention rate (%)			
	$\Delta T = 600$ °C	$\Delta T = 800$ °C	$\Delta T = 1000$ °C	$\Delta T = 1200$ °C
SiC_f/Si_2BC_3N	83.1/39.0	44.3/20.1	41.6/19.5	61.3/28.8
$SiC_f/Si_2BC_3N^*$	134.7/77.9	61.8/35.7	62.3/36.0	52.0/30.1

* SiC_f is coated with h-BN

since healing of the oxides is limited. Precipitates of mainly cristobalites distribute beneath the relative dense oxide scale (Fig. 6.75).

Ablation resistance

After 2 s of oxyacetylene flame (heat flux ~ 4.2 MW/m^2) ablation, the surface temperature of the 20 vol% SiC_f/Si_2BC_3N CMCs (SLFC1 fibers) increases rapidly to ~ 1600 °C, obtaining maximum surface temperature of ~ 2102 °C and ~ 2166 °C, respectively [21]. The backside temperature of modified composites reaches 102 °C after 12.8 s of ablation, while that of unmodified composites increases to 104 °C after 21.3 s ablation. The thermal conductivity of the unmodified composites is lower than that of the counterparts, showing better thermal insulation performance (Fig. 6.76).

The ablation resistance of the 20 vol% SiC_f/Si_2BC_3N composites is slightly improved by adopting h-BN coated SiC_f (SLFC1). Compared with the 20 vol% C_f/Si_2BC_3N composites prepared by the same HP sintering parameters, their mass and linear ablation rates are reduced by ~ 47.9% and ~ 37.8%, respectively (Table 6.17). After ablation, amorphous SiO_2 forms on the ablation surfaces, meanwhile, part of amorphous SiO_2 crystallizes to some cristobalite. Under such a violent thermal shock environment, macrocracks and deep ablation pits appear on the sample surfaces leading to the fragment. Degradation of SiC fiber and the strong bonding between

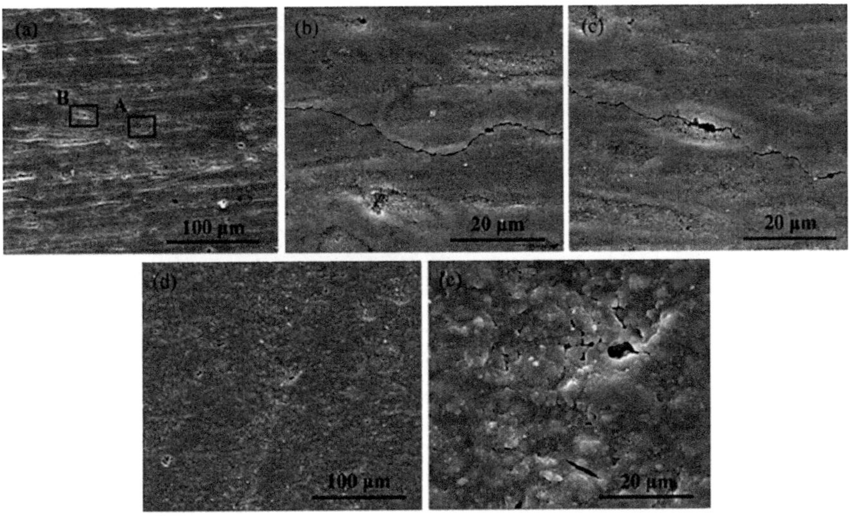

Fig. 6.74 SEM surface morphologies of the hot-pressed 20 vol% SiC_f/Si_2BC_3N composites after thermal shock at $\Delta T = 1100$ °C: **a–c** Without h-BN coating; (d-e) with h-BN coating [20]

matrix and fibers both restrict the toughening and strengthening of the short SiC_f (SLFC1) (Fig. 6.77).

Thermal shock cracks appear in the ablation pits and further spread to the edge of the specimen. Since the fibers in hot-pressed SiC_f/Si_2BC_3N CMCs are mainly distributed perpendicular to the pressure direction, thus some grooves from SiC fiber corrosion are found in the ablation center surfaces (Fig. 6.78). The transition region consists of porous zone and dense region. Due to the scouring of flowing gas-stream, SiO_2 oxides are not only flowed to the adjacent regions but evaporation or sealing pits left by eroded fibers in the center region. In the transition region, some liquid SiO_2 are also scoured to the heat-affected zone by the high-speed gas-stream. Therefore, the porous zone in transition region is constituted by porous SiO_2 with considerable broken bubbles. In dense area, dense SiO_2 with few micropores both exist on the ablation surfaces (Fig. 6.79).

In heat-affect region, the appearance of unbroken bubbles on ablation surfaces is observed owing to the increasing viscosity of SiO_2. Besides, most bubbles unite and grow into larger ones. Some white particles containing Si, B, O and C elements are produced. Thus, the borosilicate glass and/or SiO_2-B_2O_3 binary melt form in the heat-affect region. In addition, some spherical SiO_2 particles are also found on the surfaces (Fig. 6.80).

Surface spalling is not found for all the SiC_f (KD-II) reinforced MA-Si_2BC_3N composites, and the ablation areas can be clearly divided into three zones [22]. There is a white and continuous oxide scale covering the ablation pits. The ablation area enlarges with increasing fiber contents. For the thermal stress and gases scouring, all samples are split into two halves showing similar fracture behavior (Fig. 6.81). The

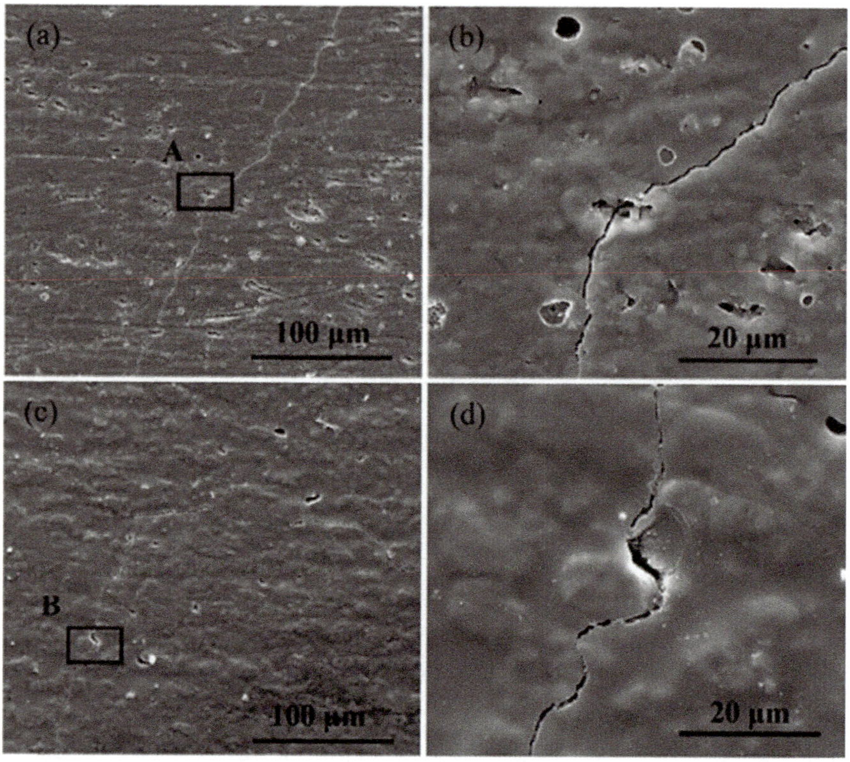

Fig. 6.75 SEM surface morphologies of the hot-pressed 20 vol% SiC_f/Si_2BC_3N composites after thermal shock at $\Delta T = 1200$ °C: **a**, **b** Without h-BN coating; **c**, **d** with h-BN coating [20]

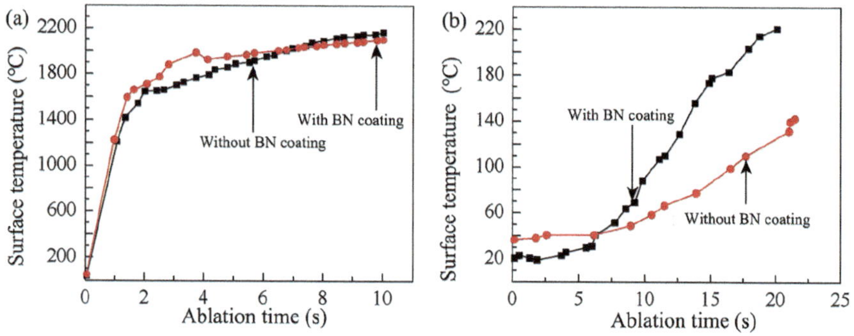

Fig. 6.76 Surface temperature curves of the hot-pressed 20 vol% SiC_f (SLFC1) reinforced MA-Si_2BC_3N composites during oxyacetylene flame (heat flux ~ 4.2 MW/m^2) ablation: **a** Surface temperature; **b** backside temperature [21]

Table 6.17 Ablation properties of the 20 vol% SLFC1 SiC_f reinforced MA-Si_2BC_3N composites after oxyacetylene flame (heat flux ~ 4.2 MW/m^2) ablation for 10 s [21]

Composition	Mass ablation rate (mg/s)	Linear ablation rate (mm/s)
$SiC_f/Si_2BC_3N^a$	9.5	0.0237
SiC_f/Si_2BC_3N	10.2	0.0274

[a] SiC_f is coated with h-BN

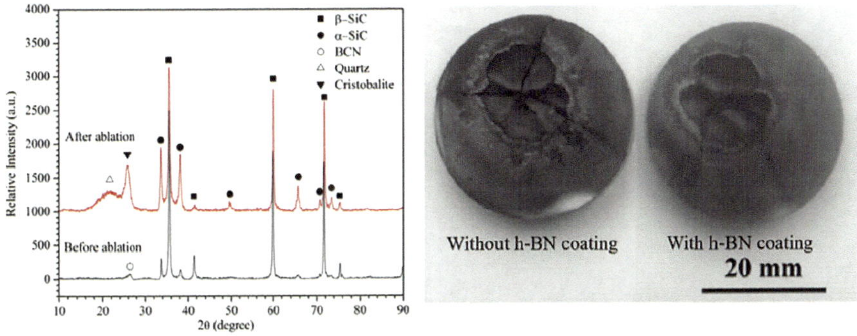

Fig. 6.77 Phase composition and macroscopic morphologies of the hot-pressed 20 vol% SiC_f (SLFC1) reinforced MA-Si_2BC_3N composites after oxyacetylene flame (heat flux ~ 4.2 MW/m^2) ablation: **a** XRD patterns; **b** optical photos [21]

MA-Si_2BC_3N monoliths exhibit the best performance, obtaining mass and linear ablation rates of only 0.61 mg/s and 0.0253 mm/s, respectively. Under the same ablation condition, composites with 5 vol% modified fibers show mass and linear ablation rates of 1.62 mg/s and 0.0412 mm/s. However, they increase rapidly when as-coated SiC fiber contents reach 10 vol%. The unmodified composites have a linear ablation rate similar to the composites with 5 vol% modified fibers, but show a higher mass ablation rate. The mass ablation rates of the SiC_f (KD-II) reinforced MA-Si_2BC_3N CMCs are much lower but the linear rates are higher in all cases. This indicates the better ablation resistance for KD-II SiC fibers *vs.* SLFC1 fibers (Table 6.18).

After ablation, except for the matrix phases α/β-SiC and BN(C), only amorphous SiO_2 are detected in the heat-affected region because of the gases scouring and SiC oxidation. The ablation center and transition region suffer the most severe thermal attack and gas scouring, thus amorphous SiO_2 tends to crystallize forming cristobalites. No peaks for B_2O_3 are detected likely because of the high evaporation rates (Fig. 6.82).

For MA-Si_2BC_3N monoliths, a continuous and dense oxide scale covers the ablative center surfaces without micropores and microcracks. XRD shows some cristobalite, however, the inserted EDS suggests the amorphous layer containing Si, C and O. In contrast, lots of bubbles and pores form in transition region shaping a porous oxide layer. These bubbles consisting mainly of Si, C and O seems like cristobalite,

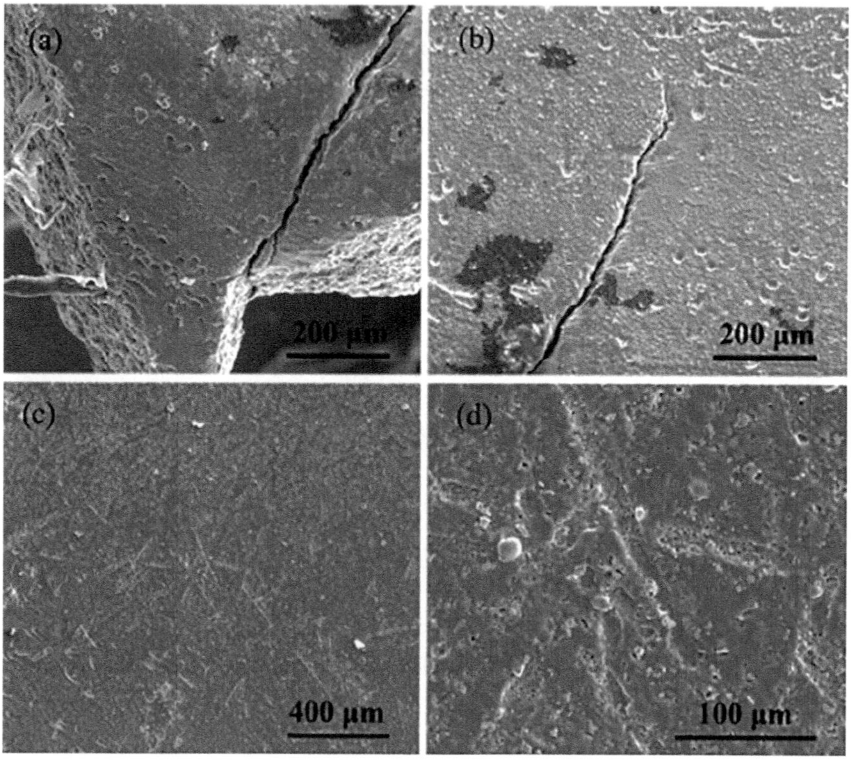

Fig. 6.78 SEM surface morphologies of the ablation center of the 20 vol% SiC_f (SLFC1) reinforced MA-Si_2BC_3N CMCs: **a**, **b** Thermal shock cracks; **c** relative dense surface; **d** fiber corrosion [21]

while the smooth areas contain Si, N and C with some O The sizes, quantities and distribution of these bubbles and pores vary with the direct distance away from the ablation center to transition area. This porous structure, on one hand, supports diffusion channels for oxidizing species to diffuse inward. On the other hand, the release of substantial high vapor pressure gases can prevent unexpected fracture during ablation. (Fig. 6.83).

A relatively dense morphology with considerable bubbles locates in heat-affected area. This oxide layer with multiple spherical bubbles likely arises from a molten surface layer trapping gas. The bubbles are hollow and some holes are observed at the bottom of the surface. The gas pressure from the gaseous products, such as SiO, CO, N_2, B_2O_3 and CO_2 likely causes the formation of hollow bubbles. Bubble sizes increase with increasing gas pressure eventually bursting. For relative low ablation temperatures in this region, the oxidation of SiC and SiC fibers cannot offer enough fused SiO_2 to fill pores and cracks. The continuous release of CO derived from SiC fiber oxidation constructs pores on oxide surface, thus corrosion in this region for fiber containing ceramics is more serious. In heat-affected area, all the investigated samples show a similar microstructure consisting of substantial bubbles of mainly

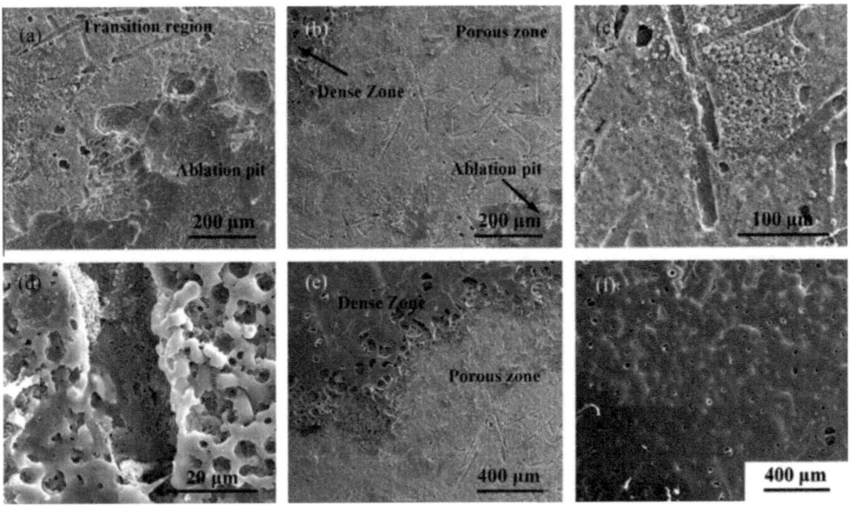

Fig. 6.79 SEM surface morphologies of the transition region of the SiC_f/Si_2BC_3N CMCs: **a** Morphology between ablation center and transition region; **b** porous zone and dense zone; **c**, **d** corrosion of short SiC fibers; **e** morphology between the porous zone and dense zone; **f** SiO_2 layer in the dense zone [21]

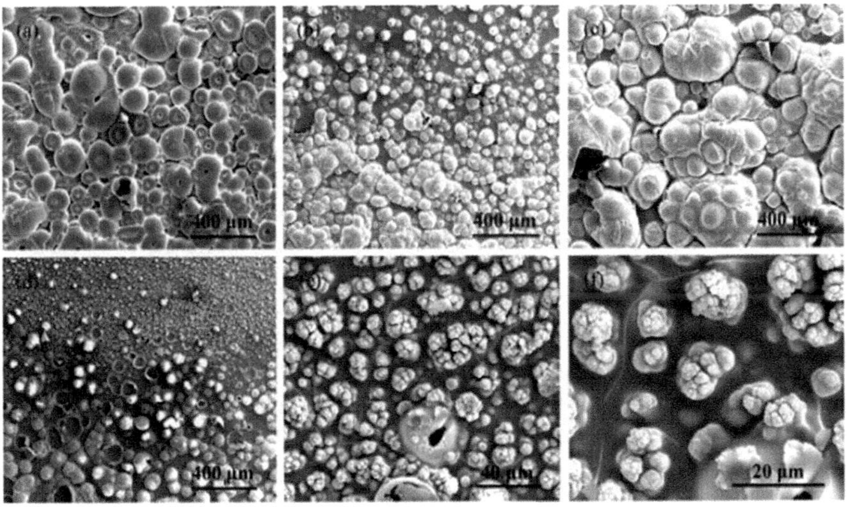

Fig. 6.80 SEM surface morphologies of the heat-affected zone of the SiC_f/Si_2BC_3N composites: **a–c** Bubbling on the surfaces; **d–f** particle precipitates [21]

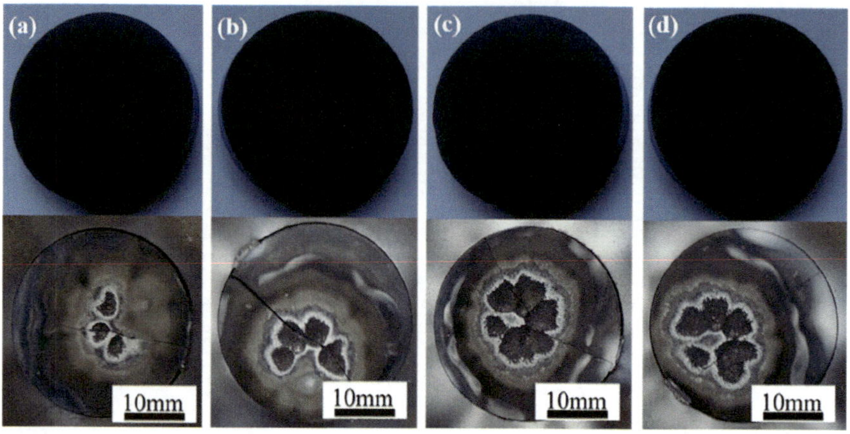

Fig. 6.81 Macroscopic morphologies of the various SiC_f (KD-II) reinforced MA-Si_2BC_3N composites after oxyethylene flame (heat flux ~ 4.2 MW/m^2) ablation for 10 s: **a** MA-Si_2BC_3N; **b** + 5 vol% modified SiC_f; **c** + 10 vol% modified SiC_f; **d** + 10 vol% unmodified SiC_f [22]

Table 6.18 Ablation properties of the various SiC_f (KD-II) reinforced MA-Si_2BC_3N composites after oxyethylene flame (heat flux ~ 4.2 MW/m^2) ablation for 10 s [22]

Composition	Mass ablation rate (mg/s)	Linear ablation rate (mm/s)
MA-Si_2BC_3N	0.61	0.0253
+ 5 vol% coated SiC_f	1.62	0.0412
+ 10 vol% coated SiC_f	4.92	0.0681
+ 10 vol% uncoated SiC_f	3.41	0.0486

SiO_2 but with different bubbles sizes that grow with increasing gas pressure. The SiC_f/Si_2BC_3N CMCs with 10 vol% modified fibers show the largest bubbles of 50 μm, while MA-Si_2BC_3N monoliths exhibit minimum sizes of~20 μm. Away from the heat-affected area, the oxide surface displays coarse morphology containing randomly positioned silica particles. Though SiC fibers still maintain their original morphologies, some deep pits are observed in ceramic matrix. One can infer that the consumption of ceramic matrix is faster than the SiC fibers. The incorporation of SiC fibers seems to reduce ablation resistance.

The oxide layer thickness for the MA-Si_2BC_3N monoliths and SiC_f/Si_2BC_3N composites is ~100 μm, 30 μm, 60 μm and 30 μm, respectively (Fig. 6.84). Their microstructure from the ablation pit to the substrate can be divided into three layers: outermost layer mainly consisting of molten SiO_2, transition region (porous structure with some SiO_2) and ceramic matrix. The outermost layer remains dense and continuous, while many micro-pores are observed in the transition region. In all cases, the oxide layers are strongly adhered to ceramic matrix.

The FIB-TEM slice shows amorphous nature of the oxide scale; however, from the B-O mapping, some B_2O_3 agglomerate in the outermost region (~25 nm). Further,

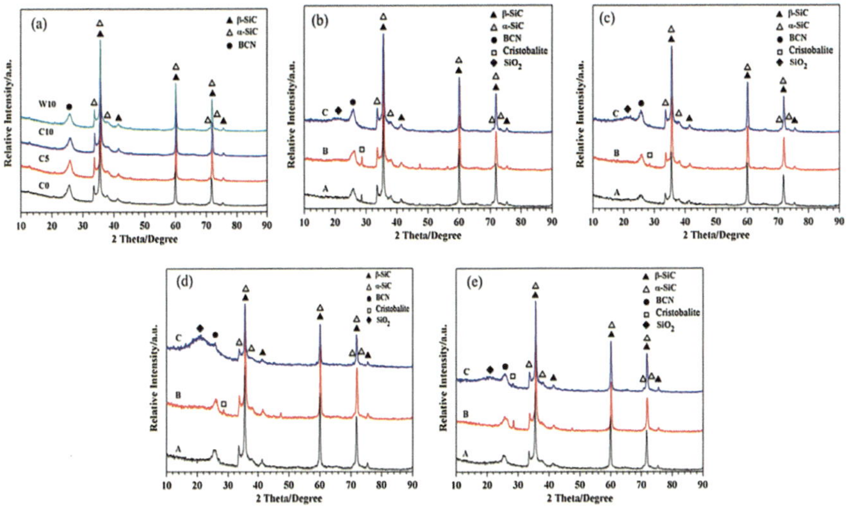

Fig. 6.82 Micro XRD patterns of the various SiC_f (KD-II) reinforced MA-Si_2BC_3N composites after oxyethylene flame ablation for 10 s: **a** MA-Si_2BC_3N; **b** + 5 vol% modified SiC_f; **c** + 10 vol% modified SiC_f; **d** + 10 vol% unmodified SiC_f (A: ablative center; B: transition zone; C: heat-affected zone) [22]

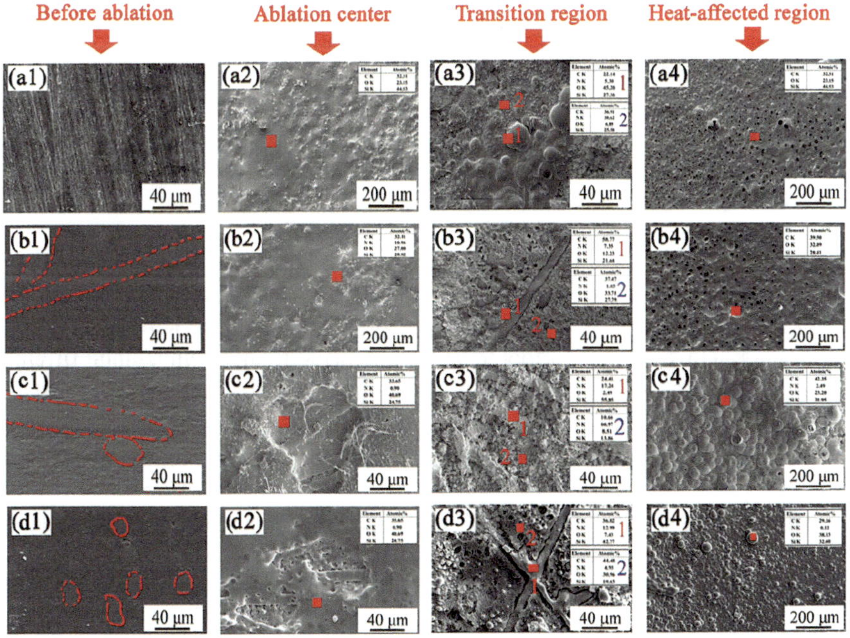

Fig. 6.83 SEM surface morphologies of the various SiC_f (KD-II) reinforced MA-Si_2BC_3N composites after oxyacetylene flame ablation for 10 s: **a** MA-Si_2BC_3N; **b** + 5 vol% modified SiC_f; **c** + 10 vol% modified SiC_f; **d** + 10 vol% unmodified SiC_f [22]

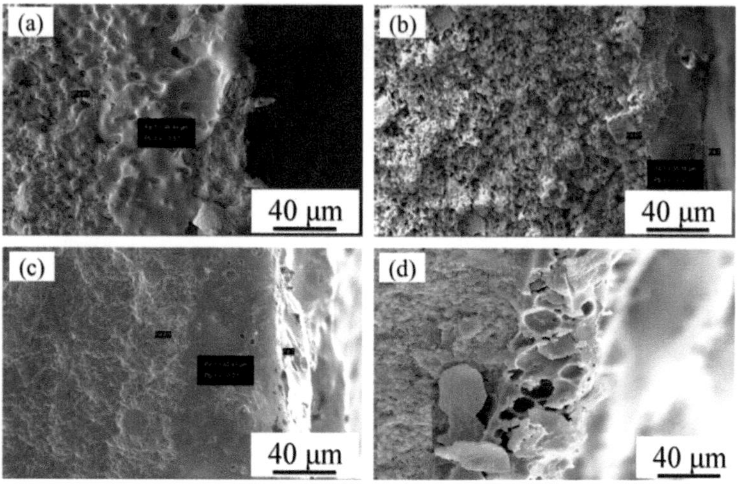

Fig. 6.84 SEM cross-sectional morphologies of the monoliths and composites after ablation in an oxyacetylene flame for 10 s. **a** MA-Si_2BC_3N; **b** + 5 vol% modified SiC_f; **c** + 10 vol% modified SiC_f; **d** + 10 vol% unmodified SiC_f [22]

Si–N and Si–C mappings indicate the outermost region is rich in C and N. Some SiO_2 are also found as a possible consequence of reacting with B_2O_3 forming borosilicate glass. Most studies imply that B_2O_3 does not volatilize completely but reacts with the amorphous silica at ~ 1000 °C to form borosilicate glass that then decomposes at ~ 1600 °C; however, some B_2O_3 have survived above 2100 °C for the as-ablated SiC_f/Si_2BC_3N CMCs. The C and N containing amorphous SiO_2 with some B_2O_3 can serve as a diffusion barrier, showing a better oxidation protection than pure SiO_2 (Fig. 6.85).

(7) *SiC_f/Si_2BC_3N-MAS CMCs*

After thermal shock at ΔT = 1000 °C, the h-BN coated KD-SA SiC_f reinforced Si_2BC_3N-MAS CMCs show rapid decrease in bending strength, but almost remain unchanged at ΔT = 1200–1300 °C [23]. Interestingly, the CMCs with 10 vol% as-coated SiC_f, show slight increases in strength as ΔT increases (Fig. 6.86 and Table 6.19). After thermal shock test, no trace of oxides is found on the sample surfaces, but showing high diffraction intensities of the matrix phases α/β-SiC, and BN(C).

After thermal shock at ΔT = 1000–1300 °C, some oxides are found covering the sample surfaces. Especially, a dense amorphous oxide scale with some nanocrystallites forms after thermal shock at ΔT = 1200 °C. Unfortunately, crack bridging, crack deflection and pull-out of SiC fibers do not respond to the thermal shock attack, and only fracturing of fibers parallel to the fracture surfaces are observed (Fig. 6.87).

SiC_f/Si_2BC_3N-MAS CMCs show similar surface temperature ~ 1250 °C within 1 s of oxyacetylene flame (heat flux ~ 4.2 MW/m^2) ablation, and finally leveled out at ~ 2000 °C. It seems that the addition of MAS doesn't favor the ablation improvement

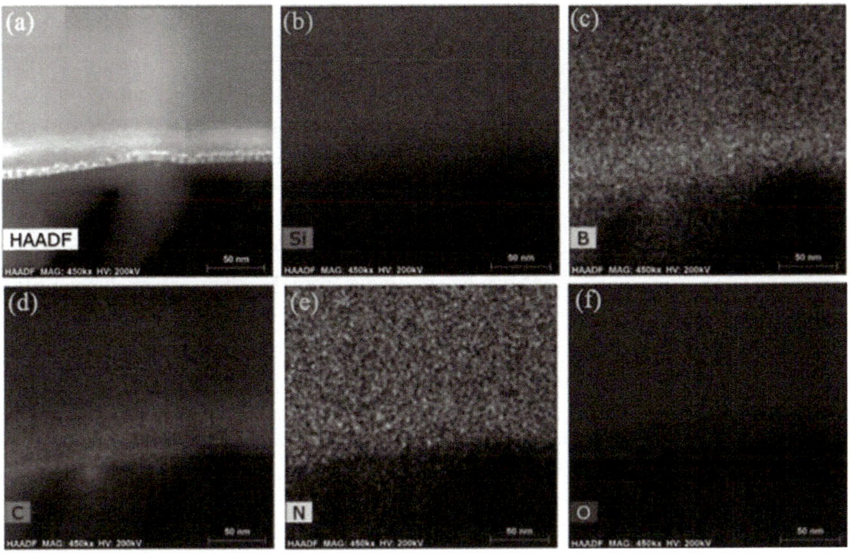

Fig. 6.85 Oxide scale microstructure of the 10 vol% modified SiC_f (KD-II) reinforced MA-Si_2BC_3N composites after 10 s of oxyacetylene flame ablation: **a** STEM morphology; (b-f) corresponding element mappings [22]

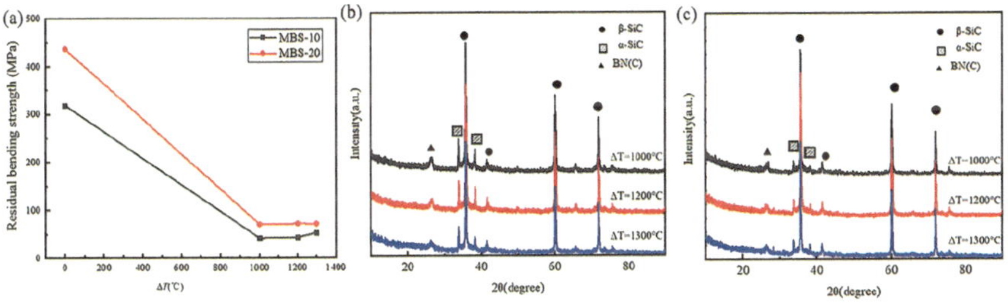

Fig. 6.86 The residual strength and phase composition of the hot-pressed SiC_f/Si_2BC_3N-MAS ceramic composites after thermal shock at different temperature difference: **a** Residual strength; **b** + 10 vol% h-BN coated SiC_f; **c** + 20 vol% h-BN coated SiC_f [23]

Table 6.19 Residual strength and strength retention rate of the hot-pressed SiC_f/Si_2BC_3N-MAS ceramic composites after thermal shock at different temperature difference [23]

Composition	Residual strength (MPa)/strength retention rate (%)		
	$\Delta T = 1000$ °C	$\Delta T = 1200$ °C	$\Delta T = 1300$ °C
+ 10 vol% h-BN coated KD-SA SiC_f	42.3/13.3	43.4/13.6	53.1/16.7
+ 20 vol% h-BN coated KD-SA SiC_f	71.0/16.2	72.6/16.6	72.2/16.5

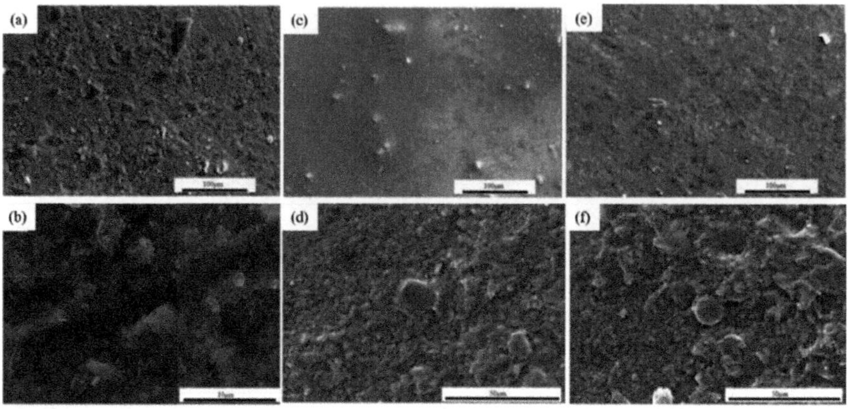

Fig. 6.87 Surface and fracture morphologies of SiC_f/Si_2BC_3N-MAS CMCs after thermal shock at different temperature difference: **a**, **b** $\Delta T = 1000$ °C; **c**, **d** $\Delta T = 1200$ °C; **e**, **f** $\Delta T = 1300$ °C [23]

of the SiC_f/Si_2BC_3N CMCs as compared with the former (Table 6.20). After ablation, matrix phases of α/β-SiC and BN(C) can be detected in all the ablation regions, but differs in their relative content. Additionally, considerable amorphous phases form on the heat-affected zone (Fig. 6.88). In all cases, the as-ablated composites show porous oxide structure on sample surfaces regardless of the SiC fiber content. In ablation center, the rod-like oxides distribute on a porous surface, but a relative dense oxide scale forms on the transition region. Substantial micropores varying from 10 to 100 μm position on the porous oxide layer in heat-affected zone. Some micropores unite and grow into bigger rupturing bubbles (Fig. 6.89).

(8) (8) *$(C_f$-$SiC_f)/Si_2BC_3N$ CMCs*

Thermal shock resistance

The residual strength of the four $(C_f$-$SiC_f)/Si_2BC_3N$ CMCs exhibit similar change trend, showing decreasing strength with the increasing ΔT [24, 25]. For example, after thermal shock at $\Delta T = 800$ °C, the residual flexural strength of the composites with h-BN coated 10 vol% C_f and 10 vol% SiC_f, first sharply decreases from 69.2 MPa to 49.4 MPa, 71.4% of the original value. And then it remains almost

Table 6.20 Ablation properties of the h-BN coated KD-SA SiC_f reinforced Si_2BC_3N-MAS composites after oxyacetylene flame (heat flux ~ 4.2 MW/m^2) ablation for 10 s [23]

Composition	Mass ablation rate (mg/s)	Linear ablation rate (mm/s)
+ 5 vol% SiC_f	15.6	0.0039
+ 10 vol% SiC_f	27.9	0.0651
+ 20 vol% SiC_f	16.2	0.0401

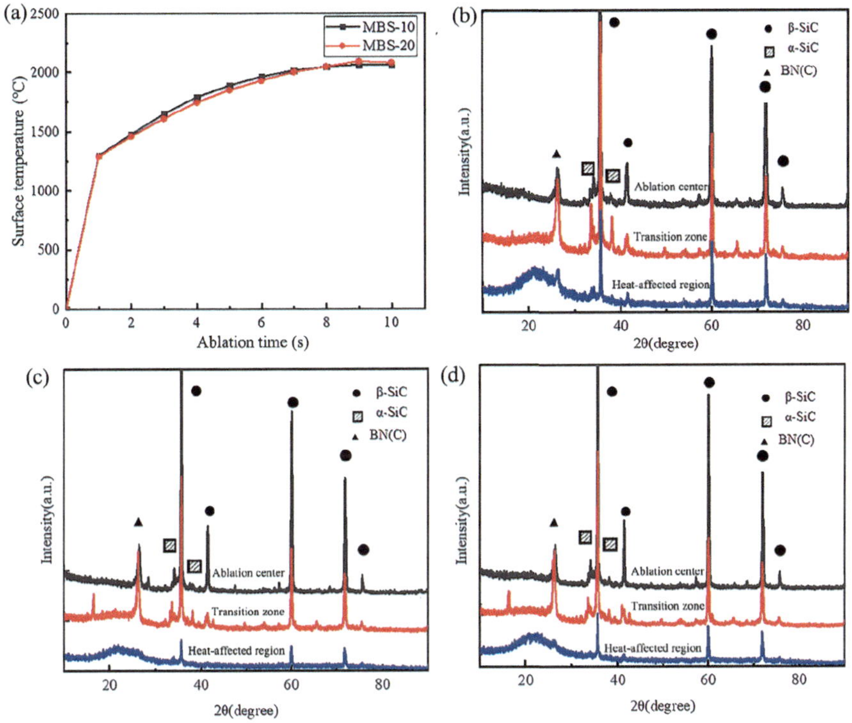

Fig. 6.88 Ablation behavior of the KD-SA SiC_f reinforced Si_2BC_3N-MAS composites: **a** Surface temperature curve; **b** + 5 vol% h-BN coated SiC_f; **c** + 10 vol% h-BN coated SiC_f; **d** + 20 vol% h-BN coated SiC_f (A: ablation center; B: transition zone; C: heat-affected zone) [23]

unchanged at $\Delta T = 1000$ °C and finally decreases to 42.8 MPa, i.e., a final retention of 61.8% with a ΔT of 1200 °C (Fig. 6.90 and Table 6.21).

The various (C_f-SiC_f)/Si_2BC_3N composites show similar phase composition and morphological/microstructural evolution after thermal shock at different temperature difference (Fig. 6.91). Typically, for h-BN coated fibers reinforced CMCs tested at $\Delta T = 800$ °C, the carbon fibers and SiC fibers are slightly damaged. A few strips shaped pits are clearly seen after oxidation of carbon fibers and the ceramic matrix. Smooth knots can be also observed resulting from the oxidized SiC_f and C_f. Meanwhile, some microcracks appear on the surfaces. Besides, crack deflection and even propagation along the fiber and the fiber debonding can also be found (Fig. 6.92). All the above is responsible for the dramatic decrease in flexural strength from 69.2 to 49.4 MPa.

At $\Delta T = 1000$ °C, the damaged surface is covered by SiO_2 and even part of B_2O_3 that do not evaporate in time. The interior material is efficiently protected from being damaged, and most defects such as microcracks and cavities are greatly reduced. Therefore, there is little change in residual flexural strength when ΔT increasing from 800 °C to 1000 °C. At $\Delta T = 1200$ °C, some microcracks and pits (or holes) after fiber oxidation and exfoliation are observed. Meanwhile, fractured fibers can be

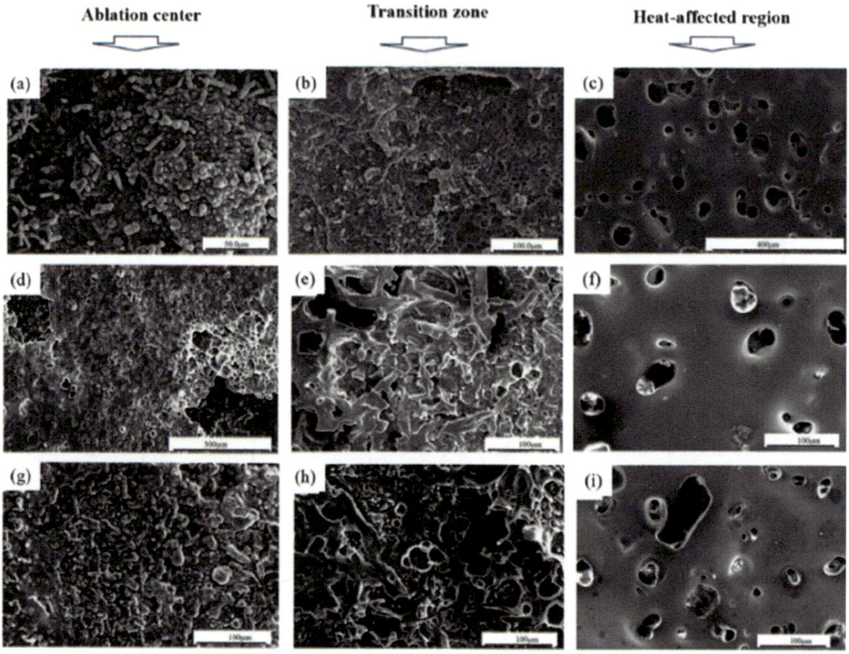

Fig. 6.89 SEM surface morphologies of the KD-SA SiC_f reinforced Si_2BC_3N-MAS composites after oxyacetylene flame (heat flux ~ 4.2 MW/m^2) ablation for 10 s: **a–c** + 5 vol% h-BN coated SiC_f; **d–f** + 10 vol% h-BN coated SiC_f; **g–i** + 20 vol% h-BN coated SiC_f [23]

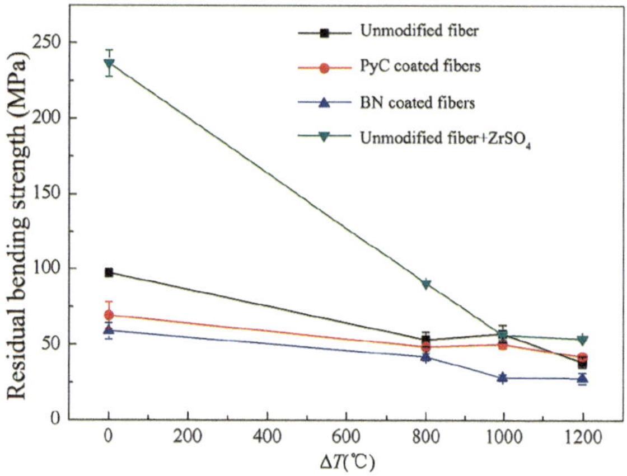

Fig. 6.90 The residual strength as a function thermal shock temperature difference of the various hot-pressed (C_f-SiC_f)/Si_2BC_3N composites [24]

also found which is ascribed to a reduced strength rather than stress due to mismatches in CTE and limited thermal conductivity. In addition, some cavities appear due to the significant gas emission and B_2O_3 evaporation caused by strong oxidation of the

Table 6.21 Residual bending strength and retention rate of the various hot-pressed (C_f-SiC_f)/Si_2BC_3N composites after thermal shock at different temperature difference [24]

Composition	Residual strength (MPa)/strength retention rate (%)		
	ΔT = 800 °C	ΔT = 1000 °C	ΔT = 1200 °C
+ Unmodified fibers	53.7/55.2	55.9/57.5	39.2/40.3
+ PyC coated fibers	43.0/72.4	28.3/47.6	29.1/49.0
+ h-BN coated fibers	49.4/71.4	49.4/71.4	42.8/61.8
+ Unmodified fibers and $ZrSiO_4$ additive	89.4/38.0	55.2/23.5	54.0/22.9

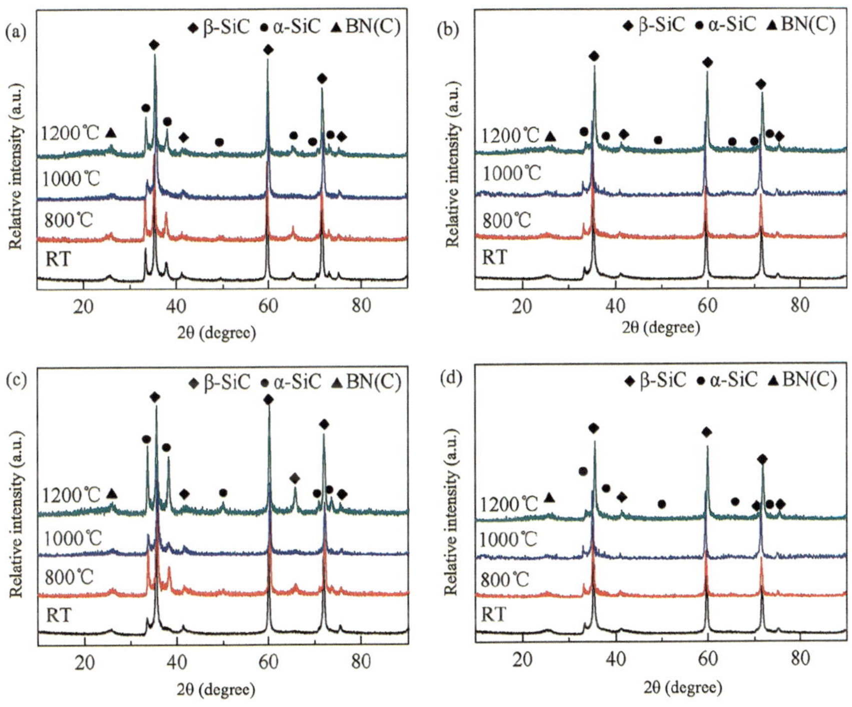

Fig. 6.91 XRD patterns of the various (C_f-SiC_f)/Si_2BC_3N composites after thermal shock at different temperature difference: **a** + unmodified fibers; **b** + PyC coated fibers; **c** + h-BN coated fibers; **d** + unmodified fibers and $ZrSiO_4$ additive [24]

carbon fibers and the ceramic matrix which plays a leading role in thermal shock, although the SiO_2 on the surfaces can be helpful for reducing damage to some extent.

The carbon fiber pull-out is still clearly observed on the fracture surfaces after thermal shock at temperature differences of 800 °C, 1000 °C and 1200 °C. With the increase in ΔT, the observed fiber pull-out decreases as well as residual stripe shaped pits after fiber oxidation or exfoliation, indicative of an increase in oxidation degrees of damage to fibers and the MA-Si_2BC_3N ceramic matrix (Fig. 6.93).

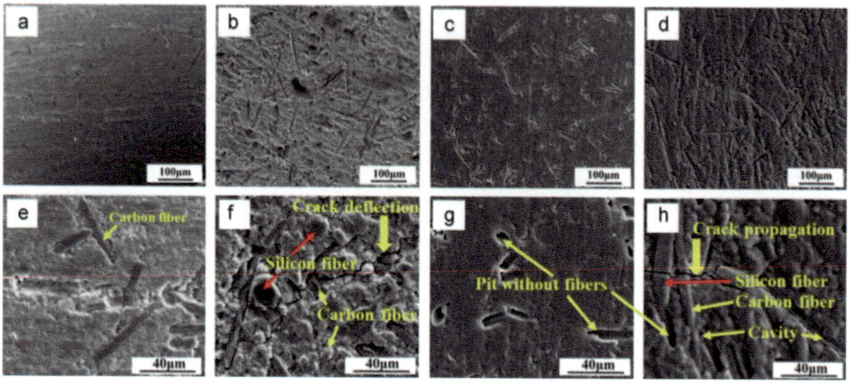

Fig. 6.92 SEM surface morphologies of the (C_f-SiC_f)/Si_2BC_3N composites with h-BN coating after thermal shock at different temperature difference: **a, e** RT; **b, f** $\Delta T = 800$ °C; **c, g** $\Delta T = 1000$ °C; **d, h** $\Delta T = 1200$ °C [25]

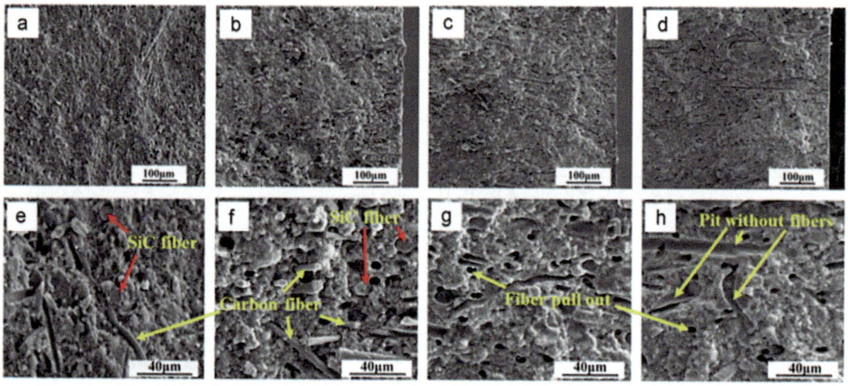

Fig. 6.93 SEM fracture morphologies of the (C_f-SiC_f)/Si_2BC_3N composites with h-BN coating after thermal shock at different temperature difference: **a, e** RT; **b, f** $\Delta T = 800$ °C; **c, g** $\Delta T = 1000$ °C; **d, h** $\Delta T = 1200$ °C [25]

The calculated R and R'''' values for the (C_f-SiC_f)/Si_2BC_3N composites are 233 °C and 941 μm, respectively, according to the room temperature mechanical properties (flexural strength of 69.2 MPa, Young's modulus of 72.9 GPa, fracture toughness of 2.6 MPa·m$^{1/2}$), the average thermal expansion coefficient of 3.05×10^{-6} °C^{-1} (25–1200 °C) and Poisson ratio of 0.25. Compared with the values of R (219 °C) and R'''' (274 μm) for the SiC_f/Si_2BC_3N CMCs [20], the R of the (C_f-SiC_f)/Si_2BC_3N composites increases by 14 °C and R'''' is 2.4 times higher, implying superior thermal stress fracture and damage resistance. Moreover, the SiC_f/Si_2BC_3N CMCs show brittle fracture, while (C_f-SiC_f)/Si_2BC_3N presents distinct pseudo-plastic fracture.

Ablation resistance

The ablation surface temperature of various hot-pressed (C_f-SiC_f)/Si_2BC_3N composites increases sharply within a few seconds and then levels out at around 2250 °C. No macrocracks and catastrophic damage in (C_f-SiC_f)/Si_2BC_3N CMCs is found under these circumstances (Fig. 6.94). The linear ablation rates of the various (C_f-SiC_f)/Si_2BC_3N composites show the same order of magnitude comparable to the C_f/Si_2BC_3N [15]. However, the mass ablation rate of the (C_f-SiC_f)/Si_2BC_3N composites decreases by one order of magnitude (Table 6.22).

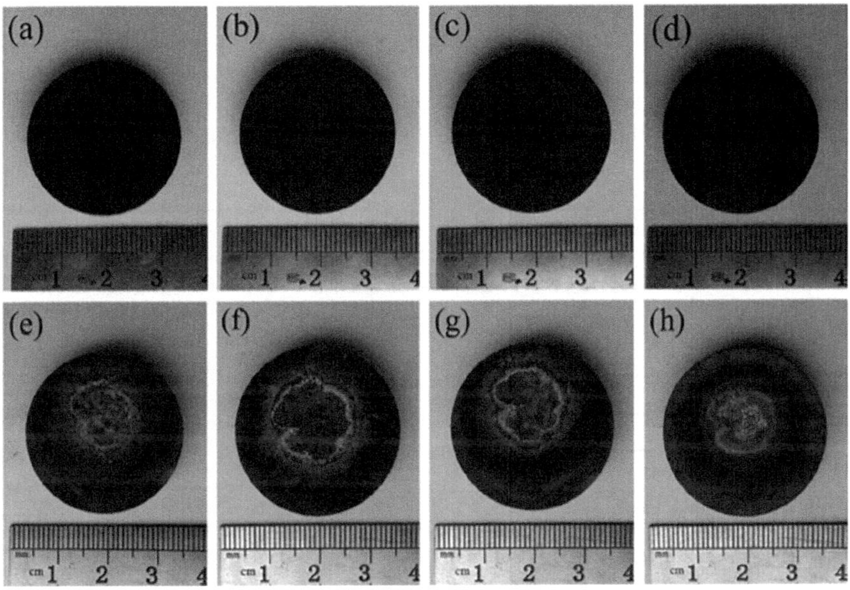

Fig. 6.94 Optical macro morphologies of the various hot-pressed (C_f-SiC_f)/Si_2BC_3N composites after oxyacetylene flame (heat flux ~ 4.2 MW/m^2) ablation for 10 s: **a**, **e** + unmodified fibers; **b**, **f** + PyC coated fibers; **c**, **g** + h-BN coated fibers; **d**, **h** + unmodified fibers and $ZrSiO_4$ additive [25]

Table 6.22 Bulk density and ablation rates of the various (C_f-SiC_f)/Si_2BC_3N composites after oxyacetylene flame (heat flux ~ 4.2 MW/m^2) ablation for 10 s [25]

Composition	Bulk density (g/cm^3)	Mass ablation rate (mg/s)	Linear ablation rate (mm/s)
+ Unmodified fibers	2.24	4.33	0.0523
+ PyC coated fibers	2.04	7.01	0.0603
+ h-BN coated fibers	2.18	4.64	0.0480
+ Unmodified fibers and $ZrSiO_4$ additive	2.58	1.36	0.0273

For the as-ablated CMCs, α/β-SiC, cristobalite and amorphous SiO_2 appear in the transition and heat-affected regions, while amorphous SiO_2 disappear and strong diffraction peaks for α-SiC appear in the central region. During ablation some SiC react with oxygen to form amorphous SiO_2. Coincidentally, a fraction of the amorphous SiO_2 crystallize into some cristobalite. The higher ablation temperatures in the central region are responsible for complete crystallization of amorphous SiO_2 and are also sufficient to transform β- to α-SiC (Fig. 6.95).

The ablation center surfaces of the as-ablated (C_f-SiC_f)/Si_2BC_3N composites roughen. Numerous strip shaped pits are clearly visible, left after the erosion of short SiC_f and/or C_f. These fibers are aligned mainly along the vertical direction of axial pressure as a result of hot pressing. In addition, due to the weaker adhesive strength between the fibers and the ceramic matrix, the fibers are stripped off the ceramic matrix during ablation. A few microcracks are also observed in the central ablation region. Besides, in the central region, some spherical particles of SiO_2 appear along the strip shaped pits inlaying in the foam-like ceramic matrix. Note that there

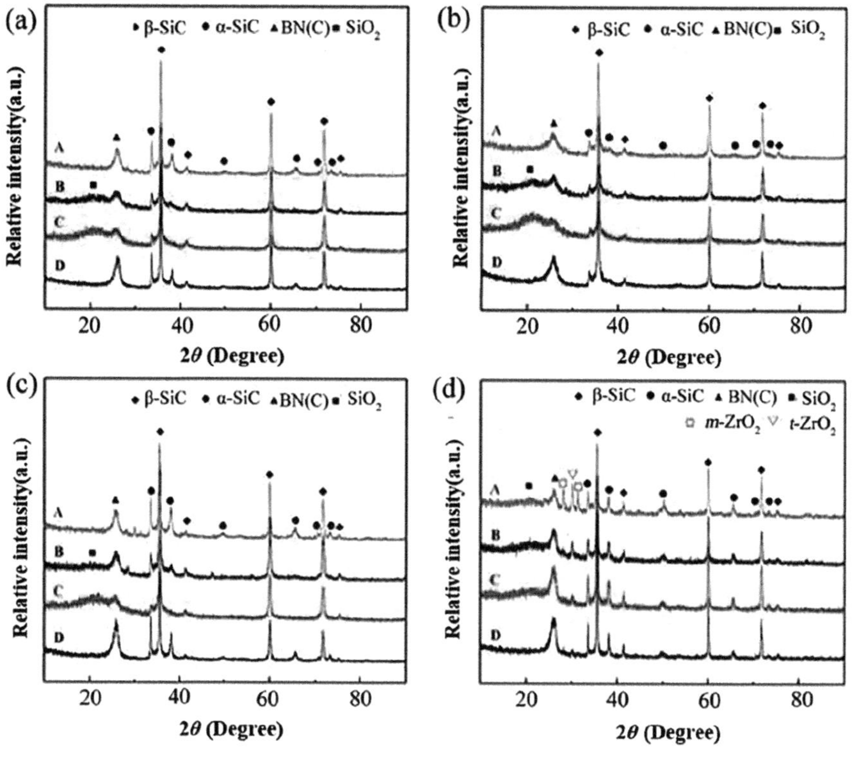

Fig. 6.95 Micro XRD patterns of the various hot-pressed (C_f-SiC_f)/Si_2BC_3N composites after oxyacetylene flame ablation for 10 s: **a** + unmodified fibers; **b** + PyC coated fibers; **c** + h-BN coated fibers; **d** + unmodified fibers and $ZrSiO_4$ additive. (A: ablation center; B: transition zone; C: heat-affected zone; D: before ablation) [25]

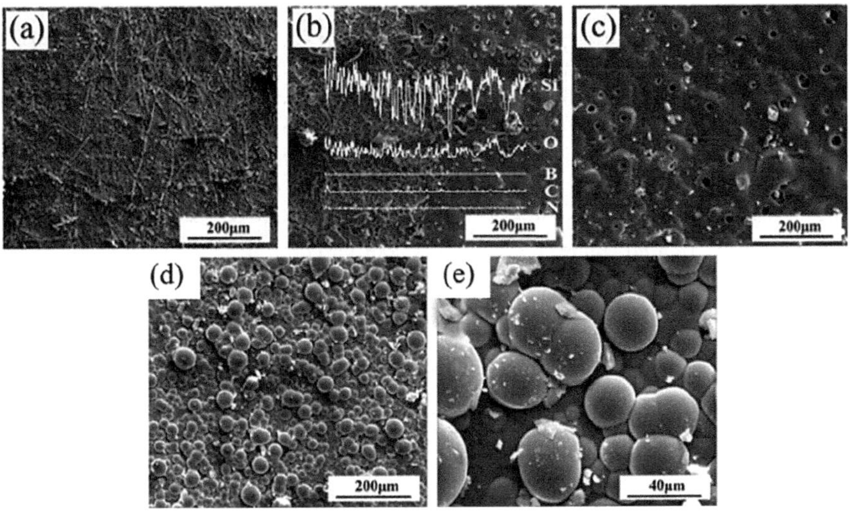

Fig. 6.96 SEM surface morphologies of the as-ablated (C_f-SiC_f)/Si_2BC_3N composites with h-BN coating: **a** Ablation center; **b** interface between center area (left) and transition region (right); **c** transition zone; **d–e** heat-affected region [25]

are no noticeable elemental changes (only for Si and O) along the direction from the central to transition region apart from the obviously different morphologies. In the transition region, the temperature ensures that SiO_2 has good fluidity. Therefore, scoured by high temperature combustion flow, a large amount of SiO_2 reaches the transition region and uniformly covers the surface. The gases emitted coupled with B_2O_3 evaporation during ablation produce pores with a diameter of ~ 20 μm (Fig. 6.96).

The bubbles with average sizes ~ 30 μm form on the surfaces of the heat-affected region, most remaining unbroken. These bubbles consist mainly of SiO_2. The SiO_2 viscosity is quite high at the low ablation temperatures in the heat-affected region limiting the forces needed to break the bubbles. Meanwhile, a few of broken bubbles can also be observed and some stripe shaped protuberances appear under the broken bubbles. The protuberances are SiC fibers in view of the observed diameter and excellent oxidation resistance. In addition, some white particles near the broken bubbles are borosilicate glass phases. B_2O_3 from the oxidation of BN does not evaporate easily at the relatively low-temperatures in the heat-affected region far from the central ablation region. The resultant B_2O_3 reacts with SiO_2 to produce borosilicate glass.

(9) *MWCNTs/Si_2BC_3N ceramic composites*

Thermal shock resistance

One of the advantages of the introducing multi walled carbon nanotubes (MWCNTs) is the improvement on the thermal shock resistance. The calculated *R* parameter suggests that both the MWCNTs and SiC coated MWCNTs can improve the crack initiation resistance [26]. Similarly, the thermal stress damage resistance can be enhanced with the introduction of MWCNTs regardless of the SiC coating (Table 6.23 and Table 6.24).

After thermal shock test at $\Delta T = 400$ °C, the residual strength of the MWCNTs/Si_2BC_3N composites does not change significantly, but the residual strength of pure MA-Si_2BC_3N is lower than that of CMCs. At $\Delta T = 600$ °C, the residual strength of all the investigated materials decreases, but showing a relatively high residual strength. Further increasing ΔT to 1000 °C, the monoliths and composites show significant strength attenuation. In addition, the elastic modulus of the composites decreases gradually with the increase of ΔT (Fig. 6.97).

Table 6.23 The calculated thermal shock damage parameters of the SPS densified MWCNTs/Si_2BC_3N CMCs [26]

MWCNTs content (vol%)	σ_f (MPa)	γ_{WOF} (J/m^2)	α (10^{-6}/K)	R (°C)	R'''' (mm)	R_{st} (K m$^{1/2}$)
0	320.1 ± 25.8	1349.3 ± 205.0	4.51	470	1.8	23.5
1	426.1 ± 35.2	2676.5 ± 450.3	4.85	615	1.9	31.3
2	322.1 ± 6.4	1449.6 ± 264.2	4.89	530	1.7	25.5

Table 6.24 The calculated thermal shock damage parameters and measured strength of the SPS densified SiC coated MWCNTs reinforced MA-Si_2BC_3N CMCs [26]

SiC coated MWCNTs content (vol%)	R (°C)	R'''' (mm)	R_{st} (K·m$^{1/2}$)	Residual strength (MPa)/strength retention rate (%)
1	432	0.21	7.3	128.2/32.8
2	579	0.24	10.4	192.6/36.2

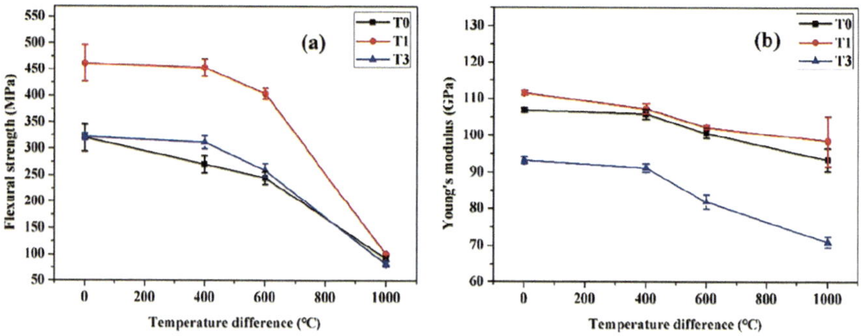

Fig. 6.97 Mechanical properties of the MWCNTs/Si_2BC_3N composites after thermal shock at different temperature difference: **a** Residual strength; **b** residual elastic modulus

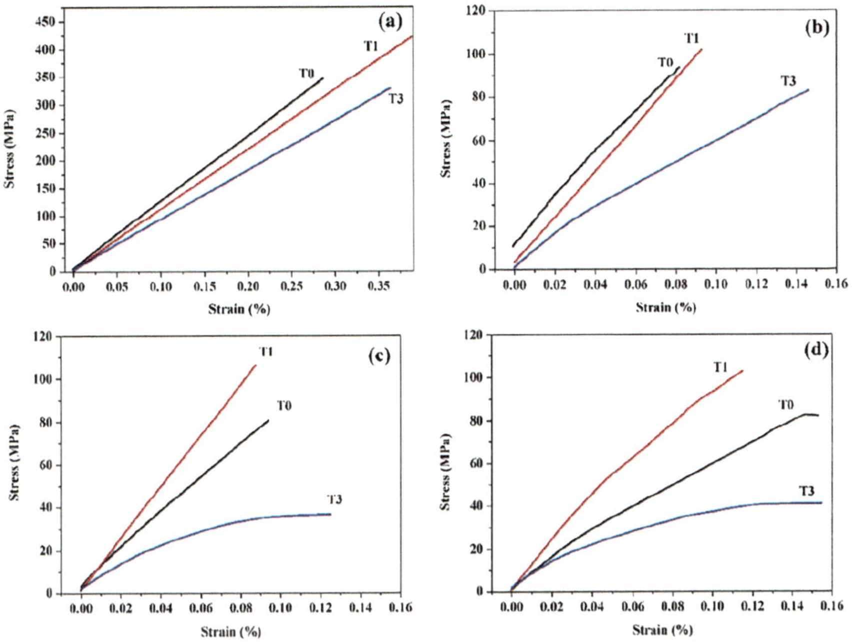

Fig. 6.98 Stress–strain curves of the MWCNTs/Si_2BC_3N composites after cyclic thermal shock tests at $\Delta T = 1000$ °C: **a** Before thermal shock test; **b** 1 cycle; **c** 2 cycles; **d** 3 cycles

The zigzag degree of the stress–strain curve represents the intensity of microcrack formation in a certain sense. After cyclic thermal shock tests at $\Delta T = 1000$ °C, the stress–strain curves of the CMCs materials change obviously. After 1 cycle test, only 1 vol% MWCNTs containing MA-Si_2BC_3N CMCs retain its linear stress–strain behavior. Therefore, more microcracks seem appear in pure MA-Si_2BC_3N and 3 vol% MWCNTs/Si_2BC_3N composites after 1 cycle test, and composites with 1 vol% MWCNTs should have better crack resistance (Fig. 6.98).

After 2 cycles thermal shock test, both the MA-Si_2BC_3N ceramics and 1 vol% MWCNTs/Si_2BC_3N CMCs show a linear stress–strain curve, while other presents a more curved stress–strain curve. The thermal shock induced surface oxidation can help to heal the microcracks thereby inhibiting the crack propagation to a certain extent. For the composites with 3 vol% MWCNTs, substantial cracks have generated after 1 cycle test, thus to achieve an effective crack-healing seems impossible. After 3 cycles test, more microcracks are formed in inner materials leading to more nonlinear mechanical behavior. In terms of strength retention rate, thermal shock resistance of MA-Si_2BC_3N cannot be improved by MWCNTs addition, once the ΔT is ≥ 1000 °C (Fig. 6.99).

After thermal shock at $\Delta T = 1000$ °C, a uniform and dense oxide layer 1–3 μm is formed on the material surfaces. In all cases, the thermal shock generated oxide

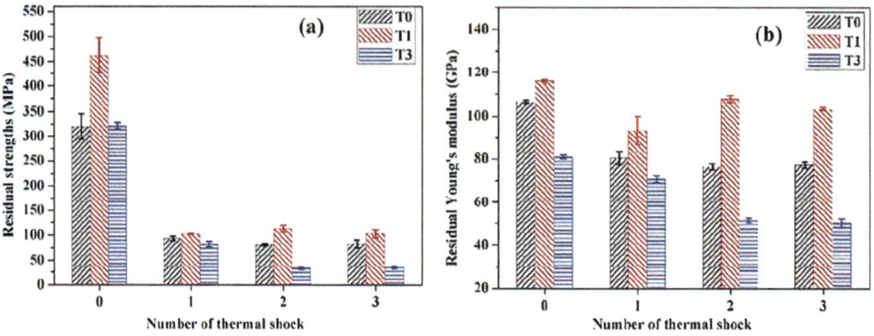

Fig. 6.99 Mechanical properties of MWCNTs/Si_2BC_3N composites after cyclic thermal shock test at $\Delta T = 1000$ °C: **a** Residual strength; **b** residual elastic modulus [16]

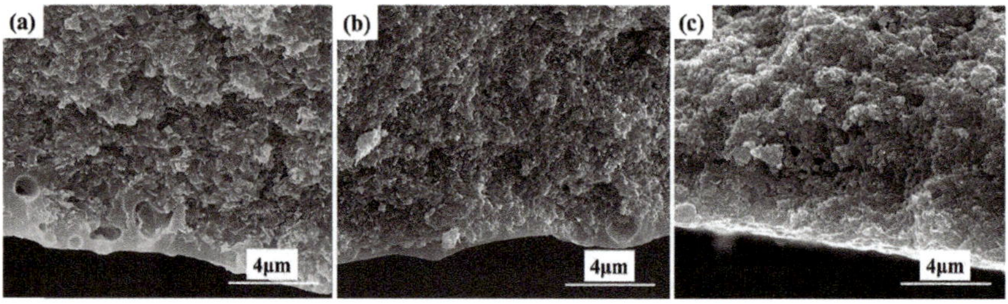

Fig. 6.100 SEM fracture morphologies of the MA-Si_2BC_3N ceramics and MWCNTs/Si_2BC_3N composites after the 1 cycle thermal shock test at $\Delta T = 1000$ °C: **a** Pure MA-Si_2BC_3N; **b** + 1 vol% MWCNTs; **c** + 3 vol% MWCNTs [16]

scales adhere tightly with the ceramic matrix, without micropores and microcracks agglomerated at the interfaces (Fig. 6.100).

Ablation resistance

After 6 s of oxyacetylene flame (heat flux ~ 4.2 MW/m^2) ablation, the surface temperature of the 1 vol% MWCNTs/Si_2BC_3N composites reaches ~ 1900 °C. After 10 s of ablation, the surface temperature exceeds over 2000 °C, and finally stabilized at ~ 2210 °C. The MA-Si_2BC_3N monoliths has been split into two halves, while fragmentations of their CMCs with multiple penetrating cracks are observed. Oxides forms on the ablation surfaces, in which the ablation area is different (Fig. 6.101). The introduction of MWCNTs weakens the ablation resistance of the MWCNTs/Si_2BC_3N composites. For example, after the introduction of 1 vol% or 3 vol% MWCNTs, the linear ablation rate of the composites increases from 0.047 mm/s to 0.064 mm/s or 0.049 mm/s, while mass ablation rate raises from 20.80 mg/s to 38.97 mg/s or 28.24 mg/s, respectively (Table 6.25).

After ablation, only some α/β-SiC are detected in ablation center and transition zone of pure MA-Si_2BC_3N ceramics, while oxides of cristobalite and amorphous SiO_2 are located in heat-affected area. With the introduction of 1 vol% MWCNTs,

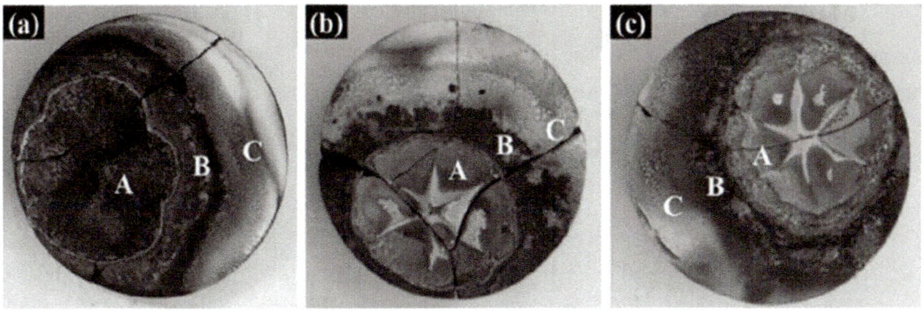

Fig. 6.101 Surface macro morphologies of SPS densified MWCNTs/Si_2BC_3N CMCs after oxyacetylene flame (heat flux ~ 4.2 MW/m^2) ablation for 30 s: **a** Pure MA-Si_2BC_3N; **b** + 1 vol% MWCNTs; **c** + 3 vol% MWCNTs. (A: ablation center; B: transition zone; C: heat-affected area)

Table 6.25 Ablation properties of the SPS densified MWCNTs/Si_2BC_3N composites after oxyacetylene flame (heat flux ~ 4.2 MW/m^2) ablation for 30 s

MWCNTs content (vol%)	Linear ablation rate (mm/s)	Mass ablation rate (mg/s)
0	0.047	20.80
1	0.064	38.97
2	0.049	28.24

the surface phase composition of the composites is similar, but substantial cristobalite and amorphous SiO_2 are found in transition zone and heat-affect region (Fig. 6.102).

After ablation, the loose and porous oxide structure of mainly SiO_2 form in the ablation center. In the transition zone, some SiO_2 cover on the oxide surfaces, and substantial bubbles and spherical particles position within the scale. Spherical blistering and holes overspread the oxide surfaces showing porous structure of heat-affected region. Bubble formation at interfaces of oxide layer and matrix in ablation

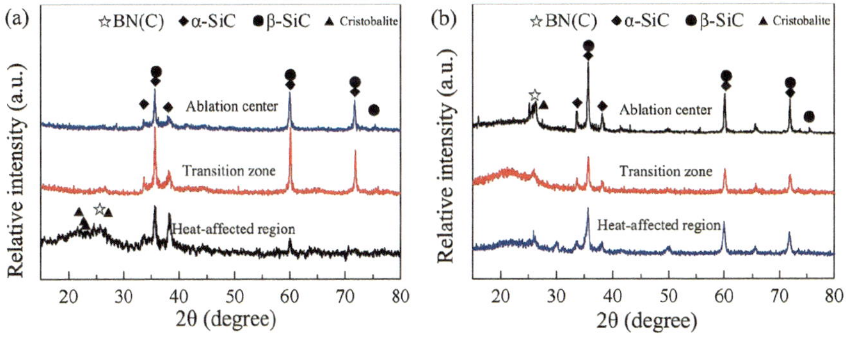

Fig. 6.102 Micro XRD patterns of the MA-Si_2BC_3N CMCs after oxyacetylene flame ablation for 30 s: **a** Pure MA-Si_2BC_3N; **b** + 1 vol% MWCNTs

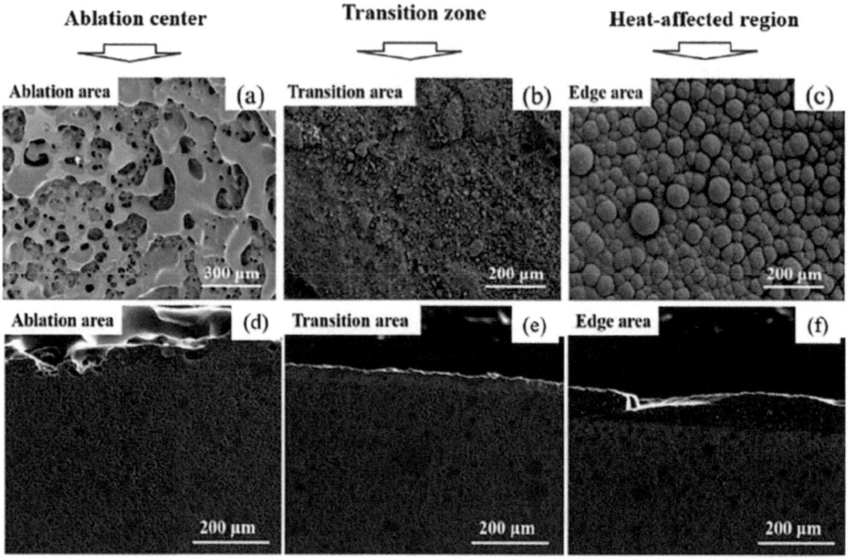

Fig. 6.103 SEM surface and cross-section morphologies of pure MA-Si_2BC_3N after oxyacetylene flame ablation for 30 s: **a–c** Surface morphologies; **d–f** cross-section morphologies

center, which flakes off indicating poor adhesion and exposing the interior to further oxidation. The oxide scale becomes dense and smooth from ablation center to the heat-affected zone in view of the cross-section morphologies, in which the oxide scale thickness is in range of 50–200 μm (Fig. 6.103).

The composites with 1 vol% MWCNTs experience the same ablation process and thus show similar rough oxide scale structure with bubbles some of which have been broken. No trace of microcracks appear in oxide scales. Coincidently, the cross-section SEM images also illustrate the oxide scales are very loose in ablation center, but change to dense adhered tightly to ceramic matrix in other ablation regions (Fig. 6.104).

(10) *GPLs/Si_2BC_3N CMCs*

Thermal shock resistance

The residual bending strength of the monolithic samples at a ΔT of 1400 °C decreases ~ 84.3% relative to its primary strength [27]. But the residual bending strength of the GPLs/Si_2BC_3N composites with different graphene platelets (GPLs) also displays the same downward trend, indicating similar thermal shock resistance. The residual bending strength of the composites with 1 vol% or 2 vol% GPLs remains 28.3 MPa or 35.5 MPa after thermal shock test at $\Delta T = 1400$ °C, respectively (Fig. 6.105 and Table 6.26).

After thermal shock test at $\Delta T = 1000$ °C, a relative dense and smooth surface is found; however, some micropores and bubbles are also distributed on the oxide

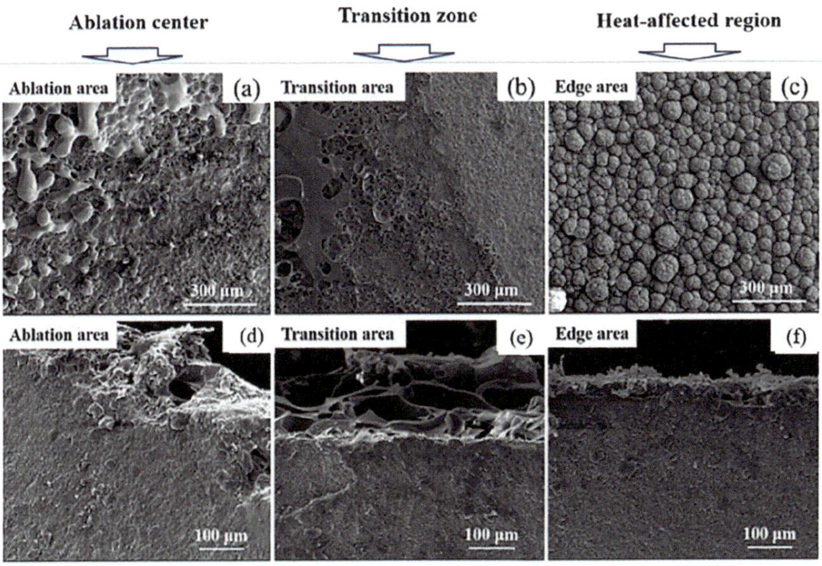

Fig. 6.104 SEM surface and cross-section morphologies of the 1 vol% MWCNTs/Si_2BC_3N CMCs after oxyacetylene flame ablation for 30 s: **a–c** Surface morphologies; **d–f** cross-section morphologies

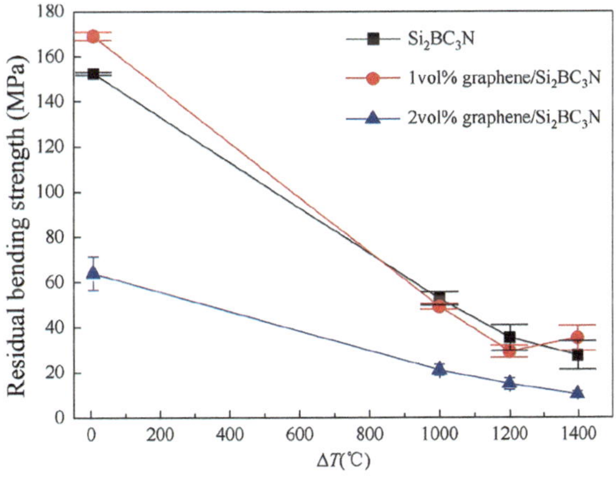

Fig. 6.105 Residual strength curves of the SPS densified GPLs/Si_2BC_3N composites after thermal shock at different temperature difference [27]

Table 6.26 Residual strength and strength retention rate of the SPS densified GPLs/Si_2BC_3N CMCs after thermal shock at different temperature difference [27]

GPLs content (vol%)	Residual strength (MPa)/strength retention rate (%)		
	$\Delta T = 1000$ °C	$\Delta T = 1200$ °C	$\Delta T = 1400$ °C
0	21.1/32.6	15.3/23.6	10.2/15.7
1	54.1/35.3	33.8/22.0	28.3/18.4
2	51.0/30.2	28.3/16.8	35.5/21.0

surfaces at $\Delta T = 1400\,°C$. Fortunately, conventional toughening mechanisms derived from fiber-reinforced and rod-like phase reinforced ceramics are still applicable in GPLs/Si_2BC_3N CMCs after thermal shock test at $\Delta T = 1000$–$1400\,°C$ (Fig. 6.106).

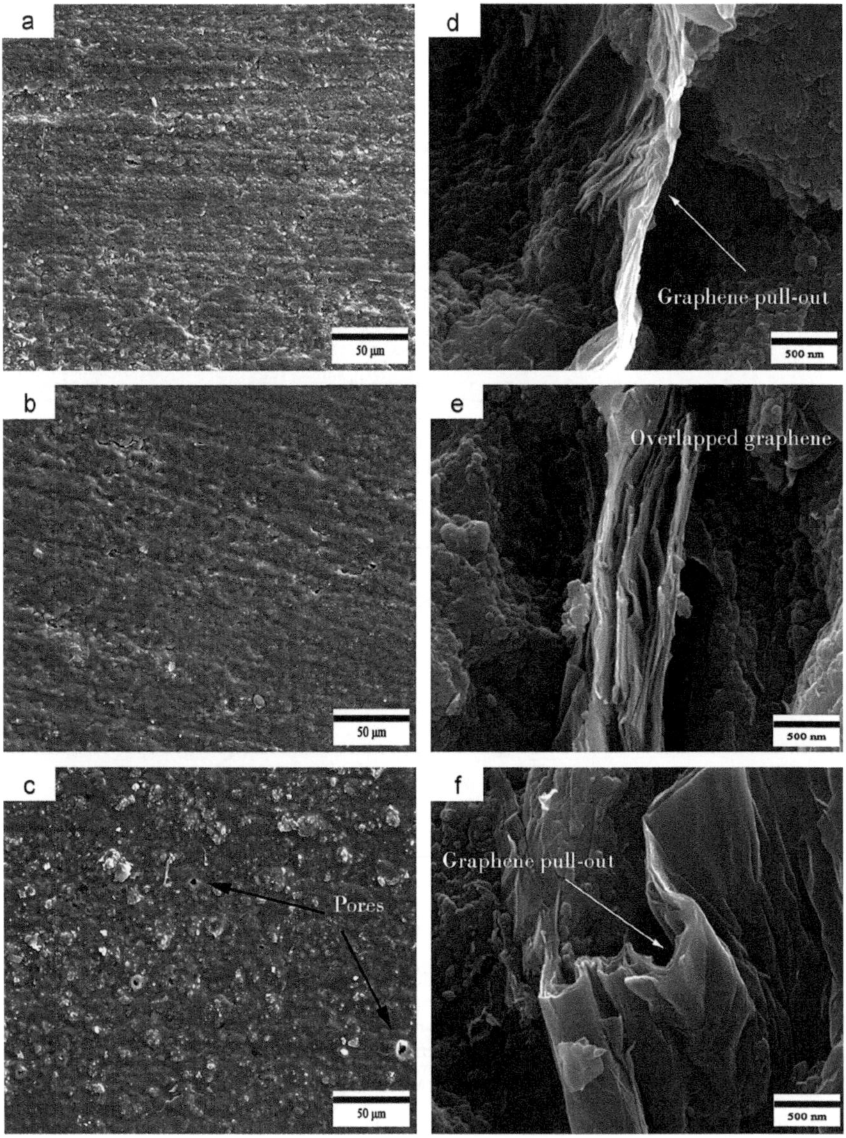

Fig. 6.106 SEM surface and fracture morphologies of the GPLs/Si_2BC_3N CMCs after thermal shock at different temperature difference: **a**, **d** $\Delta T = 1000\,°C$; **b**, **e** $\Delta T = 1200\,°C$; **c**, **f** $\Delta T = 1400\,°C$ [27]

Fig. 6.107 Crack initiation and propagation on fracture surface of the GPLs/Si_2BC_3N CMCs after thermal shock at ΔT = 1400 °C [27]

At ΔT = 1400 °C, some GPLs pile up together and form a larger area of interface which can also be helpful for toughening. GPLs piled up together with a sheet of belt-like geometry have lager interfacial contact area with the MA-Si_2BC_3N matrix and are easier to adhere at nanosized grain boundaries, for which enhances the fracture toughness even though applied at thermal shock condition. Interestingly, there still exists microcracks initiation and propagation on the fracture surfaces after thermal shock at ΔT = 1400 °C (Fig. 6.107).

Ablation resistance

After ablation in an oxyacetylene combustion flame (heat flux ~ 4.2 MW/m^2), the ablation surface of the pure MA-Si_2BC_3N has been peeled off due to the thermal stress and gas fluxes scouring [28]. In contrast, there is a white and discontinuous oxide layer covering on the ablation center for GPLs/Si_2BC_3N CMCs with intact macro structure (Fig. 6.108).

The GPLs addition enhances the ablation resistance of MA-Si_2BC_3N ceramics. The mass and linear ablation rates of the composites with 10 vol% GPLs are 12.10 mg/s and 0.0071 mm/s, while those of 5 vol% GPLs containing composites are 18.5 mg/s and 0.0092 mm/s, respectively. As compared, the 10 vol% GPLs/Si_2BC_3N composites show a similar contribution on ablation resistance as well as the SiC_f/Si_2BC_3N CMCs, but do much better than that of C_f/Si_2BC_3N. Compared with the (C_f-SiC_f)/Si_2BC_3N, the linear ablation rate of 10 vol% GPLs/Si_2BC_3N CMCs is lower than the former, but the mass ablation rate shows an opposite result (Table 6.27).

For the as-ablated MA-Si_2BC_3N monoliths, a continuous amorphous layer forms on ablation center surfaces without micropores and microcracks. This is mainly comprised of what appear to be cristobalite grains 5–50 μm in diameter. The formation of a dense SiO_2 layer is an effective way to limit inward diffusion of oxygen to MA-Si_2BC_3N matrix, thereby enhancing the oxidation resistance. However, when it comes to the transition area, the ablation surfaces are covered with rod-like structures

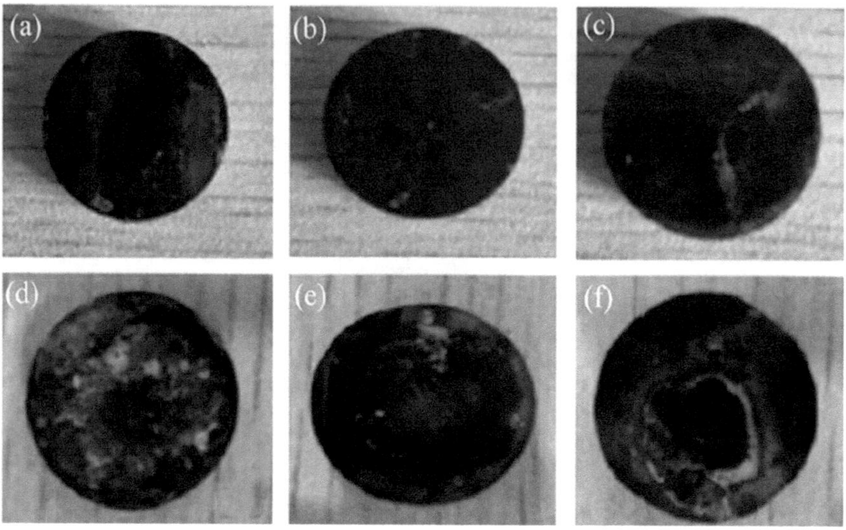

Fig. 6.108 Surface macro morphologies o of the GPLs/Si_2BC_3N CMCs after oxyacetylene flame (heat flux ~ 4.2 MW/m^2) ablation for 10 s: **a, d** Pure MA-Si_2BC_3N; **b, e** + 5 vol% GPLs; **c, f** + 10 vol% GPLs [28]

Table 6.27 The mass and linear ablation rates of the GPLs/Si_2BC_3N composites after oxyacetylene flame (heat flux ~ 4.2 MW/m^2) ablation for 10 s [28]

GPLs content (vol%)	Relative density (%)	Mass ablation rate (mg/s)	Linear ablation rate (mm/s)
0	97.3	32.60	0.0125
5	85.9	18.50	0.0092
10	86.6	12.10	0.0071

that appear to be SiO_2. The very large temperature gradient across the ablation region is likely responsible for the formation of such rod-like SiO_2 structure (Fig. 6.109).

After ablation, the GPLs containing composites exhibit a similar structure to the former. In the central area, a relatively smooth and dense oxide layer with multiple spherical structures likely arising from a molten surface layer trapping gas bubbles is formed during ablation process. The spheres are composed primarily of SiO_2 and small amounts of C likely from the residual SiC or partially oxidized GPLs. Spherical particles form to minimize surface free energy. Some open pores appear at the top of SiO_2 spheres. some spheres are hollow and some holes are located at the bottom of the spheres. The sizes of the hollow spheres grow with the increasing gas pressure.

Note that when the gas pressure in SiO_2 spheres exceeds the sum of ambient pressure and tensile stress of the oxide layer, the gas products can escape from the hollow spheres and gas pressure is released. Meantime, pores are formed at the top of the spheres and channels are left at the bottom, which leads to a continuous oxygen

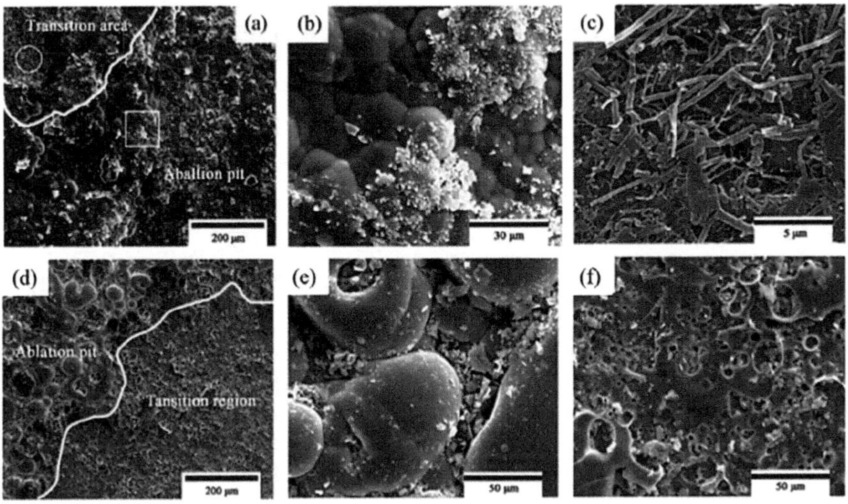

Fig. 6.109 SEM surface morphologies of the **a–c** MA-Si_2BC_3N and **d–f** 10 vol% GPLs/Si_2BC_3N CMCs in different ablation region: **a**, **d** Interfaces between ablation center and transition zone; **b**, **e** magnified ablation center; **c**, **f** magnified transition zone [28]

diffusion into inner matrix and decreases the anti-oxidation ability. The ablation center surfaces of 5 vol% GPLs incorporating composites tend to form fused and dense SiO_2 layers with small scale pores uniformly distributed on surface, while a similar surface structure at the ablation center of 10 vol% GPLs/Si_2BC_3N generates spheres ranging from 20 to 70 μm.

For the as-ablated GPLs/Si_2BC_3N composites, the boundaries between transition zone and heat-affected area are obviously (Fig. 6.110). The transition area shows porous structure while the heat-affected area keeps smooth, dense and continuous for both samples. For low surface temperature in heat-affected region, the oxidation process and oxidation rate in this zone should be slight and slow. As is known to all that gas evaporation at relatively low temperature and pores filled by high viscosity of SiO_2 ineffectively can lead to porous morphology in heat-affected area, which is not matched with the morphologies shown above. The possible reasons for this should attribute to the flowing of low viscosity of SiO_2 from ablation center by high-speed gas scouring, thus heals the pores impactful and covers on the heat-affected area surface. The glassy melt layer is comprised of SiO_2 with few residual carbons, indicating the enrichment of SiO_2 herein.

From the optical macroscopic morphologies, we can make a conclusion that the GPLs nanofillers contribute to ablation resistance improvement of the MA-Si_2BC_3N ceramic matrix. For the 100–400 nm lateral size GPLs used and oxide layer covering on the ablation surface, it is difficult to observe whether or not GPLs remains on the sample surfaces after ablation. However, the presence of multilayer graphene near to the ablation surface in fracture morphologies is confirmed (Fig. 6.111).

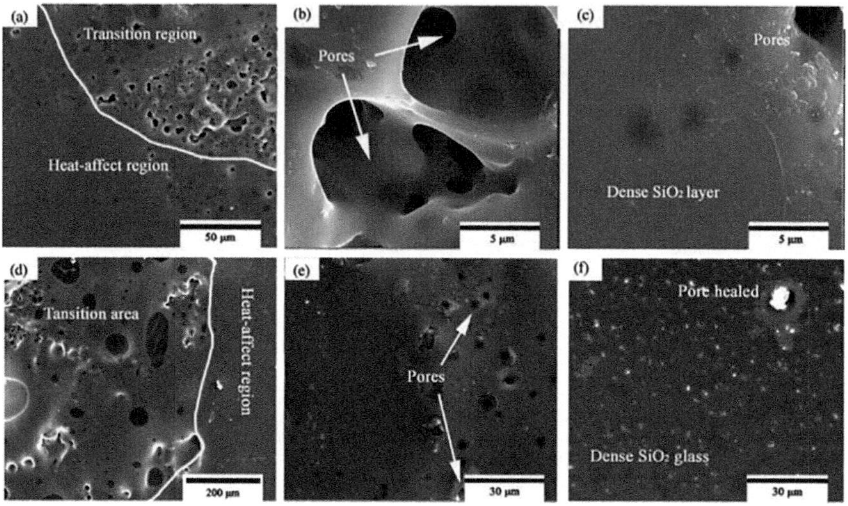

Fig. 6.110 SEM surface morphologies of the as-ablated 10 vol% GPLs/Si_2BC_3N CMC: **a**, **d** Interfaces between transition region and heat-affected area; **b**, **e** magnified transition region; **c**, **f** magnified heat-affected region [28]

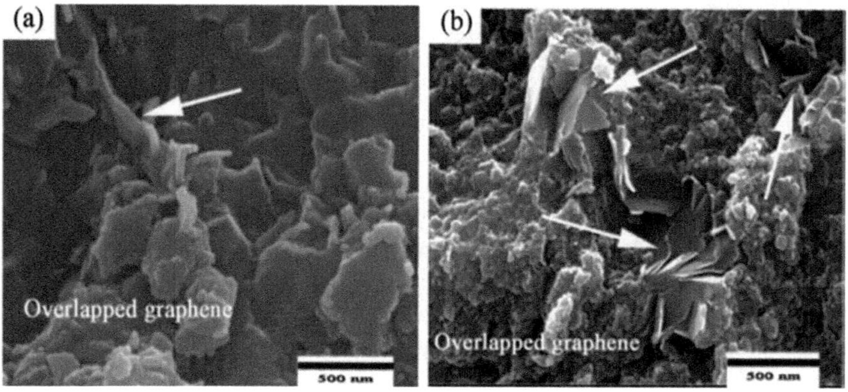

Fig. 6.111 Fracture morphologies near to the oxide layer after exposing to an oxyacetylene torch experiment for 10 s: **a** + 5 vol% GPLs; **b** + 10 vol% GPLs [28]

During ablation, the GPLs with smaller sizes are most prone to oxidation and erosion, while larger sizes and multilayer forms are much less like to be completely destroyed. Thus, the GPLs plays double roles during ablation. On one hand, GPLs shows a degree of chemical inertness, thus it not only protects the underlying matrix from being oxidizing, but also remains its structure unchanged depending on its location and size. On the other hand, pores are left behind due to the escape of gaseous products, which may accelerate oxygen diffusion rapidly into inner matrix. Consequently, the ablation resistance and anti-oxidation ability will decrease as compared with the MA-Si_2BC_3N monoliths. To a certain extent, pores or channels can be healed by filling of low viscosity SiO_2 and B_2O_3, therefore retards the oxidation rate of ceramic matrix.

6.4 Thermal Shock Damage and Corrosion Mechanisms of SiBCN Materials

6.4.1 Thermal Shock Damage Mechanisms

Thermal shock of ceramic materials is influenced by many factors such as strength, Young's modulus, fracture toughness, thermal conductivity and thermal expansion coefficient. Thermal shock resistance parameters can be used for the initial selection and screening of SiBCN based materials for a given thermal environment though recently finite element methods have been developed to determine thermal stresses which combined with failure statistics can give a better indication of the overall failure probability of a structural part.

There are two classical theories on thermal shock resistance in a brittle ceramic. One is Kingery's critical stress fracture theory where thermal shock cracks must be avoided altogether high strength is required as predicted by R, R', and R''. R can be expressed as [6]

$$R = \Delta T_c = \frac{\sigma_f(1-\nu)}{\alpha E} \tag{6.1}$$

the other is Hasselman's thermal shock damage theory where the strength is low cracks do not propagate as predicted by parameters R''', R'''', and R_{st}. R'''' is given as [29]

$$R'''' = \frac{E\gamma_f}{\sigma_f^2(1-\nu)} \approx \frac{(K_{IC}/\sigma_f)^2}{2(1-\nu)} \tag{6.2}$$

where σ_f is the flexural strength, α is the thermal expansion coefficient, ν is the Poisson ratio and γ_f is the fracture surface energy.

The thermal expansion coefficient (CET) of the MA-Si_2BC_3N composite materials usually follows the simple rule of mixtures (or more complex models), based on the CET of the respective components, their volume fraction and the volume fraction of interfacial phases. Based on these models, a PDCs-SiBCN/MA-Si_2BC_3N hybrid ceramic is expected to possess a similar CET to monolithic MA-Si_2BC_3N [(4.0–4.8) $\times 10^{-6}$ °C^{-1}]. Obviously, the chemical composition of the sintering additives, such as Zr, Zr-Al, ZrN, MgO-Al_2O_3-SiO_2, etc. will have a certain influence but should remain within the variations observed for monolithic MA-Si_2BC_3N.

However, in the case of C_f, the CET of the reinforcement phase (C_f: 1.7 × 10^{-6} °C^{-1}) is much lower than that of the MA-Si_2BC_3N matrix, implying a decrease in the composite CET. For example, the CET of 20 vol% C_f/Si_2BC_3N CMCs HP sintered at 1800 °C, reaches only (0.43–1.60) × 10^{-6} °C^{-1}, which is less than the double the value of monolithic MA-Si_2BC_3N [(4.0–4.8) × 10^{-6} °C^{-1}]. However, their measurements are performed on samples with bulk densities varying between 1.92–2.18 g/cm^3 and with residual porosity levels reaching up to 23% ~ 33%, both of which can affect the CET values. In case of SiC_f (~4.5 × 10^{-6} °C^{-1}) or (C_f-SiC_f) reinforcements, the similar CET of the 20 vol% SiC_f/Si_2BC_3N (~4.3 × 10^{-6} °C^{-1}) and 20 vol% (C_f-SiC_f)/Si_2BC_3N CMCs [(3.2–4.2) × 10^{-6} °C^{-1}] are expected.

The calculated R and R'''' values for the (C_f-SiC_f)/Si_2BC_3N CMCs are 233 °C and 941 μm, and for the C_f/Si_2BC_3N CMCs are 748.8 °C and 3515.6 μm, respectively. Compared with the values of R (219 °C) and R'''' (274 μm) of the SiC_f/Si_2BC_3N CMCs, the R of the (C_f-SiC_f)/Si_2BC_3N CMCs increases by 14 °C and R'''' is 2.4 times higher due to the low flexural strength, Young's modulus and thermal expansion coefficient. Consequently, the SiC_f/Si_2BC_3N materials shows brittle fracture, while the (C_f-SiC_f)/Si_2BC_3N and C_f/Si_2BC_3N CMCs present distinct pseudo-plastic fracture, implying a superior thermal stress fracture and damage resistance.

In addition to the CET mismatches, others factors such as thermal conductivity and relative density (or pores) also influence thermal shock resistance. The thermal conductivity of the MA-Si_2BC_3N matrix composites is controlled by the level of porosity, the relative phase content, the amount and conductivity of the reinforcement phases (metal particles, ceramic particles, MWCNTs, C_f, SiC_f, etc.), the number of glassy phases (ZrO_2, $ZrSiO_4$, MAS, MZS, etc.) and any orientation effect. For example, the thermal conductivity of the SPS densified monoliths will increase with the increasing Si/C ratio (amount of α/β-SiC, as it possesses a higher thermal conductivity than BN(C) phase). In addition, the introduction of C_f with high thermal conductivity into SiC_f/Si_2BC_3N composites contributes significantly to thermal shock resistance. An integrated effect including crack deflection caused by SiC_f and C_f, fiber pull-out and debonding, can also consume much fracture energy and thus significantly improve thermal shock resistance in (C_f-SiC_f)/Si_2BC_3N composites. Moreover, the relative density of the (C_f-SiC_f)/Si_2BC_3N CMCs ceramics is 80.8%, 8.2% lower than that of SiC_f/Si_2BC_3N CMCs, which is also beneficial to thermal shock resistance to a certain extent. Generally, the incorporation of UHTCs components or *in-situ* generated UHTCs in MA-Si_2BC_3N matrix both favor their thermal resistance (Fig. 6.112).

Note that, orientation of the short fibers may produce an anisotropic thermal conductivity behaviour during uniaxial pressure sintering, where a higher thermal conductivity observed in the direction of orientation compared with the direction perpendicular to the fiber orientation. Besides, the volume fraction of the glassy phase influences conductivity of MA-Si_2BC_3N matrix, as the sintering additives typically used possess a lower thermal conductivity and act as a thermal barrier.

Significantly, a great CET mismatch can cause stress failure, but the low thermal conductivity cannot reduce the temporary temperature gradient that arises in thermal shock. The result is residual stress buildup inside the materials (namely, compression

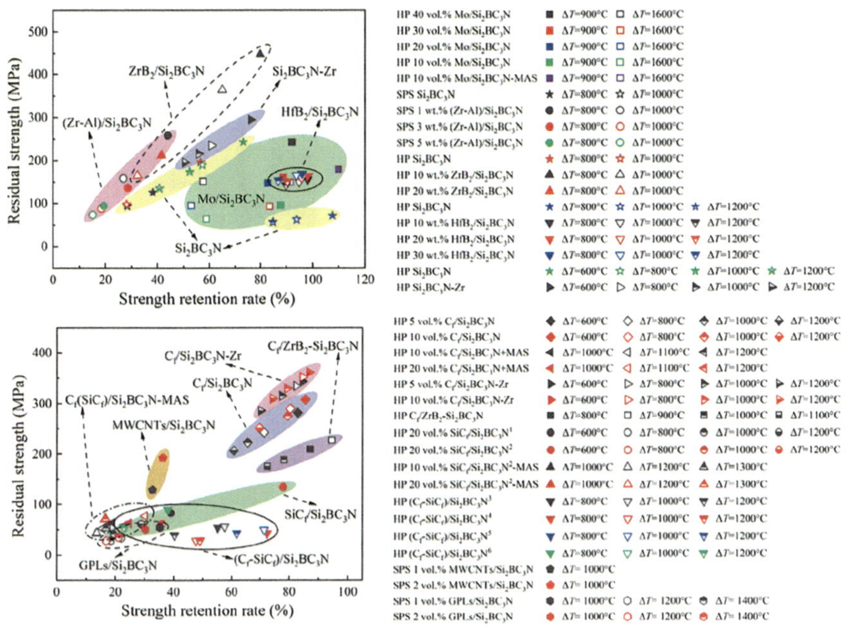

Fig. 6.112 Comparison of the thermal shock resistance of the various MA-Si_2BC_3N based materials prepared by hot-pressing or spark plasma sintering ways. ([1]: SLFC1 SiC_f without coating; [2]: SLFC1 SiC_f with h-BN coating; [3]: Unmodified (C_f-SiC_f); [4]: PyC coated (C_f-SiC_f); [5]: h-BN coated (C_f-SiC_f); [6]: Unmodified (C_f-SiC_f) with $ZrSiO_4$ additive)

stress in the hot interior but tensile stress buildup on rapidly cooled surfaces) that can induce formation of microcracks. Thus, a modest relative density can enhance resistance to thermal shock damage because pores with appropriate contents, uniform sizes and homogeneous distribution can consume much of the thermal induced elastic strain energy and relax the thermal stress. Obviously, the ability to withstand sharp temperature changes and resist catastrophic damage for a ceramic material depends on the complex combination of CET, thermal conductivity, mechanical properties and relative density (or pores).

As noted above, the C_f/Si_2BC_3N and (C_f-SiC_f)/Si_2BC_3N CMCs both show thermal shock resistance superior to SiC_f/Si_2BC_3N CMCs as supported from calculated thermal stress fracture and damage resistance parameters evaluating the thermal shock resistance. The C_f/Si_2BC_3N composite nose cone models are thus prepared to further test thermal shock resistance for thermal protection system applications (Fig. 6.113). When subjected to a heat flux of 12.0 MW/m^2 (flame temperature on sample surface of ~ 3000 °C) promptly and kept being heated for 10 s in engine bench tests using liquid oxygen and kerosene as the fuel, the nose cone samples survive successfully without failure. However, the fiber-free MA-Si_2BC_3N samples break on heating for just 4 s due to the poor thermal shock resistance. Therefore, the resulting

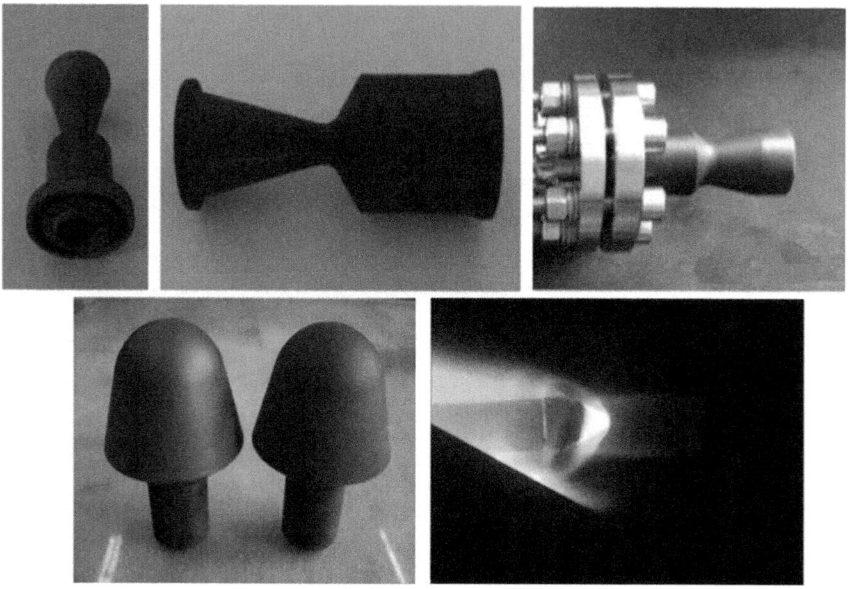

Fig. 6.113 Digital photos of combustion chamber samples of the attitude control engine for a new type of satellite, and the macro morphologies of the ward end caps before and after oxyacetylene flame ablation assessments

C_f/Si_2BC_3N composites show potential for multi-functional thermal protection applications, especially in severe thermal shock environment. The combustion chamber samples are also prepared using the MA-Si_2BC_3N CMCs. They passed successfully the simulation test on ground bench indicating attractive potential applications in the fields of thermal protection and high temperature refractory materials, especially served in air-rich atmosphere.

The PIP processing PDCs-SiBCN and PDCs-SiCN materials [30–32] also show good thermal shock resistance (Fig. 6.114). For the multilayer hydrothermal carbon coatings modified continuous C_f reinforced ZrB_2/SiBCN (C_f/ZrB_2-SiBCN) CMCs [30], the increasing coating thickness alleviates the damage of fibers in oxidation environment thus improving the critical thermal shock temperature difference (ΔT_c) of the composites, while the excessive thick coating weakens the fiber strength and limit toughening abilities of the continuous C_f. Especially, the enhancement effect is more distinct in the composites fabricated with high-density fabric. For example, the ΔT_c of the C_f/ZrB_2-SiBCN CMCs fabricated with high-density fabric is calculated to be 769 °C, 931 °C, 954 °C and 1026 °C, higher than 679 °C, 969 °C, 989 °C and 974 °C of the counterparts, respectively.

For the 3-D needle-quenched C_f reinforced PDCs-SiCN CMCs, their strength retention rate is 93.3% after thermal shock test at $\Delta T = 780$ °C. It further degrades to 14.6% when ΔT increasing to only 1180 °C. The strength deterioration of the PDCs-SiCN CMCs should be attributed to the strength degradation of carbon fibers

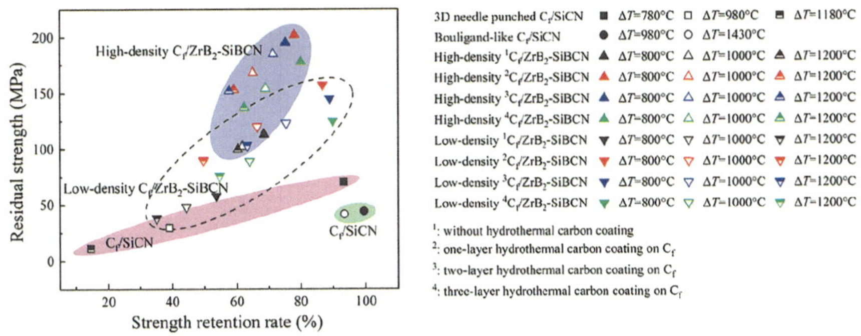

Fig. 6.114 Thermal shock performance of the PIP processing C_f reinforced PDCs-SiBCN and PDCs-SiCN materials [30–32]

and decomposition of interfacial structures [31]. The thermal shock resistance of PIP processing PDCs-SiCN composites can be significantly enhanced by the use of bio-inspired Bouligand layups. The retained strength of the Bouligand-type C_f/SiCN CMCs decreases slightly (less than 1%) as ΔT increases from 980 °C to 1430 °C, resulting from the helicoidal distributed fibers effectively resisting crack initiation and propagation. Even after seven thermal shock cycles, the Bouligand-like C_f/SiCN composites with 10° pitch angle still show the higher retained strength, suggesting an excellent thermal shock resistance [32].

6.4.2 Ablation Corrosion Mechanisms

Under a corrosive condition, ablation results in parts of a SiBCN material being removed by coupled thermo-mechanical, thermo-chemical and thermo-physical interactions due to a combustion flame with high temperature, pressure and velocity. Generally, oxyacetylene, arc-jet and plasma ablation tests are the three main methods to measure the ablation properties of ablators. In consideration of high-cost of arc-jet and plasma wind tunnel tests, oxyacetylene torch ablation is the most often used method for evaluating the performance of thermal protection materials under hypersonic re-entry conditions. If not clearly specified, the ablation tests mentioned below are performed under an oxyacetylene combustion flame. The combustion products of the oxyacetylene torch comprise of 43.02 mol% O_2, 15.38 mol% CO_2, 12.04 mol% CO, 10.75 mol% O, 8.45 mol% OH, 7.83 mol% H_2O, 1.77 mol% H and 0.76 mol% H_2, generally leads to a flame temperature exceeding ~ 3000 °C and a heat flux ~ 4.2 MW/m^2.

The linear and mass ablation rates are usually used to evaluate the ablation property of the MA-Si_2BC_3N based materials (Fig. 6.115). For the fiber-toughened MA-Si_2BC_3N CMCs, the ablation surface temperature increases sharply to about 1800 °C within a few seconds and then levels out at about 2200 °C. Consequently, severe

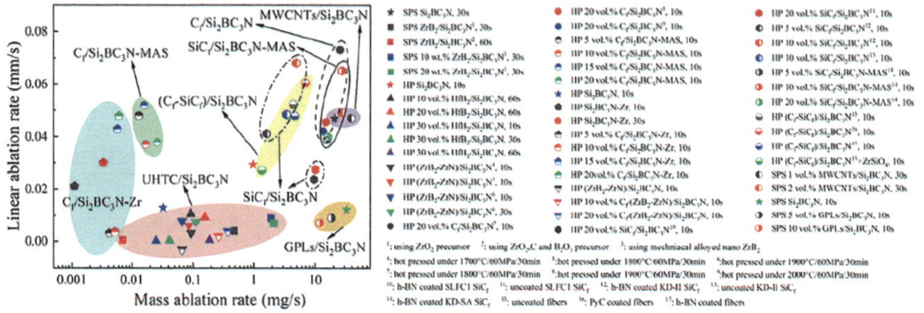

Fig. 6.115 Comparison of the ablation resistance of the MA-Si_2BC_3N based materials after oxyacetylene torch ablation for varying time

thermal shock occurs actually along with the ablation. The relatively dense SiC_f/Si_2BC_3N CMCs hot-pressed at 1900–2000 °C and 60 MPa shows good ablation properties, obtaining linear and mass ablation rates of 0.0237 mm/s and 9.5 mg/s, respectively. The CMCs samples after ablation test still maintain integrity except the shallow ablation pits on the surface indicating an excellent thermal shock resistance. The combination of thermal shock and ablation resistance is confirmed strongly by the aforementioned engine bench test of the nose cone CMCs sample, which shows a linear ablation rate of ~0.544 mm/s.

The SiC_f/Si_2BC_3N CMCs offer better ablation resistance to C_f/Si_2BC_3N. The linear ablation rate decreases by ~ 36% and mass ablation rate decreases by ~ 32% compared to C_f/Si_2BC_3N. As aforementioned, under the same ablation conditions, no catastrophic damage and macrocracks can be observed on surfaces of (C_f-SiC_f)/Si_2BC_3N materials. These results strongly demonstrate that the as-prepared (C_f-SiC_f)/Si_2BC_3N CMCs show effective integration of the superior ablation resistance of SiC_f/Si_2BC_3N with the superior thermal shock resistance of C_f/Si_2BC_3N. Just like C_f/Si_2BC_3N CMCs, the (C_f-SiC_f)/Si_2BC_3N CMCs show consequently attractive potential applications in the severe thermal shock and ablation environments.

Note that the linear ablation rate of (C_f-SiC_f)/Si_2BC_3N CMCs is higher than that of SiC_f/Si_2BC_3N CMCs, while the mass ablation rate of (C_f-SiC_f)/Si_2BC_3N decreases by one order of magnitude. The abnormal ablation rate changes can be explained as follows. Firstly, the (C_f-SiC_f)/Si_2BC_3N CMCs have a lower relative density than SiC_f/Si_2BC_3N, so the oxyacetylene combustion flow more easily penetrates toward the central materials and then reacts with the constituent phases. As a result, within the same ablation time, the (C_f-SiC_f)/Si_2BC_3N CMCs exhibit a larger linear ablation depth but lower mass loss than SiC_f/Si_2BC_3N. Secondly, during ablation, the oxidation and exfoliation of carbon fibers (10 vol%) generates a loose, porous microstructure that offers combustion flow penetration paths and resultant higher linear ablation rates of C_f-containing MA-Si_2BC_3N CMCs. Moreover, the central ablation pits or holes obstruct SiO_2(l) and B_2O_3(l) from flowing towards the specimen edge thereby diminishing the degree of ablation in the regions far from the central region but aggravating the degree of ablation in the central region. This can be also responsible for the higher linear ablation rate but lower mass ablation rate of the (C_f-SiC_f)/Si_2BC_3N CMCs.

After ablation under high-temperature flame, the as-ablated specimen surface can be divided into three regions: central, transitional and heat-affected regions. Ablation usually occurs at the interface between fibers and ceramic matrix, and the latter has a higher ablation resistance than the former. For the SiC_f/Si_2BC_3N materials, in the central region, liquid SiO_2 with good fluidity fills the left pits or holes after the exfoliation of SiC fibers thus leading to the ablation pits with smooth bottom.

The oxidation of SiC fibers and MA-Si_2BC_3N matrix plays a dominant role in the transition region resulting in a thin SiO_2 layer with some pores covering the sample surface. No bubbles are observed due to the gas emitting coupled with B_2O_3 evaporation. The temperature in the heat-affected region far away the central region is so low that high-viscosity SiO_2 shows a poor fluidity and cannot cover uniformly the entire region. Consequently, a number of bubbles form on the surface of the heat-affected region, most remaining unbroken. Additionally, the reaction of partial B_2O_3 with SiO_2 most likely leads to the formation of borosilicate glass in the heat-affected region.

In contrast, the bottom of ablation pits for the C_f/Si_2BC_3N CMCs looks very rough. Some microcracks coupled with spherical SiO_2 particles can be observed at the bottom of the central ablation pit. Significantly, the flocculent products which could be self-assembled carbon nanowires are formed during the ablation of C_f/Si_2BC_3N materials.

The above analysis of phase composition, microstructures and elements coupled with the thermodynamic calculation indicate that the ablation mechanisms of short fiber toughening MA-Si_2BC_3N CMCs include mainly oxidation of fibers and ceramic matrix, high-speed combustion flow scouring and mechanical exfoliation. Taking the SiC_f/Si_2BC_3N for example, in the initial stage of ablation process, the temperature of sample surface is relatively low and the situation is thus similar to the heat-affect region. The oxidation of SiC fibers and MA-Si_2BC_3N matrix is the main ablation mechanism. The ablation products such as silica and borosilicate form in this stage; SiO(g) and B_2O_3 gases emit from the ablation sample surface. With the increasing surface temperature, the ablation product SiO_2 is melted. But it hardly evaporates due to the relatively low surface temperature (~2200 °C) detected by the optical pyrometer. As a result, the dense liquid SiO_2 layer covers on the ablation sample surface. Pores or bubbles are left in the region full of the low-viscosity SiO_2 due to the emitting of SiO(g) and B_2O_3(g). In the final stage of ablation testing, the severe chemical erosion and mechanical exfoliation occur and the ablation pit forms on the central region by the scouring of high-speed combustion flow. Consequently, oxidation of SiC fibers and MA-Si_2BC_3N matrix, erosion of SiC fibers in the ablation center regions, evaporation of SiO(g) and B_2O_3(g), and flowing of low-viscosity SiO_2 liquid contribute to the formation of various ablation regions during the ablation process.

The refractory metals and their borides, nitrides and carbides, such as ZrB_2, HfB_2, TaB2, ZrN, HfC, TaC, and Ta_4HfC_5 show extraordinary ablation resistance. They are, therefore, often used as the secondary phase or protective coatings to improve the ablation resistance of the MA-Si_2BC_3N monoliths and their CMCs. Experimental results have clearly exhibited the ZrB_2 addition indeed improved the ablation resistance of MA-Si_2BC_3N composite ceramics. Significantly, the 15 wt% ZrB_2 incorporating C_f/Si_2BC_3N composites (2.65 g/cm^3) provide more excellent ablation resistance even after ablation of 60 s. The composite samples still kept intact without cracking after the very serious temperature increase-induced thermal shock occurring within very few seconds in the initial stage of ablation and after the subsequent quick cooling shock in air following the 60 s ablation. Their linear and mass ablation rates are 0.0013 mm/s and 0.00019 mg/(mm^2·s), respectively.

The PDCs-SiBCN and PDCs-SiCN materials also exhibit outstanding ablation resistance under different combustion flame condition [33–38]. For the PIP processing 3-D C_f/SiBCN composites ablated in a plasma flame at ~2200 °C (heat flux ~ 4.02 MW/m^2) [33], the thermal damage of internal C_f/SiBCN during ablation includes crystallization and decomposition of SiBCN matrix, as well as interphase damage between the SiBCN (enriched O) and carbon fibers & PyC interphase. The evolution process of microstructures in SiBCN matrix has showed that ablation firstly appears in the interfaces of PyC and active spots. Their mass and linear ablation rates are 0.05390 mg/(mm^2·s) and 0.0030 mm/s. However, improved ablation resistance of the 3-D C_f/SiBCN composites with $(PyC/SiC)_3$ multi-layers as interphase are observed [34], obtaining mass and linear ablation rates of 0.04270 mg/(mm^2·s) and 0.0017 mm/s, which are ~ 43% and ~ 21% lower than the formers (Fig. 6.116 and Table 6.28).

In terms of testing results, the performance of PIP processing Zr modified C_f/SiBCNZr composites [35] shows much better resistance to oxyethylene torch ablation, originating from a dense protective oxide layer of dominated *in-situ* ZrO_2 and SiO_2 on the surface effectively retarding outside oxygen inward. In another case [36], the PIP (C_f-SiC_{NWS})/SiBCN composites with spraying Ta-Mo based coating show measured linear ablation rate of 0.00014 mm/s at 1500 °C and 0.000062 mm/s at 1800 °C, and mass ablation rate of 0.000024 mg/(mm^2·s) at 1500 °C and 0.00160 mg/

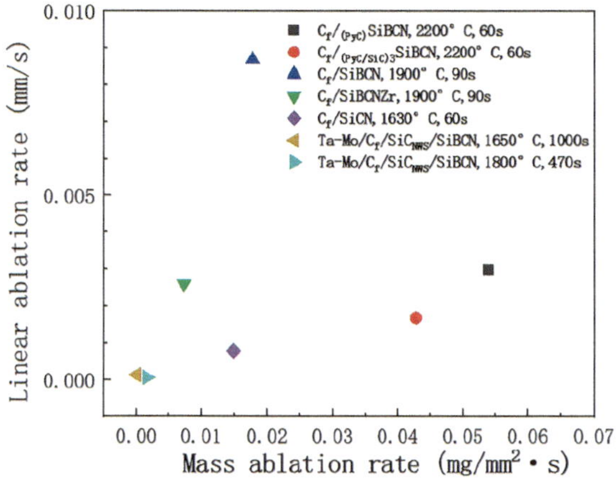

Fig. 6.116 Ablation performance of the PDCs-Si(B)CN ceramic composites under different ablation condition

Table 6.28 Ablation properties of the PDCs-Si(B)CN materials under different ablation condition [33–38]

Composition	Torch	Ablation temperature (°C)	Ablation time (s)	Mass ablation rate (mg/(mm^2·s))	Linear ablation rate (mm/s)
PyC coated C_f/SiBCN	Plasma	2200	60	0.05390	0.0030
$(PyC/SiC)_3$ coated C_f/SiBCN	Plasma	2200	60	0.04270	0.0017
C_f/SiBCN	Oxyacetylene	1900	90	0.01771	0.0087
C_f/SiBCNZr	Oxyacetylene	1900	90	0.00729	0.0026
C_f/SiCN	Oxyacetylene	1630	60	– 0.01479	0.0008
Ta-Mo/C_f/SiC_{NWS}/SiBCN	Oxyacetylene	1500	1000	0.000024	0.00014
Ta-Mo/C_f/SiC_{NWS}/SiBCN	Oxyacetylene	1800	500	0.00160	0.000062
10 vol% SiCN-HfC	Oxyacetylene	2500	60	– 0.00019	– 0.000156

(mm^2·s) at 1800 °C, respectively. Obviously, the addition of refractory metals and subsequently generated borides, nitrides, carbides, or oxides favors the ablation resistance of the PDCs-SiBCN materials.

The PIP processing C_f/SiCN gradient composites abated under an oxyacetylene flame with a heat flux ~ 2.38 MW/m^2 show the linear ablation rate of 0.000833 mm/s and mass ablation rate of -0.014791 mg/(mm^2·s), respectively [37]. The addition of only 10 vol% PDCs-SiCN greatly improve the ablation performance of the HfC ceramics, due to the formation of the triple-layer structure oxide with good thermal stability and mechanical scouring resistance [38]. After ablation under an oxyacetylene flame for 60 s, the mass and linear ablation rates of the 10 vol% SiCN containing HfC composite ceramics are -0.00019 mg/(mm^2·s) and -0.000156 mm/s, respectively. The negative mass or linear ablation rates suggest the mass gain after ablation, and substantial oxides strong adhered to the ablation surfaces.

Looking forward, more types of thermal-protective high-temperature structural parts should be developed to expand their industry applications and to promote their application process based on the development of SiBCN ceramics and their matrix composites with more extraordinary comprehensive properties as well as larger size. Herein, the relevant basic scientific topics should not be neglected during the further exploration of engineering applications of SiBCN materials. For example, it is significantly necessary to develop the highly dense and large-scale (dozens or even hundreds of centimeters) SiBCN structural parts (especially by hot pressing or HIP technology) to meet the relevant requirements. Coincidently, more types of thermal protection and high-temperature structural parts, especially short fiber toughened SiBCN CMCs, should be developed to expand their engineering applications in more fields and to promote their application process. Persistent effort should be made to develop centimeter-sized or larger-size amorphous SiBCN ceramic monoliths by high pressure sintering or post-HIP technique. Persistent effort should be also made to study the high-temperature damage (such as creep, thermal shock oxidation and ablation) mechanisms of SiBCN system materials. For example, there have been disagreements over the investigations of oxidation resistance, especially the oxide scale microstructure and oxidation kinetics at early exposing time. Therefore, their high-temperature oxidation behavior requires further study even compared with that of typical Si-based ceramics such as SiC, SiCN, etc. Additionally, the influence of chemical composition, processing routes (such as sintering methods) on the creep resistance of SiBCN materials should be paid more attention. The cyclic thermal shock with different times of thermal cycle is worthy of investigation which is necessary to develop the SiBCN structural parts, especially the aforementioned combustion chamber samples of attitude control engine.

References

1. Liao XQ. Microstructure and properties of mechanical alloyed SiBN powders and ceramics. Harbin: Harbin Institute of Technology; 2014.
2. Zhang M, Hou JN, Li DX, Liang B, Niu ZB, Zhang Q, Yang ZH, Jia DC, Cai DL, Duan XM, He PG, Zhou Y. Mo-SiBCN metal-ceramic composites with enhanced and tunable thermophysical properties and thermal shock resistance. Ceram Int. 2022;48:5744–51.
3. Matsunoshita H, Fujiwara K, Sasai Y, Kishida K, Inui H. Orientation relationships, interface structures, and mechanical properties of directionally solidified $MoSi_2/Mo_5Si_3/Mo_5Si_3C$ composites. Intermetallics. 2016;73:12–20.
4. Zhu Q, Shobu K, Tani E, Kishi K, Umebayashi S. Oxidation behavior of $Mo_{\leq 5}Si_3C_{\leq 1}$ and its composites. J Mater Sci. 2000;35:863–72.
5. Liao N, Jia DC, Yang ZH, Zhou Y, Li YW. Mechanical properties and thermal shock resistance of Si_2BC_3N ceramics with ternary Al_4SiC_4 additive. Ceram Int. 2018;44:9009–17.
6. Kingery WD. Factors affecting thermal stress resistance of ceramic materials. J Am Ceram Soc. 1955;38:3–15.
7. Hasselman DPH. Elastic energy at fracture and surface energy as design criteria for thermal shock. J Am Ceram Soc. 2006;46:535–40.
8. Liao N, Jia DC, Yang ZH, Zhou Y, Li YW. Enhanced mechanical properties, thermal shock resistance and oxidation resistance of Si_2BC_3N ceramics with Zr–Al addition. Mater Sci Eng, A. 2018;725:364–74.
9. Wang GY. Preparation and properties of SIBCN(Zr) gradient porous ceramics prepared by ice molding and their bicontinuous composites. Harbin: Harbin Institute of Technology; 2018.
10. Wang BZ. Microstructure evolution and high temperature oxidation and ablation mechanism of mechanical alloying nano-Ta_4HfC_{5p}/SiBCN powder and ceramic. Harbin: Harbin Institute of Technology; 2018.
11. Miao Y. Preparation, oxidation resistance and ablation mechanisms of ZrB_2/SiBCN composites. Harbin: Harbin Institute of Technology; 2017.
12. Liao N, Jia DC, Yang ZH, Niu B, Zhou Y, Li YW. Enhanced mechanical properties, thermal shock resistance and ablation resistance of Si_2BC_3N ceramics with nano ZrB_2 addition. J Eur Ceram Soc. 2019;39:846–59.

13. Miao Y, Yang ZH, Zhu QS, Liang B, Li Q, Tian Z, Jia DC, Cheng YB, Zhou Y. Thermal ablation behavior of SiBCN-Zr composites prepared by reactive spark plasma sintering. Ceram Int. 2017;43:7978–83.
14. Ao DF. The thermal shock and ablative properties of HfB_2/SiBCN multiphase ceramic. Harbin: Harbin Institute of Technology; 2018.
15. Niu ZB, Wang BZ, Pan LJ, Li DX, Jia DC, Yang ZH, Peng H, Ye SQ, Zhou Y. Mechanical and thermal shock properties of C_f/SiBCN composite: effect of sintering densification and fiber coating. J Am Ceram Soc. 2022;105:4321–35.
16. Pan LJ. The surface coating on C_f and the preparation and properties of C_f/SiBCN composite. Harbin: Harbin Institute of Technology; 2012.
17. Zhoutang JE. Microstructure and high temperature thermal shock, ablation properties of the C_f/SiBCN-MAS composites. Harbin: Harbin Institute of Technology, 2023.
18. Zhu QS. Fabrication and high temperature damage mechanical of (C_f-ZrB_2)/Si-B-C-N ceramic composites. Harbin: Harbin Institute of Technology; 2021.
19. Zhu QS, Wang GY, Liao XQ, Liang B, Li DX, Yang ZH, Jia DC, Zhou Y, Zhang T, Gao CG. Ablation properties and mechanisms of BN-coated C_f-reinforced SiBCNZr ceramic composites under an oxyacetylene combustion torch. Ceram Int. 2021;47:15533–41.
20. Wang JY. Mechanical, thermal shock and ablation properties of SiC_f/SiBCN ceramic composite. Harbin: Harbin Institute of Technology; 2013.
21. Wang JY, Duan XM, Yang ZH, Jia DC, Zhou Y. Ablation mechanism and properties of SiC_f/SiBCN ceramic composites under an oxyacetylene torch environment. Corros Sci. 2014;82:101–7.
22. Li DX, Wu DX, Yang ZH, Zhou YY, Wang SJ, Duan XM, Jia DC, Zhu QS, Zhou Y. Effects of *in situ* amorphous graphite coating on ablation resistance of SiC fiber reinforced SiBCN ceramics in an oxyacetylene flame. Corros Sci. 2016;113:31–45.
23. Xiao JZ. Microstructure and thermal shock ablative behavior of the SiC_f/SiBCN composites. Harbin: Harbin Institute of Technology; 2023.
24. Li YT. Mechanical properties, thermal shock and ablation resistance of (C_f-SiC_f)/SiBCN composites. Harbin: Harbin Institute of Technology; 2014.
25. Liang B, Yang ZH, Li YT, Yuan JK, Jia DC, Zhou Y. Ablation behavior and mechanism of SiC_f/C_f/SiBCN ceramic composites with improved thermal shock resistance under oxyacetylene combustion flow. Ceram Int. 2015;41:8868–77.
26. Liao N, Jia DC, Zhou Y, Li YW. Enhanced thermal shock and oxidation resistance of SiBCN ceramics through MWCNTs incorporation. Journal of Advanced Ceramics. 2018;7:276–88.
27. Li DX, Yang ZH, Jia DC, Wu DX, Zhu QS, Liang B, Wang SJ, Zhou Y. Microstructure, oxidation and thermal shock resistance of graphene reinforced SiBCN ceramics. Ceram Int. 2016;42:4429–44.
28. Li DX, Yang ZH, Jia DC, Duan XM, He PG, Zhou Y. Ablation behavior of graphene reinforced SiBCN ceramics in an oxyacetylene combustion flame. Corros Sci. 2015;100:85–100.
29. Hasselman DP. Elastic energy at fracture and surface energy as design criteria for thermal shock. J Am Ceram Soc. 1963;46:535–40.
30. Fang C, Bao Y, Hu P, Dong S, Liu DZ, Wang HL, Zhang XH. Effects of multilayer hydrothermal carbon interphases on mechanical properties and thermal shock resistance of CF/ZrB_2-SiCBN. J Eur Ceram Soc. 2022;42:4759–69.
31. Yang ZH, Li DX, Hou YW, Liao XQ, Zhu QS, Jia DC, Duan XM, Cai DL, Wang SJ, Zhou Y. Processing and mechanical performance of 3D C_f/SiCN composites prepared by polymer impregnation and pyrolysis. Ceram Int. 2019;45:17344–53.
32. Yang F, Meng SH. Cyclic thermal shock behaviors of carbon fibers reinforced SiCN ceramics with bio-inspired Bouligand-like architectures. J Eur Ceram Soc. 2023;43:3172–81.
33. Ding Q, Ni DW, Ni N, Jiang YL, Chen BW, Chen XW, Lu J, Niu YR, Zhou HJ, Dong SM. Thermal damage and microstructure evolution mechanisms of C_f/SiBCN composites during plasma ablation. Corros Sci. 2020;169: 108621.
34. Ding Q, Chen BW, Ni DW, Ni N, He P, Gao L, Wang HD, Zhou HJ, Dong SM. Improved ablation resistance of 3D-C_f/SiBCN composites with $(PyC/SiC)_3$ multi-layers as interphase. J Eur Ceram Soc 2021;41:1114–120.

35. Lyu Y, Cheng Y, Zhao GD, Wang MY, Chen GQ, Zhang XH, Han WB. Modification of SiBCN by Zr atom and its effect on ablative resistance of C_f/SiBCN(Zr) composites. Compos B. 2023;253: 110511.
36. Tang WK. Preparation and properties of CF-SiC_{NWS} reinforced SiBCN ceramic composites. Harbin: Harbin Institute of Technology; 2019.
37. Hu ZC, Meng SH, Yi FJ, Li JP, Xu CH. Oxyacetylene ablation resistance of continuous gradient ceramic-polymer composite. Corros Sci. 2021;179: 109050.
38. Ni N, Hao W, Liu TY, Zhou L, Guo FW, Zhao XF, Xiao P. Oxidation/ablation behaviors of hafnium carbide-silicon carbonitride systems at 1500 and 2500 °C. Ceram Int. 2020;46:23840–53.

GPSR Compliance

The European Union's (EU) General Product Safety Regulation (GPSR) is a set of rules that requires consumer products to be safe and our obligations to ensure this.

If you have any concerns about our products, you can contact us on ProductSafety@springernature.com

In case Publisher is established outside the EU, the EU authorized representative is:

Springer Nature Customer Service Center GmbH
Europaplatz 3
69115 Heidelberg, Germany

Batch number: 10703867

Printed by Printforce, the Netherlands